Introductory Statistics
for the
Health Sciences

Introductory Statistics
for the
Health Sciences

Lise DeShea

University of Oklahoma
Oklahoma City, Oklahoma, USA

Larry E. Toothaker

University of Oklahoma
Norman, Oklahoma, USA

Illustrations by William Howard Beasley

CRC Press
Taylor & Francis Group
Boca Raton London New York

CRC Press is an imprint of the
Taylor & Francis Group, an **informa** business

A CHAPMAN & HALL BOOK

Cover photo: "Disparity Five-World AIDS" (6'x8') for CommonGround191 art project. By Gary Simpson, used with permission. Corroded copper depicts the number of AIDS/HIV cases per country. One inch equals 12,000 humans, with some countries "off the canvas."

Data sets used in this book are available in .csv, SAS and SPSS via links at http://desheastats.com. The data, graphing code and figures also are available directly from our data repository: Beasley, W. H. (2015). *Graphing code to accompany the book "Introductory Statistics for the Health Sciences," by Lise DeShea and Larry E. Toothaker (2015).* https://github.com/OuhscBbmc/DeSheaToothakerIntroStats. doi:10.5281/zenodo.12778

CRC Press
Taylor & Francis Group
6000 Broken Sound Parkway NW, Suite 300
Boca Raton, FL 33487-2742

First issued in paperback 2020

© 2015 by Taylor & Francis Group, LLC
CRC Press is an imprint of Taylor & Francis Group, an Informa business

No claim to original U.S. Government works

ISBN-13: 978-1-4665-6533-3 (hbk)
ISBN-13: 978-0-367-78353-2 (pbk)

Library of Congress Cataloging-in-Publication Data

DeShea, Lise.
  Introductory statistics for the health sciences / Lise DeShea and Larry E. Toothaker.
    pages cm
  "A CRC title, part of the Taylor & Francis imprint, a member of the Taylor & Francis Group, the academic division of T&F Informa plc."
  Includes bibliographical references and index.
  ISBN 978-1-4665-6533-3 (hardcover : alk. paper) 1. Medical statistics--Textbooks. I. Toothaker, Larry E. II. Title.

RA409.D474 2015
610.2'1--dc23

2014043098

Visit the Taylor & Francis Web site at
http://www.taylorandfrancis.com

and the CRC Press Web site at
http://www.crcpress.com

Contents

10 One-Sample Tests and Estimates   275

11 Two-Sample Tests and Estimates   297

Preface

Welcome to *Introductory Statistics for the Health Sciences*. If you are a student, you may be approaching this book with a number of emotions: anticipation, dread, excitement, skepticism, and so on. We have taught a lot of people like you over the decades, and your needs and concerns motivated this book. Consider this preface to be our way of greeting you at the door to our home. We will do everything we can in this book to make you comfortable during your stay, even if it was not your first choice to be here. The research examples should be interesting and relevant to students interested in various health sciences, such as medicine, nursing, dentistry, and physical therapy. We have helped many fearful students who succeeded in learning statistics and told us afterward, "I can't believe that my favorite class this semester was statistics!"

The website for this book is http://desheastats.com, where you find links to many of the data sets used in this book. You also may see the data files, the book's figures and the graphing code used to create the figures in our data repository, https://github.com/OuhscBbmc/DeSheaToothakerIntroStats. Students may wish to reproduce results in the book or perform additional analyses. Statistics requires remembering many new terms and concepts, so we have created electronic flashcards for use with existing iPhone and Android cell phone apps. These modern flashcards were motivated by an old quotation from Herbert Spencer in the preface of the book, *The Data of Ethics* (1881): "... for only by varied iteration can alien conceptions be forced on reluctant minds." Everyone has a reluctant mind in some respects. We all want to feel comfortable, accomplished, and smart. Being a scholar requires pushing past the comfort zone of current knowledge, which may require trying different study techniques. Information on the flashcards is available on the website, http://desheastats.com.

If you are an instructor, this book may be used in a number of different types of statistics courses. We intended it mainly for two kinds of courses: (1) a service course for undergraduate students who must complete a statistics course before being admitted to a health sciences program or (2) a first-semester statistics course on a health sciences campus. The book also could be used by students in other disciplines, as the research examples are written by and for people who do not have training in the health sciences. The emphasis is on conceptual understanding, with formulas being introduced only when they support concepts. The text gives a limited number of symbols, mainly those that tend to appear in journal articles, so that learning is not impeded by symbols.

This book differs from other statistics texts in many ways, most notably in the first and last chapters. We begin with an overview of the context for statistics in the health sciences: different kinds of research, variables, inferences about relationships between variables, and so on. Explaining the research context allows students to connect the new material with their existing knowledge, and then they have a framework to which they can add knowledge of statistics. The material from Chapter 1 is revisited throughout the book. Chapter 15 organizes the list of tests and estimates covered in the text, and readers are led through a process of assessing research scenarios and choosing the best analysis plan from among the covered statistics. We routinely are able to cover almost everything in the book in a first-semester statistics course, so students are not buying an overstuffed book.

We are always looking for ways to improve our explanation of statistics, and inspiration can come from anywhere—from walking the dog to watching the movie *Steel Magnolias*, both of which appear in examples in this text. Do you have ideas for improving the book? Please let us hear from you at Lise-DeShea@ ouhsc.edu or LToothaker@ou.edu.

Lise DeShea and Larry E. Toothaker

Acknowledgments

A project as ambitious as this book could not succeed without the support of family, friends, and colleagues. We extend our sincere gratitude to William Howard Beasley for producing the graphs and figures in this book. We deeply appreciate his thoughtful consideration of the most effective ways of displaying information graphically; Dr. Beasley made this book better. We also are grateful to Helen Farrar, who gave us a student's perspective on the manuscript and drew our attention to subtle ways that we could improve our explanations. We thank artist Gary Simpson for sharing photographs of his art. The works in his Disparity series are part of a larger art project called "CommonGround191." The artist writes, "The series embraces the dichotomy of static numbers versus the randomness of my technique.... As with much of our observations: there is always more to see and learn for the vigilant, if time is taken." We invite you to take time to explore his website, www.commonground191.com.

Our thanks also go to the following supporters. Some of them read sections of the manuscript and provided helpful reviews and suggestions; others shared their data and insights about research; everyone cheered us along and supported the completion of this book:

David Grubbs, Marsha Pronin, and everyone else at Chapman & Hall/CRC Press; Robert A. Oster, University of Alabama at Birmingham School of Medicine; Gammon M. Earhart, Washington University School of Medicine in St. Louis; Michael J. Falvo, U.S. Department of Veterans Affairs; Bradley Price; Saeid B. Amini; Kaelyn Kappeler; Daniel J. Buysse, University of Pittsburgh; Robert A. J. Matthews; Chenchen Wang, Tufts Medical Center; Christopher Schmid and Yoojin Lee, Brown University; Virginia Todd Holeman, Asbury Theological Seminary; and Nietzie Toothaker. From the University of Oklahoma Health Sciences Center: Mark Chaffin, Barbara J. Holtzclaw, Jo Azzarello, Barbara Carlson, John Carlson, Mark Fisher, Deborah Wisnieski, John Jandebeur, Trevor Utley, Erica Perryman, Stevie Warner, Maria Cordeiro, Elizabeth Goodwin, Kristy Johnson, Kimethria Jackson, Stephanie Moore, Angela Ramey, Carol Stewart, and students in Lise DeShea's statistics courses.

Authors

Lise DeShea is the senior research biostatistician in the College of Nursing at the University of Oklahoma Health Sciences Center. She earned her MS and PhD in quantitative psychology from the University of Oklahoma, served on the faculty of the University of Kentucky, and worked as a statistician for a Medicaid agency. In addition to conducting research on emergency room utilization, bootstrapping, and forgiveness, she has coauthored two previous statistics books with Dr. Toothaker. She dedicates this work to her parents, from whom she gratefully inherited a work ethic.

Larry E. Toothaker is an emeritus David Ross Boyd Professor, the highest honor given for teaching excellence at the University of Oklahoma. He taught statistics in the OU Department of Psychology for 40 years and retired in 2008. He has seven grandchildren who currently consume much of his time. He also has a hobby of woodworking; he loves to make furniture and to teach others how to do the same. He occasionally teaches courses through OU's College of Continuing Education. He is actively involved in his church and tries to follow "In all your ways acknowledge Him..." every day of his life.

1

The Frontier Between Knowledge and Ignorance

"Science works on the frontier between knowledge and ignorance, not afraid to admit what we don't know. There's no shame in that. The only shame is to pretend we have all the answers."

—Neil deGrasse Tyson, PhD, Astrophysicist

Introduction

Why do people enjoy careers in the health sciences? Many health-care professionals find their greatest satisfaction in relationships with patients—no surprise, given that they have gravitated toward the helping professions. For example, physical therapists say that helping patients is extremely rewarding for them, and physical therapy often is listed as one of the most satisfying careers. Factors such as relationships with coworkers and job security add to the job satisfaction for nurses. Similar factors are related to job satisfaction for physicians, who also say personal growth and freedom to provide quality care are important.

We feel certain that your choice of a career in the health sciences was not motivated by a desire to study statistics.

Yet every statement in the first paragraph is a conclusion emerging from research that used statistics. As you pursue a career in the health sciences, you will need to understand research and draw conclusions as an informed consumer of the results. This chapter will help you to understand the context in which statistics are used. Some of the language will be familiar—*research, statistics, population, sample, control group*—but precise definitions are needed. Just as there are many vocabulary words with highly specific meanings within your health profession, we are introducing you to the language of quantitative research.

And, believe it or not, we chose a helping profession too: teaching. We love working with students, and we hope to make this journey into statistics enjoyable for you. We want you to be a skeptical consumer, even when you are consuming the ideas we present in this book. You might think that statistics is a cut-and-dried topic packed with facts that are widely accepted. Yet statistics textbook authors disagree on many topics, and statistical definitions can vary. We have chosen the content of this book based on what we think is most accurate, widely accepted, and helpful for your understanding. But we want you to know that our way of presenting the information is not the only way.

The Context for Statistics: Science and Research

The largest context in which statistics are used is *science*, which may be defined as an area of study that objectively, systematically, and impartially weighs empirical evidence for the purpose of discovering and verifying order within and between phenomena. Science is a huge undertaking, and people specialize within a science to address questions about narrow topics. Researchers in occupational therapy may have a wider goal of finding the best ways of helping people to participate in everyday activities, but their research may focus on whether stability balls would be helpful for children with attention and hyperactivity concerns (Fedewa & Erwin, 2011).

We just used a word that is familiar to you, but it needs a definition—*research*, or scientific structured problem solving. Each word in this definition contributes to an understanding of the term *research*. The approach must be scientific, which brings in the ideas from our definition of science—impartially weighing empirical evidence, for example. All research involves a problem or question. For the occupational therapists considering stability balls, the question is whether such balls might increase certain children's classroom attention. And those researchers approached their problem solving in a structured and ordered way. They did not choose children from their families to participate in the study; they followed institutionally approved guidelines for conducting ethical research. Also, those researchers had to address many details about conducting the study, such as how they would measure attention, whether different kinds of stability balls would be tested, how many children would be in each group, and so forth. In trying to solve this research problem, the researchers sought solutions to classroom challenges facing children with attention deficit and hyperactivity.

Research may be categorized in a number of ways. One common categorization is qualitative research versus quantitative research. *Qualitative research* refers to an approach to research involving in-depth description of a topic with the goal of obtaining a rich understanding of some phenomenon. Qualitative researchers analyze information that is mostly nonnumeric, such as spoken and written words, although some qualitative researchers may study photographs and pictures. Qualitative researchers' words also are important, as they may make written observations about *participants* or *subjects*—the people being studied. This kind of research uses qualitative methods, which may include interviews or diaries. Qualitative researchers collect rich descriptions and follow proven methods of analyzing nonnumeric information. For example, qualitative researchers have asked, "How do emergency department physicians perceive the contribution of physical therapists?" Lebec et al. (2010) conducted qualitative research to develop an understanding of physicians' views of consultations with physical therapists. These researchers did not seek to collect a little information from a wide range of physicians across many contexts; they interviewed 11 physicians to obtain detailed descriptions of the working relationships in one hospital's emergency department. By studying transcripts of the interviews, the researchers analyzed the insights from these physicians for themes that seemed to be common among all of the interviews. One theme was that physical therapists provide unique expertise that is valuable to the physician, the patient, and the emergency department itself. To understand this theme, can you see how a detailed description would be necessary?

The results of a qualitative study could inform the design of quantitative research, which is the domain of statistics. *Quantitative research* is an approach that relies on prior research to generate research questions and often to make predictions, with the results being analyzed mathematically. Both quantitative and qualitative researchers collect *data*, which is the information collected by researchers for later analysis to answer research questions. In quantitative research, the information may be facts or numbers. The facts could include the gender and diagnoses for all patients in a study. The numbers could include the patients' ages and blood pressure readings. Qualitative researchers collect mostly nonnumeric data, often in the form of descriptions, but sometimes numeric information as well; quantitative researchers collect mostly quantitative data, but sometimes nonnumeric information. Here is an important point: the kind of data—numeric versus nonnumeric—does *not* determine the kind of research being conducted. Qualitative and quantitative research are two entirely different approaches to examining phenomena.

Qualitative research sometimes is described as thick and narrow, meaning tremendous detail is collected on a limited number of participants or situations. Qualitative researchers are looking for a nuanced understanding of a complex phenomenon, such as physicians' perceptions of physical therapists in the stressful environment of an emergency department. Sometimes qualitative researchers will report quantitative information in the results. For instance, Lebec et al.

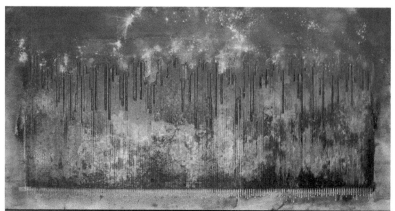

"Disparity One-Life Expectancy at Birth" (4' × 8'), by Gary Simpson, used with permission. This piece lists various countries alphabetically, along with their estimated life expectancies at birth, which are represented by vertical wooden strips.

(2010) said physical therapists in the hospital where the study was conducted usually consulted on 5–15 emergency patients per shift, with shifts lasting 8–12 hours. But the authors' focus was on the analysis of the interviews with emergency department physicians, and the quantitative information about the typical number of consultations in a shift was given simply to help their readers to understand the context of the study.

In contrast to qualitative research, quantitative research sometimes is described as thin and broad, meaning more superficial data are gathered from a wider range of participants or situations. (By the way, *data* is a plural noun.) For example, Wang et al. (2010) conducted a study of patients with fibromyalgia, a condition that amplifies pain signals to the brain. These researchers thought the patients' pain could be eased by tai chi, a martial art that has been described as a kind of meditation in motion. Wang et al. recruited 66 adults with fibromyalgia to participate in a study to compare two groups: (1) those in tai chi classes and (2) those in education classes. The researchers thought patients would have milder symptoms after 12 weeks of tai chi classes than those who participated in a 12-week series of education classes. These quantitative researchers collected nonnumeric data, such as the kinds of medications being taken, but only for the purpose of describing the participants. They mostly collected quantitative data, including numeric scores for fibromyalgia-related symptoms. Unlike qualitative researchers, these quantitative researchers made predictions in advance about the numeric results for those in the tai chi classes compared with those in the education classes. We will explain details of such quantitative research throughout this book.

One misconception about quantitative versus qualitative research is that quantitative research is objective and qualitative research is subjective. Remember when we talked about impartially weighing empirical evidence? That is a goal shared by quantitative and qualitative researchers. We would argue that all research has the potential for being subjective, because it is filtered through the viewpoints of fallible humans who have their own biases and expectations of their study's outcome. A researcher may assume that a patient survey is more objective if patients use numbers to rate their agreement with statements about how well a doctor communicates. But quantitative ratings of agreement, such as 1 (*strongly disagree*) to 5 (*strongly agree*), are no guarantee of objectivity. Surveys can be structured in a way that constrains the respondent and influences the results. For example, suppose we want to measure how well a doctor communicates and are thinking about using a survey in which patients are asked to rate their agreement with these statements:

1. I do my best to communicate well with my doctor.
2. How well I communicate with my doctor is important to my health.
3. Longer appointments cost more money.
4. Everyone needs to help lower health care costs.
5. My doctor spends enough time listening to me.

The first four questions emphasize the patient's responsibilities and the equation "time = money." After agreeing with those four statements, the patient may feel boxed into responding positively to the last question. Further, the first four questions do not tell us about the doctor's communication skills. A better set of questions might be the following:

1. My doctor talks to me in a way that I can understand.
2. My doctor listens to my concerns.
3. My doctor explains different treatment options.
4. My doctor usually asks if I have any questions.
5. My doctor spends enough time listening to me.

Now the questions focus on the doctor, not the patient. The order of the questions does not seem to force respondents to answer in a certain way. Some respondents whose doctors do not communicate well might feel uncomfortable about saying "strongly disagree" with the five statements mentioned previously. We might consider including some items that ask questions in a negative way, such as, "My doctor doesn't care about my opinion." This statement may be a chance to say "strongly agree" for the respondents who are uncomfortable with saying they disagree. Quantitative researchers should not limit their surveys to questions like the ones shown here. They may miss the opportunity to reveal an unexpected finding if their surveys fail to include open-ended questions that require written descriptive responses. When the first author created a survey for a Medicaid agency about a program that provided limited family planning benefits, she included an open-ended question asking the respondents what they would change about the program. The responses revealed that many respondents thought the program covered more services than it actually did, which led to an outreach effort to educate those in the program about its limited benefits.

All researchers should remain mindful of the fact that each person has a limited world view, which could influence the planning and interpretation of research. One way to combat this potential problem is by involving other researchers. Lebec et al. (2010) gave transcripts of interviews with emergency department physicians to two raters, who independently analyzed the transcripts for themes repeated by different physicians. These authors used an accepted, rigorous method that involves instructing raters to locate specific words, phrases, or ideas. Transcripts allow raters to focus on the words being used by the participants instead of verbal cues that the raters would hear if they listened to recorded interviews. (Other qualitative researchers may have their primary interest on those verbal cues, such as the phrases or ideas that make the respondents hesitate or stumble over their words.) The findings by different raters can be analyzed for the degree of agreement. Higher agreement among raters would support the notion that a certain theme had been identified. To check whether the tentative list of themes identified by the raters reflected the reality of life in the emergency department, the researchers shared the results with one of the hospital's main physical therapists, who confirmed the themes seemed to be realistic.

Both quantitative and qualitative research studies have their place, depending on the researchers' goal. These approaches can be combined in *mixed-methods* studies, which contain two parts: (1) an in-depth qualitative study, such as a case study or focus group, and (2) a quantitative study with the attention placed on the numeric results from a broader base of people. Such studies combine the strengths of qualitative and quantitative research. Qualitative research provides a deeper understanding of a phenomenon, and quantitative research shows results from a larger cross section of people who may be affected by the phenomenon. In either kind of research, the investigators must remember that one study is just that—one study, not the definitive final story about the phenomenon under investigation. Science requires replication across time and different situations to demonstrate that the results of one study were not a fluke.

Check Your Understanding

SCENARIO 1-A

We are nurses working on a hospital's quality improvement project involving children treated for asthma in the emergency room. We are using electronic health records to extract data on age, gender, number of emergency visits in the last 12 months, prescriptions for rescue inhalers or maintenance medications, and parental reports on the frequency of the children's use of those medications (if the information was reported). 1-1. Identify the numeric and nonnumeric data being collected. 1-2. Explain whether the scenario seems to describe quantitative or qualitative research.

(Continued)

Suggested Answers

1-1. Numeric data are the ages, numbers of emergency visits, frequency of medication use. Nonnumeric data are gender, prescribed medications. 1-2. Quantitative research. A year's worth of records are being examined for superficial facts and numbers that were chosen in advance for investigation by the quality improvement team. If the research were qualitative, the researchers might conduct extensive interviews with a small number of patients and families in pursuit of a thorough understanding of their experience in the emergency department.

Definition of Statistics

We have been using the term *statistics* without a definition. There are many ways of defining this term. We can talk about statistics as an area of research; we have conducted research in which the statistics themselves were subjected to computer simulations to see how they perform under various conditions. But *statistics* most commonly refers to numerical summary measures computed on data. Some statistics are descriptive, which is the case when we compute the average age of patients in a study. The arithmetic average, or *mean*, is a statistic, and it fits this definition because we are summarizing the patients' ages. We will talk more about descriptive statistics in Chapter 2. Other statistics are used to make decisions. You have probably heard statements such as, "The results showed a *significant difference* between the groups." The term *significant* has a specific meaning in statistics, which we will cover later.

As we have seen, statistics is a topic within quantitative research, which involves structured problem solving using scientific methods. Scientific methods can be defined in many ways; Figure 1.1 illustrates the quantitative research process as a cycle.

In Step 1, we encounter a problem or research question. Before taking a course in statistics or research methods, most students would think of research as reading articles and books about a topic; researchers call this activity a *literature review*, in which they identify the boundaries of the established knowledge on a topic and refine their research question. Research questions will be grounded in theory, and important research will add to theory. *Theory* can be defined as an organized, research-supported explanation and prediction of phenomena. For example, neuroscientists have proposed theories about how the brain organizes complex tasks. In Step 1, researchers have examined information from research literature and identified a question that needs to be answered to expand our understanding of an area of science.

After becoming familiar with published research, researchers stand on the boundary of the existing knowledge and point toward an unexplored area, and they speculate: what is happening over there? In Step 2, quantitative researchers

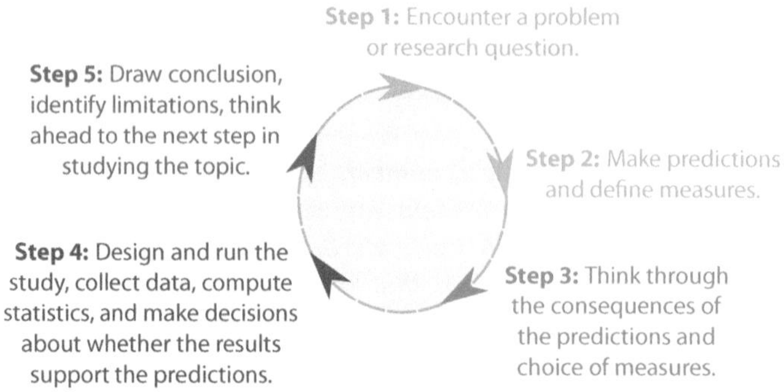

Step 1: Encounter a problem or research question.

Step 2: Make predictions and define measures.

Step 3: Think through the consequences of the predictions and choice of measures.

Step 4: Design and run the study, collect data, compute statistics, and make decisions about whether the results support the predictions.

Step 5: Draw conclusion, identify limitations, think ahead to the next step in studying the topic.

Figure 1.1

Cyclical nature of quantitative research. The process of quantitative scientific investigation begins with someone encountering a problem or research question. By the end of the process, the researcher typically has identified new research questions to be investigated.

formulate predictions and define what they will measure. How we state our predictions and which measures we choose can determine the kinds of answers we can get. If researchers studying stability balls for schoolchildren with hyperactivity do not ask the teachers how stability balls affected the children's attention span, important information might not be collected. Next, researchers must consider whether they will be able to answer their research question; this is what happens in Step 3. Sometimes they think of consequences that they did not intend, requiring them to revisit Step 2 and modify their predictions. Step 4 contains many details that people often associate with research: designing and running the study, collecting data, computing statistics, and making decisions about whether the evidence supported the predictions. When we say *designing a study*, we mean the process of making decisions about the plans for the study; for example, we may decide the study requires multiple occasions of measurement so that change across time can be assessed. Step 5 is drawing conclusions—the head-scratching about the meaning of the results, acknowledging the limitations of the results, and identifying new research questions that must be explored with a new study. Identification of new research questions makes the process cyclical because the researcher returns to Step 1: encountering a problem.

Check Your Understanding

SCENARIO 1-B

We are studying emergency room visits by young patients with asthma. A colleague says he has done prior research on asthma. He shows us several articles about ways to encourage the use of maintenance medication in young patients with asthma. 1-3. Explain whether the identification of the articles constitutes research.

(Continued)

1. The Frontier Between Knowledge and Ignorance

SCENARIO 1-C

Suppose we belong to an e-mail list for public health researchers. We receive an e-mail from an inexperienced researcher who writes, *I have some data on the number of flu shots given in each county, the number of reported flu cases, the number of prescribers per 1,000 residents, etc., and I don't know where to start. Do you have some recommendations on the statistics I should compute? I guess I could graph the data and look for patterns, but I think I should be computing some statistics as well. How should I get started?* 1-4. Where does this researcher stand in terms of the steps in scientific methods? 1-5. Based on what you have learned about scientific methods, what would you tell her?

Suggested Answers

1-3. Finding articles that other people wrote is part of the research process, but by itself this step is not research. In Figure 1.1, the literature review may be part of Step 1 or Step 3. 1-4. The researcher appears to be at Step 4, which includes collecting data and computing statistics. 1-5. Based on scientific methods, we would recommend that the researcher look back at the research questions that motivated her to collect the data in the first place. Those questions will help to determine how the data should be analyzed.

The Big Picture: Populations, Samples, and Variables

When we conduct quantitative research using human subjects, we care about these participants, but we also care about other people who may be similar to our participants. If our results will apply only to our participants, then we have not made a contribution to science. The larger group of people to whom we would like to generalize our results is the *population*. It is important to recognize that when we use the term *population*, we rarely are talking about everyone in the world or even everyone in a country. We are talking about everyone to whom our results may apply—and our results may generalize only to preadolescent children with type 1 diabetes. Sometimes researchers do care about generalizing their results to all the citizens of a country, in which case the target population carries the same meaning to you as the word *population* probably had before you started reading this book.

We will use the term *population* to refer to entities that share a characteristic of interest to researchers. These entities are not always individual people. The entity that is measured defines the *unit of analysis*. If we were studying HIV screening rates in 120 urban hospital emergency departments, the units of analysis would be the 120 emergency departments. The 120 emergency departments may provide us with information that we can generalize to all urban U.S. hospital emergency rooms, so our population may be defined as "all urban hospital emergency rooms in the United States." The 120 departments comprise the *sample*, or a subgroup of the population. Be sure to keep these two terms—sample and population—separate from each other. Students sometimes mistakenly combine them into one term. If it is a sample, it cannot be a population, and vice versa. A sample is a limited number of entities or subset of the population. Our data are collected from samples. A population, in contrast, often is large and therefore unobtainable. That is why we need samples, which are smaller and manageable. A population also may be hypothetical. In the example of preadolescent children with type 1 diabetes, we care not only about preadolescent children right now but also about the babies who eventually will be diagnosed and reach the age of the children in our study. Later in the text, we will introduce a data set on food hardship and obesity rates for the 50 American states plus the District of Columbia (DC). Do the 50 states and DC represent a population? We could argue that the rates of food hardship and obesity represent only one year. The population is all possible years of food hardship and obesity rates, and the sample is the single year's results.

Some researchers make their careers in the field of *epidemiology*, the study of the distribution and spread of health conditions and diseases in human populations. Epidemiologists may be involved in tracking cases of whooping cough across a country and provide information to health agencies on encouraging vaccinations to control its spread. In this case, the epidemiologists would hope to obtain information about all infections to provide accurate surveillance of the populace. Most of the research to be described in this book will not focus on populations, but rather samples of participants serving as representatives of everyone in the population of interest.

Let's define another term that may look familiar: *variable*. A variable is a quantity or property that is free to vary, or take on different values. Obesity rate is a variable because it is a quantity that is not constant. States vary on their obesity rates, so the obesity rate is a variable. Is "female" a variable? No, it is not. But gender is a variable; it is a property that can have different values—typically, we limit the values to "male" or "female." Gender is an example of a qualitative or categorical variable. It also sometimes is called a *discrete variable*, because it has a limited number of possible values between any two points. Categorical variables are always discrete, but the reverse is not true. A discrete variable can be quantitative. The number of times that an elderly patient has fallen in the last year is a quantitative, discrete variable; if we were looking at the number of falls last year for patients in a nursing home, and we had data from 0 to 22 falls, there is a limited number of possible values between those two numbers. No patient would have 8.769 falls. In comparison, weight could be called a *continuous variable*, because theoretically it could take on infinitely many values between any two points. If those nursing home patients were weighed, it is conceivable that a patient might weigh 138 lb, 138.5 lb, or 138.487 lb. Although a variable may be continuous, we measure using values that make sense. We usually round our numbers to a value that is precise enough to see differences among participants without getting into unmanageably large numbers of decimal places.

Check Your Understanding

SCENARIO 1-D

Is there a difference between the average stress levels of people who exercise three or more times a week, compared with people who exercise once a week or less? With the help of four family medicine clinics, we recruit 45 volunteers who exercise at least three times a week and 45 people who exercise once a week or less. We give them a survey that yields a stress score for each person, where scores range from 0 to 50 and a higher score means more stress. 1-6. What is the population? 1-7. What is the sample? 1-8. What is the unit of analysis? 1-9. Identify one continuous and one discrete variable.

Suggested Answers

1-6. The population is the larger group to which we would like to make generalizations. The first sentence of the scenario indicates that we are interested in people who exercise three or more times a week, compared with people who exercise once a week or less, so it sounds as if there actually are two populations of interest. 1-7. There actually are two samples, each containing 45 people, who differ in terms of their frequency of exercising. 1-8. In this study, the unit of analysis is the individual person. 1-9. One continuous variable may be the stress score, which ranges from 0 to 50. One discrete variable may be exercise; people belong to one of two discrete groups, depending on whether they exercise frequently or seldom.

Generalizing from the Sample to the Population

How a sample is obtained affects the strength of our generalizations from the sample to the population. Let's say we are studying preadolescent children with type 1 diabetes. Our participants, who volunteered for the study with a parent or guardian's permission, are patients at rural clinics in Mississippi. The Centers for Disease Control and Prevention says Mississippi has a high obesity rate. Should we use the results of our study to make generalizations about all American children with type 1 diabetes? In most cases we can generalize only to people who are similar to those in our sample. More specifically, the method of obtaining the sample will affect our generalization. In this example, we have volunteers. Are all preadolescents with diabetes similar to those whose parents let them participate in research? Maybe—or maybe not. We have no way of knowing. A systematic influence on a study that makes the results inaccurate is *bias*. By using volunteers in only one state, we may be biasing our results. This is an example of regional bias. There are many other kinds of bias. For example, a study relying on volunteers might have self-selection bias. The people who volunteer for research may differ substantially from those who do not volunteer.

Suppose we want to obtain a sample from the population of *all* American preadolescents with type 1 diabetes. If we want to strengthen our ability to generalize from the sample to the population, we need to consider how we are getting our sample. The sample would need to be drawn from the population using a method that does not systematically include or exclude certain people. If the process is biased, then we cannot generalize to everyone in the population of American preadolescents with type 1 diabetes. One way to reduce the bias is to conduct *simple random sampling* from the population. Simple random sampling is a process of obtaining a sample such that each participant is independently selected from the population. By selecting each person independently, the process is intended to reduce bias and increase our chances of obtaining a sample that is more representative of the population. Imagine that we could assign a number to every preadolescent child with type 1 diabetes. We could put all the numbers in a hat, mix them up and draw out one child's number for the sample ... then mix the numbers again and draw out the number of another child ... and repeat this process until we have obtained an adequate *sample size*, or number of people in the study. (The process typically would be computerized.) For this example, let's say an adequate sample size is 50 children; a journal article would use the letter N to indicate the sample size: $N = 50$. Each time we draw out a child's number, we have conducted an independent selection, and the 50 children will constitute a simple random sample.

Why would researchers go to the trouble to conduct simple random sampling? We need to look at the reason for drawing any sample: we cannot get the population, but we want to make generalizations about the population. We do not want our generalizations about all preteens with type 1 diabetes to be biased; bias could be introduced if we only studied children with type 1 diabetes living in rural Mississippi. If we want to generalize to all preadolescents with type 1 diabetes, we would be better served if we could randomly sample from that population, removing the bias introduced by studying children in rural Mississippi. In the long run, random sampling will produce samples that are representative of the population of preadolescent children with type 1 diabetes. The quality of our generalization from the sample to the population depends on the process we use to obtain the sample. With simple random sampling we have a higher quality of generalizability, or higher *external validity*. We call it *external validity* because we are taking what we observed in the sample and generalizing those results outside to the population.

The definition of simple random sampling did not say anything about the likelihood of being chosen for the sample. When we have a huge population, the chance of one person being chosen may be practically equal to the chance of any other person being chosen. But what if we work for a hospital and want to draw a simple random sample from among 256 children with type 1 diabetes who were treated in the emergency room in the last three years? For the first child randomly sampled, the chance of being chosen is 1/256. But for the second child selected, the chance of being chosen is 1/255, for the third child it is 1/254, and so forth. *Everyone having an equal chance of being chosen is not necessary for a sample to be random.* The process is the important aspect: making sure that each child is chosen independently. If each child is selected independently and bias is minimized, researchers will have greater confidence in generalizing the sample's results back to the population of interest.

Clearly we cannot assign a number to each preadolescent American child with diabetes and draw a sample randomly for our study. Most research studies use *convenience samples*—groups of participants conveniently available to researchers. These samples sometimes are called *judgment samples*, because researchers make a judgment about whether they adequately represent the population. The use of convenience samples introduces a limitation. When we ask, "To whom may we generalize these results?" we are asking about the strength of the external validity. The answer depends both on the definition of that population and the process of obtaining the sample from that population. Many authors of journal articles will not specify the population to which they would like to generalize their results; they will imply the population. It is up to skeptical readers like you to notice if the authors seem to be whispering and not drawing attention to a limitation, such as the use of a convenience sample at one location and one point in time. In the conclusions of journal articles, it may seem as if the researchers are shouting their generalizations, without regard to the limitations that may have been mentioned quite briefly earlier in the article. Good researchers know that all research has limitations, so we should be willing to include statements such as, "External validity may be limited because our convenience samples were patients at rural clinics in Mississippi." (By the way, when we say *journal articles*, we are talking about scientific journals that publish the results of research, such as *The New England Journal of Medicine*. We are not talking about magazines or online sources like Wikipedia, which publish articles that have not been reviewed by other researchers.)

We have presented two possibilities for obtaining samples: simple random sampling and convenience sampling. Many other sampling methods are possible: *snowball sampling*, where participants with specific, rare conditions or experiences may tell us about other people like themselves who could be recruited for the study; *stratified random sampling*, in which random samples are drawn from within strata, such as age groups; and so forth. The two main methods we have presented provide a context that allows contrasts to be drawn between studies with potentially strong external validity (i.e., studies with random sampling) and studies with a likelihood of weak external validity (i.e., studies with convenience sampling). In addition, a single study is insufficient for establishing that a phenomenon exists or a treatment works. As skeptical consumers of research findings, we must consider whether a study's single results are unusual or whether subsequent studies have reproduced the effects being reported. Replication across different situations and people makes research findings more trustworthy in terms of external validity.

SCENARIO 1-E

We have mentioned the study by Wang et al. (2010), who conducted a study of tai chi as a possible way of easing the pain of fibromyalgia. These researchers recruited 66 adults with fibromyalgia who had moderate to severe musculo-skeletal pain in all quadrants of the body for at least three months. People with certain other conditions (e.g., lupus) were excluded. The researchers used the Fibromyalgia Impact Questionnaire (FIQ), which measures pain, fatigue, stiffness, mental and physical functioning, and so on. Higher FIQ scores mean worse symptoms. 1-10. What is the population to which the researchers probably would like to generalize? 1-11. What is the sample? 1-12. From this limited description of the study, how would you characterize the external validity of the study? 1-13. Suppose the researchers wanted to run a second study. They identify a network of support groups for patients with the disease, and they contact every third person listed in a membership directory. Explain whether this process produces a random sample.

Suggested Answers

1-10. The researchers probably wish to generalize to all adults with fibromy-algia who do not have certain other conditions, like lupus, and who have persistent moderate to severe musculoskeletal pain in all quadrants of the body. 1-11. The sample is the 66 participants. 1-12. The external validity may be limited; there is no mention of random sampling from the population of interest. The participants probably lived in the same area of the country as the researchers. 1-13. The second study is using systematic sampling, not random sampling. The key is whether independent selection was conducted, which is necessary for a random sample. If the sample began with the third person on the list of potential participants, then everyone else's inclusion in or exclusion from the sample was decided. Selection was directly affected by the location of each name, relative to the location on the list for the first person chosen, so this sample was not randomly chosen.

Experimental Research

In this section, we will continue to take words that you have heard before and define them as they are used within quantitative research. We also will divide quantitative research into three categories: *experimental research, quasi-exper-imental research, and nonexperimental research.* The kind of research has huge implications for the degree to which we can draw causal conclusions.

Researchers are most interested in the relationship between variables, not the variables in isolation. Kamper et al. (2012) studied the relationship between pain and interference with normal movement. Patients who have suffered whiplash may avoid moving in certain ways because it hurts. It makes sense to assert a *causal relationship*, in which one variable (severity of whiplash injury) causes changes in another variable (amount of pain). These researchers proposed that disability, or interference in motor function, also may be related to fear of movement. Patients with whiplash may fear the pain and avoid movement, which then can lead to an even greater loss of ability. Perhaps patients with less pain have less fear of movement, whereas patients with more pain experience greater fear of movement. When we have a *predictive relationship* between variables, a change in one variable (amount of pain) corresponds to a change in another variable (amount of fear). But our speculation about whether the relationship is causal or predictive is not enough. We must conduct a specific kind of study to be able to claim that one variable *caused* changes in the other variable. Let's examine different ways of studying relationships between variables, then we will specify the kind of study required to assert a causal relationship exists between variables.

Suppose we think people will become more alert as long as they *believe* they are drinking caffeinated coffee and less alert if they *believe* they are drinking decaffeinated coffee. We could run a study in which we make coffee for a small office. On some days we could tell the workers that the coffee has caffeine, and on other days we could say the coffee is decaffeinated, even though we always gave them coffee with caffeine in it. Each day at 11 a.m. we could ask the coffee drinkers how alert they feel. Will the workers report being more alert on the days they thought they were drinking caffeine? In other words, is there a relationship between people's beliefs about what they are consuming and their later alertness?

There are problems with this proposed study. People might feel more stressed or tired on certain days of the week, like Monday, which also could explain any difference in alertness. Maybe people are more likely to use sleep medications later in the week than at the beginning of the week. Figure 1.2 illustrates some variables that could influence the participants' alertness. The arrows indicate a direction of influence. For example, amount of sleep may influence the differences in participants' alertness.

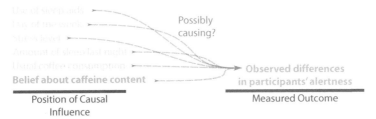

Figure 1.2

Variables that may influence an outcome. The researchers in the caffeine study were interested in people's alertness. Many variables could affect alertness. This figure shows a number of variables, including the participants' beliefs about whether they consumed caffeine, as potentially causing the differences in alertness.

This example serves as a contrast to a better approach. Depending on how a study is designed, we may or may not be able to say which variables are responsible for the results. What can researchers do to zero in on the relationship between beliefs and alertness, without the relationship being affected by variables like sleep aids or day of the week?

Researchers can use *random assignment*, a process of placing participants into groups such that the placement of any individual is independent of the placement of any other participant. This definition may sound like random sampling, but random assignment is a process that takes place after a sample has been drawn. Random assignment prevents the groups from being created in a biased way. Bias would exist if the groups were determined by the order of arrival for the study, where we could end up with early rising, more alert participants being in the first group. Before we tell the sample about whether caffeine is in the coffee, we need to ensure that the different groups are as interchangeable as possible, so we randomly assign people to groups. After that, we can make the groups different based on what we tell them about what they drank (caffeine/no caffeine). Later we can measure whether their belief affected their reported alertness.

The caffeine example was inspired by actual research. Dawkins, Shahzad, Ahmed, and Edmonds (2011) investigated whether an expectation about drinking caffeine led to greater alertness, but they also switched out the kind of coffee. The researchers had four groups. They randomly assigned participants to drinking either caffeinated or decaffeinated coffee. Then within these groups, half of the participants were randomly assigned (secretly) to being deceived or truthfully informed about the coffee's caffeine content. All groups were measured on the same day. The purpose of random assignment was to make the groups as interchangeable as possible before the participants' experience was changed by researchers. Any differences in alertness levels between groups could be the result of mere chance or the researchers stepping into the situation: whether the coffee actually contained caffeine and what the participants were told about the caffeine content. The researchers' act of changing the experience of different groups of participants is called *manipulation* or *intervention*.

Does random assignment actually make the groups interchangeable? In our experience, random assignment usually works. We have run studies in which we have randomly assigned people to groups and then compared the groups on many variables—proportion of males versus females, average age, health status—and we have found the groups to be comparable. But we also have consulted on studies in which random assignment failed to make the groups comparable in advance, and special statistical analyses were used to try to compensate for these preexisting differences in groups. Random assignment does not guarantee that any particular study will have groups that are extremely similar before the intervention. It is possible to have random assignment (also known as *randomization*) that results in one group having a slightly higher average age, more females, or a greater number of heavy coffee drinkers. But in the long run, randomization will control those interfering variables. When we say *control*, we mean we are limiting the effect of those variables that could make it harder for us to zero in on the relationship between beliefs about caffeine and alertness. (Randomization will not solve the problem of all interfering variables in all situations. A longitudinal study involves repeated measures over time, and many things can happen across time to interfere with the results.)

Why did these researchers need to study groups of people? Within a particular study, the determination of cause and effect requires measurements have been taken on person after person. Suppose these researchers had studied only one person in each condition—one person who knowingly drank caffeine, one person who drank caffeine but was deceptively told it was decaffeinated, one person who knowingly drank decaffeinated coffee, and one person who drank decaf but was deceived about the caffeine content. Any observed difference in alertness of these individuals could be an arbitrary difference in the four individuals' alertness that day; one person who had a bad night's sleep could affect the results. But if researchers study many people in each of those conditions and find a statistically noteworthy difference in their alertness, they can feel more confident that the result is not limited to a few individuals. In other words, *statistical replication*, or having a sample size greater than 1, is necessary in quantitative research. Would two participants per group be enough? Probably not. The question of whether a study has an adequate number of participants is complex and will receive further consideration later in the text. (Another issue in science is the replication of entire studies to show that a study was not a one-time arbitrary result.)

The coffee study by Dawkins et al. (2011) involved (1) random assignment of people to groups, (2) the researchers' manipulation of their experiences, and (3) statistical replication. These three details are required to say that a study is an *experiment*. The variable that the researcher manipulates is called the *independent variable*. It is independent of the results, and in fact the researcher intervenes before the outcomes are measured. Dawkins et al. manipulated two independent variables: actual caffeine in the coffee (yes or no) and whether the participants were led to believe they were drinking caffeine (yes or no). Each of these independent variables had two conditions. The possible conditions within an independent variable are called *levels* of the independent variable. The study ended up with four groups, based on all combinations of the two levels of actual caffeine and the two levels of belief about caffeine.

We want to determine whether the participants' belief about the beverage *causes* differences in the average alertness, the outcome variable. The outcome variable in an experiment is the *dependent variable*. The dependent variable *depends* on the effect of the independent variable, so it is collected after the manipulation. If the participants' belief about the beverage causes differences in their average alertness, then the scores on the dependent variable *depend* on the participants' belief. Alertness also may depend on actual caffeine—or the combined effect of caffeine content and participants' belief about the presence of caffeine.

As we have seen, many other variables could influence alertness. These potentially interfering variables are called *extraneous variables*. They also may be called *confounding* or *lurking variables*. Extraneous variables compete with the independent variable in attempting to explain any differences in the dependent variable. Figure 1.3 shows a diagram to illustrate a relationship between belief about caffeine and alertness (omitting actual caffeine for simplicity). (Figure 1.3 intentionally has not identified some extraneous variables, which use the abbreviation *EV*; we cannot predict what all of the possible lurking variables may be.)

By randomly assigning people to groups, the researchers were attempting to equate the groups in advance on many extraneous variables. Dawkins et al. (2011) wanted to show whether belief about caffeine *caused* differences in alertness. But extraneous variables try to compete with the independent variable as causal influences on alertness. This is the reason for running an experiment: to show whether a causal relationship exists between variables. These researchers had to isolate the participants' belief about the coffee in the position of causal influence, keeping extraneous variables under control so that they could zero in on the effect of beliefs on alertness. After that was accomplished, the researchers made the groups different by telling one group that it consumed caffeine and telling the other group that the coffee was decaffeinated. The group that was told the coffee was decaf would be unlikely to contain all the participants who had the least amount of sleep last night or all the participants who use sleep aids. Those sleep-deprived, medication-using participants most likely would be spread out

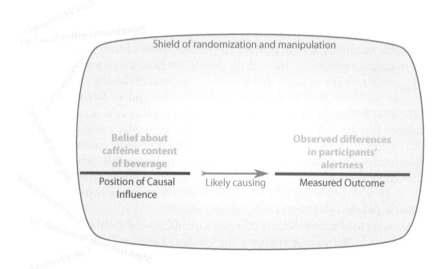

Figure 1.3

Position of cause, position of effect. An experiment in which the researcher manipulates the participants' belief about whether they consumed caffeine isolates their belief in the position of causal influence. The shield of randomization and manipulation keeps extraneous variables away from the position of being able to causally influence the participants' alertness levels. The researchers randomized the participants to groups, hoping to equate the groups on those extraneous variables. The researchers then manipulated the groups' beliefs about caffeine intake, placing that independent variable in the position of causal influence.

among the groups because of the random assignment. By randomly assigning participants to groups, the researchers controlled the effect of many extraneous variables, allowing the independent variable to dominate in the position of causal influence. Randomization and manipulation together shield the relationship between the independent variable and the dependent variable, protecting it from the extraneous variables. Then the researchers could say that differences in belief about caffeine caused differences in alertness.

Random sampling and random assignment have similar-sounding definitions, but they have different purposes. A simple random sample involves a process of choosing each participant independently from the population, whereas random assignment is a process of placing participants independently into groups. The purpose of simple random sampling is the avoidance of bias in the sampling process. If we were to rely on a convenience sample that was biased toward children with type 1 diabetes living in rural Mississippi, we could not generalize to all preadolescents with type 1 diabetes. The convenience sample would limit our external validity. The purpose of random assignment to groups is much different. Random assignment focuses our attention on the causal relationship between the independent and dependent variables. By randomly assigning people to groups, we are attempting to control extraneous variables so that the manipulated independent variable's effect on the dependent variable can be detected.

Notice that we did not describe the process that Dawkins et al. (2011) used to obtaining their sample for the caffeine study. In fact, they used a convenience sample. It was still an experiment, and the researchers may feel quite confident about asserting a causal relationship between the independent and dependent variables. But one may question whether we can generalize the results from that sample to a larger population because of the method of obtaining the sample.

Another term sometimes is used to specify a certain kind of experiment. Researchers in the health sciences often conduct *randomized controlled trials*, in which participants are randomized to groups and typically observed across time, with one group receiving a *placebo*. A placebo is a sham condition that is identical to the treatment condition except the sham condition has no effect. In medical research, a placebo may be a sugar pill or a saline injection. The group that receives a placebo or no intervention is called the *control group*. An *experimental group* receives an intervention. If an experimental group takes a nightly pill containing a drug that the researchers think will control symptoms of gastroesophageal reflux disease, the control group would take an identical-appearing nightly pill—but the control group's pill would be a placebo. The different experimental conditions sometimes are referred to as the *treatment arms* of the study. The U.S. National Institutes of Health's website provides a good overview of clinical research: http://www.nichd.nih.gov/health/clinicalresearch/. Researchers, journal editors and methods experts also have collaborated on guidelines for transparent reporting of clinical trials: http://www.consort-statement.org/.

Although randomized controlled trials are considered the gold standard of determining causality in research, the knowledge of receiving attention in a study could influence participants (Kaptchuk, 2001). That is why some studies involve a control group that receives some sort of attention. We have mentioned such a study: Wang et al. (2010), in which the researchers wanted to know whether a tai chi class would help people with fibromyalgia. The researchers did not want the control group to receive no attention at all; otherwise, at the end of the study, it might appear that people with fibromyalgia have fewer symptoms simply because they received attention, not because of the effect of tai chi. The control group participated in a class that met for 1 hour twice a week for 12 weeks, the same amount of time as the tai chi class. The control group received general health information and participated in stretching exercises. Because this control group received some attention, it would be called an *attention-control group*. By giving some attention to the control group, the researchers can make an argument for any group differences at the end of the study being attributable to the manipulated independent variable and not the mere attention paid to those in the tai chi group.

Check Your Understanding

SCENARIO 1-E, Continued

When Wang et al. (2010) conducted a study of tai chi, they randomly assigned 66 adults with fibromyalgia to one of two groups. One group took a tai chi lesson twice a week for 12 weeks. The other group participated in a twice-weekly fibromyalgia-related wellness lesson followed by stretching exercises. The researchers used the FIQ, on which higher scores mean greater severity of symptoms. 1-14. How do we know that this study is an experiment? 1-15. What kind of variable is FIQ score? 1-16. What kind of variable is the type of twice-weekly activity? 1-17. Suppose a classmate, Kay Study, says, "I'm concerned that the age of the participants might interfere with our ability to assert there is a causal relationship between the group membership and the FIQ scores." What do we tell her?

Suggested Answers

1-14. It is an experiment because the participants were randomly assigned to groups and the researcher manipulated their experience: whether they took tai chi classes or wellness classes. The study also includes statistical replication to allow the researchers to observe an effect across many participants. 1-15. FIQ score is a quantitative variable serving as the main outcome measure, so in an experiment we call it the dependent variable. 1-16. The type of twice-weekly activity (tai chi or wellness class) is a categorical or qualitative variable serving as the independent variable. 1-17. Age is a characteristic of participants that would be one of the extraneous variables controlled by randomization to groups. We would tell Kay that there is no reason to believe that one group would be different in age from the other group, so age should not impact our ability to say whether the kind of activity caused differences in FIQ scores.

Blinding and Randomized Block Design

When Dawkins et al. (2011) conducted their caffeine study, they took an additional step to control extraneous variables: *double-blinding*. A study is double-blind when the participants are in the dark about which group they are in—and so are the researchers who actually interact with the participants. Deception was involved; half of the participants drank a beverage that did not match what they were told about the beverage. Some of the researchers had to know which participants received caffeine and keep track of the results; these researchers made the coffee in a separate room and did not interact with participants. They poured the coffee in a cup that looked like any other cup used in the study, and they gave it to a "blinded" researcher. The researchers in the know told the "blinded" researcher to deliver the cup to a specific participant. In this way, the blinded researcher was prevented from unintentionally communicating any information to the participants, who also were blinded about the manipulation. Double-blinding is important because a large body of research shows that if researchers know about group membership, they *will* treat the groups differently, and this researcher bias is not controlled by random assignment. (For further reading, see Rosenthal, 2009.) Some studies are only single-blinded, meaning the researchers know which condition each participant is in. If a study has no blinding, the researchers must remember the potential effects on the behavior of those assisting with the study and on the behavior of the participants, which could influence the study's results.

Another way of controlling extraneous variables is to limit a study to participants who are all alike in some way; for example, instead of having to contend with the extraneous variable of age, we could limit a study to young adults. The trade-off for this decision is that we will be able to generalize only to a population of young adults. If our research question concerns mainly young adults, then this decision is justifiable. We also can include an extraneous variable in the study's design. If we had reason to believe that men and women may respond differently when deceived about their coffee's caffeine content, we could incorporate gender as a factor to be studied. Within each group of males and females, we could randomly assign them to one of the four conditions: actual caffeine (yes or no) combined with belief about caffeine (caffeine present or absent). The study then would have eight groups and would have a *randomized block design*. A randomized block design contains at least one variable to which participants cannot be randomly assigned, like gender. A *blocking variable* is a categorical variable with participants in naturally occurring or predefined groups. Gender can be a blocking variable. Researchers use a randomized block design when they have a reason to compare such groups. Participants within each block are randomly assigned to levels of independent variables. Then the effect of the blocking variable, like gender, can be taken into account in the statistical analysis of results.

Do researchers ever conduct studies that lack both randomization and manipulation? Yes, and it is called *nonexperimental research*. Nonexperimental research can lead to important discoveries about health risk factors and identify possible interventions for future research.

Nonexperimental Research

Now that we are aware of the impact of extraneous variables on the outcome variable, it may seem odd to think we would perform *nonexperimental research*, which does have statistical replication, but does *not* have manipulation of an independent variable or random assignment to groups. Yet nonexperimental research can identify which variables explain an outcome so that health professionals can be aware of certain risk factors or can design interventions to address those risks. Nonexperimental research also is called *observational research* or *descriptive research*. We will use all three terms interchangeably.

There are important studies in which random assignment was impossible. Field, Diego, and Hernandez-Reif (2009) reviewed a number of studies about depressed mothers and their newborns. One study focused on the babies of women who had suffered prenatal depression. These babies were less attentive than babies of nondepressed women. Attentiveness could have long-lasting effects on learning, so knowing about the relationship between the mothers' prenatal depression and babies' attentiveness could lead to interventions with pregnant women who have depression. This study is an example of nonexperimental research. Prenatal depression is a variable that is suspected of having an influence on the outcome variable, baby's attentiveness, but the researcher cannot manipulate depression or randomly assign participants to its levels.

Some researchers would read the description of this study and say that the independent variable is prenatal depression (presence/absence). We prefer to call it a *predictor variable*, which is analogous to an independent variable, except researchers cannot manipulate it or randomly assign participants to its levels. Please be aware that this book will reserve the term *independent variable* to refer to a variable manipulated by the researcher. Although we are in the minority of textbook authors to insist on this distinction, our teaching experience has shown us that using different names reinforces with students the idea that causality cannot be demonstrated with all variables. The predictor variable of maternal prenatal depression *corresponded to* or *predicted* differences in attentiveness of their babies, but the study cannot establish a causal link between prenatal depression and babies' attentiveness. (Journal articles also may refer to the predictor as an *explanatory variable*.)

The outcome variable in observational research also will have a special name in this book: *criterion variable*. Again, we are in the minority in insisting on the term *criterion variable* for the measured outcome variable in descriptive research; the researchers who conducted the study of mothers' prenatal depression may have referred to attentiveness as the dependent variable. But we are convinced that this distinction helps students to keep track of the kinds of conclusions they may draw from different sorts of studies. Using the term *predictor* can help us remember that we can establish only a *predictive* relationship with a criterion variable. (Other books may use the term *response variable* or *outcome variable* in this context.) Figure 1.4 illustrates the lack of control of extraneous variables in the study of maternal prenatal depression and babies' attentiveness.

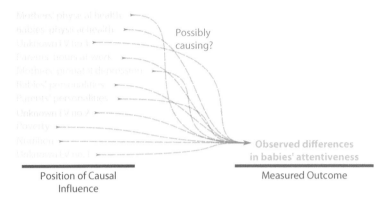

Position of Causal Influence

Measured Outcome

Figure 1.4

Observational study without control of extraneous variables. The study of maternal prenatal depression as a possible influence on babies' attentiveness lacked control of extraneous variables. As a result, those extraneous variables are free to compete with mothers' prenatal depression as a potential cause of the observed differences in the babies' attentiveness.

Many kinds of observational studies exist. Two common types of nonexperimental research that appear in health sciences research are case–control studies and cohort studies. A *case–control study* involves people with a condition (the cases) being compared to otherwise similar people who do not have that condition (the controls), then identifying risk factors that may explain why some people have the condition and others do not. A case–control study by Joshi, John, Koo, Ingles, and Stern (2011) identified multiethnic samples of patients with localized prostate cancer, advanced prostate cancer, or no prostate cancer. Those without prostate cancer (i.e., the controls) lived in the same cities as those with prostate cancer (i.e., the cases) and were similar to the cases in age, race, and ethnicity. The men were interviewed about their eating habits. An analysis showed that the men with prostate cancer tended to consume more white fish cooked at high temperatures. This summary vastly oversimplifies the results of the sophisticated data analysis, but provides an example of a case–control study.

Another kind of descriptive study is a *cohort study*, which identifies people exposed or not exposed to a potential risk factor and compares them by examining data across time, either retrospectively or prospectively. A *cohort* is a group of people who share one or more characteristics and who are studied to assess eventual disease incidence or mortality. Many cohort studies have sought to answer questions about the effect of cigarette smoking. He et al. (2001) conducted a study in which people were tracked for years to look at tobacco's effects. Clearly, the researchers could not randomly assign people to being cigarette smokers or not, then wait to see whether cigarette smokers were more likely than nonsmokers to develop certain health conditions. They collected data from people who self-determined whether they were smokers or not. Each group—smokers and nonsmokers—was a cohort. After several years of collecting data from the two cohorts, the researchers identified which people in each group eventually developed congestive heart failure. The researchers then compared the rates of congestive heart failure for the smokers and nonsmokers. This study had statistical replication (multiple people being studied), but no random assignment to groups and no manipulation of an independent variable, so it is nonexperimental research. Next, we will describe a kind of study in which there is no random assignment, but an independent variable is manipulated for existing groups.

Check Your Understanding

SCENARIO 1-F

Do people who are "night owls" (preferring to stay up late at night and rise later in the day) perform better on tasks in the evening or daytime? We recruit nursing students who say they are night owls to take a drug calculation test twice. They first take the test on a Tuesday evening and then take a similar test (same format, different numbers) the next day at noon. We compare their average score on the evening test with their average score on the daytime test. 1-18. What kind of research is this? 1-19. What kind of variable is time of day? 1-20. What kind of variable is test score? 1-21. What kind of variable is area of study within the health sciences? 1-22. Explain why we can or cannot draw causal conclusions about the effect of time of day on the test scores.

Suggested Answers

1-18. Quantitative nonexperimental research, because there is no mention of random assignment to groups or manipulation of an independent variable. (The study does have statistical replication.) 1-19. Time of day is a predictor variable in descriptive research. 1-20. Test score is a criterion variable in observational research. 1-21. Area of study within the health sciences is an extraneous variable being controlled by the researchers, who chose to study only nursing students. 1-22. No, we cannot draw causal conclusions because this is not an experiment. If we reversed the timing and gave the first test during the day and the second test at night, we might find that the nursing students always did better on the second test, no matter when it was presented. We may need to randomly assign the students to groups, then manipulate the time of day that the two groups take the test.

Quasi-Experimental Research

Sometimes researchers have no way of randomly assigning people to groups, but they can manipulate the experience of existing groups. A study that is characterized by manipulation of an independent variable in the absence of randomization is an example of *quasi-experimental research*. For example, Buron (2010) wanted to study a way to improve patient-centered care in nursing homes by increasing communication between nurses and residents. He thought communication would improve if nurses knew more about the lives that residents led before they moved into the nursing home. His idea was to work with residents to create life collages that would hang in their rooms, providing material for conversations with the nurses.

The researcher could have randomly assigned nurses to interacting with residents who received a life collage or residents who did not receive a life collage. But nurses generally serve all residents living in one area of a nursing home, so the researcher needed to separate the nurses who would see life collages from the nurses who would not see collages. His solution was to enlist the help of two nursing homes. Some residents of one nursing home received life collages, and some residents at a similar nursing home did not. The researcher began the study by collecting data on nurse-resident communication and the nurses' knowledge about the residents. Then the researcher worked with a graphic designer to create collages for residents in one of the nursing homes. After the collages had hung in the residents' rooms for a period, nurses at both nursing homes were measured again. The nurses in the "collage" nursing home showed a statistically noteworthy increase in their knowledge of the residents' work and personal lives; the nurses in the other nursing home did not show a similar increase in knowledge.

The researcher manipulated the independent variable of life collages (present or absent) for the two nursing homes, but the nurses were not randomly assigned to groups. This study was a quasi-experiment because it involved only the manipulation of an independent variable and statistical replication. In the absence of random assignment, can we draw causal conclusions about the effect of the life collages on nurses' knowledge about residents' lives? No, because extraneous variables were not controlled. Any number of factors may explain differences in the two nursing homes. One nursing home may have had nurses and residents who had known each other for many months, so no increase in knowledge could be observed. Residents of the other nursing home may have just moved in, and the nurses naturally may have been learning more about them during the course of the study. If we do not have random assignment to groups, then we cannot say whether the independent variable is responsible for the differences or changes in the measured variables. Figure 1.5 illustrates a predictive relationship between life collages (present/absent) and nurses' knowledge about residents' lives. Instead of having a shield of randomization and manipulation protecting the relationship between the two variables of interest, we have only manipulation. Life collages are in the position of cause, but without randomization, the shield is nearly ineffective in keeping extraneous variables out of the position of cause. (You might compare Figure 1.5 with Figure 1.3, where a shield of randomization and manipulation isolated an independent variable in the position of causal influence.)

Quasi-experimental research poses a vocabulary challenge for us. We use the term *dependent variable* in an experiment (characterized by randomization and manipulation) and *criterion variable* in descriptive research (which lacks those two characteristics). So what do we call the outcome variable in a

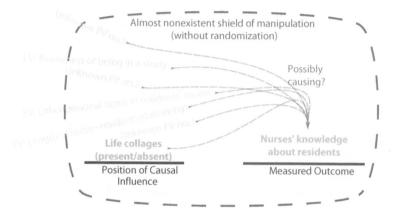

Figure 1.5

Quasi-experimental study with manipulation but no randomization. The study of life collages involved manipulating which one of the two nursing homes received life collages for some residents. But without randomization to groups, the shield that could have isolated the presence/absence of collages in the position of cause is nearly nonexistent. As a result, extraneous variables can get into the position of causal influence and affect the outcome variables, one of which was the nurses' knowledge about the residents.

quasi-experiment, which has manipulation of an independent variable but no randomization? The lack of randomization means extraneous variables compete with the independent variable to affect the outcome variable, so we cannot make causal conclusions. So what should we call the outcome variable? In this book's few examples describing a quasi-experiment, we will use both terms and remind our readers of the weakness of asserting causal relations in this sort of research.

Inferences and Kinds of Validity

Earlier we talked about external validity, which concerns our ability to generalize our sample findings externally to a larger population. External validity depends on the sampling method, or how the sample was drawn from the population. Researchers often talk about making *inferences*, which are conclusions drawn from information and reasoning. Based on our sample results, we draw conclusions about the population from which we drew the sample; that is, we infer from the sample to the population. The quality of the inference from the sample to the population is another way to define external validity.

We also draw conclusions (inferences) about the causal relationship between the independent and dependent variable; this relationship is internal to the study. *Internal validity* is the quality of inference that we can make about whether a causal relationship exists between the variables. When we randomly assign participants to groups to control extraneous variables and we manipulate an independent variable for multiple participants, we are conducting an experiment. Experimental

1. The Frontier Between Knowledge and Ignorance

research allows us to draw causal conclusions about the independent variable causing the observed changes or differences in the dependent variable. Therefore, we can say that experimental research has strong internal validity. The internal validity of nonexperimental research, with statistical replication but no random assignment to groups and little control of extraneous variables, is weak or nonexistent. Observational research still may provide us with important information, as we saw in the study of maternal depression and baby attentiveness. But we must be careful to avoid drawing causal conclusions when we have observed relationships that could be influenced by uncontrolled extraneous variables. The word *experiment* may be hiding in the term *quasi-experimental research*, but quasi-experiments are quite weak in terms of internal validity. A quasi-experiment introduces one possible causal factor, the independent variable, but many uncontrolled extraneous variables still may be responsible for any differences we observe in the outcome variable. So the internal validity of a quasi-experiment is extremely limited.

What's Next

This chapter provided an overview of the context in which we use statistics. *The rest of this book will use the information in this chapter.* Mirroring the real-life application of this information, the book will present many research scenarios for you to assess. We will ask you to identify the kind of research and variables, in addition to weighing the internal and external validity of research scenarios. In Chapter 2, we will introduce some statistics that describe data and estimate what may be going on in the population.

Exercises

SCENARIO 1-G

(Inspired by Waterhouse, Hudson, & Edwards, 2010. Details of this scenario may differ from the actual research.) We are interested in the physiological effects of music during submaximal exercise. We recruit 60 healthy young male volunteers who typically ride a bicycle for about 30 miles per week. In our study, they will ride stationary bikes while listening to music on earbuds plugged into a music player, which we will provide. They will be instructed to ride at a moderate pace. Secretly we randomly assign them to one of three conditions. The riders in Group 1 will listen to upbeat popular music. The riders in Group 2 will listen to the same music being played 10% faster than the original recording. The riders in Group 3 will listen to the same music played 10% slower than the original recording. The stationary bikes record how fast the riders are pedaling. Does the speed of the music make a difference in the pedaling speed? 1-23. What kind of research is this? 1-24. What kind of variable is speed of music? 1-25. What kind of variable is pedaling speed? 1-26. Explain why we can or cannot draw causal conclusions about the effect of music speed on pedaling speed.

(Continued)

SCENARIO 1-H

A researcher wanted to know if people whose sleep is interrupted differ from those who are able to sleep without interruptions. Specifically, she wanted to determine whether there was a difference in the amount of frustration these people reported the next morning. Ninety volunteers (45 men and 45 women) were recruited for the study, which was conducted in a sleep lab, where the researcher could observe the participants' sleep stage at any given time. The participants were randomly assigned within gender to (1) no sleep interruptions, (2) sleep interruptions during rapid eye movement (REM) sleep, or (3) sleep interruptions during non-REM sleep. After spending the night in the lab, the participant was sent to another room to pick up a survey. The room was secretly part of the research, and everyone else was secretly a research assistant. The participant encountered three lines in front of service windows. Whichever line the participant chose, the clerk closed the window as soon as the person in front of the participant had been helped. The participant had to get in a second line, and when the participant reached the front of the line, this clerk said his/her computer just crashed. So the participant had to get into the third line to pick up the survey. The first question on the survey was, "How much frustration do you feel right now?" with responses being rated from 0 (*no frustration*) to 10 (*complete frustration*). 1-27. What kind of research is this? 1-28. What kind of variable is gender? 1-29. What kind of variable is frustration? 1-30. What kind of variable is sleep interruption? 1-31. Explain whether we can draw causal conclusions about the effect of sleep interruption on frustration. 1-32. Explain whether we can draw causal conclusions about the relationship between gender and frustration.

SCENARIO 1-I

(Inspired by Carmody et al., 2011. Details of this scenario may differ from the actual research.) Researchers wanted to know whether meditating would reduce how much hot flashes bothered menopausal women. The researchers ran advertisements and recruited 110 volunteers in Worcester, Massachusetts. The women were randomly assigned to either a meditation group or a waitlist control group. Those in the meditation group attended eight weekly classes on mindfulness meditation; those in the control group were placed on a waiting list for classes that would begin after the 8-week study. During the 8 weeks, all of the women kept a diary about their hot flashes, and at the end of each day, they rated "bothersomeness," or the degree of feeling bothered by hot flashes. They used a scale of 1 (*not at all bothered today*) to 4 (*extremely bothered today*). The researchers reported that both groups showed a decrease in mean "bothersomeness" scores, but that the meditation group's "bothersomeness" scores went down more than the control group's scores did. 1-33. Is this quantitative or qualitative research—or possibly mixed methods? 1-34. How might the researchers describe their

(Continued)

population of interest? 1-35. Describe the external validity of this study. What facts in the scenario led you to this judgment? 1-36. What is the unit of analysis? 1-37. In terms of the "bothersomeness" of hot flashes, is this study an experiment, quasi-experiment, or descriptive research? How do you know? 1-38. What kind of variable is "bothersomeness"? 1-39. What kind of variable is "surgical history"? 1-40. Describe the internal validity of the study and the facts that led you to this judgment. 1-41. Suppose a classmate says, "Isn't the internal validity of this study reduced by the fact that they didn't limit the study to women who either have or have not had a hysterectomy?" Answer your classmate's question. 1-42. What extraneous variables are *not* being controlled in this study?

SCENARIO 1-J

Landon, Reschovsky, and Blumenthal (2003) compared physician satisfaction in 1997 versus 2001. Their data came from the Community Tracking Study (CTS) Physician Survey, a nationally representative telephone survey of physicians. These researchers reported that the study consisted of a "complex sample clustered in 60 randomly selected sites and a small, independently drawn, unclustered national sample" (p. 443). The physicians were asked to think about their general satisfaction with their medical career, which they then rated from 1 (*very dissatisfied*) to 5 (*very satisfied*). 1-43. Describe the external validity of this study and how you reached this conclusion. Does it matter whether this study has good external validity? 1-44. Describe the internal validity of this study and how you reached this conclusion. Does it matter whether this study has good internal validity? 1-45. Suppose we wanted to compare the career satisfaction of physicians, nurses, physical therapists, occupational therapists, and pharmacists. Explain whether it is possible to conduct an experiment to study this question.

SCENARIO 1-K

(Inspired by Walker et al., 2009. Details of this scenario may differ from the actual research.) Suppose we want to know whether a hospital can reduce readmission rates by having a pharmacist interview the patient and examine patient records before discharge. We recruit adult patients whom we identify as being at risk of medication-related complications after leaving the hospital. We restrict our sample to patients who meet certain inclusion criteria (e.g., five or more medications prescribed). We are collaborating with the pharmacist who works in a setting that alternates each month; the pharmacist works one month with hospitalists (doctors who specialize in treating patients in the hospital) and the next month with residents (doctors who are learning a specialty). While the pharmacist is working with one set of doctors, those doctors' patients who meet inclusion criteria are asked to participate; those who agree receive the intervention. The patients who are being

(*Continued*)

treated at that time by the other set of doctors also are asked to participate, but they will be part of the comparison group, which receives usual discharge instructions regarding their medications. The intervention involves the pharmacist interviewing the patients about their understanding of medications. The pharmacist also assesses the accuracy and appropriateness of the recommended prescriptions. We want to know whether the intervention leads to fewer medication discrepancies (e.g., unnecessary duplications or omitted medications) and a lower rate of readmissions, compared with the usual-care group. 1-46. Is this study an experiment, quasi-experiment, or nonexperimental research? How do you know? 1-47. What kind of variable is group (intervention vs. usual care)? What kind of variable is readmission rate? 1-48. Describe the internal validity of the study. What facts in the scenario led you to this judgment?

1-49. Begin a "statistics dictionary" for yourself by writing a list of terms in this chapter. Without looking at our definitions, write definitions of these terms in your own words. Then compare your explanation with our definition to clarify your understanding. As you read the book, you can add terms and symbols to your dictionary. (This idea came from a former student, who said, "The value was in the *making* of the dictionary; I don't think it would be that useful to have one that was made for me.")

1-50. Read the preface of this book to learn about the electronic flashcards that we have created for smartphones.

References

Buron, B. (2010). Life history collages: Effects on nursing home staff caring for residents with dementia. *Journal of Gerontological Nursing, 36,* 38–48. doi:10.3928/00989134-20100602-01

Carmody, J. F., Crawford, S., Salmoirago-Blotcher, E., Leung, K., Churchill, L., & Olendzki, N. (2011). Mindfulness training for coping with hot flashes: Results of a randomized trial. *Menopause: The Journal of the North American Menopause Society, 18,* 611–620. doi:10.1097/gme.0b013e318204a05c

Dawkins, L., Shahzad, F.-Z., Ahmed, S. S., & Edmonds, C. J. (2011). Expectation of having consumed caffeine can improve performance and mood. *Appetite, 57,* 597–600. doi:10.1016/j.appet.2011.07.011

Fedewa, A. L., & Erwin, H. A. (2011). Stability balls and students with attention and hyperactivity concerns: Implications for on-task and in-seat behavior. *American Journal of Occupational Therapy, 65,* 393–399. doi:10.5014/ajot.2011.000554

Field, T., Diego, M., & Hernandez-Reif, M. (2009). Depressed mothers' infants are less responsive to faces and voices. *Infant Behavior and Development, 32,* 239–244. doi:10.1016/j.infbeh.2009.03.005

He, J., Ogden, L. G., Bazzano, L. A., Vupputuri, S., Loria, C., & Whelton, P. K. (2001). Risk factors for congestive heart failure in U.S. men and women. *Archives of Internal Medicine, 161,* 996–1002. doi:10.1001/archinte.161.7.996

Joshi, A. D., John, E. M., Koo, J., Ingles, S. A., & Stern, M. C. (2011). Fish intake, cooking practices, and risk of prostate cancer: Results from a multi-ethnic case–control study. *Cancer Causes and Control, 23,* 405–420. doi:10.1007/s10552-011-9889-2

Kamper, S. J., Maher, C. G., Menezes Costa, L. D. C., McAuley, J. H., Hush, J. M., & Sterling, M. (2012). Does fear of movement mediate the relationship between pain intensity and disability in patients following whiplash injury? A prospective longitudinal study. *Pain, 153,* 113–119. doi:10.1016/j.pain.2011.09.023

Kaptchuk, T. J. (2001). The double-blind, randomized, placebo-controlled trial: Gold standard or golden calf? *Journal of Clinical Epidemiology, 54,* 541–549. doi:10.1016/S0895-4356(00)00347-4

Landon, B. E., Reschovsky, J., & Blumenthal, D. (2003). Changes in career satisfaction among primary care and specialist physicians, 1997–2001. *Journal of the American Medical Association, 289,* 442–449. doi:10.1007/s11606-011-1832-4

Lebec, M. T., Cernohous, S., Tenbarge, L., Gest, C., Severson, K., & Howard, S. (2010). Emergency department physical therapist service: A pilot study examining physician perceptions. *Internet Journal of Allied Health Sciences and Practice, 8.* Retrieved from http://ijahsp.nova.edu/

Rosenthal, R. (2009). Interpersonal expectations: Effects of the experimenter's hypothesis. In R. Rosenthal & R. L. Rosnow (Eds.), *Artifacts in behavioral research* (pp. 138–210). New York, NY: Oxford University Press.

Walker, P. C., Bernstein, S. J., Tucker Jones, J. N., Piersma, J., Kim, H.-W., Regal, R. E., …Flanders, S. A. (2009). Impact of a pharmacist-facilitated hospital discharge program: A quasi-experimental study. *Archives of Internal Medicine, 169,* 2003–2010. doi:10.1001/archinternmed.2009.398

Wang, C., Schmid, C. H., Rones, R., Kalish, R., Yinh, J., Goldenberg, D. L., … McAlindon, T. (2010). A randomized trial of tai chi for fibromyalgia. *New England Journal of Medicine, 363,* 743–754. doi:10.1056/NEJMoa0912611

Waterhouse, J., Hudson, P., & Edwards, B. (2010). Effects of music tempo upon submaximal cycling performance. *Scandinavian Journal of Medicine & Science in Sports, 20,* 662–669. doi:10.1111/j.1600-0838.2009.00948.x

2

Describing Distributions with Statistics: Middle, Spread, and Skewness

Introduction

Most statistics that we see in news reports are averages or percentages. These *descriptive statistics* summarize or describe information about a sample. However, averages and percentages are limited. Averages tell us generally where the middle part of the data set is located on the number line, and percentages tell us the typical response. What other characteristic of samples might be important to measure with descriptive statistics?

Consider that we have a sample of blood pressure readings from one patient who has kept track of her blood pressure for a few weeks, and we notice that on average she had slightly elevated blood pressure readings. High blood pressure is a risk factor for stroke. But does the average (mean) blood pressure reading tell us everything we might need to know about this patient's risk of stroke? Rothwell et al. (2010) reported that having a great deal of variation (spread) in blood pressure is another risk factor for stroke. The mean blood pressure is not a measure of variation; it is a measure of the location of the data on the number line. We need other statistics that will measure the spread in the blood pressure readings. We also may be interested in knowing whether some extreme readings are appearing in only one direction (i.e., a few extreme readings that are all high or all low), which may indicate the patient has episodic hypertension or hypotension. In that case, we would need statistics that measure departure from symmetry. A symmetrical data set would have the same shape below and above the middle score, as we will show in this chapter. A departure from symmetry may be clinically noteworthy and influence other statistics.

In Chapter 1, we defined *statistic* as a numerical summary measure computed on data. We also could define *statistic* as a numerical characteristic of a sample. We have just mentioned three characteristics of a data set: middle, spread, and symmetry. This chapter will cover statistics that measure these three most commonly reported characteristics of samples. Remember that statistics are computed on samples, but we are using them to get some sense for what might be happening in the population. In fact, statistics can be considered estimates of *parameters*, the numerical characteristics of the population. Because we cannot obtain scores from the entire population of interest in any given study, we use what we can get: the sample data. We will use sample statistics to make inferences about unobtainable population parameters.

We will begin with statistics that measure location on the number line, then we will talk about statistics that measure how spread out scores are, and finally, we will measure how much a data set departs from symmetry.

Measures of Location

There are many kinds of averages in statistics. We have talked about one average, which is technically called the arithmetic mean and is computed by adding up scores and dividing by the number of scores. From now on, we will simply call it "the mean." The mean is one of many measures of the general location of the middle of the data. The middle location of the data is one characteristic of a *distribution*, which is a set of scores arranged on a number line. There are advantages to using the mean: everyone understands the average, and all the scores in the sample are represented in its computation. A disadvantage is that the mean can be pulled up or down by one or more extreme scores in a data set. Suppose we asked a patient to keep track of his blood pressure for a week. He faithfully records his blood pressure every day at noon for 6 days. Then on the seventh morning, he drinks three cups of coffee before taking the reading. The coffee raises his blood pressure, and the higher reading increases his average for the week.

Researchers can avoid the problem of extreme scores by switching to a different measure of middle. The *median* is the score exactly in the middle of the distribution. Think of an interstate highway, with a median dividing the highway in half. In a similar way, the statistical median divides the numerically ordered set of scores in half; the same number of scores appears below the median and above the median. Let's consider seven days of systolic blood pressure (SBP) readings taken on the first author's left arm:

113, 116, 132, 119, 112, 120, 114

To find the median, the data must be placed in order:

112, 113, 114, 116, 119, 120, 132

The median for the seven days of readings is the score in the middle: 116. What if we had 10 days of SBP readings? Well, here they are—the same seven readings and three more days of SBPs, placed in order:

$$112, 113, 114, 116, 119, 120, 125, 127, 129, 132$$

With an even number of readings, there is not one score that divides the data set in half. By averaging the two scores in the middle of the 10-score data set, we get a number with the same number of scores below it and above it, and that is the median: $(119+120)/2 = 119.5$.

Check Your Understanding

2-1. The median for the seven days of SBP readings was 116. Compute the mean for those seven scores, then speculate on why it differs from the median.

Suggested Answers

2-1. The sum of the seven scores is 826, so the average is 826/7 = 118. The mean is slightly higher than the median, 116, which may be the result of the reading of 132, pulling the mean in that direction.

Occasionally, a journal article will report a *trimmed mean*, which is a mean computed on a data set from which some of the highest and lowest scores have been dropped. A trimmed mean sometimes is called a *truncated mean*. A trimmed mean can be used to communicate about the middle of a data set after the most extreme scores are excluded. If we wanted to compute a 10% trimmed mean for the 10 days of SBP readings, we would ignore 10% of the data on the lower end (i.e., the lowest one of the 10 scores) and 10% of the data on the upper end of the distribution (i.e., the highest one of the 10 scores). The remaining eight readings would be averaged to find the 10% trimmed mean, meaning 10% of scores were dropped from each end of the distribution. In our example of 10 days of SBP readings, we would drop the scores 112 and 132, then average the remaining eight scores to get a 10% trimmed mean of 120.375. Important information may be present in the extremes, however, so researchers must consider the impact that trimmed means may have on their conclusions.

One more measure of middle typically is introduced in textbooks, but it is rarely used with numeric data. The *mode* is the most frequently occurring score or response. The mode generally is reserved for categorical variables. If it were to be used with a numeric variable, it would be a poor choice for measuring the middle because the most frequently occurring score can happen anywhere on the number line. For example, suppose a patient takes her blood pressure every day for a month, and the most frequently occurring SBP reading is 115, occurring on 6 days. So 115 would be her mode for SBP—but if it is her lowest SBP reading, it will not be in the middle of the distribution of SBP readings. Another disadvantage of the mode is that there can be more than one mode or even no mode, if every score occurs once in the data set. If a patient tracked his blood pressure daily for a week and had three SBP readings of 118 and three readings of 122, both of these numbers would be modes. The mode tells us about the most common value for a variable, and it is well suited for use with a qualitative variable like gender. When Rothwell et al. (2010) reported on blood pressure variability between office visits, the sample consisted of 2,006 patients, with 1,438 of the patients being men. We can say that the mode for gender was male, but usually a researcher would be more precise and say that 71.7% of patients were men. This information tells us about the typical patient and helps us to judge the population to which we might wish to generalize the results.

Many other statistics exist for measuring middle, but the ones we have presented are the most common. Each of these statistics estimates a population parameter (numerical characteristic of a population). Symbols are used to represent almost every statistic and parameter, and not every textbook uses the same symbol. Most statistics textbooks use the symbol $\bar{X}$ for the sample mean; it is pronounced as "X-bar." The line or bar symbolizes the process of averaging, and the X represents each score in the data set, so the symbol is saying, "Average together the scores for this variable called X." But we find that most journals in the health sciences use the capital letter M to represent the sample mean. Therefore, we will use M.

When we introduce statistics that are used for decision making, called *inferential statistics*, we will need symbols that represent some parameters. The sample mean estimates the population mean, which has the symbol μ, the lowercase Greek letter mu, pronounced "mew." We will need the symbol when we cover statistics used for decision making about whether a sample mean (M) differs from a population mean (μ). We will use symbols and formulas only when necessary.

Check Your Understanding

SCENARIO 2-A

Marx et al. (2010) wanted to know whether nursing home residents with dementia could be encouraged to engage in positive social interactions through animal-assisted therapy. But live dogs are not always appropriate or available. The researchers wanted to know whether residents could be equally engaged by a live dog, a puppy video, a plush toy dog, a robotic dog, or a printed cartoon image of a dog that could be colored with markers. They presented one stimulus at a time, then timed how long the residents engaged with each stimulus. If they were still interacting with the stimulus after 15 minutes had elapsed, the researchers ended the session and recorded the engagement score as 15 minutes. 2-2. What statistics might the researchers compute to tell whether different stimuli resulted in different engagement times? 2-3. What characteristic of engagement scores would be measured with the statistics that you named? 2-4. Why would the researchers limit the sessions to 15 minutes?

(Continued)

2. Describing Distributions with Statistics

2-2. Descriptive statistics like the mean, median, or trimmed mean could be computed on the engagement times for each stimulus. The researchers may wish to compare the mean engagement times for each stimulus. 2-3. Location of the middle of the data on the number line. 2-4. By setting a ceiling or maximum score allowed, the researchers would limit the effect of extreme scores on the mean, while also maintaining the timely progress of the study and reducing the potential for fatiguing the participants.

Measures of Spread or Variability

We are accustomed to measuring the middle location or typical outcome using statistics like the mean. Less intuitive are statistics that measure how spread out scores are. In statistics, we use the term *variability* to refer to the amount of spread or variation within a set of scores. A colleague (Dr. Edward Kifer) used to tell his statistics students, "Never a center without a spread." He was saying that it is not enough to know the general location of the middle of a data set; we also need to know how much the scores are spread out on the number line. Suppose we had two samples of patients whose SBPs had been recorded. Sample 1 has a mean of 130, and Sample 2 has a mean of 135. These two means describe the middle of each sample's readings, and Sample 2 has a higher middle. But how spread out are the readings for each sample? Does Sample 1 represent 30-year-old men, and does Sample 2 represent 60-year-old women? If so, one sample may have much more consistent readings than the other. The means will not tell us how much consistency or inconsistency is in a set of scores. We need to measure variability.

Variability can be measured in different ways. The simplest measure of variability is the *range* statistic, which equals "high score minus low score." Let's return to the example of the first author's week of SBP readings in order:

$$112, 113, 114, 116, 119, 120, 132$$

The highest SBP was 132, and the lowest SBP was 112. The range equals 20 (i.e., $132-112 = 20$). Most journal articles do not report the range statistic, however; they simply report the maximum and minimum scores.

The problem with the range statistic is that only two scores are used in its computation, so it fails to reflect how much variation is present within all of the scores. If the second author had seven days of SBP readings, he might have recorded 112 for six days and 132 for one day, producing the same numeric value for the range statistic. But his SBP readings would have been much more consistent than the first author's readings, and we might wonder whether the 132 reading was taken shortly after a period of physical activity. We need a measure of variability that will capture the variation in all of the scores, not just the distance between the highest and lowest scores.

At this point, many statistics textbooks introduce a big, ugly formula that creates symbol shock and makes students' eyes glaze over. We would like you to follow us through an example that may seem to be leading nowhere for a while—but we do have a point that we will reach. We will verbally explain some ways of measuring variability in a set of scores. Although the following explanation may seem to wander, the process will demonstrate ways of measuring spread. We also know that if students are going to understand statistics that measure variability, they simply must understand the calculations. Later in this chapter, we will introduce a statistic and spare you from the ugly formula. But trust us—the best way to learn about measures of spread is to understand the calculations.

When considering how scores vary, we need to ask, "Compared with what?" The most common statistics for measuring variability will compare scores with the mean. Let's use the first author's seven days of SBP readings, which had a mean = 118. Perhaps we could find the average distance of each score from the mean. Subtract 118 from each of these scores:

$$112, 113, 114, 116, 119, 120, 132$$

Subtracting the mean from each score gives us distances of

$$-6, -5, -4, -2, 1, 2, 14$$

To find the average distance, we need to add up those distances: $-6 + -5 + -4 + -2 + 1 + 2 + 14$. What did you get?

You should have found a sum of zero. How can the average distance be zero? We have just encountered a characteristic of the sample mean: it is a balance point in the distribution. In other words, the distances for the scores above the mean will balance out the distances for the scores below the mean. We cannot use the average distance (or difference) from the mean as a measure of variability because the distances *always* add up to zero.

Check Your Understanding

2-5. To persuade yourself that the average distance of scores from their mean always equals zero, use the following diastolic blood pressure (DBP) readings: 79, 79, 84, 87, 78, 83, 84. Compute the mean, then find the distances by subtracting the mean from each score, then add up the distances.

Suggested Answers

2-5. The sum of scores is 574, so the mean is 82. By taking each score and subtracting 82, we get these distances: −3, −3, 2, 5, −4, 1, 2. These distances add up to zero.

2. Describing Distributions with Statistics

Let's continue to think of ways to measure the amount of spread in the scores in relation to the mean. We could ignore the negative signs to avoid the problem of distances from the mean summing to zero. But statisticians take a different approach: squaring the distances. To square a number, we multiply it by itself. A negative number multiplied by itself will give us a positive number. Returning to the seven days of SBP readings, the square of each distance is

$$36, 25, 16, 4, 1, 4, 196$$

Now let's average the *squared* distances from the mean:

$$\frac{(36+25+16+4+1+4+196)}{7} = \frac{282}{7}$$
$$= 40.285714 \approx 40.29$$

(The squiggly equals sign means "approximately equal to.") We have just computed a measure of variability called the *sample variance*, which is the average squared distance from the mean. Remember where we have come from: we wanted to measure the spread of scores around the mean, so we computed the distance of each score from the mean. We got rid of negative signs by squaring the distances, then we averaged the squared distances. We will use the term *sum of squares* to refer to a process of squaring some numbers and adding them up. Here, we are squaring distances from the mean and adding them up, getting the sum of squares of distances. The sum of squared distances is a variance's numerator (the top number in a fraction), and its denominator (the bottom number in a fraction) is the sample size because we are computing an average.

Let's step back for a moment and look at the learning process. If you are similar to 99% of our students, right now you are looking at that 40.29 and thinking, "Uh, okay ... that doesn't tell me anything." Very anticlimactic, isn't it? *Congratulations, you are having a perfectly normal reaction to computing a sample variance!* People are unaccustomed to measuring how spread out scores are, and this statistic almost always produces a "so what?" feeling. Part of the problem with understanding the sample variance seems to be that people get lost when we square those distances from the mean. The average *squared* distance from the mean is 40.29. But who in the world uses squared distances? Statisticians agree with you, and there is another statistic for measuring variability that is not expressed in squared distances.

Before explaining this next statistic that measures variability, we need to give a quick math review. If we had a *squared distance* of 4, then the *distance* would be 2. Two is the number that gets multiplied by itself, and thus it is called the *square root* of 4; in symbols it would be written as $\sqrt{4}$. The square root of 4 equals 2. Squaring the number 2 involves multiplying it by itself ("two squared" $= 2^2 = 2 \times 2 = 4$). Taking the square root of the number 4 involves finding whatever number could be multiplied by itself to result in 4 ("the square root of 4" $= \sqrt{4} = 2$).

Now let's find out how to measure variability without having squared SBP units. We computed a sample variance of about 40.29 for seven days of SBP readings. This number is in squared SBP units, because we squared the distances between each SBP reading and the mean SBP. Let's take the square root of our sample variance, 40.285714 (being more precise by using the unrounded variance): $\sqrt{40.285714} = 6.3471028 \approx 6.35$. Now we do not have a sample variance; we just computed a *standard deviation*, which is the square root of the sample variance. The standard deviation is in SBP units, not squared SBP units, and also is a measure of variability.

Are you still having that "so what?" feeling? Again, it is a typical reaction. But remember where we started: we wanted to compute the average distance of the scores from the mean as a measure of spread. But the distances always sum to zero, so we had to square the distances before averaging them, which gave us the average squared distance from the mean, also known as the sample variance. Squared distances are hard to understand, so to get rid of the squared units, we took the square root of the variance and thus computed a standard deviation. Technically, the standard deviation is *not* the average distance from the mean; we had to go the long way around by squaring the distances. But the standard deviation is the closest thing we can get to an average distance from the mean.

Check Your Understanding

2-6. Let's say you have done some statistics homework. One part required you to compute a sample variance and its corresponding standard deviation. You are checking your answers before turning in the homework, and you realize that you forgot to label each of the statistics. Your paper has two numbers for that question: 31.36 and 5.6. Which one is the standard deviation, and how do you know? 2-7. As you continue to check another part of your homework, you notice that one of your answers says, "Sample variance = −12." Why should you reconsider this answer?

Suggested Answers

2-6. The sample variance is 31.36 and the standard deviation is 5.6, because the standard deviation is the square root of the variance. Another way to look at it is: If you multiply 5.6 by itself (i.e., you square 5.6), you get 31.36. 2-7. A sample variance is computed by squaring the distance of each score from the mean, which gets rid of all negative signs, so a sample variance cannot be negative.

Just as the sample mean estimates the population mean, the sample variance estimates the *population variance*, which is a parameter for the amount of variability of scores in the population. The population variance has the symbol σ^2, a

2. Describing Distributions with Statistics

lowercase Greek letter sigma with the symbol for squaring (the superscripted 2). We say "sigma squared." The sample standard deviation estimates the *population standard deviation*, another parameter for the amount of variability in the population's original scores (i.e., not squared units of measure). The population standard deviation is symbolized by σ, or "sigma." If we could obtain numeric values for these parameters, the population standard deviation would be the square root of the population variance, and both parameters would summarize the amount of spread in the population of scores.

The sample standard deviation and the population standard deviation will be used in Chapter 4, when we talk about measuring a specific score's location in comparison with the mean. But the sample variance and standard deviation that we just computed are slightly different from the statistics reported in journal articles and computed by statistical software like SAS and SPSS. The problem with the sample variance is that it systematically underestimates the population variance. When we take the square root of the sample variance, we get a standard deviation that systematically underestimates the population standard deviation. We want our statistics to be good estimates of parameters. Imagine if you had to administer an important liquid medication to a patient in a specific dose and you had to rely on a plastic spoon for measuring the drug, then you discovered that the plastic spoon was a little too small. Your measurements consistently would be too small.

Statisticians have technical reasons for using a slightly different formula for the variance. For this new variance, we still add up the squared distances from the mean (i.e., compute the sum of squares of distances), but instead of dividing by N, which is the number of scores, we divide by $N - 1$. Using our numeric example of seven days of SBP readings, we would take the squared distances from the mean, add them up, and divide by $N - 1$:

$$\frac{(36+25+16+4+1+4+196)}{N-1} = \frac{282}{6} = 47$$

This variance statistic is called the *unbiased variance*, and it is the statistic we would see in a journal article reporting the results of a research study. The numerator, which is the sum of squares of distances, is the same as the numerator of the first variance we computed. Now, instead of computing the average squared distance from the mean, we found *almost* the average squared distance from the mean. The sample variance with N in the denominator produces a result that on average is too small. The unbiased variance is a more accurate estimate of the population variance, σ^2, but there will be occasions when each of these variance statistics will be needed, as you will see later in the book.

Now that we have another variance statistic, are we free of the squared units of measure? No, but we can take its square root to get a new standard deviation. In our numeric example, this new standard deviation would equal $\sqrt{47} = 6.8556546 \approx 6.86$. Unfortunately, for technical reasons we cannot use the word *unbiased* in front of the term *standard deviation*. To distinguish this new standard deviation from the previous one, we will use the abbreviation *SD* and describe the statistic as the *standard deviation based on the unbiased variance*. Conveniently, this statistic is the one computed by statistical software and reported in journal articles, which typically use the abbreviation *SD*.

How do we interpret these measures of variability? Suppose a patient said his daily dose of the hypertension medication lisinopril was 10 mg. This number would not make sense unless we knew what would be considered a typical daily dose. Similarly, we need some frame of reference to judge the amount of spread in a distribution. We cannot simply look at a variance or standard deviation and say whether it is big or small. By themselves, these variances and standard deviations do not tell us much. You may wonder, "Is an unbiased variance of 47 considered *big*?" Our response would be, "Compared with what?" Let's go back to one detail from the first author's SBP readings: they were taken on her left arm. On the same days, the following SBP readings were taken on her right arm:

111, 112, 117, 115, 121, 125, 125

If we computed the unbiased variance on these numbers, we would get 33.67. Which arm produced a set of SBP readings with greater variation? The left arm's readings produced an unbiased variance of 47, which is a larger number than 33.67. So there was more variability in the left arm's readings. Equivalently, we could say that the right arm's readings were more consistent and less spread out around the mean. Larger values of the unbiased variance indicate more variation or spread in the data, and smaller numbers for this statistic indicate less variation or spread around the mean. If the right arm had produced an SBP reading of 120 on seven days in a row, the mean would be 120 and there would be no spread of scores around the mean, so the unbiased variance would be zero. The same is true for the standard deviation: the smallest possible value is zero, indicating all the scores are the same (i.e., zero spread). If even one score differs from the rest, all measures of variation would be bigger than zero. As the amount of variation in a data set increases, so do the statistics measuring spread, including the unbiased variance and the *SD*.

There are more statistics for measuring variability, but the ones presented here are the most common. Let us summarize them as follows:

- Range = high score minus low score. Researchers typically report the high and low scores, leaving out the range statistic, which does not summarize all the variation in the data set.
- Sample variance = average squared distance from the mean. It will be used in Chapter 4, but is *not* the variance computed by most statistical software. It is in squared units of measure.
- Standard deviation = square root of the sample variance. It is in original units of measure, and appears in Chapter 4, but is *not* computed by most statistical software.

- Unbiased variance = *almost* the average squared distance from the mean. It has the same numerator as the sample variance (i.e., we square the distances from the mean and add them up), but the denominator is $N - 1$. This statistic is computed by most statistical software and is in squared units of measure.
- Standard deviation based on the unbiased variance = SD = square root of the unbiased variance. It is in original units, is computed by most statistical software, and is one of the most commonly reported measures of variability.

Both of the variances computed in this chapter use the same sum of squared distances for their numerator. When we took the square root of each variance, we obtained a standard deviation; these two standard deviations shared the same numerator. We could give symbols for each of these measures of variability, but the last statistic in this list is the only one with an abbreviation (SD) that commonly appears in journal articles.

One other detail about these measures of variability: one or more extreme scores can inflate these statistics. Remember that one disadvantage of the mean was that a few extremely high scores can pull the mean upward, and a few very low scores can pull the mean downward. We gave the example of the patient who took his blood pressure every day at noon for six days. Then on the seventh morning, he drank three cups of coffee before taking the reading, which increased his blood pressure and resulted in a higher mean for the week. The extremely high blood pressure reading also would increase the amount of variability in the data set, leading to higher numbers for the variance and standard deviation.

Check Your Understanding

SCENARIO 2-B

Let us compute measures of variability using the first author's left-arm DBP taken on the same days as the SBP readings. Her DBP readings were 79, 79, 84, 87, 78, 83, 84. 2-8. The sum of squared distances from the mean equals 68. Replicate this numeric result. 2-9. Compute the sample variance. 2-10. Compute the standard deviation based on the sample variance. 2-11. Compute the unbiased variance. 2-12. Compute SD, the standard deviation based on the unbiased variance. 2-13. For seven days of left-arm SBP, we found the following results: unbiased variance = 47 and SD = 6.86. Is there more variability in the SBP or DBP readings? 2-14. Do these results reflect the fact that SBP readings are higher on the number line than DBP readings?

(Continued)

Suggested Answers

2-8. As shown in Question 2-5, the mean is 82. Subtract the mean from each score to get the distances: −3, −3, 2, 5,−4, 1, 2. Square each distance: 9, 9, 4, 25, 16, 1, 4. Sum the squared distances to get 68. 2-9. The numerator of both variance statistics is 68. To get the sample variance, divide 68 by the number of scores, which is 7: 68/7 = 9.7142857 ≈ 9.71. 2-10. Take the square root of the sample variance: $\sqrt{9.7142857}$ = 3.1167749 ≈ 3.12. 2-11. To obtain the unbiased variance, take the sum of squared distances and divide it by one less than the sample size: 68/6 = 11.333333 ≈ 11.33. 2-12. To get SD, take the square root of the unbiased variance: $\sqrt{11.333333}$ = 3.3665016 ≈ 3.37. 2-13. The unbiased variance for SBP was 47, compared with an unbiased variance of 11.33 for DBP. Because 47 is greater than 11.33, there is more variability in the SBP readings than in the DBP readings. We can draw the same conclusion by comparing the results for SD: 6.86 for SBP versus 3.37 for DBP. The DBP readings were less spread out around their mean, compared with how spread out the SBP readings were around their mean. The DBP readings were more consistent than the SBP readings were. 2-14. The statistics that measure variability are not influenced by the location on the number line. It is a coincidence that the SBP readings had higher variability as well as a higher mean than the DBP readings did. The measures of variability tell us nothing about how high someone's blood pressure is; they only tell us how spread out the readings are around the mean, wherever that mean may be located.

Measure of Skewness or Departure From Symmetry

If a set of scores is perfectly symmetric, then for every score above the mean, there is another score that is the same distance below the mean. Researchers need to know about any skewness in their data, because it can indicate the presence of extreme scores that could influence other statistics, like the mean and the standard deviation. For example, suppose a patient has measured his SBP on nine straight days, with a mean = 118. We can compute a statistic that measures the degree of departure from symmetry, or *skewness*; we call it the skewness statistic. If the skewness statistic equals zero, it means the patient's nine readings have zero skewness or zero departure from symmetry; in other words, they are perfectly symmetric around the mean. For the reading of 121, there is a reading of 115; these two readings are the same distance from the mean of 118. Figure 2.1 shows an example of a symmetric distribution of nine fabricated SBP readings with a mean of 118. Each circle represents one score in this figure, which we have kept simple because we have not covered graphing yet.

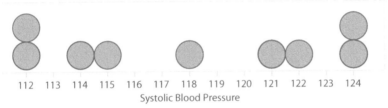

112 113 114 115 116 117 118 119 120 121 122 123 124

Systolic Blood Pressure

Figure 2.1

Symmetric distribution of blood pressure readings. This distribution of made-up systolic blood pressure readings is symmetric, meaning the left side of the distribution is a mirror image of the right side of the distribution. There is no skewness.

This set of SBP readings has a skewness statistic that equals zero because the distribution is perfectly symmetric; that is, we could fold the page along a vertical line through 118 and the lower half would be a mirror image of the upper half. When we have zero skewness, there are no extreme scores in one direction to influence the mean. In contrast, suppose we used the same patient's DBP readings on those nine days to create Figure 2.2. Now we do not have a distribution that could be folded down the middle to produce a mirror image. Most of the scores are clustered around 67 or 68, but there is one DBP reading that stands out, 79. This set of DBP readings has a skewness ≈ 2.04. The skewness statistic is positive, so we would say that the distribution is positively skewed or right skewed. The direction of the skewness gets its name from the few scores that are *different* from most of the scores. When identifying the direction of skewness, the *clue* is that the *few* name the *skew*. If there are a few big scores but mostly smaller scores, the few big scores are responsible for the skew, and we would say the distribution is positively skewed or right skewed. The one high score in Figure 2.2 acts like a weight on the right side of a seesaw, skewing the distribution to the right. Not only does the distribution look skewed, but the extreme score would act like a magnet, pulling the mean in that direction. So the skewness in the distribution can lead to some skewed statistics too.

What would a negatively skewed or left-skewed distribution look like? Suppose we took the same patient's heart rate on those nine days and found the results shown in Figure 2.3. There are two heart rates of 60 beats per minute, which

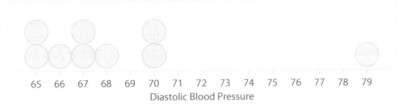

65 66 67 68 69 70 71 72 73 74 75 76 77 78 79

Diastolic Blood Pressure

Figure 2.2

Positively skewed distribution of blood pressure readings. This distribution of made-up diastolic blood pressure readings is not symmetric. In fact, it is skewed to the right or positively skewed. The location of a *few* extreme scores identifies the kind of skewness.

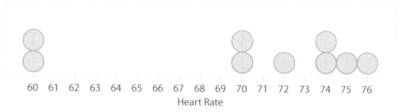

| 60 | 61 | 62 | 63 | 64 | 65 | 66 | 67 | 68 | 69 | 70 | 71 | 72 | 73 | 74 | 75 | 76 |

Heart Rate

Figure 2.3

Negatively skewed distribution of heart rates. The *few* extreme scores on the left side of the number line tell us that this distribution is skewed to the left or negatively skewed.

appear to differ from the seven other readings. The skewness statistic for this set of heart rates is −1.15, reflecting the negative skewness. The few smaller readings are skewing the distribution in the negative direction on the number line, so we would say this distribution is negatively skewed or left skewed. The mean would be skewed downward toward those lower extreme scores.

The skewness statistic has a big, ugly formula that will not help you to understand skewness. Therefore, we will not present the formula. We have explained how to interpret the numeric values of the skewness statistic. A positive skewness statistic tells us that at least one score above the mean is pulling the average in the positive direction on the number line. If the skewness statistic is a negative number, then at least one score below the mean is pulling the average in the negative direction on the number line. If the skewness statistic equals zero, then the distribution is symmetric and there is no skewness. We will come back to the concept of skewness briefly in Chapter 3, when we talk about a graph that can be used to identify extreme scores.

What's Next

The purpose of this chapter was to introduce some of the statistics that measure important characteristics of sample data: their middle, their variability or spread around the mean, and their skewness or departure from symmetry. One problem with all descriptive statistics is that they do not tell us very much about the distribution of scores. If a patient says, "I'm not feeling well," we do not have enough information to understand what is wrong. Similarly, if we observe a skewness statistic that equals 1.23, we can say there is some positive skewness in the distribution, but it does not tell us whether there may have been an anomaly in the data set. Summary statistics, therefore, are somewhat general, like the patient saying, "I'm not feeling well."

We need more information to understand a set of scores. One of the best ways to understand a data set—and one of the only decisions about data analysis that almost all statisticians could agree on—is to graph the data. We will introduce you to some graphs in Chapter 3. Certain graphs require an understanding of additional descriptive statistics, which will be introduced in the text where they will make the most sense to you.

2. Describing Distributions with Statistics

SCENARIO 2-A, Continued

This study by Marx et al. (2010) concerned nursing home residents with dementia interacting with dogs and substitutes for dogs, such as a plush toy. The researchers hoped to use these stimuli to encourage social interactions. A researcher offered one stimulus at a time to the resident. If the resident did not want to engage with the stimulus, an engagement duration score of zero was recorded. If the resident accepted the stimulus, the researcher left the participant alone for up to 15 minutes, except in the conditions involving a live dog, which was accompanied by a handler. Up to this point we have not mentioned that there were three sizes of dogs being tested: small, medium, and large. The researchers wanted to know whether the size of a dog would make a difference in its acceptability to the residents. The residents also were offered a puppy video, a plush dog toy, an activity involving coloring a picture of a dog, and a robotic dog. 2-15. Is this an example of experimental, quasi-experimental, or nonexperimental research? How do you know? 2-16. What kind of variable is type of stimulus? 2-17. What kind of variable (independent, dependent, predictor, criterion, extraneous) is engagement duration score? 2-18. What kind of variable is the participant's age? 2-19. If the researchers reported higher mean engagement with a real dog compared with the dog-coloring activity, can we say that the kind of stimulus caused the difference? 2-20. How would you describe the internal validity of the study? 2-21. How would you describe the external validity of the study?

SCENARIO 2-A, Continued

The study of dogs and substitutes for dogs involved 56 nursing home residents with dementia; 44 were women, 35 were widowed, and 47 had a high school education or more. The youngest participant was 61, and the oldest was 101, with a mean age = 87. On the Mini-Mental State Exam measuring cognitive functioning, the participants had $M = 9.1$ ($SD = 6.2$); the article says "range 0–21" for the Mini-Mental State Exam. 2-22. What is N? 2-23. Identify three modes described in the continuation of the scenario. 2-24. Compute the range statistic for age. 2-25. Given the information about age, name a reason that we might want to know the median. 2-26. What does "$SD = 6.2$" tell us? 2-27. Is "0–21" really the range?

SCENARIO 2-C

Wilkens, Scheel, Grundnes, Hellum, and Storheim (2010) wanted to know whether the dietary supplement glucosamine would reduce pain in patients with chronic lower back pain and arthritis. They ran a double-blind, placebo-controlled trial with 250 adults who were randomly assigned to taking either 1,500 mg glucosamine or an identical-looking capsule containing an inert substance every day for six months. The main outcome measure was the Roland Morris Disability Questionnaire (RMDQ), which asks patients to place a check mark next to statements about their back pain or limitations on their activities

(Continued)

(e.g., "I sleep less because of the pain in my back"). More check marks indicate the patient has greater disability related to the back problem. At baseline, the patients in the glucosamine group had a mean RDMQ = 9.2 (*SD* = 3.9), and those in the control group had a mean RDMQ = 9.7 (*SD* = 4.5). All participants were allowed to use nonsteroidal painkillers and their usual therapies (e.g., physical therapy) for their lower back pain. After six months, both groups had a mean RDMQ = 5.0; after one year the glucosamine group's mean RDMQ = 4.8, and the control group's RDMQ = 5.5 (the study did not report the *SD*s for these occasions of measurement). The researchers said that the differences between groups at each occasion were not statistically noteworthy. 2-28. Is this an example of experimental, quasi-experimental, or nonexperimental research? How do you know? 2-29. What kind of variable is treatment? 2-30. What kind of variable is RDMQ score? 2-31. What kind of variable is "usual therapies"? 2-32. Which group at baseline had more variability in its RDMQ scores? 2-33. Which group at one year appeared to have higher disability? 2-34. Can we say that glucosamine does not cause improvement in disability related to lower back pain?

SCENARIO 2-D

Suppose you are concerned about your mother's blood pressure, so you ask her to record her blood pressure and heart rate daily for six days. She records the following readings (in order: SBP/DBP, heart rate; data courtesy of the first author's mother):

> Day 1: 150/64, HR 72
> Day 2: 152/75, HR 70
> Day 3: 150/81, HR 68
> Day 4: 164/65, HR 74
> Day 5: 156/57, HR 69
> Day 6: 143/78, HR 73

2-35. Compute the mean and median for all three variables. 2-36. What do the means and medians tell you about the data? 2-37. Compute the unbiased variance for SBP. 2-38. Compute the unbiased variance for DBP. 2-39. Compute the unbiased variance for heart rate. 2-40. Compare the three values computed for the unbiased variances. What do they tell you about the data? 2-41. Compute the *SD* for each variable. Why do you suppose *SD* is reported in scientific journals more frequently than the unbiased variance? 2-42. Look at the means and the *SD*s for all three variables and discuss anything that stands out about these results.

SCENARIO 2-E

Raphael et al. (2012) reported the results of a study of nighttime teeth-grinding, called sleep bruxism. The researchers conducted a case-control study of people with and without myofascial pain associated with the temporomandibular joints (TMJ)—that is, pain related to the muscles that move the jaw. This pain also can affect the neck and shoulders. The 124 people diagnosed

(*Continued*)

with this condition were asked how many months ago they first experienced the pain; the researchers reported a mean = 126.1, median = 84, and $SD = 127.1$. 2-43. What can you say about the number of months since onset, based on the mean and median? 2-44. The researchers wanted to determine whether patients who had TMJ pain experienced more sleep bruxism (clenching or grinding of the teeth during sleep) than people without the pain. The researchers measured rhythmic masticatory muscle activity (RMMA) episodes, which are jaw movements; *masticatory* means "related to chewing." These episodes were counted and timed. The case participants' mean duration of RMMA episodes was 49.9 seconds ($SD = 69.7$), with a median = 21 seconds. We would like to have seen skewness statistics for the number of months since onset and the duration of RMMA episodes. Why?

SCENARIO 2-F

Harris et al. (1994) wanted to find out if "maternity blues," a mild form of postpartum depression, was related to changes in the hormones progesterone and cortisol after delivery. Healthy first-time mothers who carried babies to term were studied. The women responded to a scale that measured maternity blues; higher scores meant more symptoms of the blues. Saliva samples were taken two or three times a day before and after delivery for the measurement of hormones. Reporting on the results on the maternity-blues scale, the researchers reported, "The trimmed mean score (based on the middle 90% of values) fell from 5.0 on day 1 postpartum to 3.9 on day 3, rose to 5.3 on day 5, and fell to 3.7 on day 10 …" 2-45. What kind of trimmed mean was computed? 2-46. Why might these researchers have chosen to compute a trimmed mean?

2-47. Look over this chapter and make a list of the statistics and their symbols (if any). Write a brief explanation of each statistic, using your own words without looking at the book's definition. (If you are following the suggestion from Chapter 1 to create your own statistics dictionary, then you could include these symbols and explanations in your growing document.) 2-48. Compare your words with our definitions and think about whether there are meaningful differences in the two explanations.

References

Harris, B., Lovett, L., Newcombe, R. G., Read, G. F., Walker, R., & Riad-Fahmy, D. (1994). Maternity blues and major endocrine changes: Cardiff puerperal mood and hormone study II. *BMJ, 308*, 949. doi:10.1136/bmj.308.6934.949

Marx, M. S., Cohen-Mansfield, J., Regier, N. G., Dakheel-Ali, M., Srihari, A., & Thein, K. (2010). The impact of different dog-related stimuli on engagement of persons with dementia. *American Journal of Alzheimer's Disease and Other Dementias, 25*, 37–45. doi:10.1177/1533317508326976

Raphael, K. G., Sirois, D. A., Janal, M. N., Wigren, P. E., Dubrovsky, B., Nemelivsky, L., ... Lavigne, G. J. (2012). Sleep bruxism and myofascial temporomandibular disorders: A laboratory-based polysomnographic investigation. *Journal of the American Dental Association, 143,* 1223–1231. doi:10.14219/jada.archive.2012.0068

Rothwell, P. M., Howard, S. C., Dolan, E., O'Brien, E., Dobson, J. E., Dahlöf, B., ... Poulter, N. R. (2010). Prognostic significance of visit-to-visit variability, maximum systolic blood pressure, and episodic hypertension. *Lancet, 375,* 895–905. doi:10.1016/S1474-4422(10)70066-1

Wilkens, P., Scheel, I. B., Grundnes, O., Hellum, C., & Storheim, K. (2010). Effect of glucosamine on pain-related disability in patients with chronic low back pain and degenerative lumbar osteoarthritis. *Journal of the American Medical Association, 304,* 45–52. doi:10.1001/jama.2010.893

3

Exploring Data Visually

Introduction

People are so accustomed to seeing graphs on television, in publications, and on the Internet that a chapter on graphing may seem unnecessary. Graphs are easy to make with widely available software. But visual displays of data can be misleading, and the graph with the prettiest colors may not communicate the results of a study accurately and effectively. In addition, one of the most valuable aspects of graphing data is that the researcher is able to explore the data visually and achieve a greater understanding of the phenomena being studied. In this chapter, we explain the advantages and disadvantages of different kinds of graphs. We will show some bad graphs to serve as a contrast with better graphs. Making graphs is easy, but making graphs well can take careful thought and sometimes several attempts. The reasons for creating graphs include developing an understanding of research results, summarizing the data, and communicating the results quickly and accurately. These reasons may remind you of the purpose of descriptive statistics. But as you will see, graphs can reveal details about our data that are missed by descriptive statistics.

Why Graph Our Data?

After agreeing to analyze a data set for a colleague, the first author once leapt into calculating statistics without graphing the data. The key word in the previous sentence is "once," because she has never made that mistake again.

The colleague had taken systolic and diastolic blood pressure readings on participants immediately after they had gone through different experiences that were expected to provoke emotional reactions. The first author failed to notice something important in the data set: a research assistant had typed in zeroes whenever a participant's automatic blood pressure cuff failed to register a reading. For example, if the blood pressure cuff didn't work right for three participants, then for each of those three participants, a zero was entered in the data set. What is the meaning of a zero for a blood pressure reading? A zero would mean the person was dead. Instead of using zeroes, the research assistant should have used a special code to indicate missing scores. What effect would the zeroes have on the mean systolic blood pressure for a group of participants? That's right; the mean would be pulled down, leading the researchers to think that the average blood pressure was lower than it actually would have been, if only valid scores were included in the mean. If your first author had taken the time to graph the data *before* she computed statistics, she would have noticed many zeroes, realized the mistake that had been made in data entry, and avoided wasting time on useless statistical analysis. As written by the inventor of certain graphs, "There is no excuse for failing to plot and look" (Tukey, 1977).

There are many kinds of data entry errors that can be discovered by exploring the data with graphs. For example, in a study of undergraduate students, we have seen an age typed as 91 years instead of 19 years. The 91 would have led to an erroneously high average for participants' ages. We also have seen data sets in which many people had either low or high scores, but no one had scores in the middle, leaving a gap in the distribution. Such gaps can provide important information to researchers about their research topic and the participants, and no statistic measuring center or spread would tell the researchers that such a gap existed. Finding errors and looking for unexpected patterns in the data are two reasons that we recommend graphs as the first step of any data analysis. You will learn about a graph that can detect extremely high or low scores. These extreme scores may be data entry errors, or they may be legitimate, accurate measurements. They still can skew the statistics, and they may indicate to the researcher that the sample includes some participants who differ systematically from others in the sample. Suppose a researcher is studying health literacy, which is the ability to understand and use health information, and the researcher administers a questionnaire that measures health literacy. While analyzing the data, the researcher identifies a few extremely high scores. On further investigation, the researcher discovers that these participants are health-care professionals, who do not belong in the study. Without looking at the data, the researcher may have included those extreme scores and reported skewed statistics.

When summarizing a study, we can give readers a quick understanding of the results by showing graphs. Our goals should be to find and correct inaccuracies in the data, then to look for patterns to help us understand the data and explain the results to others.

3-1. List as many reasons as you can for graphing data.

Suggested Answers

3-1. Your list may include these reasons: to become familiar with the data set before computing statistics; to reveal possible data entry errors; to see unexpected patterns in the data; to understand the phenomenon being measured; to identify extreme scores; to provide a quick summary of the data for inclusion in a report.

Pie Charts and Bar Graphs

Graphs often clump together participants who are similar in some way. Some variables are categorical, such as gender. If we were graphing the genders of babies born in a certain hospital last year, the graph would clump the males together and the females together, then show the *frequency,* or number of occurrences in each category. In other words, we would need a graph that works well with a nonnumeric variable such as gender. Even though frequencies are numbers, the focus here is on the variable that is creating the clumps, and gender is a categorical variable.

Other graphs can clump together participants who have the same score or similar scores on a quantitative variable. If we asked mothers of newborns to tell us how long they were in active labor, we might create a graph that clumped together women according to two-hour blocks of time: 0–2 hours, 2–4 hours, 4–6 hours, and so forth, with a decision being made about where to include the dividing line (e.g., exactly two hours goes into the first grouping of 0–2 hours and not into the next grouping, 2–4 hours). Number of hours in active labor is a quantitative variable, and similar subjects are being clumped according to the hours of labor. Even though the graph might have blocks of time, we know we have a quantitative variable because the order of the blocks has meaning. It would not make sense to present the information in this order: 4–6 hours, then 0–2 hours, then 8–10 hours. In contrast, the categorical variable's clumps could be reordered without changing the meaning of the results. It would not matter if we gave the number of males in a study before the number of females, or if the order were reversed.

Categorical variables, such as gender, are most often graphed using a pie chart or a bar graph. A *pie chart* is a circle with wedges representing relative frequencies in categories; they are *relative* frequencies because they are shown as proportions or percentages of the total number of observations, such as the percentage of participants who were male. If a sample of 80 people included 20 males, we could create a pie chart that showed 25% of the participants (20 out of 80) were male, and 75% were female. An advantage of a pie chart is that it shows how much of the whole is contained in each category. A *bar graph* also can represent the frequencies in each category or the percentage of participants in each category. The frequencies or percentages are represented by different bar heights (for a vertical bar graph) or lengths (for a horizontal bar graph).

Which is better, a bar graph or a pie chart? It depends on the data and the message that needs to be conveyed. What follows are several graphs of the same data. We will show some truly awful graphs, then we will show better ones. The data come from a real research study. Price, Amini, and Kappeler (2012) conducted a randomized controlled trial of the effect of exercise on the health of pregnant women and their babies; the researchers graciously agreed to share their data with us (see http://desheastats.com). They obtained a sample of sedentary women who were 12–14 weeks into their pregnancy. The women were randomly assigned to either participating in an exercise program or remaining inactive. All participants' aerobic fitness and muscular strength were measured on five occasions, with the first occasion being before the intervention began and the last occasion being after delivery. The researchers also collected information about delivery method, babies' birth weight, and the mothers' speed of recovery from the delivery. Suppose we want to graph how many participants had each kind of delivery method. The graph would clump together participants in the categories of cesarean section or vaginal delivery, counting how many women were in each category. The data, therefore, are categorical. Figure 3.1 is a *terrible* pie chart showing the proportions of deliveries that were cesarean section or vaginal delivery.

What makes Figure 3.1 so terrible is the unnecessary use of three dimensions, which distorts the proportions in each category. A weakness of a pie chart is the difficulty of comparing the sizes of wedges in a circle, especially when there are many categories with small differences in the proportions. Figure 3.1 has only two categories, and clearly a minority of participants delivered by cesarean section. But the use of three dimensions makes it extremely difficult to estimate the percentages of participants in each category. Figure 3.2 is so much easier to understand without the third dimension getting in the way.

The pie chart in Figure 3.2 takes up a lot of space relative to the amount of information being communicated: 16 out of 62 women in the study (25.8%) had cesarean deliveries. Data analysts disagree on the value of pie charts (e.g., see http://tinyurl.com/9uvs5pl and http://tinyurl.com/m2l8ot2, or search online for

Cesarean

Vaginal

Figure 3.1

A terrible pie chart. The use of three dimensions can obscure the meaning of wedges in a pie chart like this one, showing the proportion of cesarean versus vaginal deliveries in the study of exercise during pregnancy. (Data from "Exercise in pregnancy: Effect on fitness and obstetric outcomes—a randomized trial," by B. B. Price, S. B. Amini, and K. Kappeler, 2012, *Medicine & Science in Sports & Exercise, 44*, 2263–2269.)

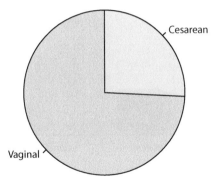

Figure 3.2

A better pie chart. A graph can be much clearer in two dimensions, compared with three dimensions. (Data from "Exercise in pregnancy: Effect on fitness and obstetric outcomes—a randomized trial," by B. B. Price, S. B. Amini, and K. Kappeler, 2012, *Medicine & Science in Sports & Exercise, 44,* 2263–2269.)

"bad pie charts"). Some of the bad pie charts on the Internet actually combine percentages from different variables. Instead of taking one variable such as delivery method and creating a pie chart, a bad pie chart would try to say that 52% of participants in a study were female, 38% were African American, 14% had type II diabetes, and 41% were taking medication for high blood pressure. So, do these numbers represent 145% of respondents? Of course not. There could be four separate pie charts for these variables (gender, race, diabetes, and medication), but it makes no sense to try to force these numbers into one pie chart.

Quite a bit of research has been done on how people perceive information visually. The brain has to decipher information that has been translated into a visual image. The various ways of displaying the information are not equally easy to decipher (see Cleveland's hierarchy of graphic elements in Wilkinson, 2005). Research shows it is easier to compare two points on a number line than it is to compare two volumes; for example, imagine trying to judge whether a short, wide beverage glass versus a tall, narrow glass would hold more water. The idea behind graphing data is to find the best way to understand and communicate the story of a research project's results. We would recommend graphing your data in multiple ways, then choosing the graphs that help to explain relationships among variables.

If we wanted to see the frequencies in each category instead of the proportions of the whole, we could create a bar graph. Would three dimensions be acceptable in a bar graph? Let's look at Figure 3.3 (and we hope your skepticism is primed to critique this graph).

Figure 3.3 is awful too. The use of three dimensions again is problematic. Determining the number of women in each category based on the bars' heights would be impossible without the lines drawn across the background of the graph. The graph also does not tell us that the numbers represent frequencies in each category. Let's look at a simpler bar graph that is easier to understand (Figure 3.4).

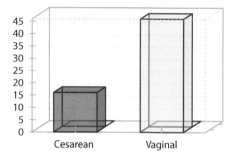

Figure 3.3

A terrible bar graph. This graph tries to communicate about the kinds of deliveries in the study of exercise during pregnancy, but use of three dimensions makes the graph harder to understand. (Data from "Exercise in pregnancy: Effect on fitness and obstetric outcomes—a randomized trial," by B. B. Price, S. B. Amini, and K. Kappeler, 2012, *Medicine & Science in Sports & Exercise, 44,* 2263–2269.)

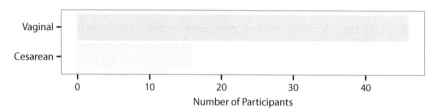

Figure 3.4

A better bar graph. Without a needless third dimension, the bar graph becomes easier to interpret. (Data from "Exercise in pregnancy: Effect on fitness and obstetric outcomes—a randomized trial," by B. B. Price, S. B. Amini, and K. Kappeler, 2012, *Medicine & Science in Sports & Exercise, 44,* 2263–2269.)

By using only two dimensions in Figure 3.4, we remove some unnecessary complexity and make the graph easier to understand. Having horizontal bars and labels often can make the reader's task easier, at least in cultures where we are accustomed to reading from left to right. Figure 3.4 also takes up a fraction of the space required by the three-dimensional plots to convey the same amount of information. Frequencies sometimes are harder to understand than percentages. By dividing each frequency by the total N, we can create a bar graph that shows percentages, similar to the information conveyed by the pie chart in Figure 3.1. Figure 3.5 is like Figure 3.4, except now we have inserted the percentages in each category of delivery method.

As in all bar graphs, the bars in Figure 3.5 do not touch each other, and the order of the bars could be reversed without changing the meaning of the results. Even though Figure 3.5 is better than the earlier graphs, it might raise an important question: how much space does this graph require, relative to the amount of information being conveyed? One could argue that the same information can be conveyed in the following sentence: "Sixteen out of the 62 participants (25.8%) had cesarean deliveries." This statement is simple, takes up less space

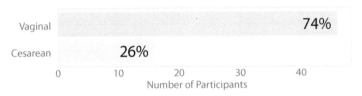

Vaginal				**74%**
Cesarean		**26%**		
0	10	20	30	40

Number of Participants

Figure 3.5

An even better bar graph. By adding the percentages, we have made the relative lengths of the bars even more understandable. Compared with Figure 3.4, what other improvements can you see? (Data from "Exercise in pregnancy: Effect on fitness and obstetric outcomes—a randomized trial," by B. B. Price, S. B. Amini, and K. Kappeler, 2012, *Medicine & Science in Sports & Exercise, 44*, 2263–2269.)

on the page, and shows that a graph is not always necessary. The information about delivery method could be placed in a small table containing many other details, such as descriptive statistics on length of pregnancy and babies' birth weights. Researchers should use graphs when they enhance the written text of their reports and explain the results better than words could.

Check Your Understanding

3-2. Suppose your instructor tells you to delete any graphic elements that are unnecessary in a graph. Why might you hear this recommendation?
3-3. What makes a bar graph better than a pie chart? What advantage does a pie chart have over a bar graph?

Suggested Answers

3-2. Unnecessary details, such as the use of three dimensions, can interfere with the reader's understanding of the data. This recommendation reminds us to focus on the purpose of the graph and make graphing decisions that increase clarity of the communication of information. 3-3. Comparing the heights or lengths of bars can be easier than comparing the sizes of wedges in a circle. But the wedges can show which category contains the greatest proportion or percentage of participants.

Two Kinds of Dot Plots

There are two graphs that have the same name: dot plot, a term that sometimes is written as one word. You already saw one kind of dot plot in Chapter 2. Figure 3.6 reproduces Figure 2.1 showing nine systolic blood pressure readings that were invented for the purpose of illustrating a symmetric distribution.

Figure 3.6 is an example of what we call a *simple dot plot* (Wilkinson, 1999). The graph uses data for one quantitative variable, systolic blood pressure. The scores are not being clumped together in a simple dot plot. Each dot

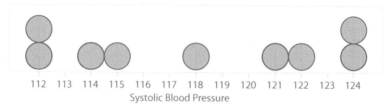

| 112 | 113 | 114 | 115 | 116 | 117 | 118 | 119 | 120 | 121 | 122 | 123 | 124 |

Systolic Blood Pressure

Figure 3.6

Simple dot plot. Each dot represents one systolic blood pressure reading. The data were fabricated and previously shown in Chapter 2.

represents one systolic blood pressure reading, and the dots are stacked when the same blood pressure reading occurred more than once. Now imagine introducing a second variable to a dot plot. Suppose we have obesity rates for 17 states in the Southern United States. We could form a simple dot plot for those 17 rates, or we create another kind of dot plot that allows us to add a second variable, the state name. Figure 3.7 shows this kind of dot plot, which we call a *multi-way dot plot* (Cleveland, 1993). "Multi-way" means there is more than one variable being graphed. This kind of dot plot displays frequencies, percentages, or rates according to a categorical variable such as location.

The multi-way dot plot in Figure 3.7 lifts the dots from the horizontal axis, and a line is drawn from each dot to the vertical axis, where the state's abbreviation is listed. The obesity rate was the percentage of adults in a representative sample from each state who met criteria to be categorized as obese. The multi-way dot plot is similar to a bar graph. Instead of having bars representing percentages, this graph has a dot and a line for each percentage. The rates for these 17 Southern states (with "Southern" defined by the U.S. Census Bureau) were placed in order,

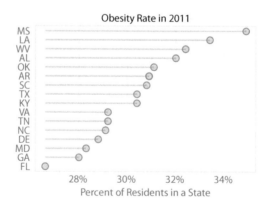

Figure 3.7

Multi-way dot plot. The 2011 obesity rates for 17 Southern states are shown in rank order, with Mississippi having the highest rate. (Data from "Adult obesity facts," by the Centers for Disease Control and Prevention, 2012, August 13, retrieved from http://www.cdc.gov/obesity/data/adult.html.)

3. Exploring Data Visually

and a dot's location relative to the number line at the bottom of Figure 3.7 indicates the state's obesity rate. Participants were clumped according to the state; that is, people who met the criteria to be considered obese were counted, then this number was divided by the total number of people who responded to the survey for that state. The states could have been listed alphabetically along the vertical axis, but then the numeric rank order of the obesity rates would have been lost. A multi-way dot plot also can be organized by groups. That is, if we expanded Figure 3.7 to include all 50 states and the District of Columbia, we could use one color to represent Southern states' obesity rates and another color to show non-Southern states' rates. (Your instructor may wish to have you create graphs with these data. Visit http://desheastats.com.)

Scatterplots

The multi-way dot plot in Figure 3.7 allowed us to compare the 17 Southern states' obesity rates. We can imagine a similar graph being created for a different variable: food hardship. This variable has been defined as the percentage of adults who said they lacked money to buy food for their families or themselves on at least one day in the last year. The Food Research and Access Center annually estimates the rates of food hardship for each state, based on representative samples. It would not be a surprise to find that the 50 states and the District of Columbia would vary in terms of food hardship rates, just as they vary on obesity rates. We might ask: Are states where people are more likely to lack money to feed their families also the states with lower obesity rates? Or is there something about lacking money for food that results in poorer nutrition and therefore higher obesity rates? Or will there be no discernible relationship between the two variables?

We can graph two variables together in a *scatterplot*, a graph that plots scores on two quantitative variables, with each variable having its own number line or axis. Figure 3.8 shows a scatterplot of food hardship and obesity rates for the United States, including the District of Columbia. We have put the food hardship scores on the horizontal (*X*) axis and obesity rates on the vertical (*Y*) axis.

Figure 3.8 uses one circle per location. We prefer to use circles instead of solid dots in a scatterplot because they can help us to see overlapping data points. The points that are toward the left side of the graph correspond to lower rates of food hardship; the point that is farthest to the left represents North Dakota, which had a food hardship rate of 10%, coupled with an obesity rate of 27.8%. (We had to look at the data set to determine which state had these rates.) The points that are toward the right side of the scatterplot correspond to higher rates of food hardship; the point that is farthest to the right represents Mississippi, which had a food hardship rate of 24.5%. As you probably noticed, Mississippi's circle also is highest vertically on the scatterplot. That's because Mississippi had the country's highest rate of obesity (34.9%), which we saw in the multi-way dot plot (Figure 3.7). Which state had the lowest rate of obesity? Obesity is shown on the vertical axis, so we are looking for the state represented by the circle that is lowest on the graph vertically—by looking at the data set, we learned it is Colorado, with

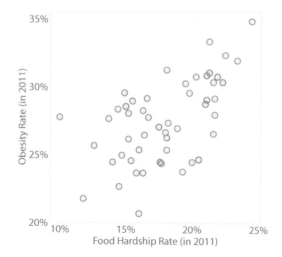

Figure 3.8

Scatterplot of food hardship and obesity rates. This graph uses scores on two variables. Each circle represents the food hardship rate and the obesity rate for one location, with the data representing the 50 states and the District of Columbia. (Obesity data from "Adult obesity facts," by the Centers for Disease Control and Prevention, 2012, August 13, retrieved from http://www.cdc.gov/obesity/data/adult.html; food hardship data from "Food hardship in America 2011: Data for the nation, states, 100 MSAs, and every congressional district," by the Food Research and Action Center, 2012, February, retrieved from http://www.frac.org.)

a rate of 20.7%. But Colorado appears to be closer to the middle of the pack in terms of food hardship at 16%. So there is not a perfect relationship between food hardship and obesity, where the states would be in the same order on food hardship as they are on obesity. But there appears to be a trend, where lower numbers on food hardship tend to be paired with lower rates on obesity, and higher scores on food hardship generally go along with higher numbers on obesity. (Do these results surprise you? The Food Research and Access Center's website, www.frac.org, can help you to understand the relationship between food hardship and obesity.) We will examine many scatterplots in Chapters 5 and 13 when we talk at length about a relationship between two quantitative variables.

Scatterplots are useful because they can help us to understand how two quantitative variables may be related. Scatterplots show one point for each unit of analysis; the two scores for each location are clumped together and represented by a single point (or circle, in the case of Figure 3.8) on the graph. We can look for trends in the relationship between the states' food hardship rates and obesity rates by studying the shape of the *point cloud*, or the collection of dots on the scatterplot. We must remember that scatterplots like Figure 3.8 are limited to only two variables. Obesity is related to many factors, and to understand this health issue would require a more detailed analysis, taking into account factors such as poverty and more objective ways of measuring obesity, rather than asking people to self-report their weight.

3. Exploring Data Visually

3-4. Figure 3.9 shows another scatterplot of food hardship and obesity rates, except now we have replaced the circles with the abbreviations for the 50 states and the District of Columbia. Further, we have used a purple font for the Southern states' abbreviations and an orange font for the non-Southern states' abbreviations. What can we observe about the relationship between food hardship and obesity by examining Figure 3.9?

Suggested Answers

3-4. The most obvious detail in Figure 3.9 is that Southern states tend to have the highest rates in both food hardship and obesity. When all the states are taken together, there appears to be an upward trend as we read the graph from left to right: lower food hardship rates tend to appear with lower obesity rates and higher food hardship rates generally are paired with higher obesity rates. But if we focused only on the non-Southern states shown in orange font, the trend does not seem as strong. This scatterplot illustrates the importance

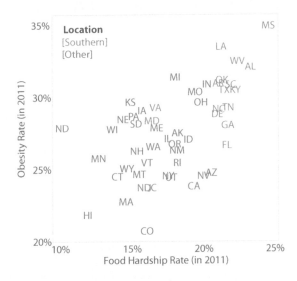

Figure 3.9

Scatterplot with state abbreviations as markers and color representing regions. This graph shows the same data as Figure 3.8, except the circles have been replaced with state abbreviations, and two colors are being used to distinguish between Southern and non-Southern states. (Obesity data from "Adult obesity facts," by the Centers for Disease Control and Prevention, 2012, August 13, retrieved from http://www.cdc.gov/obesity/data/adult.html; food hardship data from "Food hardship in America 2011: Data for the nation, states, 100 MSAs, and every congressional district," by the Food Research and Action Center, 2012, February, retrieved from http://www.frac.org.)

(Continued)

of exploring the data and making multiple graphs. The use of different colors for Southern and non-Southern states allowed us to add information about the region of the country. (Interestingly, a recent study said these obesity rates may not be trustworthy because they are based on self-reports. Le et al. (2013) suggested that people in certain regions of the United States may be more honest about their weight than people in other regions. Further, a study by Ezzati, Martin, Skjold, Vander Hoorn, and Murray (2006) examined the bias in self-reported weights and heights—overestimation of height by men and underestimation of weight by women.)

Histograms

Scatterplots allowed us to look at two quantitative variables at once. Now we will talk about some graphs that clump together participants with similar numeric scores on one variable. We return to the study by Price et al. (2012) in which sedentary pregnant women were randomly assigned to either participating in a supervised aerobic training program or remaining inactive. The pregnant participants were measured once before the intervention began (12–14 weeks gestation). After the researchers began the intervention, the participants were measured three more times during pregnancy and a fifth time about 6–8 weeks after they gave birth. To measure strength, the researchers recorded the number of times that each expectant mother could lift a 7-kg (15.4-lb) medicine ball in a minute. Suppose we would like to graph the number of lifts for the participants at Time 5. For simplicity, we will disregard their assigned groups (treatment or control). By looking at the raw data, we see that the lowest score was 5 lifts in a minute and the highest number of lifts was 43 in a minute (wow!). The strength scores clearly must remain in numeric order on the graph. We can create a *histogram*, a graph that looks somewhat similar to a bar graph, except a quantitative variable is being graphed and the bars touch each other. Figure 3.10 is a histogram of the number of lifts at Time 5.

Recall that a bar graph clumped participants into categories such as cesarean section versus vaginal delivery, so a bar graph is used with a categorical variable (such as delivery method). Now participants are clumped according to a quantitative variable, strength score. By looking at the data set, we learned that the lowest score was 5 and the highest score was 43; these scores are represented in Figure 3.10 by the small bars at each end of the distribution. On the number line near where 10 lifts would be, there is no bar. There also is no bar around 40 lifts. The lack of a bar conveys meaning in a histogram: there is a gap in the distribution where no participant had a score. Being able to see gaps in the distribution is an advantage of a histogram. Depending on the variable, the gaps may or may not be meaningful to the researchers. The heights of the bars show us which scores were most common and which scores were less common. Here is a way to remember the difference between a histogram and a bar graph. You can think of the

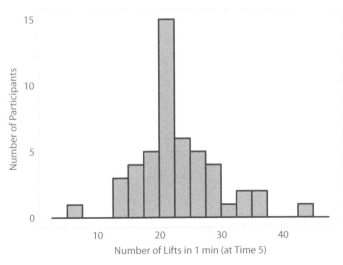

Figure 3.10

Histogram. This graph shows the number of lifts of a medicine ball by the participants in the study of exercise during pregnancy. These scores were collected after the women had given birth. (Data from "Exercise in pregnancy: Effect on fitness and obstetric outcomes—a randomized trial," by B. B. Price, S. B. Amini, and K. Kappeler, 2012, *Medicine & Science in Sports & Exercise, 44,* 2263–2269.)

bars touching in a *histogram* for a continuous variable, similar to the way *history* is continuous. In contrast, the bars do not touch in a bar graph because the bars represent separate categories, so we are *bar* hopping in a *bar* graph.

At first glance you may not realize something about Figure 3.10: the histogram clumped together different scores in the same bar. The bars can obscure the exact values being represented, but we still can get a good idea of the shape of the distribution. The tallest bar is close to 20, and 15 scores are represented in that bar. By looking at the raw data, we figured out that those 15 participants lifted the medicine ball 20, 21, or 22 times. So each bar in this histogram represents three possible scores. Look at the bar just to the left of the tallest bar; it would represent 17, 18, or 19 lifts, but without looking at the data set, we cannot exactly know what those five scores were. The bar could represent five 17's ... or two 17's and three 19's ... or any other combination of five scores in that range. If you create the same graph with a different statistical program, you might get a histogram that looks slightly different from ours.

Check Your Understanding

3-5. We showed a multi-way dot plot of the obesity rates for 17 Southern states. The obesity rates for all 50 states and the District of Columbia were used to create the histogram in Figure 3.11. What does this graph tell us about adult obesity rates?

(*Continued*)

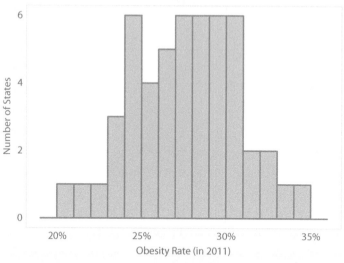

Figure 3.11

Histogram of obesity rates. The 2011 obesity rates that were used in the scatterplots with food hardship now are displayed in a histogram. (Data from "Adult obesity facts," by the Centers for Disease Control and Prevention, 2012, August 13, retrieved from http://www.cdc.gov/obesity/data/adult.html.)

Suggested Answers

3-5. *Many states appear to have obesity rates between about 24% and 31% of adult residents because that is where the tallest bars are located. The shortest bars indicate that there are a few states with lower obesity rates (about 20% to 22%) and a few other states with higher obesity rates (about 34%).*

Time Plots (Line Graphs)

The researchers studying exercise during pregnancy measured the strength of expectant mothers in the two groups on five occasions, not just the single time shown in the histogram. They collected data four times during pregnancy (at 12–14, 18–20, 24–26, and 30–32 weeks) and once postpartum (6–8 weeks after delivery). We could summarize the results for the number of lifts of the medicine ball in 1 minute by graphing the means for each group at each occasion. Prepare yourself for a colorful, hideous graph (Figure 3.12).

We hope Figure 3.12 convinces you that three-dimensional graphs tend to hinder understanding of data. This graph makes it almost impossible to compare the means for the two groups across time. Let's look at a much simpler, clearer display of the same results. A *time plot*, also known as a *line graph*, connects

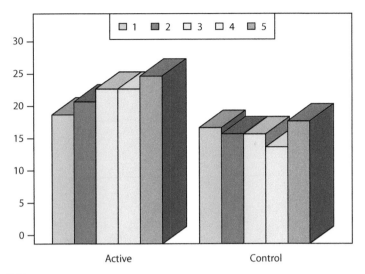

Figure 3.12

A terrible graph of sample means. Again, the use of three dimensions obscures the message about the means for the two groups in the study of exercise during pregnancy. (Data from "Exercise in pregnancy: Effect on fitness and obstetric outcomes—a randomized trial," by B. B. Price, S. B. Amini, and K. Kappeler, 2012, *Medicine & Science in Sports & Exercise, 44,* 2263–2269.)

observations or means across time. Figure 3.13 shows the same 10 means, but now without clutter.

A time plot's strength is in its name: it displays change over time. How are participants being clumped in this graph? We have more than one way of clumping here. Participants are clumped according to whether they were in the experimental group or the control group. Further, their strength scores were clumped into occasions of measurement (Time 1, Time 2, Time 3, Time 4, and Time 5). What's more, a mean was computed for each group on each occasion of measurement, and 10 means are shown on the graph. The independent variable, group, is categorical. Previously, we saw an example of a bar graph (Figure 3.4), where the order of the bars (cesarean vs. vaginal delivery) could have been swapped, and the meaning of the graph would be the same. In Figure 3.13, the two lines cannot be moved around on the graph because each point is decided by two axes: the mean number of lifts is represented on the vertical axis and the occasion in time is shown on the horizontal axis. A line graph, therefore, is useful for showing trends across time for quantitative variables, such as strength scores (i.e., the number of lifts of the medicine ball in 1 minute).

Let's think about the meaning of Figure 3.13. It appears that the active group's average strength increased through Time 3, then leveled off, then increased again after delivery. Meanwhile, the control group showed an apparent decline in mean number of lifts by Time 4, then an uptick after delivery. A disadvantage of this graph is that we are seeing the measures of the center (means) without

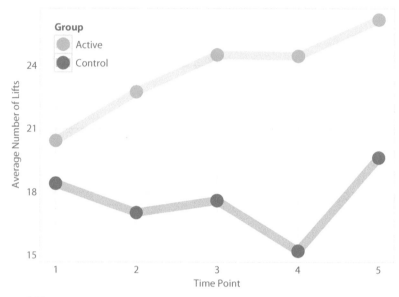

Figure 3.13

Time plot or line graph. This graph shows dots representing the mean number of medicine ball lifts on five occasions for the two groups in the study of exercise during pregnancy. By connecting the dots, we can have a better understanding of the performance of the two groups in the study of exercise during pregnancy. (Data from "Exercise in pregnancy: Effect on fitness and obstetric outcomes—a randomized trial," by B. B. Price, S. B. Amini, and K. Kappeler, 2012, *Medicine & Science in Sports & Exercise, 44,* 2263–2269.)

the measures of spread (standard deviations). In other words, we cannot tell from this graph how much variation existed in the scores for each group at each occasion. For either group at any given time, were the participants' numbers of lifts very similar to each other, or were the scores quite spread out around the mean? Were some of the active women's scores as low as the scores for members of the control group? This plot of the 10 means cannot say. But what if we added one line per participant in the study? Figure 3.14 adds these lines.

The participants in the active group have light green lines and the participants in the control group have lines that are a kind of rust color. There are many advantages to adding these lines. We can see a tendency for the active participants to have higher numbers of lifts than the control participants. The individual trajectories also illustrate the variation in scores; not all participants were close to their group mean at all occasions. Some people in the experimental group had lines that were lower than the means for the control group and a few people in the control group had higher scores than the mean in the treatment group. You may have noticed that the bold lines in Figure 3.14 do not look exactly like the lines in Figure 3.13, even though both graphs show the same 10 means. Figure 3.14 had to show a larger range on the vertical (*Y*) axis for the number of lifts because

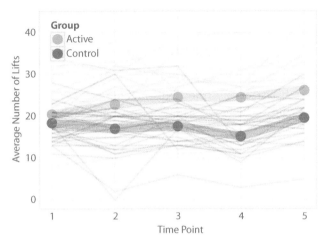

Figure 3.14

Time plot with lines for individuals. Now we have added fine lines to represent each participant in the study of exercise during pregnancy. This graph allows us to see how much variation exists within the two groups. (Data from "Exercise in pregnancy: Effect on fitness and obstetric outcomes—a randomized trial," by B. B. Price, S. B. Amini, and K. Kappeler, 2012, *Medicine & Science in Sports & Exercise, 44*, 2263–2269.)

there were participants with higher-than-average numbers of lifts and other participants with lower-than-average numbers of lifts. It is as if Figure 3.13 was zoomed in for a close-up of the means, and Figure 3.14 zoomed out to show all the variation in the scores. (You may notice that some of the individual lines appear to stop at Time 4; some participants did not complete the last occasion of measurement.)

Whenever we examine a time plot, we must remember that each occasion's mean may not have the same number of participants represented; it is common for longitudinal studies to lose participants over time—people move away, they lose interest, they have health complications that keep them from participating, and so forth. We also must remember to be careful about saying whether there is a difference between any two means. A difference may look big on a graph, but it might be statistically negligible. Some of the statistics that we will cover later in the book can be used to say whether an observed difference is noteworthy.

Boxplots

To summarize the shape of a distribution of numeric data, researchers sometimes use a kind of graph that you probably have never seen before. *Boxplots*, or *box-and-whisker plots*, are graphs that show the locations of scores and the amount of spread in a data set, while also providing a way to define scores as notably extreme. A boxplot consists of a box with two lines (called *whiskers*) extending from it; in addition, sometimes there are symbols beyond the lines. We used data from the Centers for Disease Control and Prevention's (CDC) Behavioral Risk Factor Surveillance System (BRFSS). Representative samples are taken in the 50 U.S. states and the District of Columbia. Among other health-related questions, the BRFSS survey asks people 18 years and older about their smoking habits, if any. The adult smoking prevalence is defined as the percentage of adults who have smoked at least 100 cigarettes in their lives and currently smoke every day or some days. (*Prevalence* is the proportion of people with a condition, usually expressed as a percentage or a rate, such as the number of people with the condition per 1,000 people.) Figure 3.15 is an example of a boxplot of the 2009 adult cigarette smoking rates for the 50 U.S. states plus the District of Columbia.

In a boxplot, three numbers can divide a data set into four parts that contain roughly the same number of scores. Think of folding a piece of paper in half, from top to bottom, then folding it in half again in the same direction. When you unfold the piece of paper, you will have three fold marks that divide the page into four roughly equal parts. That is what a boxplot does—the four parts will contain roughly the same *number* of scores. But unlike a folded piece of paper, the four parts of a boxplot often are not the same size because the boxplot displays how spread out the scores are, as well as where they are located on the number line.

The box part of the boxplot in Figure 3.15 represents about the middle half of the data set. The box has a line dividing it into two parts; that line represents the median, so the same number of scores are represented below the median

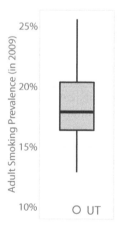

Figure 3.15

Box-and-whiskers plot, or boxplot. The adult cigarette smoking rates for the 50 states plus the District of Columbia are displayed. The circle at the bottom of the graph represents the adult smoking rate for the state of Utah. (Data from "Current cigarette smoking among adults," by the Centers for Disease Control and Prevention, 2013, April 25, retrieved from http://www.cdc.gov/tobacco/data_statistics/fact_sheets/adult_data/cig_smoking/.)

3. Exploring Data Visually

and above the median. A separate analysis of the 2009 adult smoking rates showed the median adult smoking rate was 17.9%. Look at Figure 3.15 again and make sure you understand that the horizontal line in the middle of the box appears roughly at 17.9%. When we looked at the data, we learned that three states had an adult smoking rate of 17.9%, 24 locations had rates less than 17.9%, and 24 locations had rates greater than 17.9%. The top quarter of the data is above the box, represented by the upper whisker. The bottom quarter of the data is below the box, represented by the lower whisker and the circle marked "UT," the abbreviation for Utah. The median divides the box into two parts, and each part of the box contains about one-fourth of the data. We say "*about* one-fourth" because different software packages and calculators use different rules for calculating the numbers for the bottom and top of the box. Because you may not have the same statistical software that was used to create our graphs, you might use our data and create a boxplot that looks slightly different from Figure 3.15, so we will be a bit vague about the calculations underlying our graph.

How does Figure 3.15 communicate information about variability? The length of different parts of the boxplot tells us about spread. For example, the top half of the box itself appears taller than the bottom half of the box. That means the data are slightly more spread out in the top half of the box than the bottom half. Let's compare the top quarter of the data set with the bottom quarter by looking at the whiskers: in which of these quarters of the data set do we have more spread in the scores? At a glance, we might think the top quarter because the upper whisker is longer than the lower whisker. But remember that the circle for Utah is part of the bottom quarter of the data set, so we have to include it in our assessment of spread in the top and bottom quarters. Taking into account the circle for Utah, there is a little bit more spread in the scores in the bottom quarter than the top quarter. The width of the box often carries no meaning; in our examples we could make the box extremely narrow or extremely wide, without changing its meaning. (Some complicated boxplots, which we will not cover, do modify the width and shape of the box to convey additional information.)

Let's talk about the circle for Utah. This circle identifies an *outlier*, an extreme score that meets certain criteria to be defined as notably different from the rest of the data. In this data set, Utah was the state with the lowest rate in 2009; 9.8% of adults in Utah's representative sample were categorized as current smokers. We now have identified one of the main advantages of a boxplot: it gives us a way of defining outliers (although different statistical software packages have slight differences in the exact definition of an outlier and other details of a boxplot). What was the highest smoking rate in the United States? We can answer this question by looking at how far the upper whisker reaches in Figure 3.15. It appears to be something slightly above 25%. When we look at the raw data (available via http://desheastats.com), we find that both Kentucky and West Virginia had the highest rate, 25.6%. But the rates for those two states were not extreme enough to be detected by our software package as outliers, and that's why there is not a circle representing that rate.

Before we go into more details about this kind of graph, let's summarize the parts of a boxplot:

- A whisker reaches down from the bottom of the box. That line and any symbols beyond that line represent about a quarter of the scores. Figure 3.15 has a circle below the lower whisker, representing Utah, which is an outlier with the lowest adult smoking rate. (Your statistical software might use a different symbol, such as an asterisk.)
- About one-fourth of the scores are in the lower part of the box itself.
- About one-fourth of the scores are in the upper part of the box itself.
- And about one-fourth of scores are represented by the upper whisker and any symbols beyond that line. Figure 3.15 does not have any symbols above the upper whisker.

There are more than a dozen ways of defining the statistics that determine where to draw the top and bottom lines for the box, which then affects the definition of an outlier (Langford, 2006). Different statistical programs and calculators use different rules for drawing the box and defining outliers. Your instructor may ask you to download our data and create a boxplot of the adult smoking rates, and you could get a graph that does not look exactly like Figure 3.15, which was created using the software called R. Small differences among the various rules for creating boxplots could lead to different software packages identifying different numbers of data points as outliers. If the goal is to explore the data and identify points that could skew the statistics, these small differences in the rules for boxplots may or may not be important. By creating different kinds of graphs, researchers can become familiar with possible extreme scores and judge them based on the variables being studied. If boxplots were used to make crucial decisions about whether to include or exclude certain patients in a treatment, the small differences in software packages may need to be taken into account.

We want to give you a working understanding of how outliers are defined, but because of the different rules used by different software, we need to be a bit vague. We also want you to have an idea about why there are different rules. In the process, we will talk about the percentage of scores in different parts of a distribution—and that could be confusing if we continue to use the example of adult smoking rates, which are percentages! Therefore, let's change examples and go back to the study of exercise and pregnancy by Price et al. (2012) and the number of lifts of a medicine ball in 1 minute. When the women first were measured at 12–14 weeks of pregnancy, the median number of lifts was 18, meaning that the number of scores above 18 is the same as the number of scores below 18. To give you an idea of why the rules for boxplots differ, let's think about how many scores would be below 18. The sample size was $N = 61$ participants. We know the median is the middle score, and because we have an odd number of people, we would expect that the median = 18 means there are 30 people with scores below 18 and 30 people with scores above 18. But here is one more detail: five participants lifted the medicine ball 18 times. Would all five scores of 18 be the median? Would those five scores be counted in the upper half of the distribution or the lower half of the distribution—or both? Or neither?! Such details can lead to different rules for boxplots. We could say that for our sample of $N = 61$ participants, having a median = 18 lifts tells us that at least half of the scores are equal to 18 or less and at least half of the scores are equal to 18 or more.

Now suppose we want to find the median of only the *bottom half* of the data set. That number would cut the bottom half of the data set into two pieces, giving us the bottom two quarters of the data set. That number also would define the bottom of the box itself in the boxplot. Do we include the five scores of 18 in that computation or not? It depends on which rules are being used. Obviously we cannot explain all the rules for various software packages, so we will keep our explanation in general terms.

To help us to complete our explanation of boxplots, let's start with a term that you may have heard before, but you may not know exactly what it means. A *percentile* is a score that has a certain percentage of the distribution below it. We might say that the median is the 50th percentile for the number of lifts of the medicine ball. The score that has 25% of the distribution below it is called the 25th percentile; this score is also called the *first quartile* because one-fourth of the data set is below this score. The score that has 75% of the distribution below it is called the 75th percentile; this score is also called the *third quartile* because three-fourths of the data set is below this score.

Are the 25th and 75th percentiles the same things as the bottom and top of the box in a boxplot? Not exactly. If we had a very large data set, then the difference between "25th percentile" and "the score represented by the bottom of the box" effectively would disappear. But the conflicting rules complicate the situation, so we will say the line forming the bottom edge of the box has *about* 25% of scores below it and the line forming the top edge of the box has *about* 75% of scores below it. Despite the various rules, we use the term *interquartile range* to describe the distance between the top and bottom of the box. The interquartile range by itself is rarely interesting, but it is used to decide whether an extreme score is an outlier. Different definitions of the top and bottom of the box can lead to differences in the number of outliers detected by various software packages. To illustrate the importance of the interquartile range, let's look at the boxplot in Figure 3.16, which was created using the number of medicine ball lifts at Time 1 for the participants in the study of exercise during pregnancy.

About 25% of scores are represented by the lower whisker and the circle. The box contains the middle half of the scores. The upper whisker and circles represent about 25% of the scores, roughly the top quarter. The bottom of the box is about 16 in this graph, and the top of the box is about 21.5; we already said the median was 18. What determines the length of the whiskers, and why are some scores represented by circles? This is where the interquartile range comes in. It can be used to determine whether an extreme score is markedly different from

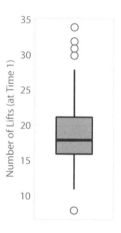

Figure 3.16

Boxplot of medicine ball lifts. This graph shows the number of times that partici-pants lifted a medicine ball in 1 minute at the beginning of the study of exercise during pregnancy. The interquartile range is approximately the length of the box. (Data from "Exercise in pregnancy: Effect on fitness and obstetric outcomes—a randomized trial," by B. B. Price, S. B. Amini, and K. Kappeler, 2012, *Medicine & Science in Sports & Exercise, 44,* 2263–2269.)

the rest of the distribution. One way of identifying an extreme score as an outlier is to check whether it is more than a certain distance from the box. According to rules from the mathematician who invented boxplots (Tukey, 1977), this certain distance is equal to 1.5 times the length of the box (i.e., 1.5 times the interquartile range). This certain distance acts like a yardstick. If a score *exceeds* that yard-stick's distance from the top or bottom of the box, then the score is called an outlier. Now we can define the whiskers too: a whisker reaches to the score that is farthest away from the box but still within the yardstick's distance. Think of the box as a house and the more extreme scores as trees. Some of the trees are on our property, and others are beyond our property line. We are preparing to have a birthday party, and we want to string a cord decorated with flags from the house to the tree that is farthest from the house but still on our property. The whisker (or cord) is drawn from the box (house) to the score (tree) that is farthest from the box but that is not an outlier (in other words, it is not a tree outside of our property line).

We know extreme scores can skew the mean and other descriptive statis-tics. Until now, we were not able to say definitely whether a score that seemed to stand out actually was an outlier. Thanks to boxplots, we can define outliers in an objective way; we do not have to base the definition on a subjective judgment about a score's appearance in a histogram. Outliers can provide important infor-mation about participants. The researchers in the pregnancy study would want to know if some participants entering the study had a great deal of trouble lifting the medicine ball more than a couple of times. Their difficulty could indicate that the researchers should monitor their activity with extra care. The researchers

3. Exploring Data Visually

also would want to know if some participants were much more athletic than the rest and able to lift the medicine ball many times. If all the athletic participants were randomly assigned to the same group, their performances could affect the statistics for that group.

Let's look again at Figure 3.16. The lower whisker goes from the bottom of the box to the lowest score that is not an outlier. By looking at the sorted data on the number of medicine ball lifts, we know that the lowest number of lifts was 8 and the next lowest number of lifts was 11. The lower whisker reaches down to the score of 11. Below the whisker there is a circle representing the score of 8, which was detected as an outlier. (Some software packages will label the outliers with the row number from a spreadsheet of the data.) Now let's look at the top whisker and the outliers at the upper end of the distribution. By looking at the raw data, we found that the top score was shared by two participants who lifted the medicine ball 34 times in a minute. The next three participants (in descending order) had scores of 32, 31, and 30. Notice that Figure 3.16 only shows four circles at the upper end of the distribution; our graph did not tell us that two people's scores were represented by the top circle. The next score in the distribution, 28, is not an outlier; the upper whisker reaches from the box to this score of 28. How might the researchers in the study of pregnancy and exercise have used this information? They may have checked to make sure that the expectant mother whose score was an outlier at the bottom of the distribution did not have a condition such as asthma that made her different from the other mothers. They also might have checked whether the five women whose high scores were outliers had reported a history of weight-lifting or physically active careers.

What can Figure 3.16 tell us about the variability in the number of lifts? We have about the same *number* of scores in each quarter of the data set, but we can see that the quarters are not equally spread out. For example, the top whisker and upper outliers are more spread out than the bottom whisker and lower outlier, indicating more variability in the top quarter of the distribution than the bottom quarter. As we have implied, the greatest strength of a boxplot is its ability to define outliers in an objective manner, but there are many definitions of outliers. There also are other ways of detecting outliers, which we will not cover in this book. A minor weakness of boxplots is that the only gaps in the distribution that we can see in a boxplot are the ones around outliers and their closest whisker.

Boxplots can be modified and combined in many ways. They can be displayed horizontally, in which case our description of the "top of the box" would have to be changed to the "right side of the box" and the "bottom of the box" would become the "left side of the box." Let's look at two vertical boxplots in the same graph, which will allow us to make comparisons between groups. The researchers who randomized sedentary pregnant women to remaining inactive or participating in physical activity tracked the participants after delivery and recorded the babies' birth weights. What can we learn about the birth weights of babies whose mothers were active, compared with the birth weights of babies whose mothers

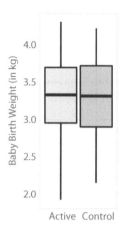

Figure 3.17

Side-by-side boxplots. This graph allows us to compare the birth weights of babies born to active versus sedentary mothers in the study of exercise during pregnancy. (Data from "Exercise in pregnancy: Effect on fitness and obstetric outcomes—a randomized trial," by B. B. Price, S. B. Amini, and K. Kappeler, 2012, *Medicine & Science in Sports & Exercise, 44*, 2263–2269.)

remained sedentary? Figure 3.17 shows side-by-side boxplots of the birth weights for babies from mothers in the two groups.

To simplify the appearance of the graph, the weights are shown in kilograms; to get from grams to kilograms, we divide the number of grams by 1,000. We immediately notice in Figure 3.17 that the medians are almost identical. A separate analysis showed that the median birth weight for the active mothers' babies was 3,330 grams (3.33 kg), and the median for the inactive/control mothers' babies was 3,317 grams (3.317 kg). Let's compare the two boxes. The box for the control group is slightly longer, indicating a bit more spread in the middle half of the data, compared with the active group's birth weights. The whiskers are longer for the active group's birth weights, showing a greater range in birth weights. Neither group had any outliers. Price et al. (2012) compared the means (not shown in Figure 3.17) and said the difference was statistically negligible.

Check Your Understanding

3-6. Figure 3.18 shows boxplots of birth weights, but this time we have grouped them by method of delivery (vaginal vs. cesarean), ignoring whether the mothers were in the active group or the sedentary group. Describe everything you can about the data based on these graphs.

(Continued)

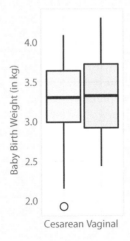

Figure 3.18

Side-by-side boxplots, different delivery methods. What can we learn about the birth weights when we compare the babies born to mothers with different delivery methods? (Data from "Exercise in pregnancy: Effect on fitness and obstetric outcomes—a randomized trial," by B. B. Price, S. B. Amini, and K. Kappeler, 2012, *Medicine & Science in Sports & Exercise, 44,* 2263–2269.)

Suggested Answers

3-6. The medians for the two groups appear to be about the same. The box for the cesarean group is slightly smaller, indicating that the middle half of the birth weights for that group was less spread out than the middle half of the birth weights for the vaginal delivery group. The upper whisker for the vaginal delivery group reaches higher than the upper whisker for the cesarean group, indicating the heaviest baby was born by vaginal delivery. The vaginal delivery group does not have any outliers. The cesarean group has one outlier on the lower end of the distribution (although a graph we created with a different software package showed two outliers). By comparison, the lower whisker for the vaginal delivery group does not reach as low on the number line, so the babies with the lowest birth weights were delivered by cesarean. What we cannot learn from this graph is whether the same number of birth weights are represented in each group; the pie charts in this chapter would tell us the answer is no.

Graphs Can Be Misleading

Sometimes we have to make several graphs to understand a data set and discover the best way to describe our results. Feel free to create as many graphs as you need to understand your data. Be cautious, however, about leaping to conclusions based only on graphs. Graphs can lie, especially when the software's default settings determine the appearance of the graph. Let's look at an example of a misleading graph. In the study of sedentary versus active pregnant women, one outcome variable was a measure of strength. The researchers taught the pregnant women to safely lift a 7-kg medicine ball as many times as they could in a minute. Figure 3.19 shows the mean number of lifts for each group at the beginning of the study. The bar graph clumps together the scores by group, and each bar represents a mean.

Figure 3.19 appears to show a difference in means, with the control group having a much lower average number of lifts. But look at where the horizontal axis begins: 18 lifts. Graphing software sometimes zooms in on any possible differences and magnifies them, somewhat like an overeager employee boasting about a minor accomplishment to impress a boss. By changing the scale on the horizontal axis, we can see a more accurate picture (Figure 3.20).

The difference in means appears smaller in Figure 3.20, compared with Figure 3.19; in fact, the researchers reported that these two means were statistically indistinguishable. It is important to be skeptical about graphs and not to

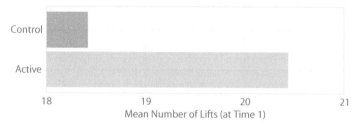

Figure 3.19

A deceptive bar graph of means. Each bar represents the mean number of lifts of a medicine ball at the beginning of the study of exercise during pregnancy. But is the difference in means actually as large as the graph seems to portray it? (Data from "Exercise in pregnancy: Effect on fitness and obstetric outcomes—a randomized trial," by B. B. Price, S. B. Amini, and K. Kappeler, 2012, *Medicine & Science in Sports & Exercise, 44*, 2263–2269.)

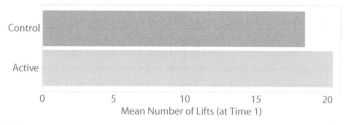

Figure 3.20

A more accurate bar graph of means. When the horizontal axis is changed so that both bars begin at zero, the difference in the mean number of medicine ball lifts is portrayed more accurately. (Data from "Exercise in pregnancy: Effect on fitness and obstetric outcomes—a randomized trial," by B. B. Price, S. B. Amini, and K. Kappeler, 2012, *Medicine & Science in Sports & Exercise, 44*, 2263–2269.)

3. Exploring Data Visually

exaggerate the results. Even if the results do not turn out as we expected, we need to report our findings as accurately as possible so that readers can judge the study for themselves. Figure 3.20 probably would not be needed because the same information could be stated in one sentence or added to a table listing statistical results for all five occasions of measurement.

Beyond These Graphs

Graphing data can be fun, and creating graphs can help researchers to under-stand the variables in a study. The collection of graphs in this chapter is far from comprehensive. People are always finding new ways of graphing data; interactive and animated graphs online can put the data into motion. An animated graph can show changes across time. For example, take a look at this animated graph that is based on data from a multiyear study of home-based services provided to families in the child welfare system in Oklahoma: http://tinyurl.com/ltzas4k. Details of the study are available in Chaffin et al. (2012).

Obviously, we cannot show moving graphs in a printed book, but we can show a couple of advanced graphs that we think are cool. When we introduced statis-tics measuring the center and spread of the data set, we quoted a colleague (Dr. Edward Kifer) as saying, "Never a center without a spread." Our colleague meant that it is not enough to report a statistic like the mean without also reporting how much variability is present in the data. This concept can be expanded to graphs. Figure 3.21 is similar to Figure 3.20, which showed the two group means for the number of lifts of the medicine ball at Time 1 in the exercise/pregnancy study.

Figure 3.21 still shows the means, represented by the right ends on the bars. Now circles have been added to represent the individuals in each group. Each circle shows the number of lifts by one person. At many places along the number line, there are people with the same number of lifts. For example, look at the number 20 on the number line. It appears that six people (four in the control group, two in the active group) lifted the medicine ball 20 times at Time 1 in the

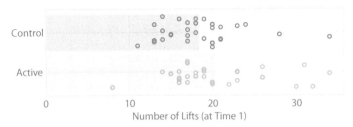

Figure 3.21

Bar graph of means with individual scores shown. This bar graph shows the mean number of medicine ball lifts by each group at Time 1 in the study of exercise during pregnancy. The circles represent each score that went into those means, represented by the bars. (Data from "Exercise in pregnancy: Effect on fitness and obstetric outcomes—a randomized trial," by B. B. Price, S. B. Amini, and K. Kappeler, 2012, *Medicine & Science in Sports & Exercise, 44*, 2263–2269.)

study. The scores have been shaken or *jittered* to add some random variability vertically; otherwise, all six people's score of 20 would overlap and appear as one circle. Figure 3.21 tells us so much more about the data than Figure 3.20 did. We still can see that the two means appear to be quite close, but now we also can tell that each group had a lot of variability in the number of lifts. The small difference in the sample means at Time 1 becomes even less impressive when we see the overlap in the scores for individuals in the two groups. Individual scores also can be added to boxplots. Figure 3.22 was created using the same data as Figure 3.21.

Again, the circles show the individual data points. To compare Figure 3.22 with Figure 3.21, look again for the people who lifted the medicine ball 20 times in a minute. You will see six little circles representing those people. We even could add symbols to indicate the location of the means, as shown in Figure 3.23.

The diamond shape in Figure 3.23 represents the mean for each group. Notice that each group's mean is slightly higher than its median, which is especially noticeable for the active group. Each boxplot has a score beyond the whisker,

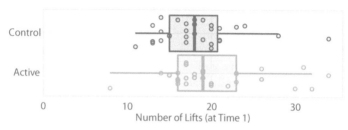

Figure 3.22

Boxplots with jittered data. The same data shown in Figure 3.21 now are shown in boxplots. By adding the circles to represent the individual scores, we now can learn more about the variability within each group. (Data from "Exercise in pregnancy: Effect on fitness and obstetric outcomes—a randomized trial," by B. B. Price, S. B. Amini, and K. Kappeler, 2012, *Medicine & Science in Sports & Exercise, 44,* 2263–2269.)

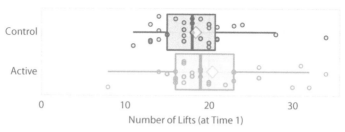

Figure 3.23

Boxplots with symbols representing the means. Diamonds represent each group's mean, adding more information to the graph of the number of medicine ball lifts at Time 1. (Data from "Exercise in pregnancy: Effect on fitness and obstetric outcomes—a randomized trial," by B. B. Price, S. B. Amini, and K. Kappeler, 2012, *Medicine & Science in Sports & Exercise, 44,* 2263–2269.)

3. Exploring Data Visually

meaning the score is an outlier. Having a high outlier could explain why each group's mean was higher than its median. This graph may require too much explanation for some purposes, such as presenting results to an audience untrained in data analysis. A complex graph may not meet researchers' needs, so we do not want to give you the impression that more complexity is always helpful.

What's Next

"Always graph your data" probably is the only rule of data analysis that would win universal approval from statisticians. It is crucial for understanding quantitative results and communicating them effectively. But *how* we should graph our data is debatable! Entire books have been written about graphing data, and new ways of graphing data are being developed. But new does not mean better. The more time we must spend staring at a graph and trying to understand it, the poorer the graph. (Search online for the term *bubble graph* and try to find one that is worth the time required to understand the data.) An effective graph gives a quick summary of data and can reveal patterns that otherwise would escape our notice. Graphs can help us to understand the entire distribution of scores, both in terms of location and spread on the number line. Location and spread will be important concepts in Chapter 4 where we will show you how to measure the relative location of any score in a distribution.

Exercises

SCENARIO 3-A

As described in Scenario 2-A in Chapter 2, researchers asked nursing home residents with dementia to interact with dogs and substitutes for dogs (Marx et al. 2010). The stimuli were used to encourage social interactions. All participants were given the chance to interact with three sizes of real dogs, plus a puppy video, a plush dog toy, an activity involving coloring a picture of a dog, and a robotic dog. The researchers measured the number of minutes that each resident engaged with each stimulus. If the resident did not want to engage with the stimulus, an engagement duration score of zero was recorded. As we saw in the exercises for Chapter 2, this research is observational, with the kind of stimulus being the predictor variable and engagement duration as the criterion variable. 3-7. Is the predictor variable numeric or nonnumeric? 3-8. Is engagement duration score numeric or nonnumeric? 3-9. Describe a graph that could compare the means for the different stimuli conditions. 3-10. Describe a graph that could compare the variability of the times measured for the participants while interacting with different stimuli. 3-11. What would be a good way to graph the number of residents who refused to interact with the different stimuli? Explain your answer. 3-12. Suppose the researchers wanted to examine the distribution of engagement duration scores for the puppy-video condition. Name two

(Continued)

graphs that would serve this purpose, and identify each graph's strength and weakness. 3-13. Can histograms be used to identify outliers? Explain.

SCENARIO 3-B

This chapter referred repeatedly to the research by Price et al. (2012) on exercise during pregnancy. Sedentary pregnant women were randomly assigned to groups, then the researchers manipulated their experiences: either remaining sedentary or following a program of physical activity. This data set is available to you via http://desheastats.com. The data set includes an additional dependent variable that we describe for the first time here. Cardiorespiratory fitness was the primary outcome in the study, and it was measured in a way that took into account the fact that women were gaining weight as pregnancy progressed. Participants walked or ran at a comfortable pace, and the power produced during the walk/run was calculated as follows: power = (weight × distance)/time. To help you understand the power score, let's consider two women of the same weight who walk the same amount of time. The woman who walks a greater distance will have a higher power score, indicating better cardiorespiratory fitness. Power scores were collected on five occasions. 3-14. What kind of research was the pregnancy study, and how do you know? 3-15. In terms of the kind of research, what kind of variable is power score? What kind of variable is the number of previous pregnancies? 3-16. If we wanted to examine the distribution of power scores for all participants at the beginning of the study before the intervention began, what would be two appropriate graphs to create? 3-17. Create those graphs with whatever statistical software is being used in your statistics class. 3-18. What might be an appropriate graph for visualizing the means for each group and each of the five occasions of measurement? 3-19. Use your statistical software to create the graph you named in the previous question. 3-20. What can we learn from a pie chart? 3-21. What is another graph that could communicate similar information as a pie chart? 3-22. Why do your authors seem to hate three-dimensional pie charts and bar graphs?

SCENARIO 3-C

This chapter referred to a data set containing rates of adult smoking for the 50 U.S. states and the District of Columbia. If you are learning statistical software, download our data set (see http://desheastats.com) and do the following exercises. 3-23. Create a histogram of adult smoking rates and compare it to Figure 3.15. What do you see in the histogram that was not obvious in Figure 3.15? What does Figure 3.15 reveal that you could not learn from the histogram? 3-24. Create a histogram and a boxplot of youth smoking rates. Compare them and describe what you have learned about youth smoking in the United States. 3-25. Create a scatterplot with adult smoking rates on the horizontal axis and youth smoking rates on the vertical axis. What do you see? 3-26. Create a scatterplot with state excise taxes on the horizontal axis and adult smoking rates on the vertical axis. What do you see?

References

Centers for Disease Control and Prevention. (2012, August 13). *Adult obesity facts*. Retrieved from http://www.cdc.gov/obesity/data/adult.html

Centers for Disease Control and Prevention. (2013, April 25). *Current cigarette smoking among adults*. Behavioral Risk Factor Surveillance System Prevalence and Trends Data, 2011. Retrieved from http://www.cdc.gov/tobacco/data_statistics/fact_sheets/adult_data/cig_smoking/

Chaffin, M., Hecht, D., Bard, D., Silovsky, J., & Beasley, W. H. (2012). A statewide trial of the SafeCare home-based services model with parents in Child Protective Services. *Pediatrics, 129*, 509–515. doi:10.1542/peds.2011–1840

Cleveland, W. S. (1993). *Visualizing data*. Lafayette, IN: Hobart Press.

Ezzati, M., Martin, H., Skjold, S., Vander Hoorn, S., & Murray, C. J. L. (2006). Trends in national and state-level obesity in the USA after correction for self-report bias: Analysis of health surveys. *Journal of the Royal Society of Medicine, 99*, 250–257.

Food Research and Action Center. (2012, February). *Food hardship in America 2011: Data for the nation, states, 100 MSAs, and every congressional district*. Washington, DC: Food Research and Action Center. Retrieved from http://frac.org/pdf/food_hardship_2011_report.pdf

Langford, E. (2006). Quartiles in elementary statistics. *Journal of Statistics Education, 14*(3). Retrieved from http://www.amstat.org/publications/jse/v14n3/langford.html

Le, A., Judd, S. E., Allison, D. B., Oza-Frank, R., Affuso, O., Safford, M. M., … Howard, G. (2013). The geographic distribution of obesity in the US and the potential regional differences in misreporting of obesity. *Obesity, 22*, 300–306. doi:10.1002/oby.20451

Marx, M. S., Cohen-Mansfield, J., Regier, N. G., Dakheel-Ali, M., Srihari, A., & Thein, K. (2010). The impact of different dog-related stimuli on engagement of persons with dementia. *American Journal of Alzheimer's Disease and Other Dementias, 25*, 37–45. doi:10.1177/1533317508326976

Price, B. B., Amini, S. B., & Kappeler, K. (2012). Exercise in pregnancy: Effect on fitness and obstetric outcomes—a randomized trial. *Medicine & Science in Sports & Exercise, 44*, 2263–2269. doi:10.1249/MSS.0b013e318267ad67

Tukey, J. W. (1977). *Exploratory data analysis*. Reading, MA: Addison-Wesley Publishing.

Wilkinson, L. (1999). Dot plots. *The American Statistician, 53*, 276–281. doi:10.2307/2686111

Wilkinson, L. (2005). *The grammar of graphics*, 2nd ed. New York, NY: Springer Publishing.

4

Relative Location and Normal Distributions

Introduction

Researchers commonly report statistics describing the location of a distribution of sample data, as well as the amount of spread in the scores, and statisticians always encourage graphing to reveal patterns in the data. Statistics that measure the center and spread can be combined to measure where a particular score is located in a sample distribution of scores. For example, patients often want to know whether their results on a lab test are average; if the results are not average, patients want to know how far above or below average their results are. Comparing a score to a mean is part of measuring *relative location*, typically defined as a score's position on the number line in relation to the mean. The mean might be a sample mean or the mean of a specific population, such as people who are the same age, sex, and weight as a patient. Some variables in the population have distributions with a special shape, which we will discuss toward the end of the chapter.

Few researchers regularly use the specific statistics in this chapter. Yet the concept of comparing something to its mean is quite common in statistics and will serve as a building block for complex concepts to be introduced later.

This chapter will help you to get used to thinking about the relative location of numbers within distributions, as well as introduce you to a special family of mathematical distributions.

Standardizing Scores

Whenever we want to understand a set of numbers, we need to know something about the possible values so that the numbers make sense to us. For example, many researchers study how sleep is related to health outcomes. Buysse, Reynolds, Monk, Berman, and Kupfer (1989) developed a scale known as the Pittsburgh Sleep Quality Index (PSQI). This 19-item questionnaire asks patients how often they have had trouble getting to sleep in the last month and how often they wake up in the night for various reasons. Suppose you score a 17 on the PSQI. What does that score tell us?

To interpret the meaning of the score, the first thing we need to know is whether a high score means better sleep or worse sleep. The PSQI generates higher scores to reflect more frequent sleep disturbances, which means worse sleep quality. The second thing we need to know is whether 17 is high or low. Does it help you to judge the meaning of 17 if we tell you that PSQI scores can range from 0 to 21?

We can tell that a score of 17 is closer to 21 than it is to zero, but we do not know whether 17 is a typical score. Comparing a score to the mean, we can tell whether the score is higher or lower than average. The distance between the score and the mean for sleep quality needs to be measured in terms of some standard yardstick that will tell us whether the distance is big or small. If we can standardize this measurement of distance, then we could tell whether a person's sleep quality differed greatly from the mean or differed just a little bit from the mean. Further, we could take a different variable, like sleep duration, and ask, "Does this person sleep for fewer hours than the average person?" To answer this question, we can compute the difference between the person's sleep duration and an average for a sample, then use a measure of spread as a yardstick to judge whether the distance is big or small. We might find that the person with a PSQI score of 17 has much worse sleep quality than average, but in terms of sleep duration, the same person may be quite close to average.

We have a way of standardizing the measurement of a score's location relative to a mean. A *z score* is a mathematical calculation that measures the location of a raw score relative to the mean, with distance measured in units of standard deviations. The *z* score tells whether the score is above or below the mean and how many standard deviations fit in the gap between the score and the mean. Here is a verbal definition of a *z* score:

(Something minus its mean) divided by its standard deviation

The parentheses contain the part of the computation that we must do first. Memorize this verbal definition because we will expand on this concept later in the book. Here, "something" is the PSQI score of 17. Notice that "something" comes before the mean in the computation; this order is important. If 17 is below the mean, then "something minus its mean" will give us a negative *z* score. A negative *z* score therefore tells us that the score is in the negative direction on the number line, relative to the mean. For sleep quality, lower scores mean better sleep quality, so a negative z score would mean the person has better-than-average sleep quality. If 17 is above the mean, then we will have a positive z score, which tells us that compared with the mean, our score is in the positive direction on the number line. Because a higher score on the PSQI means more sleep disturbances, a positive z score would mean worse-than-average sleep quality. The verbal definition says *its* mean and *its* standard deviation. The word *its* refers to the number 17, which is part of a group of numbers for which we know the mean and standard deviation. The group of numbers could be a sample or a population.

A z score is also called a *standard score* because we are computing the relative position from the mean in units of standard deviations. Using a standard score frees us from having to know the numeric value of a typical score on a possibly unfamiliar measure like the PSQI for sleep quality. The z comes from the word *standardize*. When we standardize the data, we are obtaining scores that are in units of standard deviations.

Check Your Understanding

4-1. Suppose you have taken your first quiz in statistics class and scored 25 out of 30 points. You learn that the class mean was 22.5, the median was 24, and the standard deviation was 2.5. Compute your z score and explain what it means.

Suggested Answers

4-1. We do not need the median to answer this question; that statistic was given as a distractor because it is important to know when we need and do not need various statistics. The numerator of your z score will be (something minus its mean) = 25 − 22.5 = 2.5. The denominator will be 2.5. So z = 2.5/2.5 = 1. Your z = +1 means your quiz score was one standard deviation above the mean.

Question 4.1 described a quiz with a sample mean of 22.5 and a standard deviation of 2.5. The standard deviation is the unit of measure for the z score; in the quiz example, 2.5 points equals 1 standard deviation, similar to the way 100 cm = 1 m. How many of these standard deviations fit in the gap between your quiz score of 25 and the class mean of 22.5? One standard deviation, which equaled 2.5 points. Because the z score is +1, your quiz score is one standard deviation *above* the mean. Be sure not to confuse the meaning of a z score with the meaning of a standard deviation. The z score is measured in units of standard deviations, just as we could measure height in meters.

Computing a z Score in a Sample

Let's use real data to illustrate z scores for sleep quality. Wang et al. (2010) wanted to know whether a meditative practice involving gentle movement could be beneficial for people with fibromyalgia, a complex condition that impairs many facets of daily life. People with fibromyalgia can suffer tremendous musculo-skeletal pain, are easily fatigued, and have trouble sleeping. Aerobic exercise can be difficult for many patients with this syndrome. Wang et al. wanted to know whether tai chi might help to ease the symptoms of fibromyalgia. Originally a martial art, tai chi involves meditation, deep breathing, and slow, gentle move-ment. Adults suffering from fibromyalgia agreed to be randomly assigned to a treatment group or a control group. Participants in the treatment group attended two 1-hour tai chi classes per week for 12 weeks. The control group also had 12 weeks of activities. They attended twice-weekly sessions that involved 40 minutes of discussion of topics related to fibromyalgia, followed by 20 minutes of stretch-ing. Both groups were encouraged to continue their activities after the 12 weeks, and a follow-up assessment was conducted 24 weeks after the study began.

One of the dependent variables was sleep quality, measured by the PSQI. (These researchers graciously agreed to share part of their data, which you may obtain via http://desheastats.com.) When we examined the data, we noticed that four participants in the control group had a baseline PSQI score of 17, the num-ber that we chose arbitrarily when we began writing about sleep quality in this chapter. Figure 4.1 shows the distribution of PSQI scores for the control group at the beginning of the study (baseline).

By examining Figure 4.1, we can see the general location of the four scores of 17. We computed the mean PSQI at baseline to be approximately 13.45. Now we know that a score of 17 will be above the mean—but by how much? We need to know how many standard deviations fit in the gap between 17 and the mean, and this is where computing a z score is useful.

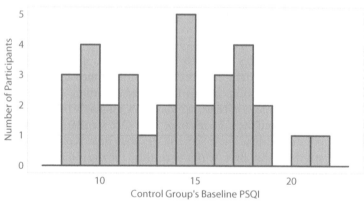

Figure 4.1

Control group's sleep quality scores at baseline. This histogram shows the control group's distribution of scores on the Pittsburgh Sleep Quality Index at the beginning of the study of tai chi and fibromyalgia. (Data from "A randomized trial of tai chi for fibro-myalgia," by C. Wang, C. H. Schmid, R. Rones, R. Kalish, J. Yinh, D. L. Goldenberg ... T. McAlindon, 2010, *The New England Journal of Medicine, 363,* 743–754)

4. Relative Location and Normal Distributions

What is the standard deviation for these data? This question presents a small problem: in Chapter 2 we presented two ways of calculating a standard deviation. Let's review: each standard deviation was the square root of a variance. The first variance that we presented was the sample variance, which was the average squared difference from the mean (with N in the denominator). The second variance that we presented was the unbiased variance, which was computed with $N - 1$ in the denominator. Most statistical software computes SD, the square root of the unbiased variance, but the z score requires the standard deviation based on the sample variance. The good news is that we can take the unbiased variance computed by statistical software and do a bit of math to get the sample variance:

$$\text{Sample variance} = \text{unbiased variance} \times \frac{N-1}{N}$$

For the 33 PSQI scores in the control group, our statistical software computed an unbiased variance = 13.443182. The number that will be multiplied against the unbiased variance is $(N - 1)/N$:

$$\frac{33-1}{33} = \frac{32}{33} = 0.969697$$

So the sample variance will equal

$$13.443182 \times 0.969697 = 13.035813$$

To get the standard deviation that we need for our z score, we take the square root of the sample variance: $\sqrt{13.035813} = 3.6105142 \approx 3.61$. We waited until the last possible calculation before rounding so that the final answer would be more accurate. For large sample sizes, the number $(N - 1)/N$ will be close to 1, which means there will be little difference between the sample variance and the unbiased variance.

Back to the participants with a PSQI score = 17: what is the relative position of this score in the control group at baseline? The mean was $M = 13.45$, and the standard deviation was 3.61. The z score is computed as follows (we will use the rounded mean and standard deviation for simplicity, although using the unrounded figures would be more accurate):

$$\text{(Something minus its mean) divided by its standard deviation}$$

$$= \frac{(17-13.45)}{3.61}$$

$$= \frac{3.55}{3.61}$$

$$= 0.9833795 \approx 0.98$$

This $z = 0.98$ tells us that the PSQI score of 17 is almost one standard deviation greater than the mean for the control group at baseline.

To help you to visualize what we are doing with this z score, let's take a look at two number lines. Figure 4.2 shows the number line for PSQI and the number line for z scores.

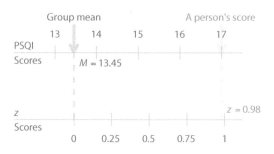

Figure 4.2

Number line for sleep quality, number line for z scores. A person's sleep quality score of 17 corresponds to a z score of 0.98. If you had a sleep quality score equal to the control group's mean at baseline, then your z score would be zero. (Data from "A randomized trial of tai chi for fibromyalgia," by C. Wang, C. H. Schmid, R. Rones, R. Kalish, J. Yinh, D. L. Goldenberg ... T. McAlindon, 2010, *The New England Journal of Medicine, 363*, 743–754.)

A *z* score of zero corresponds to the mean PSQI; that is, if your PSQI score exactly equaled the mean, then your *z* score would be zero. One standard deviation in baseline PSQI scores for the control group equaled 3.61 points. Between the person's score of 17 and *M* = 13.45, there is a gap that is not quite as wide as one standard deviation.

If we took a set of scores, like the tai chi study's baseline PSQI scores, and if we computed a *z* score for everyone in the control group, we could average all those *z* scores, and the mean will be zero. This is true for any set of scores: the corresponding set of *z* scores will average out to zero. Why? As you may recall, the sample mean is the balance point in a distribution, meaning the distances of scores above the mean balance out the distances of scores below the mean. Whenever we compute a *z* score, the numerator is computing each score's distance from the mean. So if we average together all the *z* scores, we would add up all of those distances from the mean. The positive *z* scores would balance out the negative *z* scores, and the mean of the *z* scores would be zero. Further, the standard deviation of all those *z* scores would equal one (an explanation that we will omit for the sake of space). Both the mean and standard deviation of a set of *z* scores are standard: the mean always equals zero and the standard deviation is always one. Just to clarify what we mean by "a set of *z* scores," a *z* score can be computed for every raw score in a sample, so after we do that computation for every raw score, we will have a set or collection of *z* scores. Because the set of *z* scores always will have a mean = 0 and a standard deviation = 1, we call *z* a standard score.

What will the distribution of a set of *z* scores look like if we standardize the baseline PSQI scores for everyone in the control group? Although the mean and standard deviation are standard, the shape is *not standard*. Figure 4.3 displays a histogram of these z scores below the distribution of raw scores, which were previously shown in Figure 4.1.

The distribution of z scores has the same shape as the original distribution of PSQI scores. The most frequently occurring score is still close to the middle, and the other peaks and gaps are still in the same places relative to each other. The distribution simply has been moved down to the number line so that it is centered on zero, and its spread corresponds to a standard deviation of one. Computing a standard score provides *only* a standard for the mean and standard deviation; it does *not* provide a standard shape for the distribution.

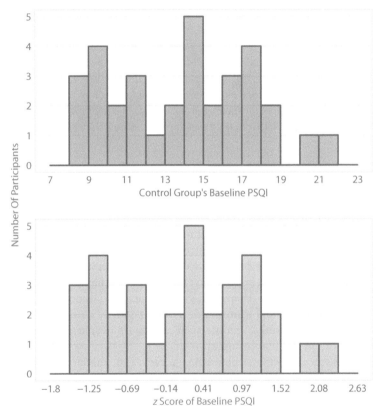

Figure 4.3

Distribution of sleep quality, distribution of corresponding z scores. Computing z scores does not change the shape of the distribution. The control group's baseline sleep quality scores have the same shape as their corresponding z scores. (Data from "A randomized trial of tai chi for fibromyalgia," by C. Wang, C. H. Schmid, R. Rones, R. Kalish, J. Yinh, D. L. Goldenberg ... T. McAlindon, 2010, *The New England Journal of Medicine, 363,* 743–754.)

4-2. Let's calculate z scores for the members of the control group with the lowest and highest PSQI scores at baseline. The lowest PSQI was 8, and the highest PSQI was 21, which also is the maximum possible score, indicating the worst sleep quality. Using the control group's mean (13.45) and standard deviation (3.61), compute the z scores for 8 and 21, and explain what they mean.

Suggested Answers

4-2. For the person with the lowest PSQI: $(8 - 13.45)/3.61 = -5.45/3.61 = -1.5096953 \approx -1.51$. For the person with the highest PSQI: $(21 - 13.45)/3.61 = 7.55/3.61 = 2.0914127 \approx 2.09$. The lowest PSQI was one and a half standard deviations below the mean for the control group's sleep quality scores at baseline, and the highest PSQI was more than two standard deviations above the mean. You might see if you can find the approximate location of these two z scores and their corresponding PSQI scores of 8 and 21 in Figure 4.3. (If you use statistical software to analyze the data downloaded via our website, you can get more decimal places for the mean and standard deviation, and your final answer will be more accurate than these approximate z scores.)

Computing a z Score in a Population

We can measure the location of a score relative to the mean of a sample or a population. We dearly wish that we knew the mean PSQI for a population of adults! If someone compiled the PSQI results for thousands of adults, we might have a sense for an approximate population mean and standard deviation. Then we could compute another z score to see how many population standard deviations are between 17 and a population mean. The PSQI has been translated into dozens of languages, and perhaps someday the data will be compiled from many sources so that means and standard deviations could be calculated for various populations—young adults, middle-aged adults, or people diagnosed with a chronic condition like fibromyalgia.

In the absence of known population means and standard deviations, let's pretend for a while and use some made-up numbers to illustrate a z score for a sleep quality score's location relative to a population mean. Our made-up parameters will have some basis in reality. Buysse et al. (2008) conducted a study of sleep quality in a racially diverse sample of men and women with a mean age of 59.5 years. By comparison, the sample in the tai chi study by Wang et al. (2010) consisted of men and women with a mean age of about 50 years. The major difference between the participants in these two studies is the diagnosis of fibromyalgia. It might make sense to ask how a score of 17 for our patient with fibromyalgia compares with the mean of a population like the one studied by Buysse et al.. Here is the point where we will begin to pretend. We will use the Buysse et al. study as a springboard into fantasy land, where we will pretend we know that the mean PSQI for a population of adults is 6, and the population standard deviation is 3. How does our patient with fibromyalgia who scored a 17 on the PSQI compare with this population mean?

Now our verbal definition is "(something minus its *population* mean) divided by its *population* standard deviation." We could substitute the names of these parameters with their corresponding symbols, introduced in Chapter 2, and the definition of z would be "(something minus μ) divided by σ." Using the pretend numbers for μ and σ, our computation is

$$z = \frac{17-6}{3}$$

$$= \frac{11}{3}$$

$$= 3.666666... \approx 3.67$$

Our patient with fibromyalgia who scored a 17 on the PSQI had a z score that was more than three population standard deviations above the population mean for adults, according to our pretend figures. Let's compare this result with the previous z score for the location of the score of 17 within a sample. Compared with the mean sleep quality for people with fibromyalgia who were randomly assigned to a control group and measured before any intervention, our patient's score of 17 was almost one standard deviation above the mean. So compared with other people with fibromyalgia, this patient's sleep quality score is higher than average, meaning worse sleep quality. But compared with adults in general (based on our made-up parameters), this patient's sleep quality was much worse than the population mean. Therefore it is crucial to know about the group to which a person is being compared.

Check Your Understanding

4-3. Let's continue using our made-up parameters for PSQI scores in the population. The control group in the tai chi study included baseline PSQI scores from a low = 8 to a high = 21. Using the made-up parameters (population mean = 6, population standard deviation = 3), calculate the relative positions of the scores 8 and 21.

Suggested Answers

4-3. For PSQI = 8, the z score = (8 − 6)/3 = 2/3 = 0.666666... ≈ 0.67. This z means the person's PSQI score was two-thirds of a population standard deviation above the population mean for adults in general (based on the made-up parameters). For PSQI = 21, the z score = (21 − 6)/3 = 15/3 = 5. The person with the worst possible sleep quality score had a PSQI score that was five population standard deviations above the population mean for adults.

Comparing z Scores for Different Variables

An advantage of z scores is that we could take scores on two different measures of sleep quality that exist on different numeric scales and transform both scores to z scores. Then the two scores would be on the same scale; they both would be in units of standard deviations. For example, Buysse et al. (2008) explored the relationship between scores on the PSQI and the Epworth Sleepiness Scale (ESS). The PSQI can produce scores on seven components of sleep quality, but our examples have used the global score, which ranges from 0 to 21, with higher numbers meaning worse sleep quality. The eight-item ESS asks respondents to rate the "likelihood of dozing or falling asleep" during various situations. Scores on the ESS can range from 0 to 24. The scales measure different aspects of sleep. The PSQI measures recent trouble with sleeping at night, while the ESS measures a patient's likelihood of falling asleep during the day. We could measure a patient using both scales and calculate z scores for the patient's PSQI and the ESS scores.

Previously we saw the patient with fibromyalgia who had a PSQI score = 17, which we compared to the made-up population mean of 6. Using the made-up population standard deviation of 3, we found $z \approx 3.67$, meaning the patient had a PSQI score that was more than three population standard deviations above the population mean. Let's pretend that the same patient with fibromyalgia had an ESS score = 12. Further, we will use the Buysse et al. (2008) study again as a springboard into fantasy land, where we will find an ESS population mean for adults = 8 and an ESS population standard deviation = 4. Now we can compute the patient's z score for ESS, relative to adults in general:

$$z = \frac{12-8}{4}$$

$$= \frac{4}{4}$$

$$= 1$$

The patient has an ESS score that is one population standard deviation above the population mean. Now that we have two z scores, one for PSQI and another for ESS, we can compare the patient's z scores for the two different measures of sleep quality. Remembering that higher scores on both scales mean worse outcomes, is the patient worse in terms of the ESS (the measure of daytime sleepiness) or the PSQI (the measure that focuses on recent trouble with sleeping)? The z score for the ESS was 1, and the z score for the PSQI was about 3.67. The patient's score for daytime sleepiness is higher than the made-up population average for adults in general by exactly one population standard deviation. But we were able to fit more than three population standard deviations in the gap between the patient's score on the measure of recent sleep trouble and the population mean for adults. Because the z score for the PSQI was higher than the z score for the ESS, the patient's score for nighttime sleep trouble is much worse than the population mean on that measure, compared with the patient's score for daytime sleepiness, relative to its population mean.

Check Your Understanding

4-4. Let's continue using our made-up parameters for ESS scores in the population. We previously saw that the person with the lowest PSQI = 8 in the tai chi study had a $z = 0.67$ when compared with the fictitious population mean. Imagine this person had an ESS score = 3. Using the made-up parameters for the ESS ($\mu = 8$, $\sigma = 4$), compute the z score for this person's ESS score. Then compare this person's z scores for PSQI and ESS.

Suggested Answers

4-4. For ESS = 3, the z score = $(3 - 8)/4 = -5/4 = -1.25$. For daytime sleepiness (ESS), this patient's z score is 1.25 population standard deviations below the made-up population mean. But for nighttime sleeping trouble, the person's PSQI score is two-thirds of a population standard deviation above the made-up population mean. As you will recall, higher numbers on the PSQI meant worse sleep problems, and higher numbers on the ESS meant greater likelihood of falling asleep during daytime activities. So this person has an ESS score that is better than the average for adults in general and a PSQI score that is worse than the population mean.

A Different Kind of Standard Score

A z score is not the only kind of standard score. Another standard score is called a *T score*. (Later we will introduce statistics that are symbolized by a lowercase t, so it is important to capitalize this T score.) Typically T scores are used in education to compare a child's score on an academic achievement test with the mean of similar children, usually based on age or educational level. When a large-scale academic test is created, it is given to thousands of children who are representative of the population that will use the test. These thousands of children serve as a reference group, and the process of gathering scores and assessing the numerical results from a large reference group is called *norming* (Petersen, Kolen, & Hoover, 1993). The reference group's mean and standard deviation often are called *norms* and are used for comparison purposes. In educational and other settings, T scores have a mean = 50 and a standard deviation = 10. A child whose test score is one standard deviation below the mean would have a T score = 40. (It would be much more pleasant to explain a T score of 40 to a parent or guardian, compared with having to say the words "negative one" in explaining the child's comparable z score.)

In the health sciences, however, T scores are most commonly used to report the results of a bone mineral density test, and for reasons that we cannot explain, this T is scaled like a z score. Unlike most T scores, the one for bone mineral density has a mean of zero and a standard deviation of one, just like a z score. Bone mineral density often declines as people age, and a dual-energy X-ray absorptiometry (DXA) test is used to diagnose osteoporosis, a risk factor for fractures in older adults. A T score for bone mineral density uses a reference group of the mean bone density for 30-year-olds, the age at which most people's bones are healthiest. A patient also receives DXA results in the form of a z score, which has a reference group of people who are the same age and gender (and sometimes the same race/ethnicity and weight). A patient could have a z score $= 0.5$, meaning that the patient's bone density is one-half of a standard deviation above the mean for people who are demographically similar to the patient, while having a T score $= -1$, meaning that the bone mineral density was one standard deviation below the mean for young healthy adults. This comparison of z and T scores for the same numeric measure of bone density illustrates why it is important to know the comparison group for any standard score.

Different DXA systems compute z scores in different ways. Age and gender are the most common reference groupings, but imagine two DXA systems being used to compute z scores for the same patient. One system uses the mean and standard deviation for a reference group that is the same age and gender as the patient. The other system uses the mean and standard deviation for a reference group that is the same age, gender, race/ethnicity, and weight. The patient's z score from the first system may differ from the z score from the second system. There is no agreed-upon standard for different reference groups (Carey et al., 2007). Again, knowing the comparison group is crucial to interpreting standard scores.

Check Your Understanding

4-5. Suppose we have three patients who are older men, and we are examining the following z and T scores for bone mineral density. Al Dente: $z = 0$ and $T = 0$. Beau Tocks: $z = 1$ and $T = 0$. Cy Stolic: $z = 0$ and $T = -1$. Which patient's results would cause the greatest concern? Which patient's results indicate the best bone density?

Suggested Answers

4-5. We would be most concerned about Cy's bone density; although his bone density score is equal to the mean for people who are demographically similar to him ($z = 0$), his bone density score is one standard deviation below the mean of healthy young adults ($T = -1$). The two other patients had results that were average or higher. Beau has the best bone mineral density because his density score is one standard deviation above the mean for people in his demographic group ($z = 1$) and the same as the mean of healthy young adults ($T = 0$). Al's

(Continued)

*result of z = 0 means his bone density is the same as the average for his com-
parison group, and his T = 0 means the density also is the same as the mean
for healthy young adults. Notice that Al, Beau, and Cy all could have had different demographic comparison groups, so all of their z scores may have been
computed with different population means and standard deviations*

Distributions and Proportions

Take another look at Figure 4.1, which shows the distribution of baseline PSQI
scores for the control group in the tai chi study. We could look at parts of the distribution within certain ranges. A *proportion* is a fraction expressed in decimals.
We might ask: What proportion of the participants had scores higher than 17?

Out of the 33 participants in the control group, there were four participants
with PSQI scores greater than 17. Four divided by 33 gives us a proportion
approximately equal to 0.12. We can get a percentage by multiplying the proportion by 100: about 12% of participants in the control group scored higher than 17
on the PSQI at baseline.

Computing a proportion of a sample is not very interesting. If enough data
were available and we could describe the shape of the population, however,
then we could ask some interesting questions. For example, we might wonder
whether a PSQI score = 17 is unusually high. We could ask, "Compared with
other adults in the population of patients with fibromyalgia, what is the proportion of adults with a PSQI score greater than 17?" Or we could ask, "Compared
with healthy adults without any diagnosed chronic illnesses, what is the proportion of adults with a PSQI greater than 17?" *If we knew the shape of the
distribution of PSQI scores in those populations*, then we could estimate these
proportions.

Unfortunately, we do not know anything about the shape of those populations
of PSQI scores. Sometimes, however, we read about variables that are measured so
often that we can venture a guess about the shape of the distribution of scores in a
population. The National Health and Nutrition Examination Survey (NHANES)
uses a complex sampling procedure to select thousands of people for representative samples and measures them in many ways. Physicians take measurements
on children of all ages and consult World Health Organization growth charts
provided by the Centers for Disease Control and Prevention (CDC) to determine
whether children appear to be developing as expected. Sophisticated analyses
have been conducted to produce these growth charts, and we recommend caution in making assumptions about the shape of any population of scores.

Let's build a case for one possible shape of a population distribution. Imagine
that you are a ticket-taker at a sporting event in the United States, and you
decide to pay attention to the heights of men entering the arena or stadium.

(The first author is a huge fan of certain sports, so nothing sexist is implied in this example of men attending a sporting event!) As you pay attention to how frequently you see men of different heights, do you have a large proportion or a small proportion of the men being shorter than 5 ft (about 1.5 m)? If we are focusing on adults, the fraction of males under this height would be small. What proportion of men would be shorter than 5 ft 9 in. (about 1.75 m)? Quite a large fraction of the men would be that height or shorter. What proportion of men would be taller than 6 ft 2 in. (about 1.9 m)? Probably fewer than half of the men are that height or taller. What proportion would be taller than 6 ft 6 in. (about 2 m)? Now we would be looking at a small proportion of the men entering the sporting event.

The heights that are about average would be most common, and the shorter heights and taller heights would be less frequently occurring. If we created a histogram of the heights of the men at the sporting event, the bars probably would be tallest in the middle, shortest at the extremes, and the heights of the bars probably would get progressively shorter from the middle toward the extremes. Adult height for a given person is not constant. Young men who have a late-onset adolescent growth spurt might be shorter at age 18 than at age 20. People who are advancing in age may experience some decline in bone health and become shorter. Perhaps the distribution of heights for younger men would be shaped differently and centered higher on the number line than the distribution of heights for older men.

The reason we are drawing out this discussion of heights is that we want you to think about the meaning of any global statements about populations of scores. Many variables may follow this pattern of more frequent scores around the mean, less frequent scores at the extremes, and a tapering off between the mean and the extremes. But this pattern is not enough to say that a variable follows a specific mathematical curve called a *normal distribution*. Normal distributions are defined mathematically. You probably have heard of the normal curve, which some describe as a bell curve because it looks like the silhouette of a bell sitting on a table. There are many possible normal distributions because they can be centered anywhere on the number line, and they can have little spread or a lot of spread. Figure 4.4 shows four normal distributions, differing in terms of their middle location on the number line and their spread.

A normal distribution is an example of a *theoretical reference distribution*, a distribution that is defined by a mathematical formula and describes the relative frequency of occurrence of all possible values for a variable. Mathematicians have known about theoretical reference distributions for centuries, long before statistics came along as a discipline. With a normal distribution, all numeric values are theoretically possible, so the curved line would go on forever in each end of the distribution, never actually touching the horizontal number line. Of course, our graphs are finite, so the distributions will look like they are sitting on the horizontal axis. We are avoiding big formulas in this book, but if you search online for "formula for a normal distribution" and look for images, you will see an ugly formula containing the number pi, the natural number e, fractions, exponents, and square roots.

Figure 4.4

Four normal distributions. A normal distribution can be located anywhere on the number line, and different normal distributions also can have different amounts of spread.

By specifying a population mean and standard deviation, the smooth curve can be drawn for any location on the number line or amount of spread. All normal distributions are symmetric. The mean equals the median, so half of the distribution is above the mean and half is below it. If we drew a vertical line through the median and folded the distribution in half, the bottom half would be a mirror image of the top half. All of the scores are under the curve: 100%, which corresponds to a proportion = 1. Because of the known shape defined by a mathematical formula, all normal distributions also have other known proportions. Within one standard deviation either direction from the mean, we will find about 68% of the scores, and within two standard deviations either direction from the mean, we will find about 95% of the scores. Let's illustrate those areas by pretending we have a normal distribution for men's heights, with a mean = 69 in. and a standard deviation = 3. A height of 66 in. is one standard deviation below the mean, and a height of 72 in. is one standard deviation above the mean. Between these two heights, we would find about 68% of men's heights in our pretend population, as shown in Figure 4.5. What if we go two standard deviations below and above the mean? Now we are talking about the heights from 63 to 75 in. If men's heights in our pretend population are normally distributed with a mean = 69 and a standard deviation = 3, then about 95% of men's heights will be within two standard deviations of the mean, as shown in Figure 4.5.

Is it realistic to say that any variable is truly normally distributed in the population? We don't know because populations usually are large and therefore unobtainable, and sometimes they are even hypothetical, as discussed in Chapter 1. Samples are never really normal; they have limited numbers of scores, so they are lumpy and may have gaps, unlike a smooth normal curve. For now, we will focus on the possibility of populations being normally distributed. Demographers like A'Hearn, Peracchi, and Vecchi (2009) have examined men's heights. These researchers looked

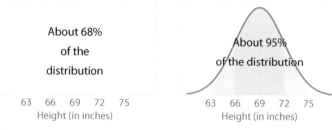

About 68%
of the
distribution

63 66 69 72 75
Height (in inches)

About 95%
of the distribution

63 66 69 72 75
Height (in inches)

Figure 4.5

Approximate areas under a normal curve. Normal distributions can be divided into different areas. The left-hand distribution shows that about 68% of the population is within one standard deviation (3 in.) of the mean, which is 69 in. The right-hand distribution shows that about 95% of adult male heights are contained within two standard deviations of the mean.

at height data from the Italian military for the years 1855–1910 and questioned whether a normal distribution is an appropriate mathematical description of those heights. Specifically affecting the distribution of men's heights in military records are the younger ages of men being measured and the lack of data for men who did not meet minimum height requirements. So this distribution of men's heights in military records might not be well matched with a theoretical normal distribution.

If we can *assume* that scores are normally distributed in a population, we gain the ability to answer questions about the proportions of people with scores higher or lower than a score that interests us at the moment. McDowell, Fryar, Ogden, and Flegal (2008) wrote a National Health Statistics Report that describes results from the NHANES study, an ongoing longitudinal project for understanding health in the United States. Let's use information from this report to illustrate the potential use of normal distributions. The heights of thousands of men have been measured for the NHANES, including hundreds of men in their 30s from various racial/ethnic groups. The report says the mean height for all men in this age group is 69.4 in. Based on other information in the report, a reasonable estimate of the standard deviation of these men's heights is 3.1 in. (Our example in Figure 4.5 rounded these numbers for simplicity.) If we can assume these heights are normally distributed, then half of the men are shorter than 69.4, and half of them are taller than 69.4. By converting heights to z scores, we would not need a normal distribution that is centered on 69.4 with a standard deviation = 3.1. We can use the *standard normal distribution*, which is the only normal distribution with a mean = 0 and a standard deviation = 1. The standard normal is a convenient distribution because z scores also have a mean = 0 and a standard deviation = 1.

A word of caution: *please* remember that the distribution of z scores is shaped the same as the distribution of the original scores. Students sometimes mistakenly believe that computing z scores magically changes their data into a normal distribution. We began with the notion that men's heights were normally distributed, an idea that we cannot confirm. *If* these heights have a distribution that is approximately normal, then the z scores for the heights also would be approximately normal. Computing z scores changes nothing about the shape of the distribution. We cannot go back to the example of sleep quality scores measured by the PSQI and talk about the proportion of adults with PSQI scores lower than 17 because we do not know the shape of the scores in that population. Dr. Daniel J. Buysse, the lead developer of this scale, says PSQI scores are definitely *not* normally distributed (D. J. Buysse, personal communication, April 3, 2013). Interestingly, Dr. Buysse says researchers believe sleep duration (how long people sleep at night) *is* approximately normally distributed in the population. Think about what that would mean: generally speaking, most people's sleep durations would be piled up around a mean number of hours, which would be the same as the median and the most frequently occurring sleep duration (the mode). As we moved down the number line for the number of sleep hours, we would find fewer and fewer people. Looking at people with above-average sleep duration, the farther we went above the mean, the fewer people we would find. But just because we could compute z scores on a variable does not give us free rein to using a standard normal distribution; we first must be confident that the original variable is normally distributed.

Check Your Understanding

4-6. Using the population mean = 69.4 and population standard deviation = 3.1 for American men's heights, compute the z scores for the following two American men in their 30s: Al Buterol, who is 72.5 in. tall, and Bo Dacious, who is 66.3 in. tall. 4-7. If American men's heights are approximately normally distributed, what is the proportion of men shorter than Al? Shorter than Bo?

Suggested Answers

4-6. Al's z score would be (72.5 − 69.4)/3.1 = 1. Bo's z score would be (66.3 − 72.5)/3.1 = −1. 4-7. We know that 68% of scores are within one standard deviation in either direction from the mean. About 68% of men would have heights between Bo's 66.3 in. and Al's 72.5 in. That leaves 32% of the men's heights beyond one standard deviation from the mean. Because of the symmetry of a normal distribution, half of the 32% would have a z score less than −1. Therefore, about 16% of men in their 30s are shorter than Bo. All of those men also would be shorter than Al, along with the 68% of men whose heights are between Al and Bo. That means about 84% of men are shorter than Al.

Areas Under the Standard Normal Curve

The last Check Your Understanding question may have been challenging for you. It required the information given earlier about the proportion of scores that are found within certain distances from the mean of a normal distribution. As it turns out, we could treat a normal distribution like the bridegroom's cake in the old movie *Steel Magnolias*, which was remade in 2012. The film took place in the U.S. South, and the groom's aunt made a red velvet cake in the shape of an armadillo and covered it with gray icing; another character called it a bleeding armadillo cake. We could cut off a portion of the cake on one end, or we could slice off a portion in the middle, or we could figure out what proportion of the cake remained after part of the cake had been served. Unlike the cake, the standard normal distribution is two-dimensional. Our slices of the standard normal distribution will be made vertically at different points along the number line. The z scores are on the number line. If we make a vertical slice through the cake at $z = 0$, then we are cutting the standard normal distribution in half. To the left of the mean ($z = 0$), we have negative numbers for places where we could slice; these locations (z scores) are distances from the middle. To the right of the mean, we have positive numbers on the number line for places where we could slice; these too are distances from the mean. The size of the cake slices will always be positive. We could cut no cake and serve 0% of the cake to someone. But if we slice from either end of the bleeding armadillo cake and serve someone, there will be a positive percentage of cake being served.

We can think of the two halves of a normal distribution as being areas. If we are looking at the entire distribution, we are considering 100% of the scores, which corresponds to a proportion = 1. Half of the distribution would be a proportion of .5 or a percentage of 50%. Are we slicing the cake at $z = 1$? If so, then about 84% of the cake (or a proportion of about .84) will be to the left of $z = 1$, and almost 16% of the cake will be to the right of $z = 1$. What if we slice the cake at $z = -1$? Then almost 16% of the cake will be below $z = -1$, and about 84% of the cake will be above $z = -1$. As we said, cake is always positive; that is, we always have a positive number for area, proportion, or percentage of a normal distribution. Only the z scores (i.e., locations on the number line) can be negative.

Let's be more exact with the proportions and percentages. Table A.1 in the back of the book lists the areas as proportions under the standard normal distribution for different sections that are defined by values of z. Table A.1 only lists positive values of z because the top half of the standard normal distribution is a mirror image of the bottom half, so we can save space by giving the numbers for only the top half. Figure 4.6 shows that the table is arranged in sets of three columns. The first three columns go together, and they continue as the second set of three columns. They wrap around to the top of the page again to make a third set of three columns. Figure 4.7 shows the images at the top of each set of three columns.

The first column gives values of z, starting at zero (0.00) and increasing in size down that column. The column is labeled "z or -z" because we can focus only on positive values of z (as shown in the first column) or we can think about the mirror image of those values in the negative direction from the mean. The values in the first column represent a z that we are interested in

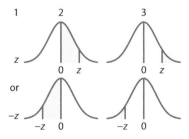

1	2	3	1	2	3	1	2	3
0.00	.0000	.5000	0.20	.0793	.4207	0.40	.1554	.3446
0.01	.0040	.4960	0.21	.0832	.4168	0.41	.1591	.3409
0.02	.0080	.4920	0.22	.0871	.4129	0.42	.1628	.3372
0.03	.0120	.4880	0.23	.0910	.4090	0.43	.1664	.3336
0.04	.0160	.4840	0.24	.0948	.4052	0.44	.1700	.3300
0.05	.0199	.4801	0.25	.0987	.4013	0.45	.1736	.3264
0.06	.0239	.4761	0.26	.1026	.3974	0.46	.1772	.3228
0.07	.0279	.4721	0.27	.1064	.3936	0.47	.1808	.3192
0.08	.0319	.4681	0.28	.1103	.3897	0.48	.1844	.3156
0.09	.0359	.4641	0.29	.1141	.3859	0.49	.1879	.3121
0.10	.0398	.4602	0.30	.1179	.3821	0.50	.1915	.3085
0.11	.0438	.4562	0.31	.1217	.3783	0.51	.1950	.3050
0.12	.0478	.4522	0.32	.1255	.3745	0.52	.1985	.3015
0.13	.0517	.4483	0.33	.1293	.3707	0.53	.2019	.2981
0.14	.0557	.4443	0.34	.1331	.3669	0.54	.2054	.2946
0.15	.0596	.4404	0.35	.1368	.3632	0.55	.2088	.2912
0.16	.0636	.4364	0.36	.1406	.3594	0.56	.2123	.2877
0.17	.0675	.4325	0.37	.1443	.3557	0.57	.2157	.2843
0.18	.0714	.4286	0.38	.1480	.3520	0.58	.2190	.2810
0.19	.0753	.4247	0.39	.1517	.3483	0.59	.2224	.2776

Figure 4.6

Excerpt from Table A.1. This table in the back of the book shows different values of z and different areas under the standard normal curve.

Figure 4.7

Column headings in Table A.1. Column 1 in Table A.1 shows values of z. Column 2 shows a "middle" area, between z = 0 and the z that is listed in the first column. Column 3 shows a tail area.

at the moment; we will call it our "z of interest." The next two columns show areas for the top half of the distribution that are related to our z of interest. The top image above the second column shows a shaded area between $z = 0$ and a positive z of interest; this area would correspond to the size of the cake slice between the mean and the positive z of interest. The bottom image above the second column shows the mirror image of that shaded area; now the area is between $z = 0$ and the negative z of interest. These shaded slices are examples of what we call "middle areas" in the distribution. Now let's look at the third column. The top image above the third column in Figure 4.7 shows a shaded area beyond the positive z of interest. The bottom image above the third column shows a shaded area beyond the negative z of interest. The areas in the third column represent what we call "tail areas." For any value of z in the table, positive or negative, we can find either tail area or either middle area. These areas represent proportions (or percentages) of scores in various parts of a standard normal distribution.

If $z = 0$, then the score equals the mean. A z greater than zero represents a score greater than the mean; as z gets bigger, it represents a greater distance between a score and its mean. The first "z of interest" in Figure 4.7 is $z = 0$, which is shown in the first row of the table. If $z = 0$, then there is no distance between a score and its mean. As we just said, the second column represents the proportion of scores between zero and the "z of interest"—but now zero and "z of interest" are the same thing, so the second column has no area to shade, so the proportion is .0000. All of the area in the upper half of the distribution is above $z = 0$, and the third column gives us the area beyond our "z of interest," which is a proportion $= .5000$, or .5. Imagine our armadillo-shaped cake sitting on the number line with the middle sitting on $z = 0$. Cutting vertically through the cake at $z = 0$ would result in the cake being cut in half, with half of the cake above $z = 0$. But no middle area is being cut; the slice at $z = 0$ does not give us any cake. We would need to slice again at a value of z that differs from zero.

Let's jump down the first column to $z = 0.7$, representing a score that is seven-tenths of a standard deviation above its mean. The second column now will be more meaningful because we can get a slice of cake. The second column tells the proportion of the standard normal distribution that is between $z = 0$ and $z = 0.7$. We find an area $= .2580$ (or, equivalently, .258). This number means that 25.8% of the area in the standard normal distribution is found between $z = 0$ (the mean) and $z = 0.7$ (our "z of interest"). If half of the area under the curve is above the mean, then the area beyond $z = 0.7$ should be $.5 - .258 = .242$. In fact, that is the number in the third column (shown as .2420). So 24.2% of scores in a standard normal distribution are greater than $z = 0.7$.

The left-hand distribution in Figure 4.8 shows these proportions relative to $z = 0.7$. The right-hand distribution in Figure 4.8 illustrates the symmetry in a normal distribution. Thanks to symmetry, we can use the same entries in Table A.1 to learn about the areas associated with $z = -0.7$. The proportion or area

Figure 4.8

Example of $z = 0.7$ and areas under the standard normal curve. If our "z of interest" is 0.7, then we can find four parts or areas of a standard normal distribution. The left-hand distribution shows the area (in green) between $z = 0$ and $z = 0.7$ and the area (in blue) that is above $z = 0.7$. The mirror images of those areas are shown in the right-hand distribution.

between the mean and -0.7 is .258; it is equivalent to say that 25.8% of the distribution is in that slice. If we slice off the lower tail of the distribution at $z = -0.7$, the proportion in the tail area would be .242, or 24.2%. By looking at this one line in the table and remembering the symmetry of a normal distribution, we can find information about four areas:

- The area below $z = -0.7$ is .242
- The area between $z = -0.7$ and the mean ($z = 0$) is .258
- The area between the mean and $z = 0.7$ is .258
- The area above $z = .7$ is .242.

These areas add up to 100% of the distribution. Again, the areas are always positive. (To help you remember, think of it this way: if we were computing the area of a rug, we would never find a negative value for the area.)

Check Your Understanding

SCENARIO 4-A

Cal Q. Layte is a 32-year-old American man who is 74 in. tall (6 ft 2 in.). Let's return to fantasy land where we know that the heights of men in their 30s in the United States are approximately normally distributed with a mean = 69.4 in. and a standard deviation = 3.1. 4-8. Can you refer Cal's height directly to Table A.1 to find the proportion of men in their 30s whose heights are greater than Cal's height? If you need to calculate something first, do so now. 4-9. Now find the proportion of men in their 30s with heights greater than Cal's height. 4-10. What is the proportion of heights less than Cal's height?

(Continued)

Suggested Answers

4-8. No, we first must calculate Cal's z score: $(74 - 69.4)/3.1 = 4.6/3.1 = 1.483871 \approx 1.48$. 4-9. We need to find areas under a standard normal curve, so we recommend sketching a standard normal distribution; your authors still sketch when answering questions like this one, and we have decades of experience! Draw a curve and divide it in half with a vertical line. The point where the vertical line and the horizontal number line intersect is $z = 0$. We always write the numbers for z below the horizontal line to remind ourselves that the z scores are on the number line, and we write the proportions above the horizontal number line, making arrows to point to the corresponding areas. Our "z of interest" is 1.48, which will be on the right side of $z = 0$ on the horizontal number line. Find $z = 1.48$ on the number line and draw a vertical line through the point. You were asked about the proportion of heights of men taller than Cal; shade that area. You should shade the small area in the upper tail of the distribution to the right of the vertical line through 1.48. Look at Table A.1 and find 1.48 in the column labeled "z or −z," and find $z = 1.48$. The third column, corresponding to a tail area, shows the answer to this question: a proportion of .0694. 4-10. Sketch another standard normal distribution, marking $z = 0$ in the middle and $z = 1.48$ on the right side of the distribution. Shade the area representing the men with heights less than Cal's. You should shade the entire lower half of the distribution and the area between $z = 0$ and $z = 1.48$. The answer can be found in multiple ways. Remember, 100% of heights are represented under the curve. We already found the top tail area = .0694. If we subtract this proportion from the total proportion, we would have $1 - .0694 = .9306$. Another way to find the answer is to look in Table A.1 for the area between the mean and Cal's height. Beside $z = 1.48$, the second column gives us a "middle" area (between $z = 0$ and $z = 1.48$); this proportion is .4306. The area below $z = 0$ is .5 (the bottom half of the distribution). To find the same answer for the proportion of heights below Cal's height, we add $.4306 + .5 = .9306$. So 93.06% of men in their 30s are shorter than Cal.

Students sometimes wonder about whether to include the value of the "z of interest" when figuring the proportions above or below that value. Consider these two ways of asking a question about a proportion:

- What is the proportion of scores greater than $z = 0.15$?
- What is the proportion of scores greater than *or equal to* $z = 0.15$?

Is there any difference in these two questions? No. We cannot get an area for a point on the number line. Imagine two knives cutting through different places on

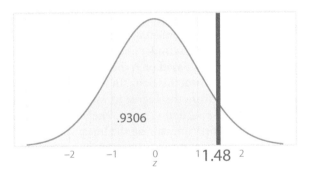

.9306

−2   −1   0   ¹1.48 ²
          z

Figure 4.9

Proportion of American men shorter than 74 in. in height. If Cal Q. Layte has a z score = 1.48 for his height, then we can calculate the proportion of American men shorter than Cal. This proportion, .9306, is shaded in blue.

the bleeding armadillo cake. The size of the cake (area) gets smaller and smaller as the two knives get closer. If they could cut the same spot on the cake at the same time, there would be no cake in between them—so it makes no difference whether we include or exclude $z = 0.15$ when calculating a proportion of the standard normal distribution.

A z score of 1.48 was calculated for Cal Q. Layte in the last Check Your Understanding section. Let's extend that example to connect it with percentiles, which were covered in Chapter 3. If we say that the heights of American men are approximately normally distributed in the population, then every man's height could be converted to a z score, and the distribution of z's would look like a standard normal distribution. Figure 4.9 shows a vertical line cutting through Cal's $z = 1.48$.

We asked in the Check Your Understanding section for the proportion of men with heights shorter than Cal's height. That area is shown in blue, and it was calculated to be .9306. We know that we can convert this proportion into a percentage by multiplying it by 100: 93.06% of men are shorter than Cal, if we are correct in saying that the population of men's heights approximates a normal distribution. In Chapter 3, we learned that a percentile is a score with a certain percentage of the distribution below it. So Cal's height of 74 in. is approximately the 93rd percentile.

What's Next

Physicians sometimes use the CDC's growth charts to determine whether children are growing as expected. These charts for height, weight, and body mass index (BMI) are available from the CDC in special software and increasingly are being used in electronic medical records. Underlying these growth charts are volumes of data collected from large representative samples, and we will use some examples in the exercises. The idea of "(something minus its mean) divided by its standard deviation" is a fundamental concept to which we will return later in the book. Similarly, finding areas under a standard normal curve is giving you some practice with proportions based on a theoretical reference distribution, a skill that you will use repeatedly in this book. In the next chapter, we will see how statistics can be used to measure the degree to which variation in one variable corresponds to variation in a second variable. The verbal definition of z scores will help us to explain a statistic measuring the linear relationship between two variables.

Exercises

4-11. Why would someone want to compute a standard score?

4-12. Explain the difference between the statistic called *standard deviation* and the statistic called a *z score*.

SCENARIO 4-B

This chapter referred repeatedly to Wang et al. (2010), in which people with fibromyalgia were randomly assigned to groups. The researchers determined that one group would participate in a twice-weekly class involving tai chi, and the other group would participate in a twice-weekly class involving educational talks, followed by stretching. One of the outcome variables was sleep quality, measured by the PSQI. 4-13. What kind of quantitative research did Wang et al. conduct, and how do you know? 4-14. How would you characterize the internal validity of this study? 4-15. What kind of variable is the class (tai chi versus education and stretching)? 4-16. What kind of variable is sleep quality?

SCENARIO 4-B, Continued

Download the tai chi data set via http://desheastats.com and create a histogram of the treatment group's baseline PSQI scores. 4-17. Describe the shape of the distribution. 4-18. If a z score is computed for each of those PSQI scores, then all the z scores for the treatment group are arranged in a distribution, what will be the shape of the distribution? 4-19. If you computed an unbiased variance = 9.808712, what would be the numeric value of the standard deviation needed to compute a z score for someone in the treatment group? 4-20. Suppose we have calculated the following statistics for the treatment group's baseline PSQI scores: mean = 13.94, median = 14, and standard deviation based on the sample variance = 3.08. Compute a z score for the person with the lowest score in the treatment group, which was 6. 4-21. Compute a z score for the person with the highest score in the treatment group, which was 19. 4-22. Explain the meaning of each of the z scores that you computed.

(Continued)

SCENARIO 4-C

This chapter referred to a U.S. governmental report on a longitudinal health study called NHANES (McDowell et al., 2008). Included in the report are results of weight measurements taken on children at various ages. The report says 1-year-old boys have a mean weight = 25.5 lb; based on information in the report, we estimate the standard deviation of these boys' weights is 3.7. Suppose your 1-year-old nephew, Alan Rench, weighs 30.5 lb. 4-23. Compute the z score for Alan's weight. 4-24. Does this z score compare Alan's weight to a sample mean or a population mean? 4-25. Let's pretend we feel confident that the weights of 1-year-old boys in the United States are approximately normally distributed. What is the proportion of boys who weigh less than Alan? 4-26. What is the proportion of boys who weigh less than Alan but more than the mean?

SCENARIO 4-C, Continued

You also have a 1-year-old niece named Anna Lyze. We consult the NHANES results and determine that 1-year-old girls in the United States have a mean weight of 24.1 lb with a standard deviation of about 3.51. We weigh Anna and find that she weighs 21.5 lb. 4-27. Compute the z score for Anna's weight. 4-28. Let's extend our make-believe session and say that the weights of 1-year-old girls in the United States are distributed approximately like a normal distribution. What is the proportion of girls who weigh less than Anna? 4-29. What is the proportion of girls who weigh more than Anna but less than the mean?

SCENARIO 4-C, Continued

Let's say that a year has passed since we weighed Alan, and now he is 2 years old. We consult the U.S. study and find that 2-year-old boys have a mean weight of 31.1 and a standard deviation of 3.73. We weigh Alan and find that he now weighs 35 lb. 4-30. Compute the z score for Alan's weight at age 2. 4-31. Again pretending that the weights of 2-year-old boys are known to approximate a normal distribution, what proportion of boys weigh less than Alan? 4-32. Can any of these calculations tell us whether Alan's change in weight since age 1 is normal?

References

A'Hearn, B., Peracchi, F., & Vecchi, G. (2009). Height and the normal distribution: Evidence from Italian military data. *Demography, 46*, 1–25. doi:10.1353/dem.0.0049

Buysse, D. J., Hall, M. L., Strollo, P. J., Kamarck, T. W., Owens, J., Lee, L., … Matthews, K. A. (2008). Relationships between the Pittsburgh Sleep Quality

Index (PSQI), Epworth Sleepiness Scale (ESS), and clinical/polysomno-graphic measures in a community sample. *Journal of Clinical Sleep Medicine, 4*, 563–571. Retrieved from http://www.aasmnet.org/jcsm/

Buysse, D. J., Reynolds, C. F., Monk, T. H., Berman, S. R., & Kupfer, D. J. (1989). The Pittsburgh Sleep Quality Index (PSQI): A new instrument for psychiatric research and practice. *Psychiatry Research, 28*, 193–213. Retrieved from http://www.journals.elsevier.com/psychiatry-research/

Carey, J. J., Delaney, M. F., Love, T. E., Richmond, B. J., Cromer, B. A., Miller, P. D., … Licata, A. A. (2007). DXA-generated z-scores and T-scores may differ substantially and significantly in young adults. *Journal of Clinical Densitometry, 10*, 351–358. doi:10.1016/j.jocd.2007.06.001

McDowell, M. A., Fryar, C. D., Ogden, C. L., & Flegal, K. M. (2008, October 22). Anthropometric reference data for children and adults: United States, 2003–2006. *National Health Statistics Reports, 10*, 6–16. Retrieved from www.cdc.gov/nchs/data/nhsr/nhsr010.pdf

Petersen, N. S., Kolen, M. J., & Hoover, H. D. (1993). Scaling, norming, and equating. In R. L. Linn (Ed.), *Educational measurement*, 3rd ed. (pp. 221–262). New York: Macmillan Publishing Co.

Wang, C., Schmid, C. H., Rones, R., Kalish, R., Yinh, J., Goldenberg, D. L., … McAlindon, T. (2010). A randomized trial of tai chi for fibromyalgia. *The New England Journal of Medicine, 363*, 743–754. doi:10.1056/NEJMoa0912611

Bivariate Correlation

Introduction

Most of the graphs and statistics covered so far have dealt with one variable at a time. For example, we saw a histogram of the obesity rates of the 50 U.S. states and the District of Columbia. The obesity rate was defined as a state's percentage of adults surveyed by the Behavioral Risk Factor Surveillance System (BRFSS), who had a body mass index (BMI) of 30 or higher, based on the participants' self-reported height and weight. Obesity rates became more interesting when graphed along with a second variable, food hardship. The Food Research and Access Center defined food hardship as the percentage of adults who said they lacked money to buy food for their families at some point in the previous year. Figure 3.8 showed a scatterplot of food hardship and obesity rates for the 50 U.S. states plus the District of Columbia; it is reproduced here as Figure 5.1.

In Figure 5.1, the horizontal axis shows food hardship rates, with lower numbers on the left side and higher numbers on the right. The variable on the horizontal axis

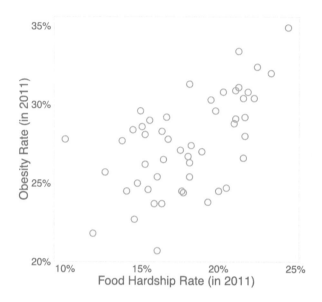

Figure 5.1

Scatterplot of food hardship and obesity rates. The predictor variable, food hardship rate, appears on the horizontal (*X*) axis, and the criterion variable, obesity rate, appears on the vertical (*Y*) axis. (Food hardship data from "Food hardship in America 2011: Data for the nation, states, 100 MSAs, and every congressional district," by the Food Research and Action Center, 2012, February, retrieved from http://frac.org. Obesity data from "Adult obesity facts," by the Centers for Disease Control and Prevention, 2012, August 13, retrieved from http://www.cdc.gov/obesity/data/adult.html.)

is typically labeled as the *X* variable; because this is not an experiment or quasi-experiment, we would use the term *predictor variable*. The vertical axis shows obesity rates, with lower numbers toward the bottom of the graph and higher numbers toward the top. The variable on the vertical axis is typically labeled as the *Y* variable, or criterion variable. There is spread in both of the variables: food hardship rates vary, and so do the obesity rates. But something else is happening: there is a trend in the relationship between these two variables. As we examine the graph from left to right, the point cloud appears to go uphill in the direction of higher numbers on both variables. These variables are not only varying one at a time; they are varying in correspondence with each other, like two dancers stepping in the same direction with each other. As the point cloud goes up toward higher rates of food hardship, it also generally goes up in terms of obesity rates.

In experimental research, we look for the effect of the manipulated independent variable on the dependent variable. But an experiment often is impossible, as in the case of food hardship and obesity. Instead of comparing groups, we can focus instead on the degree to which the variation in one quantitative variable corresponds to similar or opposite variation in another variable. As we examine the higher scores on one variable, are we seeing generally higher values for the second variable at the same time? Or are higher scores on the first variable usually paired with lower values on the second variable? Or do the two variables seem to be dancing to different tunes, and does the variation in the first variable seem to have no connection to the variation in the second variable? A statistic that can help us to answer these questions is the focus of this chapter.

Pearson's Correlation Coefficient

We will describe a certain kind of relationship between two variables: a *linear relationship*, or straight-line relationship. As we look at Figure 5.1, we can imagine drawing a line through the point cloud to summarize the relationship between food hardship and obesity. A linear relationship between two variables is commonly measured with a statistic called *Pearson's correlation coefficient*. This statistic goes by other names as well: *zero-order correlation* or *product-moment correlation*. It is symbolized by the lowercase letter *r*, as in *relationship*. Pearson's *r* will tell us how strong of a straight-line relationship that it can "see" and whether the point cloud is going uphill or downhill (as we scan the graph from left to right). Many kinds of correlation-type statistics exist; Pearson's *r* is the most common one. Like other descriptive statistics, the correlation coefficient estimates a parameter, specifically the *population correlation*, which is symbolized by ρ, the lowercase Greek letter rho. The population correlation is the degree of linear relationship between two variables in a population.

To begin to understand Pearson's *r*, let's consider its possible numeric values. With many statistics, a smaller number means there is less of something and a larger number means there is more of something. But Pearson's *r* is unusual. Its numeric value can be as small as −1, and its largest value is +1. But these extremes (−1 and +1) *both* represent the strongest values for Pearson's *r*. Its weakest value is in the middle, $r = 0$. When $r = 0$, it means there is no linear relationship between the two variables. As the *r* statistic takes on values in either direction from zero, it is indicating an increasingly strong linear relationship, until it reaches one of its strongest values of −1 or +1.

What does it mean if $r = +1$? It means there is a perfect positive linear relationship between two variables. Suppose we are buying liter-sized refill bottles of hand sanitizer for a clinic, and we are considering a product that costs $20 per liter. Let's ignore factors such as taxes, shipping costs, and discounts for large purchases. If we buy one bottle, we will owe $20. If we buy two bottles, we will owe $40. If we buy three bottles, we will owe $60. Figure 5.2 displays the prices of purchases up to 10 bottles. It was created using the pairs of numbers in Table 5.1.

Table 5.1 highlights an important detail: Pearson's correlation coefficient requires scores on each of two variables arranged in pairs. This detail is linked to this chapter's title: Bivariate Correlation. *Bivariate* means "related to two variables." In Table 5.1, the number 1 in the first column has a meaningful link to the number 20 in the second column. The numbers must be kept in those pairs because we are talking about 10 possible purchase scenarios, and each point on the scatterplot depends on one pair of numbers. If we used the pairs of numbers

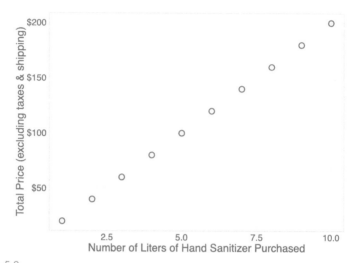

Figure 5.2

Liters of hand sanitizer and total price. For each additional liter purchased, the total price goes up $20. This graph represents a perfect positive linear relationship (fabricated data).

Table 5.1 Number of Liters of Hand Sanitizer Purchased and Total Price ($)

Number of Liters	Total Price
1	20
2	40
3	60
4	80
5	100
6	120
7	140
8	160
9	180
10	200

in Table 5.1 and computed Pearson's correlation coefficient between the number of liters purchased and the total price, we would find $r = +1$. The correlation is positive, because the 10 points in Figure 5.2 form an uphill line, as we scan the graph from left to right. Further, we know that +1 is one of the two strongest values for r. This value means there is a perfect positive linear relationship between the number of liters and the price. We can perfectly predict the total price if we know how many liters we need.

Based on what you have just read, you probably can guess what $r = -1$ means, but let's run through another scenario. Suppose our clinic is in a neighborhood with a large number of African American adults, who tend to have a higher risk of high blood pressure or hypertension. Untreated hypertension can increase the risk of heart problems. Suppose we receive a grant to reach out to the neighborhood and conduct screenings for hypertension. Our grant includes a budget of $6,000 to pay for gift cards in the amount of $30 each. We send letters to adults in the neighborhood, asking them to call our clinic to schedule a free screening and to receive a $30 gift card for their time. Consider the relationship between the number of adults screened and the amount of money *remaining* in the gift-card budget. What are the pairs of numbers? If we do one screening, we will have $5,970 left in the gift-card budget. If we perform two screenings, we will have $5,940 left in the budget. So the pairs of numbers are 1 and 5,970, 2 and 5,940, and so on. The relationship between the number of screenings and the amount of money left over is shown in Figure 5.3.

If the budget began with $6,000 and we were paying $30 per screening, then 200 screenings are possible. This graph shows 200 pairs of scores. With so many pairs of scores, we do not even see the gaps between the points, which form a downhill line. If we computed Pearson's r for those 200 pairs of numbers, we would get $r = -1$. Figure 5.3 is an example of a perfect negative linear relationship. We can perfectly predict how much money will be left in the gift-card budget, based on how many people are screened.

Correlation coefficients in research studies do not represent perfect linear relationships. Look again at Figure 5.1, where instead of a line of points, we see a point cloud that appears to have an uphill trend. When we ran a correlation analysis for the food hardship and obesity rates, we found Pearson's $r = .581$, indicating a positive linear relationship between food hardship rates

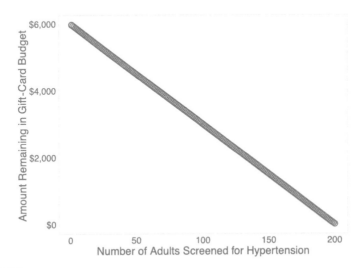

Figure 5.3

Number of screenings and money left in the budget. For each additional person screened for high blood pressure, we spend $30 from our budget for gift cards. This graph represents a perfect negative linear relationship (fabricated data).

and obesity rates for the United States plus the District of Columbia. (You can download the food hardship data set via http://desheastats.com.) The value of Pearson's r has no units of measure, so we do not have .581 "somethings." The correlation coefficient is just a number existing within a continuum: from −1 for the strongest negative linear relationship ... to zero, meaning no linear relationship ... to +1 for the strongest positive linear relationship.

When r is close to zero, the scatterplot does not have any apparent uphill or downhill linear trend. Let's look at an example. Figure 5.4 shows a scatterplot of 2010 birth rates and death rates for the 50 U.S. states plus the District of Columbia. The birth rates are the number of live births per 1,000 people in the state, and the death rates are the number of deaths per 100,000 people, with the rates adjusted for age (without the adjustment, states with older populations would appear to have unusually high rates). These numbers were reported by the Centers for Disease Control and Prevention (CDC). The birth rates and age-adjusted death rates for 2010 had a correlation of .053, which is close to zero. We recommend searching online for "guessing correlations" and visiting one of many websites that show a variety of scatterplots with a choice of values of r. You can match the values of r with the scatterplots and get familiar with different degrees of linear relationships.

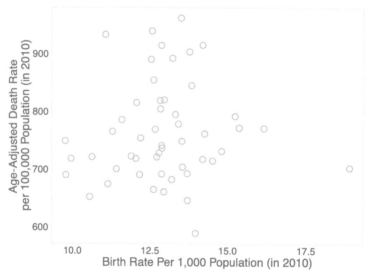

Figure 5.4

U.S. birth rates and age-adjusted death rates. It appears as if there is no linear relationship between these two variables. This graph is one example of a data set with a correlation close to zero. (Birth rates from "Birth data: National Vital Statistics System," by the Centers for Disease Control and Prevention, 2013a, April 26, retrieved from http://www.cdc.gov/nchs/births.htm; age-adjusted death rates from "Mortality tables: National Vital Statistics System," by the CDC, 2013c, April 26, retrieved from http://www.cdc.gov/nchs/nvss/mortality_tables.htm.)

5. Bivariate Correlation

SCENARIO 5-A

We found data online for 42 states' percentages of high school students who smoke cigarettes and the states' excise taxes per pack of cigarettes (rates were unavailable for eight states and the District of Columbia). We computed $r = -.372$. 5-1. Explain the meaning of this correlation. (The data set is available via http://desheastats.com.)

Suggested Answers

5-1. This correlation is negative, which means there appears to be a negative linear relationship between the excise tax rates per cigarette pack and the rate of youth smoking in the states. States that have higher taxes on cigarettes tend to have lower rates of youth smoking, and states that have lower cigarette taxes generally have higher youth smoking rates.

Verbal Definition of Pearson's *r*

In our introduction to correlation, we talked about how each variable in a data set may have variation or spread. Further, pairs of variables like food hardship and obesity may vary together in correspondence with each other. The states tended to have higher scores on both food hardship and obesity at once and lower scores on both variables at once. This shared corresponding variation between a pair of variables is the idea behind *covariance*. We know that variance is a measure of spread, specifically, how much scores vary around their mean. When we graph two variables on a scatterplot, our focus shifts away from the amount of spread in each variable, and we switch to talking about how the variables *covary*. When variables covary, then the variation in one variable has a corresponding variation in another variable. Food hardship and obesity covary; because $r = .581$, we know there is a positive linear relationship between food hardship and obesity. As food hardship goes up, obesity generally goes up; and as food hardship goes down, obesity tends to go down. They vary together, or covary. Please note that when we talk about scores increasing or decreasing, we are not talking about changes across time for any state. The data sets used in this chapter represent only one point in time. When we say rates go up or down together, we are describing a trend in the scatterplot. The point cloud in Figure 5.1 tends to go uphill from left to right. The variation in the direction from lower to higher food hardship rates is roughly mirrored in the concurrent variation from lower to higher obesity rates.

There is a statistic called a *covariance* that measures the degree to which two variables covary, but the covariance statistic is affected by the units of measure for the variables in the analysis, making it hard to interpret. For example, the strength of a covariance for food hardship and obesity could not be compared with the strength of the covariance between cigarette tax rates and youth smoking rates, because the units of measure are different. Luckily, we can standardize the covariance and remove the units of measures—and that is exactly what Pearson's r does. The correlation coefficient r is a standardized covariance. Its numerator is the covariance between the two variables, and its denominator is a big, ugly formula with a square root over it. The big, ugly formula is related to the variances of each variable, so the denominator is functioning to standardize the covariance. After standardizing the covariances, we have different correlation coefficients that can be compared with each other because they all are limited to the same continuum from −1 to +1. We can compare the correlation between food hardship and obesity (r = .581) with the correlation between cigarette tax rates and youth smoking rates (r = −.372). The food hardship/obesity correlation is stronger, because it is farther from zero than is the correlation between taxes and youth smoking rates.

We are avoiding the formula for the correlation coefficient because it is big and ugly, and we prefer small, cute formulas. To help you understand the statistic, we now will explain a verbal definition of Pearson's r, which statistical software can calculate for you. Pearson's r can be verbally defined as

r = the average product of z scores for the two variables

Now, this verbal definition by itself is not useful, so let's step through an explanation. A z score is a measure of a score's location relative to its mean. We could compute a z score for every state's food hardship rate and determine whether a state is above or below the mean of all states. We also could compute a z score for every state's obesity rate and see whether the state has a rate that is higher or lower than average. A product is found when we multiply two numbers, such as the z score for Ohio's food hardship rate times Ohio's z score for obesity rate. Figure 5.5 shows the scatterplot of food hardship and obesity for the 50 U.S. states plus the District of Columbia, this time with gray lines to show where the mean of each variable is located.

The mean for food hardship is nearly 18% and is represented by the vertical gray line. The points to the left of the vertical gray line are below the mean on food hardship, so they would have negative z scores on that variable. The points to the right of the vertical gray line are above the mean on food hardship, so they would have positive z scores on that variable. The mean for obesity is 27.6%, represented by the horizontal gray line. The points above the horizontal gray line would have positive z scores for obesity rates, and the points below the horizontal gray line would have negative z scores for obesity.

Each state's z score for one variable at a time is not our main interest. What we care about is the *combination* of the state's z score on food hardship and its z score on obesity. We can combine each state's two z scores by multiplying them. If two z scores are positive and we multiply them, we will have a positive product. If two z scores are negative and we multiply them, we will again have a positive product. But if one z score is positive and the other z score is negative, the product will be negative. Will most states have a positive or negative product of z scores? The gray lines in Figure 5.5 divided the graph into quadrants. Let's consider one quadrant at a time, starting with the quadrants with the fewest points:

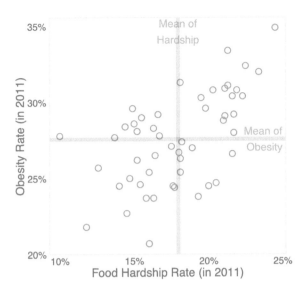

Figure 5.5

Means added to the food hardship and obesity scatterplot. The vertical gray line is the mean of the food hardship rates. The horizontal gray line is the mean of the obesity rates. Our focus will be on the four quadrants and the z scores that could be computed on each of the two variables. (Food hardship data from "Food hardship in America 2011: Data for the nation, states, 100 MSAs, and every congressional district," by the Food Research and Action Center, 2012, February, retrieved from http://frac.org. Obesity data from "Adult obesity facts," by the Centers for Disease Control and Prevention, 2012, August 13, retrieved from http://www.cdc.gov/obesity/data/adult.html.)

- The top left quadrant shows states with lower-than-average food hardship but higher-than-average obesity, so the products of z scores will be negative.
- The bottom right quadrant shows states with higher-than-average food hardship and lower-than-average obesity; each of these states also will have a negative product of z scores.

Now let's consider the two other quadrants, where two-thirds of the states are represented:

- In the bottom left quadrant we have states that are lower than average on both food hardship and obesity rates; their two negative z scores will have a positive product (because a negative times a negative is a positive).
- The top right quadrant shows the states that are above average on both food hardship and obesity; their two positive z scores will have a positive product.

The scatterplot in Figure 5.5 is dominated by positive products of z scores. If we were to go through the process of computing all of those z scores on both variables, computing the product of z scores for each state, then averaging the products, we would get Pearson's $r = .581$. Statistical software uses a different calculating process that is algebraically equivalent to the average product of z scores. This process was explained only to provide you with an informal understanding of the measurement of bivariate linear relationships.

Check Your Understanding

SCENARIO 5-A, Continued

This scenario involved cigarette taxes and youth smoking rates for 42 states. 5-2. Explain the correlation $r = -.372$ between these variables in terms of the verbal definition of Pearson's r.

Suggested Answers

5-2. If we computed each state's z score on cigarette tax and each state's z score on youth smoking rate, and then multiplied each state's two z scores, we would get a product of z scores. If we average the products for all states, we would get $r = -.372$. This number would indicate that on average, states tended to have positive z scores on one variable paired with negative z scores on the other variable. If we graphed the data, we would not be surprised to find many points in the top left quadrant and the bottom right quadrant of the scatterplot.

Judging the Strength of a Correlation

The linear relationship between food hardship and obesity ($r = .581$) is stronger than the correlation between cigarette tax rates and youth smoking rates ($r = -.372$), because .581 is farther from zero. There is a better way to judge the strength of correlations.

Before we explain this better way, let's contrast the strongest possible linear relationships with the case where we have no linear relationship. If we have a perfect linear relationship, then one variable is perfectly predictable from another, and $r = 1$ or $r = -1$. The variation in the Y variable is 100% explainable by the variation in the X variable (or vice versa, although we traditionally think of predicting the Y variable from the X variable). We illustrated $r = +1$ with the example of the number of bottles of hand sanitizer and the total cost of the purchase. We can perfectly predict the cost based on the number of bottles. Now consider the case where $r = 0$. If there is no linear relationship between two variables, then the X variable provides no ability to perform a linear prediction of the variation in the Y variable; the two variables have no shared variation.

The better way to assess the strength of a correlation is to compute the proportion of variance that is shared by the two variables; this proportion is computed by squaring Pearson's r. This new statistic, r^2, is called the *coefficient of determination*. (Please note there are many statistics that are called coefficients, including Pearson's correlation coefficient.) For cigarette tax rates and youth smoking rates, we have $r = -.372$; we compute the coefficient of determination by multiplying the correlation by itself:

$$r^2 = r \times r$$

$$= -.372 \times -.372$$

$$= .138384 \approx .138$$

How do we interpret this number, $r^2 = .138$? It is a proportion, or fraction, of the variation in one variable that is explained by the other variable. It will always be a positive number. It can be changed into a percentage by multiplying by 100: $.138384 \times 100 \approx 13.8\%$. This is the percentage of the variation in youth smoking rates that is shared with cigarette tax rates. Is that a lot of shared variance? Compared with $r^2 = 1$ (corresponding to 100% of the variance in the cost of hand sanitizer being explained by the number of bottles ordered), it is a smaller number. Compared with $r^2 = 0$ (corresponding to 0%), an $r^2 = .138$ might be statistically the same as zero, or it could be statistically noteworthy (we will explain how to make that determination later in the text). For now we can say that this coefficient of determination means the two variables covary in such a way that 13.8% of the variance in youth smoking rates can be explained by differences in cigarette taxes.

Let's take this explanation a bit further. Many factors lead to the states having different rates of youth smoking. Some states are tobacco-producing states, some states have stronger laws limiting where people can buy cigarettes, and some states have put more money into campaigns to encourage people to quit smoking. Out of all of the factors that may be related to differences in youth smoking rates, cigarette taxes is one of these factors. There is a negative linear relationship between cigarette tax rates and youth smoking rates ($r = -.372$). We can say that 13.8% of the variation in youth smoking rates is tied up in this variable's covariation with cigarette taxes. This amount may be noteworthy when we think about trying to keep people from becoming lifelong smokers.

How does the strength of the correlation for youth smoking and cigarette taxes compare with the linear relationship between food hardship and obesity? We found $r = .581$, so the coefficient of determination is

$$r^2 = r \times r$$

$$= .581 \times .581$$

$$= .337561 \approx .338$$

We can conclude that 33.8% of the variance in obesity rates is accounted for by its relationship with food hardship. Out of all of the reasons that different states have different obesity rates, we have found one potentially noteworthy variable, food hardship. If we consider all the variation in obesity rates, about one-third of that variation can be explained by the way obesity varies along with food hardship. Other factors would explain the remaining two thirds of the variation in obesity rates. Imagine if a state could lower its food hardship rate and the possible effect on obesity rates and related health outcomes.

We have looked online and found the youth smoking rates and adult smoking rates for 42 states (rates were unavailable for eight states and the District of Columbia) and computed Pearson's $r = .654$. 5-3. Explain the meaning of this number, then judge the strength of the correlation. 5-4. Let's pretend that we have found a correlation of .06 between youth smoking rates and the rate of vehicle ownership by teenagers for these 42 states. Compute something to show why we might be justified in saying the correlation is almost zero.

Suggested Answers

5-3. There is a positive linear relationship between adult smoking rates and youth smoking rates. States that have lower adult smoking rates also generally have lower youth smoking rates, and states with higher adult smoking rates tend to have higher youth smoking rates. To judge the strength of the correlation, we need to compute the coefficient of determination: $r^2 = .654 \times .654 = .427716 \approx .428$. This proportion corresponds to 42.8%. We can conclude that 42.8% of the variance in youth smoking rates can be explained by adult smoking rates. We chose to consider youth smoking rates as predicted by adult smoking rates, because of the influence of environment on younger people. But because 42.8% is the percentage of shared variation between the two variables, we also could interpret this result as youth smoking's variation corresponding to 42.8% of the variation in adult smoking. 5-4. If $r = .06$, then $r^2 = .0036$, meaning that about one-third of 1% of the variance in teen ownership of vehicles was related to youth smoking rates. There is virtually no shared variance between the variables. The variation in youth smoking rates is not mirrored by corresponding variation in the rates of vehicle ownership by teens.

What Most Introductory Statistics Texts Say About Correlation

Almost all introductory statistics textbooks caution readers that correlation does not imply that a causal relationship exists between two variables. Even when two variables have a strong correlation, it could be purely accidental. Matthews (2000) illustrated this point for teaching purposes. He compiled data from 17 countries on the number of breeding pairs of storks and the number of live human births to explore the notion that storks deliver babies. He found a positive correlation, $r = .62$. Countries with higher numbers of breeding pairs of storks also had more births, and countries with fewer stork pairs had fewer births. So, do storks deliver human babies, as the myth goes? What could explain the linear relationship? Matthews' data set included a variable that might explain the fact that these two variables covary: the area of the country in square kilometers appears to be a lurking variable. Area was correlated with the number of stork pairs ($r = .579$) and strongly correlated with the number of births ($r = .923$). Bigger countries have more people giving birth and, coincidentally, more room for storks.

Turning to a health-related example of correlation not implying causation, several studies have suggested a link between sleep problems and obesity. Some studies have found that people who are overweight tend to sleep less than people who were not overweight. Others studies have indicated that people who were overweight tend to sleep *more* than people who were not overweight. But many factors can influence sleep. People with chronic conditions, such as fibromyalgia, sometimes gain weight as a side effect of medications or as a result of pain preventing them from exercising. The same medications for pain could affect their sleeping patterns. Other conditions such as sleep apnea or depression could lead to more sleep in patients who happen to be overweight. Whenever we interpret a correlation, we must keep in mind that we may be oversimplifying reality, and we cannot leap to a causal conclusion without conducting experimental research.

The rest of this chapter will describe factors that should be considered when calculating and interpreting correlations.

Pearson's *r* Measures Linear Relationships Only

The first factor to keep in mind while interpreting Pearson's *r* concerns the idea of *linear* relationships. There are many possible kinds of relationships between two variables. Figure 5.6 shows a scatterplot of the expected average life span of people born in 2011 and the number of maternal deaths per 100,000 births in 2010 for 159 countries. We downloaded the maternal mortality rates from the U.S. Central Intelligence Agency's World Factbook and the life expectancy estimates from the World Health Organization.

There appears to be a fairly strong downhill trend in the point cloud in Figure 5.6, indicating that countries with longer expected life spans also tend to have lower maternal mortality rates. But we see a bend in the point cloud. A straight line could be drawn through the point cloud where most points seem to be located, but the line would be relatively far away from the points on the top left side of the graph. A curved line probably would fit the data better than a straight line. The term *linear relationship* refers only to *straight* lines. For these data we computed $r = -.855$, indicating a negative linear relationship. This numeric value of Pearson's *r* only tells us about the direction and degree of a linear relationship and cannot reveal that a curved line might do a better job than a straight line to describe the relationship. This example reinforces the importance of always graphing our data.

Correlations Can Be Influenced by Outliers

A second factor to consider is that extreme scores can affect the numeric value of Pearson's *r*. Every point is taken into account when *r* is calculated. If a few points are far away from the rest of the points, they could increase the strength

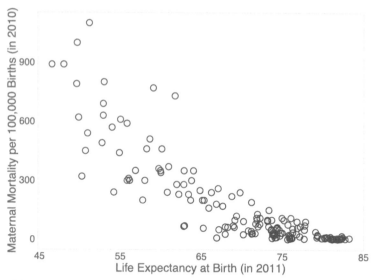

Figure 5.6

Life expectancy at birth and maternal mortality rates. Sometimes variables share a relationship that could be best described with a curved line, not a straight line, as shown in this scatterplot of expected life spans and the number of maternal deaths per 100,000 births for 159 countries. (Life expectancy data from "Life expectancy tables," by the World Health Organization, retrieved from http://www.who.int/gho/mortality_burden_disease/life_tables/en/, 2013; maternal mortality rates from "Maternal mortality rates per 1,000 population in 2012," World Factbook, 2013b, retrieved from https://www.cia.gov/library/publications/the-world-factbook/.)

of the apparent linear relationship, pushing r toward −1 or +1. Or, depending on the location of the extreme points, they could push r toward zero and deflate the apparent strength of the relationship. Figure 5.7 shows a scatterplot created using the data in the Matthews (2000) article on storks and human births in 17 countries. The correlation between the number of breeding pairs of storks and the number of births was $r = .62$, so 38.44% of the variance in the number of births was accounted for by the number of stork pairs.

Figure 5.7 shows the number of human births in thousands, so the number 400 on the vertical axis represents 400,000 births that year. Notice the two points on the right side of Figure 5.7, which we are displaying in red to draw your attention. What would happen if these two extreme points were ignored? We recomputed the correlation without those points and found $r = .163$, indicating only 2.66% of the variance in the number of births was explained by the number of stork pairs. Figure 5.8 shows the stork data without the two extreme points. We will come back to this example shortly. The two red points strengthened the correlation coefficient.

Now let's consider an example of extreme values that weaken Pearson's r. We downloaded 2012 estimates of the crude birth and death rates from the World Factbook, produced by the U.S. Central Intelligence Agency. The crude birth rate is an estimate of the annual number of births per 1,000 people in a country, and the

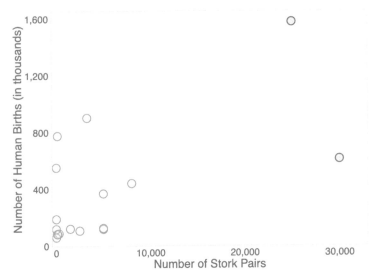

Figure 5.7

Number of breeding stork pairs, number of human births. Matthews (2000) illustrated how correlations can lead people astray. He found a strong correlation, $r = .62$, between the number of breeding pairs of storks and the number of human births in 17 countries. Notice the two extreme points. How do they affect the correlation? (Data from "Storks deliver babies ($p = .008$)," by R. Matthews, 2000, *Teaching Statistics, 22*, 36–38.)

crude death rate is an estimate of the annual number of deaths per 1,000 people. (This example should be viewed with skepticism because the rates should be age adjusted. As mentioned earlier, unadjusted rates can make regions with older populations appear to have much higher death rates. We are using the crude rates here for teaching reasons.) Figure 5.9 shows birth and death rates from a selection of 178 countries.

The correlation between the birth and death rates represented in Figure 5.9 was $r = -.401$, meaning that about 16.1% of the variance in death rates was explained by the birth rates. Countries with lower birth rates tended to have higher death rates, and countries with higher birth rates generally had lower death rates. Notice the five points floating above the main part of the point cloud in Figure 5.9. These countries—South Africa, Namibia, Botswana, Lesotho, and Swaziland—appear to have higher-than-average birth and death rates. What would happen to the correlation if we ran the analysis without those five countries in southern Africa? We removed those countries and recalculated Pearson's r. We found that the remaining 173 countries had a correlation of $r = -.555$, meaning that 30.8% of the variance in death rates was accounted for by birth rates, which is almost twice as much explained variance as we computed when those five countries were included. The presence of the extreme data points weakened the correlation. Graphing our data can help to identify observations that appear to differ from the bulk of the observations and may be influencing descriptive statistics.

You may have wondered why we used "a selection of 178 countries." We have a reason, which we will explain in the next section.

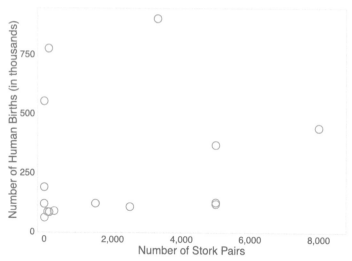

Figure 5.8

Stork data after deleting two extreme points. What happens to the correlation when two extreme points are deleted? Without those points, the correlation between the number of breeding pairs of storks and the number of human births dropped from $r = .62$ to $r = .163$. (Data from "Storks deliver babies ($p = .008$)," by R. Matthews, 2000, *Teaching Statistics, 22*, 36–38.)

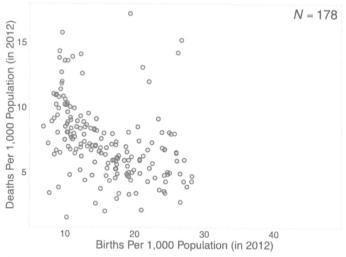

Figure 5.9

Birth and death rates for 178 countries. What kind of linear relationship appears to exist between the crude birth and death rates for this selection of 178 countries? (Birth data from "Crude birth rates per 1,000 population in 2012," World Factbook, 2013a; death data from "Crude death rates per 1,000 population in 2012," World Factbook, 2013b; all data retrieved from https://www.cia.gov/library/publications/the-world-factbook/.)

SCENARIO 5-B

An experiment discussed in Chapter 3 involved sedentary pregnant women who were randomized to groups, and then they were instructed to either remain sedentary or participate in a program of physical activity. On five occasions the women's strength and ability to perform physical activity were measured. Strength was defined as the number of times that the participants could lift a medicine ball in 1 min. Physical ability was summarized in a power score, which involved walking or running 3.2 km (2 miles). To account for increasing weight during pregnancy, power was calculated as (weight × distance)/time. We calculated the correlation between strength and power scores at the beginning of the study for the women in the study and found $r = .042$. (Data from Price, Amini, & Kappeler, 2012, can be downloaded via http://desheastats.com, if you would like to compute this correlation.) 5-5. Explain the meaning of $r = .042$. 5-6. Do something to interpret the strength of the correlation. 5-7. What factors could have affected the correlation? 5-8. The study was an experiment. Explain whether we can infer a causal relationship between these two variables.

Suggested Answers

5-5. The correlation is positive, but it appears to be close to zero. We will defer further interpretation until we answer the next question. 5-6. We can compute the coefficient of determination: $r^2 = .042 \times .042 = .001764 \approx .0018$. A fraction of 1% of the variance in power scores is accounted for by strength. We therefore would conclude that the observed correlation is very close to zero, indicating no linear relationship between the two variables. 5-7. There could be a nonlinear relationship between the two variables, or there could be extreme scores suppressing the value of r—or there simply might be no linear relationship between the two variables. Perhaps the researchers chose two important, uncorrelated variables that individually explain different aspects of physical health. Graphing the data would help us to better understand this correlation. 5-8. No. The causality that can be inferred relates to the effect of the independent variable (group: sedentary versus active) on the dependent variables. This analysis combined the data from all participants at baseline.

Correlations and Restriction of Range

Another factor that must be kept in mind when interpreting correlations is the range of the data. Restricting the range of the data can change the value of Pearson's *r*. We intentionally limited a previous example to countries with lower crude birth rates (i.e., fewer than 30 births per 1,000 inhabitants), so we could demonstrate that outliers can weaken a correlation. Now let's consider the effect of the decision to limit the range of the data. Figure 5.9 ($N = 178$ countries) showed the scatterplot for a limited range. In contrast, Figure 5.10 shows the scatterplot for the entire range of birth rates ($N = 222$ countries), with the additional countries shown in purple.

The vertical grayish purple line marks a cutoff point of 30 births per 1,000 population, which is the value we used to select the 178 countries for Figure 5.9. Now that we have graphed the data for all 222 countries in our data set, the point cloud in Figure 5.10 appears to form a U shape, and the points for those five countries in southern Africa do not seem to stand out as much. When the entire range of values is examined, the birth and death rates do not seem to have a straight-line relationship. The relationship between birth and death rates for the 222 countries in Figure 5.10 appears to be *curvilinear*, meaning a curved line could be drawn through the middle of the point cloud to summarize its shape. Countries that have the lowest estimated birth rates tend to have higher estimated death rates. In the middle of the graph, where countries have slightly higher birth rates, we see generally lower death rates. On the right side of the graph, we see that the countries with the highest birth rates generally have higher death rates as well. (Can you think of possible reasons for this relationship?) When we computed a correlation for the 178 countries with rates under 30 births per 1,000 inhabitants, we found $r = -.555$. When we computed the correlation for the entire data set, we found Pearson's $r = +.202$. The range of the data affects the r statistic. Pearson's r is not a good statistic for this data set because

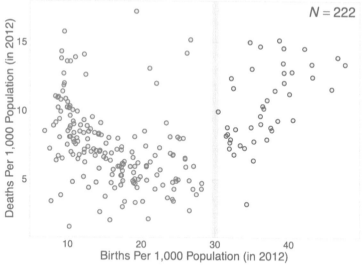

Figure 5.10

Birth and death rates for 222 countries. Now we have graphed crude birth and death rates from 222 countries. Figure 5.9 showed only the dark green part of the scatterplot. When the range of data is restricted, we may not be seeing the full picture of the relationship between two variables. (Birth data from "Crude birth rates per 1,000 population in 2012," World Factbook, 2013a; death data from "Crude death rates per 1,000 population in 2012," World Factbook, 2013b; all data retrieved from https://www.cia.gov/library/publications/the-world-factbook/.)

5. Bivariate Correlation

these crude birth and death rates do not appear to have a straight-line relationship. Pearson's *r* can only tell us the degree to which it sees a straight line. If we limit the range of the data being used, we could be influencing the correlations. This example again reminds us of the main message in Chapter 3: always graph your data.

Many health-related variables can share a curvilinear relationship. For example, Tiffin, Arnott, Moore, and Summerbell (2011) studied the association between obesity and psychological well-being in English children. They used the Strengths and Difficulties Questionnaire, which produces higher scores to indicate more psychological adjustment problems. The researchers found that the adjustment scores were lower for children close to the average BMI for their height, weight, age, and gender. Children who were underweight *or* overweight had more adjustment problems, whereas average-weight children had fewer such problems. If a data set is limited to a certain range of possible scores for a variable like BMI, statistics like *r* may be affected.

Restricting the range also can lead to a value of *r* that is closer to zero than the correlation we would have computed for the entire range of values. We previously saw an example of the correlation between cigarette taxes and youth smoking rates in the United States (the CDC did not have sufficient data to calculate youth rates for eight states and the District of Columbia in 2009). Figure 5.11 shows the

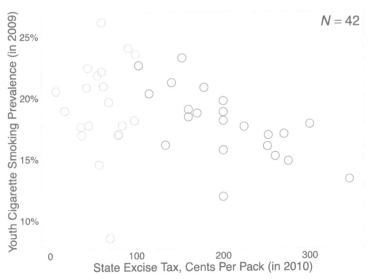

Figure 5.11

Cigarette taxes and youth smoking rates. This scatterplot represents 42 states. The light blue line divides the scatterplot into two parts. The states with excise taxes of less than $1 per pack of cigarettes are shown in light green, and the states with excise taxes of $1 per pack and higher are shown in light blue. (Youth smoking data from "Youth and tobacco use," by the Centers for Disease Control and Prevention, 2013d, April 24, retrieved from http://www.cdc.gov/tobacco/data_statistics/fact_sheets/youth_data/tobacco_use/; tax data from "State cigarette excise tax rates and rankings," by the Campaign for Tobacco-Free Kids, 2013, April 24, retrieved from https://www.tobaccofreekids.org.)

Figure 5.12

Youth smoking rates for states with lower taxes on cigarettes. When we look only at the states with excise taxes less than $1 per pack of cigarettes, there appears to be almost no linear relationship between taxes and youth smoking rates. Restricting the range can make a correlation appear to be stronger or weaker than it would be if the full range of data were analyzed. (Youth smoking data from "Youth and tobacco use," by the Centers for Disease Control and Prevention, 2013d, April 24, retrieved from http://www.cdc.gov/tobacco/data_statistics/fact_sheets/youth_data/tobacco_use/; tax data from "State cigarette excise tax rates and rankings," by the Campaign for Tobacco-Free Kids, 2013, April 24, retrieved from https://www.tobaccofreekids.org.)

cigarette taxes and youth smoking rates ($r = -.372$). Let's focus only on the locations with a tax of less than $1 (100 cents) per pack of cigarettes; those 20 states are represented by pale green circles to the left of the vertical pastel line in Figure 5.11. Will the apparent negative linear relationship between cigarette taxes and youth smoking prevalence be found among the states with lower taxes? Figure 5.12 shows that portion of the data set. The axes were redrawn to zoom in on those few states; the pastel vertical line remains at $1 per pack of cigarettes.

Restricting the range so that we see only the states with an excise tax of less than $1 per pack of cigarettes weakened the correlation. For the 42 states whose rates were reported by the CDC, youth smoking rates shared a correlation of $r = -.372$ with excise taxes; for the 20 states with cigarette taxes under $1 per pack, we computed $r = .021$ for youth smoking rates and excise taxes. By squaring .021, we get .000441, meaning that only four-hundredths of 1% of the variance in youth smoking rates is related to cigarette taxes for these 20 states. The correlation is essentially zero.

Combining Groups of Scores Can Affect Correlations

Every observation included in the computation of r can have an influence on the statistic, which is related to the fourth factor that influences our interpretation

Figure 5.13

Food hardship and obesity rates, with state abbreviations as markers. We already saw that food hardship and obesity rates share a positive linear relationship. What if we analyzed the data for Southern states and non-Southern states separately? The abbreviations for the Southern states are shown in purple. (Food hardship data from "Food hardship in America 2011: Data for the nation, states, 100 MSAs, and every congressional district," by the Food Research and Action Center, 2012, February, retrieved from http://frac.org. Obesity data from "Adult obesity facts," by the Centers for Disease Control and Prevention, 2012, August 13, retrieved from http://www.cdc.gov/obesity/data/adult.html.)

of a correlation coefficient. Combining groups of scores can increase or decrease the strength of *r*. Figure 5.1 showed the scatterplot of the data on food hardship and obesity rates for the 50 U.S. states plus the District of Columbia. Using regional definitions from the U.S. Census Bureau, we identified which states were in the South and which were not. Then we created Figure 3.9, which appeared in Chapter 3. We reproduce this graph as Figure 5.13.

We computed three Pearson's correlation coefficients for food hardship and obesity rates: one for the Southern states, one for the non-Southern states, and one for all states put together. Here's what we found:

- For the Southern states, food hardship and obesity shared a positive linear relationship, *r* = .518, meaning that about 26.8% of the variance in obesity rates was accounted for by food hardship.
- The correlation for the non-Southern states was *r* = .192, so about 3.7% of the variance in obesity rates was explained by food hardship in those states.

- When we computed the correlation for all states, we found $r = .581$, indicating that about one-third of the variance in obesity rates was related to food hardship.

Combining the Southern states with the non-Southern states led to a different result for Pearson's r, compared with the separate correlations for Southern and non-Southern states. The correlation was strongest when the data from Southern states were combined with the data from non-Southern states. When using this statistic, it is important to think about whether different groups could have different degrees of linear relationships for the variables being studied and how the overall correlation could be affected by combining data from different groups.

Check Your Understanding

SCENARIO 5-C

(Inspired by Robbins, Mehl, Holleran, & Kasle, 2011. Details of this scenario may differ from the actual research.) Suppose we read a study suggesting that the frequency of sighing in daily life can be an indicator of depression level for patients with rheumatoid arthritis. The study found that participants who sighed more times per hour (based on unobtrusive recordings) also had higher depression levels, whereas participants who sighed less frequently had lower depression scores. The researchers acknowledged that their small sample size made their results preliminary, but knowledge of the relationship could give health-care providers an auditory cue of possible depression in patients with this condition. We decide to run a larger study and look for the relationship among older adult patients. We recruit volunteer participants from two medical offices, each of which serves patients who may become depressed from dealing with pain or a chronic condition. One office is run by a group of orthopedic surgeons specializing in joint replacement. The other office is run by a group of heart surgeons. 5-9. What kind of linear relationship was found in the small study described at the beginning of Scenario 5-C? 5-10. If we collect data from patients served by these two particular medical offices and want to correlate the number of sighs with depression scores, what factor should we consider when analyzing the data?

Suggested Answers

5-9. Positive linear relationship. 5-10. Patients in one of the offices are being treated for heart conditions. Shortness of breath is common in people with heart conditions. The relationship between sighing and depression may not
(Continued)

be observable in people with heart conditions if they also have trouble breathing. Combining the data from these two clinics might lead to a different conclusion about sighing and depression than we might reach if we analyzed the data from the two clinics separately.

Missing Data Are Omitted From Correlations

A fifth factor that researchers must remember when computing correlations is that statistical software computes Pearson's r only on the complete cases. If a participant is missing a score on one of the two variables, that participant is not included in the computation of r. We have mentioned the correlation between youth smoking rates and state excise taxes for cigarettes. The analysis included only 42 states because the rates of smoking by teenagers were unavailable for eight states and the District of Columbia. Pearson's r was based only on the rates for those 42 states with complete data. Be sure to pay attention to the number of observations that were correlated so that you can determine whether data were missing.

Pearson's r Does Not Specify Which Variable Is the Predictor

A final consideration when computing correlations is that Pearson's r does not specify a direction of influence. Researchers often interpret correlations based on logically assumed directions; for example, we suspect that adult smoking rates have an influence on youth smoking rates, because younger people receive important messages from adults about what is acceptable or unacceptable behavior. But the statistic itself, Pearson's r, does not determine what is the predictor or criterion variable. It only tells us the degree of linear relationship shared by the two variables.

What's Next

How do we know whether a correlation is strong enough to be noteworthy? The answer depends on many factors, including the area of research and the number of participants in the study. In Chapter 6, we will introduce the concepts of probability and risk, which is the beginning of our transition from descriptive statistics to inferential statistics. Inferential statistics are used in making decisions about statistical noteworthiness. Later we will explain how to determine whether a correlation is markedly different from zero and how to use the evidence of a linear relationship to make predictions.

SCENARIO 5-D

(Inspired by Wansink, Painter, & North, 2005. Details of this scenario may differ from the actual research.) Researchers wanted to know whether visual cues about how much food was available would influence the amount eaten. The researchers randomly assigned 54 volunteers to one of two groups. Members of the control group were served 12 oz of soup in a normal bowl. Those in the treatment group ate from "bottomless bowls," which initially held 12 oz of soup but secretly were rigged to refill themselves from the bottom. So the participants in the treatment group never received a visual cue that they were finishing the soup. Everyone ate alone to control for the extraneous variable of social interaction. After the participant finished eating and left the room, the researchers recorded how much soup was consumed. Participants completed a questionnaire, asking them to estimate how many calories of soup they ate. The researchers correlated the participants' estimated number of calories consumed with their actual number of calories consumed. For all 54 participants, they found $r = .31$. For those who used a normal soup bowl, they found $r = .67$. For those who were secretly given a bottomless bowl, they computed $r = .12$. 5-11. What kind of research is this? 5-12. What kind of variable is the type of bowl? 5-13. What kind of variable is the participants' estimated number of calories consumed? 5-14. What kind of variable is the actual number of calories consumed? 5-15. Interpret the meaning of the three correlation coefficients. 5-16. Explain whether the participants' estimated number of calories has a causal relationship with the actual number of calories consumed. 5-17. Why was it important for the researchers to compute correlations separately for the two groups? 5-18. Compute something to judge the strength of the three linear relationships, then explain your numeric results, using variable names.

SCENARIO 5-E

Chapter 4 referred repeatedly to a study by Wang et al. (2010), who wanted to know whether tai chi would help people with fibromyalgia. The dependent variables included the Pittsburgh Sleep Quality Index (PSQI) and the Fibromyalgia Impact Questionnaire (FIQ). Higher scores on the PSQI indicate more sleep troubles, and higher scores on the FIQ mean more disruption of everyday life as a result of the pain and other effects of fibromyalgia. Data from Wang et al. (2010) may be downloaded via http://desheastats.com. The researchers measured all participants before they were randomly assigned to either a tai chi group or an education group. We computed Pearson's $r = .471$ between the baseline FIQ and baseline PSQI scores ($N = 66$). 5-19. What kind of linear relationship was found between these variables? 5-20. Explain the meaning of this correlation, using variable names. 5-21. Judge the strength of this correlation. 5-22. Explain whether we should be concerned about combining the data from the treatment and control groups for this correlation. 5-23. Why should we create a scatterplot of these data? 5-24. If you have downloaded the data, create the scatterplot and describe what you see, including anything that could be influencing the value of r.

(Continued)

SCENARIO 5-F

Falvo and Earhart (2009) reported the results of a study of patients with Parkinson's disease (PD). Impairments in balance and gait can affect the walking ability of these patients. The researchers wanted to study the distance that PD patients could walk in six minutes, a common assessment of walking capacity. Prior research showed that for healthy older adults, much of the variance in six-minute walk distance was explained by age, sex, height, and weight. Falvo and Earhart proposed that some of the variance in six-minute walk distance could be explained by (among other things) the patients' Unified Parkinson Disease Rating Scale (UPDRS) motor score. Higher UPDRS motor scores mean greater effects of Parkinson symptoms such as rigidity and tremor. Falvo and Earhart also measured Timed Up and Go (TUG), a test in which patients are asked to get up from a chair, walk to a mark 3 meters from the chair, turn around, and return to a seated position. Timed in seconds, the task begins when the person doing the assessment says "go" and stops when the patient has returned to a completely seated position. The researchers reported the following correlations: for UPDRS motor scores and six-minute walk distances, $r = -.27$; for UPDRS motor scores and TUG scores, $r = .19$; and for TUG scores and six-minute walk distances, $r = -.64$. 5-25. What kind of research is this, and how do you know? 5-26. Which correlation is strongest, and how do you know? 5-27. Explain each of the three correlations using the variable names and judging its strength with the coefficient of determination. 5-28. Why should we examine scatterplots for the three correlations?

References

Campaign for Tobacco-Free Kids. (2013, April 24). *State cigarette excise tax rates and rankings*. Retrieved from https://www.tobaccofreekids.org/

Centers for Disease Control and Prevention. (2012, August 13). *Adult obesity facts*. Retrieved from http://www.cdc.gov/obesity/data/adult.html

Centers for Disease Control and Prevention. (2013a, April 26). *Birth data: National Vital Statistics System*. Retrieved from http://www.cdc.gov/nchs/births.htm

Centers for Disease Control and Prevention. (2013b, April 25). *Current cigarette smoking among adults. Behavioral Risk Factor Surveillance System Prevalence and Trends Data, 2011*. Retrieved from http://www.cdc.gov/tobacco/data_statistics/fact_sheets/adult_data/cig_smoking/

Centers for Disease Control and Prevention. (2013c, April 26). *Mortality tables: National Vital Statistics System*. Retrieved from http://www.cdc.gov/nchs/nvss/mortality_tables.htm

Centers for Disease Control and Prevention. (2013d, April 24). *Youth and tobacco use*. Retrieved from http://www.cdc.gov/tobacco/data_statistics/fact_sheets/youth_data/tobacco_use/

Falvo, M. J., & Earhart, G. M. (2009). Six-minute walk distance in persons with Parkinson disease: A hierarchical regression model. *Archives of Physical Medicine and Rehabilitation, 90,* 1004–1008. doi:10.1016/j.apmr.2008.12.018

Food Research and Action Center. (2012, February). *Food hardship in America 2011: Data for the nation, states, 100 MSAs, and every congressional district.* Washington, DC. Retrieved from http://frac.org/pdf/food_hardship_2011_report.pdf

Matthews, R. (2000). Storks deliver babies ($p = .008$). *Teaching Statistics, 22,* 36–38.

Price, B. B., Amini, S. B., & Kappeler, K. (2012). Exercise in pregnancy: Effect on fitness and obstetric outcomes—A randomized trial. *Medicine & Science in Sports & Exercise, 44,* 2263–2269. doi:10.1249/MSS.0b013e318267ad67

Robbins, M. L., Mehl, M. R., Holleran, S. E., & Kasle, S. (2011). Naturalistically observed sighing and depression in rheumatoid arthritis patients: A preliminary study. *Health Psychology, 30,* 129–133. doi:10.1037/a0021558

Tiffin, P. A., Arnott, B., Moore, H. J., & Summerbell, C. D. (2011). Modelling the relationship between obesity and mental health in children and adolescents: Findings from the Health Survey for England 2007. *Child and Adolescent Psychiatry and Mental Health, 5,* 31. doi:10.1186/1753-2000-5-31

Wang, C., Schmid, C. H., Rones, R., Kalish, R., Yinh, J., Goldenberg, D. L., … McAlindon, T. (2010). A randomized trial of tai chi for fibromyalgia. *New England Journal of Medicine, 363,* 743–754. doi:10.1056/NEJMoa0912611

Wansink, B., Painter, J. E., & North, J. (2005). Bottomless bowls: Why visual cues of portion size may influence intake. *Obesity Research, 13,* 93–100. doi:10.1038/oby.2005.12

World Factbook. (2013a). *Crude birth rates per 1,000 population in 2012.* Washington, DC: Central Intelligence Agency. Retrieved from https://www.cia.gov/library/publications/the-world-factbook/

World Factbook. (2013b). *Crude death rates per 1,000 population in 2012.* Washington, DC: Central Intelligence Agency. Retrieved from https://www.cia.gov/library/publications/the-world-factbook/

World Factbook. (2013c). *Maternal mortality rates per 100,000 births in 2010.* Washington, DC: Central Intelligence Agency. Retrieved from https://www.cia.gov/library/publications/the-world-factbook/

World Health Organization. (2013). *Life expectancy tables. Global Health Observatory.* Retrieved from http://www.who.int/gho/mortality_burden_disease/life_tables/en/

6

Probability and Risk

Introduction

People make decisions every day by guessing the likelihood of events. Your assessment of the likelihood of rain today may determine whether you carry an umbrella. Patients with chest pain judge how likely it is that they are having indigestion versus a heart attack, and as a result they make decisions about whether to seek immediate help. Life is full of uncertainty, and in statistics we try to quantify that uncertainty. Statistics requires an understanding of *probability*, which can be defined as a relative frequency of occurrence. Probability may sound like a scary, highly mathematical term, but in fact we know from decades of teaching statistics that students intuitively know quite a bit about probability without realizing it. We will build upon what you know.

Relative Frequency of Occurrence

Let's say that you are taking a class with 100 students, who have been divided into 10 groups. The teacher says each group of 10 students must have a leader, whose name will be chosen randomly. You and the nine other students in your group write your names on slips of paper, which are placed in a hat and shuffled thoroughly to mix them up, and one name is drawn. Everyone in the group has a chance of becoming the leader. What are the chances of being chosen? You should say that each person in the group has a 1 in 10 chance of becoming the leader. That is a probability, which we defined as a relative frequency of occurrence: 1/10. Only one leader can be chosen out of the total number of students who were eligible to be the leader of your group. What is the probability of *not* having to serve as the leader? Now we are interested in the nine people who would not be the leader, out of the 10 people who are eligible to avoid the leadership role. So the probability of *not* being selected as leader is 9/10, which can be written as .9. (Health-care professionals are trained to write a zero before a decimal point to avoid confusion, so this number might be written as 0.9. We wrote ".9" without a zero because of the style rules we are following in this book.)

These probabilities were computed by identifying two numbers: a numerator and a denominator. The numerator is the number of outcomes that specifically interest us at the moment. The denominator is the number of options available, or the pool from which we are choosing. When we computed the probability of 1/10 for being selected as leader, the numerator is 1 because we are specifically interested in selecting one leader. In contrast, when we computed the probability of 9/10, we were interested in the 9 people who would not be named the leader. In both probabilities, the denominator was the 10 people who were in the pool for possible selection.

Probabilities are numbers that occur only within a certain range: zero to one. The smallest probability is zero, meaning no chance of something happening. What is the probability of a student not belonging to a group? Zero, because no one in the class of 100 was left out of a group. The highest probability is one, meaning a guaranteed event. If every student in the class belongs to a group, then 100 out of 100 students belong to a group, so the probability of being in a group equals 1. Notice that the probability of a group member's name being pulled from the hat depended on a random process, the mixing of the names. If the teacher chooses the team leaders based on their class performance and personalities, then we cannot know the probability of being selected as team leader. In this chapter, we repeatedly ask questions about the probability of randomly selecting someone or something from a group of people or objects.

Check Your Understanding

SCENARIO 6-A

In 2009, the Behavioral Risk Factor Surveillance Survey (BRFSS) surveyed 7,769 adults in Oklahoma about smoking and asthma. The sample included 769 people who said they had asthma and 1,291 people who said they smoked daily. 6-1. What is the probability of randomly selecting a person

(Continued)

in this sample who had asthma? 6-2. How likely is it to randomly choose a person in this sample who smoked daily?

Suggested Answers

6-1. We need to identify how many people are of interest and put that number in the numerator. Then we put the total number eligible to be considered in the denominator. The question asks about the occurrence of asthma; 769 people said they had asthma, so that is the numerator. The denominator is the number of people eligible, or 7,769 survey respondents—that is, 769 people with asthma and 7,000 people without asthma. The probability of randomly selecting someone with asthma is 769/7769 = .0989831 ≈ .10. 6-2. This question asks about daily smoking, reported by 1,291 people, so that is the numerator. The denominator again is the 7,769 people surveyed (1,291 daily smokers and 6,478 people who said they did not smoke daily). The probability of randomly selecting someone who smoked daily is 1291/7769 = .1661733 ≈ .17.

Conditional Probability

Suppose your teacher decided that there should be the same number of male and female group leaders. Your group consists of six women and four men, and your teacher decides your group will be led by a man. What is the probability of a randomly chosen person being a leader, given that the group must be led by a man? Now we have restricted the number of people eligible to be leader; only the four men are eligible. The probability of a person being chosen as the leader is no longer 1/10 because 6 of the 10 people cannot be considered for the leadership role. Given that the leader must be male, one of the four men will be chosen, so the probability is 1/4. This is an example of a *conditional probability*, which is a relative frequency based on a reduced number of possible options. The teacher placed a condition on the situation: the leader of this group must be male. So we are being *given* the four males, and we must randomly choose one leader from among them. We have reduced the number of options to four men. The teacher decided that a woman would not be chosen as leader for this group, so those six group members' names would be taken out of the hat, and they would have no chance of being randomly selected.

People intuitively use the concept of conditional probability all the time. Your judgment about the likelihood of rain is not made in the absence of information. Given that you see dark clouds moving in your direction, you may think rain is more likely today, compared with a day in which the skies are clear in the morning. The patient with chest pains may weigh the likelihood of a heart attack to be higher, given the patient's age and the fact that two close relatives recently died of heart attacks. But the main concept to remember about *computing* a conditional probability is that we limit the denominator to only part of the total sample. In

	Diagnosis from Dermatopathologist	
Result from App #1	Melanoma	Benign
"Problematic"	42	74
"Okay"	18	48

Figure 6.1

Shown here are the estimated frequencies for App #1. (Data extrapolated from "Diagnostic inaccuracy of smartphone applications for melanoma detection," by J. A. Wolf et al., 2013, *JAMA Dermatology, 149*, 422–426.)

the classroom example, the condition was that the leader had to be male, so we were given only part of the total group from which to choose a leader.

Let's look at another example. Wolf et al. (2013) conducted a study of smartphone apps designed to identify a skin lesion as melanoma or not. The study tested four apps that allowed the user to upload a photo of a lesion. The research team chose photos showing lesions that had been tested by a dermatopathologist, an expert who had performed skin biopsies on the lesions. The dermatopathologist had verified whether the lesions were cancerous (specifically, melanoma) or benign. If an app is accurate, then it should reach the same conclusions as the dermatopathologist. Working backward from the statistics reported in the study, we found App #1's results, shown in Figure 6.1.

The app evaluated each image and gave one of three responses: "problematic," "okay," or "error"; we show only the 182 responses that were problematic or okay. The researchers treated the "problematic" response as indicating a cancerous lesion requiring medical attention. The "okay" response was treated as an indication of a benign lesion. Each cell (or square) in Figure 6.1 contains a frequency. Out of the 60 lesions that the expert said were melanoma, the app reported that 42 of them were problematic and 18 were okay. Forty-eight lesions that were diagnosed as benign received an "okay" result from the app. But 74 lesions that the app said were problematic had been diagnosed by the expert as benign.

Now that we understand the data, let's add some row totals, column totals, and total sample size. Figure 6.2 shows the results.

The expert identified 60 cancerous lesions and 122 benign lesions, whereas App #1 said 116 lesions were problematic and 66 were okay. Now let's ask some questions about probability. What is the probability that a randomly chosen

	Diagnosis from Dermatopathologist		
Result from App #1	Melanoma	Benign	Row Totals
"Problematic"	42	74	116
"Okay"	18	48	66
Column Totals	60	122	N = 182

Figure 6.2

Now we have inserted the row totals and column totals. (Data extrapolated from "Diagnostic inaccuracy of smartphone applications for melanoma detection," by J. A. Wolf et al., 2013, *JAMA Dermatology, 149*, 422–426.)

6. Probability and Risk

lesion in the study was identified as problematic? This question does not limit the pool of images to a subset; that is, the question is not asking for a conditional probability. Out of all 182 images, 116 of them were labeled problematic by the app. Therefore, the probability of randomly choosing an image that had a problematic result is 116/182. We can write this probability as .6373626 ≈ .64. (For the rest of these examples, we will round to two decimal places.) Here is a similar question: what is the probability of a randomly chosen image being labeled as okay by the app? The answer is 66/182 ≈ .36. It makes sense—the app classified about two thirds of the images as problematic (probability ≈ .64), so the remaining images were labeled as okay (probability ≈ .36). What is the probability that a randomly chosen image was diagnosed by the dermatopathologist as melanoma? Now we are looking at the results divided up by columns. The dermatopathologist said 60 out of the 182 lesions were cancerous, so the probability is 60/182 ≈ .33. The expert said the rest of the lesions were benign, so the probability of a randomly selected image being benign is 122/182 ≈ .67.

You may have noticed that all of the probabilities in the previous paragraph had the entire sample size of 182 in the denominator, and only one fact about the sample was pertinent for finding each numerator. Conditional probabilities are far more interesting, however. What is the probability that a randomly chosen image was found to be problematic by the app, *given that* the dermatopathologist said it was melanoma? Now two facts are pertinent. When you read "given that," you should think to yourself, "Which portion of the data set am I being *given*?" This question is requiring the condition that the expert said it was cancer. So we limit our answer only to the 60 images showing confirmed cases of melanoma, and we ignore everything in the Benign column of Figure 6.2. (If we were teaching you in person, we would tell you to cover up the part of the sample that we are excluding from consideration so that we could see only the column for melanoma cases.) Now that we have limited ourselves to the Melanoma column, out of those 60 images how many did the app label as problematic? There were 42. So the conditional probability of a randomly chosen image from this study being labeled as problematic, *given* the condition that the image shows a verified case of melanoma, is 42/60 = .7. Out of all the expert-tested lesions that were definitely cancer, the app caught 70% of them and said they were problematic. In the section on "Special Names for Certain Conditional Probabilities," we will come back to this probability.

Let's look at another conditional probability. What is the probability that a randomly chosen image in the study was benign, *given that* the app said it was okay? Now we are being *given* 66 images of lesions that the app reported were okay. We confine our focus to the second row of the figure and ignore the first row's results entirely. Out of those 66 "okay" lesions, 48 were benign, according to the dermatopathologist. So the conditional probability of a randomly chosen image being benign, given that the app said it was okay, is 48/66 ≈ .73. That is, 73% of the benign lesions were correctly classified as "okay" by the app.

Whenever we are computing a conditional probability, we always make sure we understand which part of the sample has been given to us and which part of the sample has been excluded. It may feel as if you are working backward, but the only way to make sure that the denominator is correct is to find it first. The denominator is the "given" number, a subset of the total sample. Then we find the numerator; within the identified subset, it is the number of observations having the desired characteristic.

Check Your Understanding

6-3. What is the probability that a randomly selected image in the study was melanoma, given that the app said it was problematic? 6-4. What is the probability that a randomly chosen image was reported by the app to be okay, given that the expert diagnosed it as benign?

Suggested Answers

6-3. This question is giving us the app's problematic lesions, or the row total of 116. Out of that total, 42 were diagnosed as melanoma, so the conditional probability is 42/116 ≈ .36. Notice that this question is different from a previous question. The probability of melanoma, given a problematic result from the app, is 42/116; but the probability of the app saying it is a problematic result, given it is verified to be melanoma, is 42/66. 6-4. We are being given the 122 images of lesions that the expert said were benign. Out of those images, the app said 48 were okay. The conditional probability is 48/122 ≈ .39. This is another probability with a special name, to be explained next.

Special Names for Certain Conditional Probabilities

We have been talking about conditional probabilities, and our implied concern is App #1's accuracy in classifying lesions. For every image showing a diagnosed case of melanoma, a really good app would say the lesion was problematic. And for every image of a confirmed benign lesion, a really good app would say the lesion was okay. But Wolf et al. (2013) reported the app said "okay" when evaluating 18 cases of melanoma and "problematic" when looking at 74 benign lesions. For many diagnoses, health professionals have a *gold standard*, or the best, most widely accepted diagnostic tool; in the present example, it is the test result from the dermatopathologist. A new or quicker test, such as the smartphone app, is evaluated in comparison to the gold standard.

When we compute the conditional probability of a positive diagnosis by the new test, given that the gold standard gave a positive diagnosis, we are computing a statistic known as the new test's *sensitivity*. Sensitivity focuses on numbers in the first column of Figure 6.2. A good test has high sensitivity, meaning it gives a positive diagnosis for most or all of the cases that were positive based on the gold standard. The sensitivity of App #1 is the probability of identifying a lesion as problematic, given that the expert said it was melanoma; this conditional probability is 42/60 ≈ .7. Sensitivity typically is reported as a percentage, which we can find by multiplying the probability by 100. So the sensitivity of App #1 was

70%, meaning that the app was sensitive for detecting 70% of the known melanoma cases. The more cases of cancer that the app misses, the lower its sensitivity. A widely used mnemonic for remembering what *high* sensitivity means is SnNout, which stands for "Sensitivity: Negative test rules out a possible diagnosis" (Akobeng, 2007). If a test for some disease had a sensitivity of 100%, then a diagnosis of that disease could be ruled out for anyone who tested negative. App #1 had sensitivity = 70%, so 70% of melanoma cases were correctly identified, and the app missed 30% of melanoma cases. Dermatologists probably would be uncomfortable with patients making medical decisions based on the relatively low sensitivity of App #1.

A second statistic often reported when evaluating the accuracy of a new test is the *specificity*, which is the conditional probability of a negative diagnosis by the new test, given negative results according to the gold standard. Specificity focuses on numbers in the second column of Figure 6.2. A good test has high specificity, meaning the app produces a negative diagnosis for most or all of the negative diagnoses according to the gold standard. To compute the specificity of App #1, we find the conditional probability of an "okay" result, given that the expert said the lesions were benign. We already computed this probability as 48/122. The specificity, typically reported as a percentage, is

$$\frac{48}{122} \times 100 \approx 39.3\%$$

A widely used mnemonic for remembering the meaning of *high* specificity is SpPin, which stands for "Specificity: Positive test rules in a possible diagnosis" (Akobeng, 2007). If a test for a disease had specificity = 100%, then everyone *without* the disease would receive a negative test result. In other words, every time a test with 100% specificity came up negative, we could trust that it was a true negative result, and a positive result should be taken seriously. App #1's specificity for detecting benign lesions is low because the negative results were identified as "okay" for only about 4 out of every 10 images of benign lesions (39.3%). The rest of the benign images were labeled problematic. The more mistakes that the app makes about negative test results based on the gold standard, the lower the specificity.

Check Your Understanding

SCENARIO 6-B

Sheeler, Houston, Radke, Dale, and Adamson (2002) tested the accuracy of a rapid test for streptococcal pharyngitis, or "quick strep" test. The gold standard for strep is to swab the back of a patient's throat and to culture the swabbed cells. The researchers reported that the quick strep

(Continued)

test had sensitivity $= 91\%$ and specificity $= 96\%$. 6-5. Explain the meaning of these two statistics. 6-6. Restate the sensitivity and specificity statistics as questions about conditional probabilities.

Suggested Answers

6-5. *The sensitivity is the percentage of the confirmed cases of strep (according to the gold standard) that were diagnosed by the quick strep test. The quick strep test detected 91% of the cases of strep, meaning that given the patients who had a positive throat culture for strep infection, 91% had a positive result on the quick test and 9% had a negative result on the quick test. The specificity is the percentage of the negative throat cultures that also were negative on the quick test. The quick strep test was negative for 96% of the patients who had negative throat cultures. This result means that given the patients who had negative throat cultures, 4% were incorrectly diagnosed by the quick test as having a strep infection. 6-6. For sensitivity, what is the probability that a randomly chosen patient in the study tested positive on the quick strep test, given that we are limiting ourselves to the patients with positive throat cultures? For specificity, what is the probability that a randomly chosen patient in the study tested negative on the quick strep test, given that we are considering only those patients with negative throat cultures?*

Statistics Often Accompanying Sensitivity and Specificity

Sensitivity gave us the percentage of the positive biopsies that were positively identified by the smartphone app. Let's look at the results in a different way: Out of all of the smartphone app's positive cases (116 problematic results), what percentage of them were found to be melanoma in a biopsy? That is a different way to look at the data. Instead of focusing on one column at a time, we use numbers in one row at a time. The results are reproduced for your convenience in Figure 6.3, with the rows highlighted.

Sensitivity focused on the Melanoma column; 42 out of the 60 cases of melanoma resulted in a problematic answer from App #1. Now let's look at the problematic results in the first row. A statistic called the *positive predictive value* is the percentage of positive results on the new test that were positive according to the gold standard. In other words, the positive predictive value depends on the conditional probability of diagnosed melanoma, given that we are limited to the app's reports indicating possible cancer. Here we had 116 problematic results from App #1, and out of those results, there were 42 confirmed cases of melanoma. So the positive predictive value of App #1 was

$$\frac{42}{116} \times 100 \approx 36.2\%$$

	Diagnosis from Dermatopathologist		
Result from App #1	Melanoma	Benign	Row Totals
"Problematic"	42	74	116
"Okay"	18	48	66
Column Totals	60	122	N = 182

Figure 6.3

The positive predictive value and negative predictive value place the focus on the rows of this figure. (Data extrapolated from "Diagnostic inaccuracy of smartphone applications for melanoma detection," by J. A. Wolf et al., 2013, *JAMA Dermatology, 149,* 422–426.)

If a new test has a high positive predictive value, then a large percentage of its positive results would turn out to be positive according to the gold standard. App #1's positive predictive value of 36.2% means that out of all the times App #1 said the lesion was problematic result, only 36.2% of the images showed a confirmed case of melanoma. About 63.8% of the images with the problematic report from the app actually were benign, according to the dermatopathologist. In other words, given the 116 images that the app suggested might be cancer, the app made a correct prediction 36.2% of the time. Ideally, every single time that the app said "problematic," those images would be melanoma. Instead, out of the problematic results from the app, one third of them were cancer and two thirds of them were benign. As a predictive device for cancerous lesions, the app is not doing a very good job.

One more statistic that often accompanies the sensitivity and specificity statistics is called the *negative predictive value*. When we calculated specificity, we looked at the Benign column and computed the percentage of results from the gold standard that the new test said were not cancer ("okay"). Negative predictive value focuses on numbers in the second row of Figure 6.3, showing the negative results from the app; it is computed based on the conditional probability of results being negative according to the gold standard, given that the new test says the results are negative. If a new test has a high negative predictive value, then a large percentage of its negative results would turn out to be negative according to the gold standard. App #1 had 66 "okay" results, and 48 of those images showed lesions that were biopsied and found to be benign. So the negative predictive value was

$$\frac{48}{66} \times 100 \approx 72.7\%$$

Almost three fourths of the app's negative results were predictive of the expert's negative result, meaning that about one-fourth of the negative results from App #1 (18/66, or ≈ 27.3%) actually were cases of melanoma. So out of all the times that the app said the lesion was okay, one fourth of the lesions were cancerous.

All of these conditional probabilities can be hard to keep straight. Figure 6.4 summarizes sensitivity, specificity, positive predictive value, and negative predictive value.

SCENARIO 6-B, Continued

In the quick strep study by Sheeler et al. (2002), a throat culture served as the gold standard. The researchers reported that the quick strep test had a positive predictive value of 96% and a negative predictive value of 90%. 6-7. Explain the meaning of these two statistics. 6-8. Restate these two statistics as questions about conditional probabilities.

Suggested Answers

6-7. The positive predictive value means that 96% of the quick strep test's positive results ended up having positive throat cultures, so 4% of patients with positive results on the quick strep test had a negative throat culture. The negative predictive value means that 90% of the quick strep test's negative results came from patients who also had a negative throat culture, so 10% of those with a negative quick strep test received a positive throat culture and a diagnosis of strep infection. 6-8. The positive predictive value can be restated as, "What is the probability that a patient chosen at random from among those with a positive quick strep test would test positive with a throat culture?" The negative predictive value can be restated as, "What is the probability that a patient randomly chosen from among those with a negative quick strep test received a negative throat culture?"

New Test's Results	Gold Standard Results		Row Totals
	Positive	Negative	
Positive	a = number of true positives	b = number of false positives	$a + b$
Negative	c = number of false negatives	d = number of true negatives	$c + d$
Column Totals	$a + c$	$b + d$	

$$Sensitivity = \frac{a}{a+c} \times 100$$

$$Specificity = \frac{d}{b+d} \times 100$$

$$Positive\ Predictive\ Value = \frac{a}{a+b} \times 100$$

$$Negative\ Predictive\ Value = \frac{d}{c+d} \times 100$$

Figure 6.4

Summary: sensitivity, specificity, positive predictive value, negative predictive value.

| | Diagnosis from Dermatopathologist | | |
Result from Perfect App	Melanoma	Benign	Row Totals
Cancer	116	0	116
Not Cancer	0	66	66
Column Totals	116	66	N = 182

Figure 6.5

Illustrating the statistics for a perfect skin-cancer detection app.

Let's pretend for a moment that we have developed the perfect skin cancer detection app for smartphones. If our app is perfect, it will give exactly the same result as a skin biopsy for melanoma. Figure 6.5 shows the numeric results for this fictitious perfect app. Let's compute the sensitivity, specificity, positive predictive value, and negative predictive value for this fictitious perfect app, showing again how each of these statistics relies on conditional probability. Sensitivity uses the conditional probability of the app saying "cancer" given that the biopsy said "melanoma." Out of the 116 images in the first column that had biopsies showing they were melanoma, all 116 of them are detected by the app as cancerous, so sensitivity is 100%. Specificity uses the conditional probability of the app saying "not cancer," given that the biopsy said "benign." Out of the 66 images in the second column that are confirmed to show benign lesions, all 66 are detected by the app as "not cancer," so specificity is 100%. The positive predictive value uses the conditional probability of a positive biopsy for melanoma, given that the app said "cancer." Out of the 116 images in the first row that the app said were cancer, all 116 of them have positive biopsies for melanoma, so the positive predictive value is 100%. The negative predictive value uses the conditional probability of a benign biopsy, given that the app said "not cancer." Out of the 66 images in the second row that the app said are not cancer, all 66 of them have negative biopsies for melanoma, so the negative predictive value is 100%. Scientists continually try to develop quicker or less expensive diagnostic tests that meet or exceed the gold standard. Unless we have developed a new gold standard, we would not expect any quick diagnostic test to have results resembling our fictitious perfect app.

Two Other Probabilities: "And" and "Or"

Many kinds of probabilities may be presented in statistics textbooks, and as we have done in previous chapters, we present only the material that we think will be most beneficial to you. Our experience is that students who understand the next two kinds of probabilities sometimes use the knowledge to reason through research scenarios. Suppose we are planning a study in which we will extend the research by Wang et al. (2010) on the effect of tai chi for helping people with fibromyalgia to deal with their chronic pain and sleep problems. Let's say that we have reason to believe that tai chi's gentle movements will provide relaxation and pain relief for people with rheumatoid arthritis and heart conditions without imposing a physical burden. With this scenario in mind, consider this question: Who will be the participants in the new study?

This question may puzzle you because we just said we would study patients with rheumatoid arthritis and heart conditions, so let's restate our question: Will we study patients who have rheumatoid arthritis *and* heart conditions? Or will we study patients who have either one of these conditions (rheumatoid arthritis *or* heart conditions)?

The use of one different word—*and* compared with *or*—can make a huge difference. Let's label these options as Sample #1 (rheumatoid arthritis and heart conditions) and Sample #2 (rheumatoid arthritis or heart conditions). Which sample do you think would be easier to recruit? We think it would be easier to find people for Sample #2, because only one of those diagnoses is necessary; we could recruit people with rheumatoid arthritis, we could recruit people with heart conditions, and we could recruit people with these comorbid conditions. In contrast, Sample #1 would require every participant to have both rheumatoid arthritis *and* a heart condition, which may make participants harder to find. There is a probability that can be computed for the "and" situations, and there is a different probability that can be computed for the "or" situations.

Let's apply the "and" concept to probability. Returning to the example of the cell phone app and the expert's test of skin lesions, we could ask, "What is the probability that a randomly selected image in the study was melanoma *and* received a problematic result from App #1?" This is an example of a *joint probability*, in which two facts must be true in order for the observations to be included in the numerator; the denominator contains the entire sample size. The researchers found that 42 out of the 182 images evaluated by App #1 had a problematic result *and* were diagnosed as melanoma, so the joint probability is $42/182 \approx .23$.

Now let's apply the "or" concept to probability. The *"or" probability* does not have a special name that is commonly used in statistics. In the example of the smartphone app, we could ask, "What is the probability that a randomly selected image in the study was diagnosed melanoma *or* received a problematic result from App #1?" Because only one of these facts must be true (the image shows melanoma *or* the image produced a problematic app result), the image could have been in any of the following cells of Figure 6.3:

- The cell representing the 42 images that were labeled problematic and were diagnosed melanoma
- The cell representing the 18 images that were labeled okay and were diagnosed melanoma
- The cell representing the 74 images that were labeled problematic and were diagnosed benign

We would add up these three numbers and find that the probability of melanoma or problematic was $(42 + 18 + 74)/182 = 134/182 \approx .74$.

This chapter so far has focused on the first word in its title: probability. Next we turn to the concept of risk, which is closely related to probability, and explain one way that health researchers evaluate disease risk.

6-9. Returning to Figure 6.3, what is the probability of randomly choosing an image of a skin lesion that was reported as problematic and tested benign by the expert? 6-10. What is the probability of randomly selecting an image that was reported to be problematic or tested benign by the expert?

Suggested Answers

6-9. As an "and" probability, this answer will require the number of images for which both facts are true: the image was problematic and tested benign. There were 74 such images out of 182 images evaluated by App #1, so the probability is 74/182 ≈ .41. 6-10. An "or" probability's numerator counts all images for which only one of the two facts must be true. We can count the 116 images in the "problematic" row of Figure 6.3, but then we also need to count the 48 benign images that were in the "okay" row. So the probability of randomly choosing an image that was problematic or benign is 164/182 ≈ .90. (Notice that we cannot add the "problematic" row total and the "benign" column total because then we would be counting the 74 images in the problematic-and-benign cell twice.)

Risk and Relative Risk

Hardly a week goes by without a news report about how people's behavior affects their chances of some disease. Websites are available for estimating the risk of different diseases for people like you, based on demographics and health history. *Risk* is most simply defined as a probability of an undesired outcome, such as diabetes or heart attack. Aven and Renn (2009) prefer to define risk as "uncertainty about and severity of the consequences (or outcomes) of an activity with respect to something that humans value." People tend to value good health, and many activities can have severe consequences to their health—but not all the time, which is where the uncertainty comes from. For example, a person mowing a lawn runs a risk of suffering a foot injury as a result of an accident with a power lawn mower. To get a rough idea of this risk, we could estimate how many people use power mowers every summer, and out of that number, how many people suffered foot injuries in the process of mowing. The risk of a power-mower injury for any given individual is not the same as the overall risk for people in general; your first author has an immeasurably small risk of being injured by a power mower because she hires people to mow her lawn. A person working for a professional lawn service may run a higher risk of injury simply because of the number of hours spent using the power mower every week.

Disease risk, which is the field of study for epidemiologists, often requires large-scale studies. For example, non-Hodgkin's lymphoma is a form of blood cancer. How common is it? Some websites say that the annual numbers of cases make it the fifth most common cancer in the United States and the sixth most

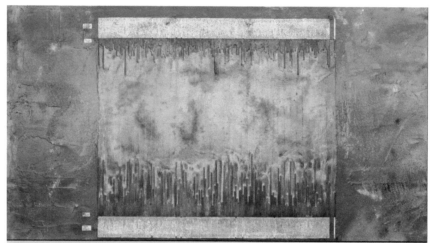

"Disparity Four-Birth/DeathRates" (4' x 8'), by Gary Simpson, used with permission. The lengths of copper rods reflect numbers for birth and death rates per thousand.

common cancer in the United Kingdom. To calculate the risk or probability of getting non-Hodgkin's lymphoma, an epidemiologist could determine the number of people who were diagnosed with the disease, out of every 100,000 residents in our country. National and global organizations conduct *disease surveillance*, or monitoring of disease incidence and trends for all inhabitants in a country or area, because risk cannot be estimated from small samples. You may know three people in one small town who had non-Hodgkin's lymphoma, which may lead you to ask two questions: How common is this disease? And given the population risk for the disease, do we have a relatively large number of cases for a small town? We may not be able to judge the risk for an entire population based on the small town's results, just as we would not estimate the risk of melanoma by using the results of the study of smartphone apps. These cases of non-Hodgkin's lymphoma may be an accidental group of people with the same disease. To determine whether that is true, epidemiologists must know quite a bit about the incidence of non-Hodgkin's lymphoma.

Disease risk usually is not the focus of research that makes the news. Usually we read reports about *risk factors*, or variables that affect the chances of a disease. Researchers compare people who have the risk factor with people who do not have the risk factor. This comparison may involve the calculation of *relative risk*, a statistic that quantifies how people with a risk factor differ from people without a risk factor. Lack of exercise is a risk factor for many diseases. The Susan G. Komen Breast Cancer Foundation (http://komen.org) says that about 500 out of every 100,000 inactive women will develop breast cancer in the next year, so the risk of breast cancer for inactive women is 500/100,000 = .005. The risk of active women developing breast cancer in the next year is 400/100,000 = .004. Notice that .004 and .005 are conditional probabilities. The probability of developing breast cancer in the next year, given that we are looking at women who are physically active, is .004. The probability of developing breast cancer in the next year, given that we are looking at women who are sedentary, is .005. The relative risk is a ratio of these two risks. We put the risk for those with the risk factor (inactivity) in the numerator, and we put the risk for the women without the risk factor (i.e., the risk for the active women) in the denominator. So the relative risk of an inactive woman developing breast cancer in the next year, compared with the active woman, is .005/.004 = 1.25. (This is an over-simplified example; actual risk of developing breast cancer involves numerous complex factors.)

How do we interpret a relative risk statistic of 1.25? Let's compare it to a number that would indicate there was no impact of inactivity. If inactivity had no impact on the chances of a woman developing breast cancer in the next year, then the risk for inactive women would equal the risk for active women, and the relative risk would be equal to 1.0. A number greater than 1 would indicate that the risk for inactive women is bigger than the risk for active women (i.e., the numerator is bigger than the denominator). For our relative risk of 1.25, we can say that the risk of breast cancer in the next year for inactive women is 25% higher than the risk for active women. A "25% higher risk" may sound extremely alarming. But notice that both risks are low—500 out of every 100,000 inactive women may get breast cancer, compared with 400 out of every 100,000 active women.

A relative risk can be less than 1.0. Consider the positive effect of exercise. We reverse these two groups and treat the inactive women as the comparison group. What is the effect of physical activity on the risk of breast cancer? The relative risk would be .004/.005 = 0.8, meaning that the risk of developing breast cancer in the next year for active women is 20% lower than the risk for inactive women. You may be surprised by this numeric result. Differences in percentages require a special calculation. Suppose we have a sample of children with diabetes who have been doing a poor job of testing their blood glucose levels, and only 20% of them test every day. We conduct an intervention, and the daily testing rate increases to 30%. We might be tempted to say there was a 10% increase. In fact, compared with the original rate of 20%, we have observed a 50% increase over the old rate. That is,

$$\text{Percentage change} = \frac{\text{New \% - Old \%}}{\text{Old \%}} \times 100$$

$$= \frac{30 - 20}{20} \times 100$$

$$= \frac{10}{20} \times 100$$

$$= .5 \times 100$$

$$= 50\%$$

The percentage change, 50%, is a positive increase. Relative to the original rate, the new rate is 50% higher. Now let's change the example but use the same numbers. Suppose we had a sample of children, and 30% of them go home after school and eat junk food. We conduct an intervention to encourage healthy eating, and later we find that only 20% of them go home after school and eat junk food. How much has the rate gone down? We must compare the new rate with the original rate:

$$\text{Percentage change} = \frac{\text{New \% } - \text{Old \%}}{\text{Old \%}} \times 100$$

$$= \frac{20 - 30}{30} \times 100$$

$$= \frac{-10}{30} \times 100$$

$$\approx -.333 \times 100$$

$$\approx -33.3\%$$

The negative percentage change means that relative to the original rate of 30%, junk-food eating after school has gone down by 33.3%. We started with a rate that was higher, so relative to that rate, the 10% difference in the numerator is a smaller percentage change. In the example of breast cancer and activity, we get different results, depending on the reference group for the relative risk. If we assessed the risk of inactivity using active women as the comparison group, we found relative risk = 1.25. When we treated activity as a positive factor and examined the risk of breast cancer in the next year for active women, compared with inactive women, we found relative risk = 0.80. So it is important to pay attention to the basis of comparison when interpreting relative risk.

One other note about disease risk: Medical professionals often cite five-year survival rates in discussing a patient's prognosis. A physician might say that someone diagnosed in the earliest stage of breast cancer has a five-year overall survival chance of 93%. What does this figure mean? A five-year survival rate means that out of all patients with the same diagnosis at least five years ago, 93% of them are alive today. We said "at least five years ago" because this figure of 93% came from the Susan G. Komen Breast Cancer Foundation's website, komen.org, which said the survival rate was for women diagnosed in 2001 and 2002. Those women underwent treatment based on the knowledge of cancer in 2001 or 2002. We cannot know the five-year survival rate for a woman diagnosed today, but we assume it would be at least that high.

SCENARIO 6-C

Larsson and Wolk (2007) combined the results of 16 studies of 21,720 people with non-Hodgkin's lymphoma to assess the risk factor of obesity. They reported that participants who were overweight had a 7% greater risk of non-Hodgkin's lymphoma compared with individuals of normal weight; and participants who were obese had a 20% greater risk than persons of normal weight. 6-11. What values of relative risk would correspond to these two percentages (7% and 20%)?

Suggested Answers

6-11. For participants who were overweight compared with participants who were normal weight, the relative risk would be 1.07. For participants who were obese compared with normal-weight individuals, the relative risk would be 1.20.

Other Statistics Associated With Probability

Many other statistics quantify the likelihood of risk factors. For example, a *hazard ratio* is a complex statistic that involves a nonlinear mathematical calculation, but it is interpreted like a relative risk. A hazard ratio substantially greater than 1.0 indicates a greater risk for people with the risk factor, compared with people who do not have the risk factor.

Another term related to probability and risk that you probably have heard in association with gambling is *odds*. Informally, people talk about the odds of something happening, but it is not the same thing as the probability of something happening. The odds can be computed by taking the probability of something happening and dividing it by the probability of that same thing *not* happening:

$$\text{Odds} = \frac{\text{probability of something happening}}{\text{probability of that same thing not happening}}$$

If the probability of something happening is .3, then the probability of that same thing not happening is $1 - .3 = .7$. So we can define the odds as follows:

$$\text{Odds} = \frac{\text{probability of something happening}}{1 - \text{probability of something happening}}$$

$$= .3 / .7 \approx .429$$

For example, the Komen Foundation (komen.org) says the lifetime probability of a woman getting breast cancer is .12; that is, about 12% of women get breast cancer at some point in their life. Therefore, the probability of *not* getting breast cancer would be .88; in other words, 88% of women do not get breast cancer. We compute the *odds* of breast cancer by taking the probability of getting breast cancer divided by the probability of *not* getting breast cancer:

$$= \frac{.12}{1 - .12}$$

$$= \frac{.12}{.88} \approx .136$$

How do we interpret the odds? Let's use a simple weather example, then return to the odds of breast cancer. When meteorologists say there is a 50% chance of rain, they are saying it rained on half of the days with weather like today's conditions. That means half of the time, there was no rain. If 50% of days had rain and 50% of days did not have rain, the odds would equal $.5/.5 = 1$. In other words, we have an even likelihood of rain versus no rain. If the probability of rain is .8, then 80% of days like today had rain, and 20% of days like today did not have rain. The odds of rain would be $.8/.2 = 4$, meaning the probability of rain today is four times higher than the probability of not having rain today. Returning to the example of .136 as the odds of breast cancer, a woman's lifetime risk of breast cancer is much lower than the probability of *not* getting breast cancer. Although it is good that the odds of breast cancer are not $.5/.5 = 1$, the odds can be difficult for most people to understand.

By the way, the odds and an *odds ratio* are not the same thing. Because we compute the odds by putting two probabilities into a fraction or ratio, it would be easy for someone to mistakenly assume that *odds* meant the same thing as *odds ratio*. But an odds ratio is a fraction that has odds in both the numerator and the denominator. We will come back to the odds ratio in Chapter 14.

What's Next

As we have seen, probability is an important concept in understanding disease risk and the assessment of the accuracy of diagnostic tools. Probability also will be crucial for us to determine whether research results are noteworthy. For example, we may ask whether the relationship between a sample's smoking behavior and its incidence of asthma is *statistically significant,* a phrase that will be explained later and will require the interpretation of probabilities. We have begun laying the foundation for inferential statistics, a process that will continue in Chapter 7.

As we proceed through this book, we quantify the uncertainty and variability in observed research results. Although uncertainty is pervasive in life, what are the chances that you will see this chapter's concepts again in this book? The probability equals 1.

Exercises

SCENARIO 6-A, Continued

We downloaded the data in Figure 6.6 from the Oklahoma State Department of Health, which contributes to an ongoing national survey called the Behavioral Risk Factor Surveillance Survey (BRFSS). Part of the survey asked a large sample of Oklahoma adults in 2009 about their experience (if any) as a smoker and whether they had asthma.

6-12. What kind of research is this? 6-13. What kind of variable is asthma? 6-14. What kind of variable is smoking behavior? 6-15. If we find that smokers are more likely than nonsmokers to have asthma, what can we say about a causal link between smoking behavior and asthma? 6-16. Compute the probability that a randomly chosen respondent never smoked. 6-17. Compute the probability that a randomly chosen respondent did not have asthma. 6-18. Give an example of a conditional probability, using variable names and numbers. 6-19. Compute the probability that a randomly selected respondent never smoked and had asthma. 6-20. Compute the probability that a respondent chosen at random smoked some days or had asthma. 6-21. Compute the probability of a randomly selected respondent having asthma, given that the respondent never smoked. 6-22. Explain the meaning of the probability 242/2199. 6-23. Explain the meaning of the probability 1957/2199. 6-24. Explain the meaning of the probability 1957/7000. 6-25. Explain the meaning of the probability 7306/7769.

SCENARIO 6-D

Rosenstein (2002) reported the results of a survey about U.S. hospital professionals' perceptions of disruptive behavior by physicians. The study defined disruptive physician behavior as "any inappropriate behavior, confrontation, or conflict, ranging from verbal abuse to physical and sexual harassment." Data came from nurses, physicians, and hospital executives at 84 hospitals around the country, ranging from small, rural not-for-profit hospitals to large, urban academic centers. One question asked whether they

Do You Have Asthma?	Smoking Behavior				
	Smokes Daily	Smokes Some Days	Formerly Smoked	Never Smoked	Row Totals
Yes	167	54	242	306	769
No	1,124	366	1,957	3,553	7,000
Column Totals	1,291	420	2,199	3,859	Total N = 7,769

Figure 6.6

Smoking behavior and asthma incidence in Oklahoma, 2009. (Data from "Asthma and cigarette smoking, Behavioral Risk Factor Surveillance system, 2009," by the Oklahoma State Department of Health, 2012, October 2, retrieved from http://www.health.state.ok.us/stats/index.shtml.)

(Continued)

**Nonpunitive Reporting Environment for
Nurses Witnessing Disruptive Behavior?**

Respondent's Role	Yes	No	Row Totals
Physician	136	17	153
Nurse	438	123	561
Executive	22	2	24
Column Totals	596	142	N = 738

Figure 6.7

Hospital professionals' opinions about nonpunitive reporting environment. (Data from "Nurse-physician relationships: Impact on nurse satisfaction and retention," by A. H. Rosenstein, 2002, retrieved from http://journals. lww.com/ajnonline/pages/default.aspx.)

felt their hospital provided a "nonpunitive reporting environment for nurses who witness disruptive behavior." Figure 6.7 contains the results.

6-26. Compute the probability that a randomly selected respondent

a. Was a nurse.
b. Answered yes.
c. Was a physician and answered yes.
d. Answered yes, given that the respondent was a physician.
e. Was a physician or answered yes.
f. Was a nurse and answered yes.
g. Answered yes, given that the respondent was a nurse.
h. Answered yes, given that the respondent was an executive.
i. Was an executive, given that the respondent answered yes.
j. Was an executive, given that the respondent answered no.

SCENARIO 6-E

This chapter drew extensively on the results from App #1 in Wolf et al. (2013), the study of smartphone apps and skin lesions. Now let's look at the results for App #2, which gave an answer of either "melanoma," which the researchers treated as a positive indication of cancer, or "looks good," which the researchers treated as a negative result. The app was able to evaluate 185 of the 188 images. Based on the statistics reported in the study, we extracted the frequencies for App #2 (Figure 6.8).

6-27. How many images could be considered true positives? 6-28. How many images could be considered true negatives? 6-29. How many images were false positives? 6-30. How many images were false negatives? 6-31. Compute the sensitivity of App #2. 6-32. Explain the meaning of this sensitivity. 6-33. Compute the specificity of App #2. 6-34. Explain the meaning of this specificity. 6-35. Compute the positive predictive value of App #2. 6-36.

(Continued)

	Diagnosis from Dermatopathologist		
Result from App #2	Melanoma	Benign	Row Totals
"Melanoma"	40	80	120
"Looks Good"	18	47	65
Column Totals	58	127	N = 185

Figure 6.8

App #2 and actual diagnosis of skin cancer. (Data extrapolated from "Diagnostic inaccuracy of smartphone applications for melanoma detection" by J. A. Wolf et al., 2013, *JAMA Dermatology, 149,* 422–426.)

Explain the meaning of this positive predictive value. 6-37. Compute the negative predictive value of App #2. 6-38. Explain the meaning of this negative predictive value.

SCENARIO 6-F

Dental sealants are thin plastic coatings placed in the pits and grooves of the chewing surface of children's teeth. The Centers for Disease Control and Prevention says sealants prevent tooth decay if they stay on the teeth. Griffin, Gray, Malvitz, and Gooch (2009) wanted to compare the risk of decay in cases where the sealants had come off, compared with teeth that never had been sealed. The researchers' concern was that before a sealant comes off, it may be loose enough to trap bacteria underneath it, leading to a greater chance of tooth decay (i.e., dental caries). Children vary widely in their dental care and eating habits, so the researchers found existing studies in which sealants were placed on the teeth on one side of a child's mouth but not the other. Teeth were studied in pairs; for example, a sealed back molar on the left side was compared with a never-sealed back molar on the right side. For each previous study identified, the researchers calculated the percentage of sealed teeth that lost the sealant and developed decay, as well as the percentage of never-sealed teeth that developed decay. These percentages were used to compute the relative risk of caries for the formerly sealed teeth. The researchers averaged the relative risks across studies, taking into account the number of participants in each study so that a small study would not have the same weight as a large study. One year after the sealants were placed, the average relative risk of caries for the formerly sealed teeth was reported as 0.998, and four years after placement of sealants, the average relative risk was 0.936. 6-39. What kind of research is this? 6-40. Why did the researchers study teeth in pairs? 6-41. Explain the meaning of the result of 0.998. 6-42. Explain the meaning of the result of 0.936. 6-43. Suppose 30% of dental sealants in one study fell off during a five-year period. Compute the odds of a participant in that study losing a sealant.

References

Akobeng, A. K. (2007). Understanding diagnostic tests 1: Sensitivity, specificity and predictive values. *Acta Paediatrica, 96*, 338–341. doi:10.1111/j.1651-2227.2006.00180.x

Aven, T., & Renn, O. (2009). On risk defined as an event where the outcome is uncertain. *Journal of Risk Research, 12*, 1–11. doi:10.1080/1366987080248888

Griffin, S. O., Gray, S. K., Malvitz, D. M., & Gooch, B. F. (2009). Caries risk in formerly sealed teeth. *Journal of the American Dental Association, 140*, 415–423. Retrieved from http://jada.ada.org/content/140/4/415.full.pdf

Larsson, S. C., & Wolk, A. (2007). Obesity and risk of non-Hodgkin's lymphoma: A meta-analysis. *International Journal of Cancer, 121*, 1564–1570. doi:10.1002/ijc.22762

Oklahoma State Department of Health. (2012, October 2). *Asthma and cigarette smoking. Behavioral Risk Factor Surveillance System, 2009*. Retrieved from http://www.health.state.ok.us/stats/index.shtml

Rosenstein, A. H. (2002). Nurse-physician relationships: Impact on nurse satisfaction and retention. *American Journal of Nursing, 102*, 26–34. Retrieved from http://journals.lww.com/ajnonline/pages/default.aspx

Sheeler, R. D., Houston, M. S., Radke, S., Dale, J. C., & Adamson, S. C. (2002). Accuracy of rapid strep testing in patients who have had recent streptococcal pharyngitis. *Journal of the American Board of Family Medicine, 15*, 261–265. Retrieved from http://www.jabfm.org/

Wang, C., Schmid, C. H., Rones, R., Kalish, R., Yinh, J., Goldenberg, D. L., … McAlindon, T. (2010). A randomized trial of tai chi for fibromyalgia. *The New England Journal of Medicine, 363*, 743–754. doi:10.1056/NEJMoa0912611

Wolf, J. A., Moreau, J., Akilov, O., Patton, T., English III, J. C., Ho, J., & Ferris, L. K. (2013). Diagnostic inaccuracy of smartphone applications for melanoma detection. *JAMA Dermatology, 149*, 422–426. Advance online publication. doi: 10.1001/jamadermatol.2013.2382

7

Sampling Distributions and Estimation

Introduction

This book began with an overview of the context for statistics—kinds of research and variables, relationships among variables, generalizations from a sample to a population, and so forth. Generalizing from a sample to a population is not a trivial matter. Any sample, even a representative one that was drawn without bias, will provide data that will differ somewhat from the data that we could have gotten from a different sample. For example, think about randomly sampling 150 young adults who joined the military last year. We want to measure height, and to control for the extraneous variable of gender, we sample only men. We could compute the mean height of these 150 men. Question: If we draw another random sample of 150 other men who joined the military last year, will we get the same mean height?

Probably not. Every time we compute the mean height for a different group of 150 new recruits, we will have a list of different heights, and the mean probably will be different. This is the idea behind *sampling variability*, the tendency for a statistic to vary when computed on different samples from the same population. Each sample is a snapshot of a portion of the population. It is as if each snapshot is taken from a slightly different vantage point and contains different scores from the same population. As another analogy, think of fish living in a lake. Fish are being born and others are dying, and fish are growing all the time, so the population is large and changing constantly. Theoretically there is an average weight of fish—a population mean. We cannot see all of the fish or catch them all at once, but we can take a sample. Any sample of fish could be weighed and a sample mean computed. But a different sample of fish would have a different mean, and we cannot know whether any particular sample mean weight is close to the population mean. But we can use information about sampling variability to judge how much variation in sample means we could expect across repeated samples.

The fish are hidden beneath the surface of the lake, making it hard to know the population or a parameter like the mean weight of the population. Sampling variability is another murky, hard-to-see concept. This chapter covers the most theoretical, abstract concepts in the book. Yet this material is crucial for understanding inferential statistics, the general topic for most of the rest of the book. There will be many times in this chapter that you may wonder if we are restating the same thing you just learned. The answer is yes! Expect some repetition as we try to help you become familiar with the abstract concepts.

Quantifying Variability From Sample to Sample

We just introduced a definition, but its location in the chapter's introduction unfortunately may have downplayed its importance. The definition deserves repeating: *sampling variability* is the tendency for a statistic to vary when computed on different samples from the same population. Sampling variability is different from the variability that we measure by computing a sample variance, which measures the amount of spread in a sample's *scores*. Sampling variability is the variation that we could expect in the many numeric values of a *statistic* that could be computed on repeated samples from the same population. If we compute the mean height on one sample of 150 people, we cannot expect the mean height to be the same if we switch to a different sample, even from the same population. Note that it is the mean height, a statistic, that we expect to have variability.

Consider the mean systolic blood pressure of middle-aged adults. The population of middle-aged adults is extremely diverse in terms of health. Amount of exercise, weight, and chronic health conditions all could affect systolic blood pressure. We cannot obtain the population mean, μ, for the systolic blood pressure of middle-aged adults because we cannot measure the whole population. But we can get a sample mean, M, and we know that our M will differ from other M's of systolic blood pressure from other samples. We also may not know the proportion of people in this population who have diabetes, but we can compute the proportion of sample members who have diabetes. And we know that our proportion will differ from other proportions of people with diabetes that could be computed for other samples. We cannot know the population variance of blood sugar readings for all middle-aged adults. But we can compute an unbiased variance on a sample of blood sugar readings. And we know that our unbiased variance will differ from other unbiased variances of blood sugar readings from other samples. Each of these statistics—sample mean blood pressure, sample proportion of people with diabetes, unbiased variance of blood sugar readings— would be computed on data from only one sample apiece, yet what do we do with this knowledge that different samples would give us different results? In other words, how can researchers account for sampling variability in their statistics when typically they will study only one sample at a time?

There are many approaches to analyzing data and making inferences about sample results. This book will demonstrate a traditional approach to dealing with sampling variability. Some statisticians would argue that other approaches are better. But our approach may be the most common. We will review the kinds of distributions that have been discussed so far in the book, and then we will introduce you to the main topic of this chapter: sampling distributions.

Check Your Understanding

7-1. What is the difference between *sample* variability and *sampling* variability?

Suggested Answers

7-1. The term "sample variability" refers to the amount of spread in scores in a sample; the scores have variation in the sample. The term "sampling variability" refers to the amount of spread in the numeric values of a statistic computed on many samples from the same population; the numeric values of the statistic computed on multiple samples would vary from each other.

Kinds of Distributions

Let's review two kinds of distributions that we have discussed, using some new terminology. Figure 7.1 shows a histogram of the number of times that a pregnant women could lift a 7-kg medicine ball in 1 min at Time 1 (12–14 weeks of pregnancy) in the study by Price, Amini, and Kappeler (2012).

A more general term that we could apply to the graph in Figure 7.1 is *sample distribution*, a group of scores from a sample arranged on a number line. This definition may seem obvious, but it will serve as a contrast to the other distributions to be discussed in this chapter. We can measure characteristics of a sample distribution, such as its center and its spread. These characteristics are measured with statistics like the median or mean (measuring the distribution's middle location on the number line) and standard deviation (measuring the spread of scores).

We can imagine the population of women who are 12–14 weeks pregnant. Further, we can imagine the population of all of their scores for the number of lifts of the medicine ball in 1 min. If we could obtain the population of scores, we could arrange the scores along a number line; the result would be a *population distribution*. We might guess that some women would be able to lift the ball only

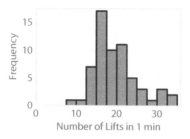

Figure 7.1

Number of medicine-ball lifts at Time 1. This histogram shows the baseline data on the number of times that pregnant women lifted a medicine ball in 1 min. This figure is an example of a sample distribution. (Data from "Exercise in pregnancy: Effect on fitness and obstetric outcomes—a randomized trial," by B. B. Price, S. B. Amini, and K. Kappeler, 2012, *Medicine & Science in Sports & Exercise, 44*, 2263–2269.)

a couple of times and that some women could lift the ball dozens of times, and that most women would be in the middle. But we cannot measure every pregnant woman, so we cannot know for sure what the population distribution would look like. If we could obtain the population of scores, we could figure out its numerical characteristics: parameters. An example of one parameter would be the population mean number of lifts.

Let's compare and contrast what we know about a sample and a population as we build toward explaining a third kind of distribution. Both samples and populations have distributions of scores. We can graph the sample distribution because we can obtain a sample. We rarely graph the population distribution because we usually cannot obtain a population. Both the sample and population have numerical characteristics. The sample's numerical characteristics are statistics, and the population's numerical characteristics are parameters. We can compute statistics because we can get samples, but we often do not know the numeric values of parameters because of the unobtainability of the population of scores. We want to make generalizations from the sample to the population. For example, Price et al. (2012) obtained their sample from a population of sedentary pregnant women, and then after the sample was studied, the researchers wanted to generalize the results back to the population. After all, if they found a safe, beneficial exercise regimen for pregnant women, they would want other women like those in their sample to benefit from the findings. When we make a generalization from the sample to the population, we need a way to account for sampling variability—the fact that different samples would provide different scores and therefore different numeric values for a sample statistic like the mean number of medicine-ball lifts.

How can we know how much variation could be expected across repeated samples from the same population? We could recruit many samples of sedentary pregnant women, compute the mean number of medicine-ball lifts for each sample, and then arrange all those means on a number line. Then we could compare our original sample mean to that arrangement of means and ask, "Is our original sample mean similar to the rest of the sample means?" But to keep performing the same study repeatedly to obtain those repeated sample means is beyond what many researchers are able or willing to do.

Here is the point where we can benefit from the work of mathematical statisticians, who have performed their own research on the statistics themselves and can provide answers about sampling variability. Given certain conditions, statisticians can tell us about the sampling variability for the mean number of medicine-ball lifts across repeated samples—or the sampling variability for any other statistic.

Let's restate what was just said. The mean will vary across different samples. We can imagine creating a distribution of this statistic, with different numeric values of the mean based on different samples. Every statistic has a distribution that could be formed in this way. By studying what would happen with repeated samples drawn on the same population and a statistic computed on every sample, mathematical statisticians can tell us a lot about those many, many values of a statistic. We are describing a *sampling distribution*, which is a distribution of a statistic that could be formed by taking all possible samples of the same size from the same population, computing the same statistic on each sample, then arranging those numeric values of the statistic in a distribution along a number line.

The definition of a sampling distribution is huge. Let's break it down:

- *"A sampling distribution is a distribution of a statistic..."* Which statistic? Any statistic. For example, the median is a statistic with a mathematical operation that can be performed on many different samples of numbers. Instead of the median, we could choose the mean as the statistical operation to be performed repeatedly. Or we could choose the unbiased variance as the statistic to be calculated over and over.
- *"...that could be formed..."* We are not actually going to do this process, except through computer simulations.
- *"...by taking all possible samples of the same size from the same population..."* Again, we can do this process only on a computer with fake numbers. Who has time to draw all possible samples from any large population? But it is theoretically possible to have same-sized samples drawn repeatedly from one population.
- *"...computing the same statistic (e.g., the sample mean) on each sample..."* If we have many same-sized samples and we compute the mean on each sample, we will have a big pile of sample means. If we compute a median on each sample, we will have a big pile of medians. If we compute an unbiased variance on each sample, we will have a big pile of unbiased variances.
- *"...then arranging those numeric values of the statistic in a distribution along a number line."* This big pile of statistics would need to be put in some kind of order, so we could put them in numeric order and create a graph like a histogram. The distribution could consist of an organized pile of sample means on a number line, like a histogram of means. If we computed repeated values of the median, the distribution would consist of all the sample medians. If we computed the unbiased variance on each sample, the distribution would be made up of all the unbiased variances.

A sampling distribution differs from a sample distribution or a population distribution. By definition, a sampling distribution contains values of a *statistic* computed on scores, whereas a sample distribution and a population distribution both contain *scores*. We do not actually have to go through this process of repeated sampling; the mathematical statisticians of the past have rescued us from that work. They are going to tell us what we could expect a distribution of statistics (e.g., many values of the sample mean) would look like—without our having to create that distribution. This is great news! We can measure one sample and take our single sample mean for the number of medicine-ball lifts and we can know something about how much the mean might vary across repeated samples. Later, with more information and knowledge, we might be able to say that our single sample mean is typical in comparison with a hypothesized value for the population mean number of lifts—or we might discover that our single sample mean is very different from some hypothesized value for the population mean.

Check Your Understanding

7-2. Suppose we had a sample size of 68 people and we were measuring the correlation between the number of hours spent watching television and the number of hours spent exercising every week. Theoretically, how could we create a sampling distribution for Pearson's *r*? 7-3. For each of the three kinds of distributions (sample distribution, population distribution, and sampling distribution), identify (a) what would be graphed and (b) whether the distribution is hypothetical or obtainable.

Suggested Answers

7-2. Theoretically, we could take all possible samples of 68 people per sample from whatever population we identified. On each sample we could ask each person about his or her number of hours spent watching TV and the number of exercise hours weekly. Then on every sample we could compute Pearson's r between TV time and exercise hours. After we had amassed a pile of correlations, we could arrange them in a distribution that would be a sampling distribution for Pearson's r. (We encourage you to think through the same question with a different statistic and perhaps to memorize the definition of a sampling distribution as the first step toward understanding the concept.) 7-3. (a) A sample distribution would be a graph of scores in one sample. A population distribution would be a graph of all scores in one population. A sampling distribution would be a graph of the many numeric values of a statistic computed on repeated samples from a population. (b) A sample distribution is obtainable because we can get a sample of scores. A population distribution of scores usually is unobtainable because we cannot measure everyone in most populations. A sampling distribution usually is unobtainable because we cannot take the time to repeatedly draw samples from the same population and compute a statistic on each sample.

Why We Need Sampling Distributions

We have implied why we need sampling distributions, but let's be more explicit about the reasons given so far, using the sample mean, M, as an example of a statistic with a sampling distribution:

- We often want to generalize from our known sample mean, M, to a hypothesized value of the population mean, μ.
- We know that statistics have sampling variability. The values for a sample mean (e.g., mean number of medicine-ball lifts) will be different for the many different samples.
- We want to know how typical our particular sample mean is, compared with all possible sample means we could have computed.
- If we wish to compare our sample mean to a hypothesized population mean, then we need to take into account the statistic's sampling variability.

Let's take this explanation a little further. Price et al. (2012) cared about the women in the sample at 12–14 weeks gestation and their mean number of lifts of the medicine ball, but they also wanted to use the sample mean as an estimate of the unknown population mean. Specifically, the sample mean is called a *point estimate*, because it is a single number or point on the number line being used to estimate the parameter. The researchers might need to quantify how likely it is to obtain a sample mean at least as large as theirs, if the sample came from a population that they hypothesized to have a specific population mean. Suppose their years of research on exercise during pregnancy led them to guess that on average, women who were 12–14 weeks into pregnancy could lift the medicine ball 21 times in 1 min. In other words, they may have hypothesized that the population mean, μ, was 21. Then the researchers recruited their sample, measured 61 women with the medicine ball, and computed a mean number of lifts = 19.41 (data available via http://desheastats.com). We might ask, "If this sample comes from a population with a mean = 21, how likely is it to get a sample mean at least this different from 21?" The previous sentence packs a lot of information:

- *"If this sample comes from a population with a mean = 21"* describes a *hypothesis*, or a testable guess, about the population mean. We usually cannot know exactly the parameters of the population from which our samples are drawn. Researchers can hypothesize about a parameter like the population mean.
- *"... how likely is it to get a sample mean at least this different from 21?"* This part of the sentence is describing a probability. The distribution of repeated sample means can tell us something about the frequency of finding means within certain ranges of the number line. We can see there is a gap between 19.41 and the hypothesized population mean = 21—but is

that a small difference or a large difference? What proportion of the "all possible samples of the same size" would produce sample means that far or even farther from 21? We will consider this kind of question at greater length later in the text.

Sampling distributions, therefore, will tell us something about how frequently we would find sample means that are at least as far from the hypothesized population mean as our specific sample mean. If we would find a large proportion of our repeated sample means being the same distance or even farther from the hypothesized μ, then our sample mean may be typical for a population that has a mean = 21. But if we find a small proportion of our repeated sample means being the same distance or even farther from 21, then our sample mean may differ from the hypothesized μ in a noteworthy way.

Check Your Understanding

7-4. What will sampling distributions give us the ability to do?

Suggested Answers

7-4. Sampling distributions will give us the ability to compute probabilities, which we will use to test hypotheses. Then we will be able to generalize from a sample to a population. Without sampling distributions, we would be unable to know something about typical values across repeated samples; that is, we could not quantify sampling variability.

Comparing Three Distributions: What We Know So Far

Figure 7.2 summarizes what we know and do not know at this point about the three kinds of distributions for the example of the number of lifts of a medicine ball in 1 min when the women were 12–14 weeks pregnant.

As shown on the left in Figure 7.2, we have a sample distribution of number of lifts, and we have a single sample mean computed on the 61 lift scores for the first occasion of measurement in the pregnancy/exercise data set. We could obtain any number of other statistics measuring characteristics of the sample distribution—its variability, its skewness, and so on. The center of Figure 7.2 shows only one number that is part of the sampling distribution of the mean: our sample mean = 19.41. Right now we do not know the shape or average of the sampling distribution of the mean, how spread out it is, or whether 19.41 is actually in the middle or far from the middle of the sampling distribution. On the right side of Figure 7.2, we have a hypothesized population mean, $\mu = 21$. Most likely, we will never know the shape of the population distribution of scores because we usually cannot obtain the entire population.

Sometimes we lose sight of the big picture when we are dealing with a topic as theoretical as this one. Let's summarize what we have learned and try to regain

Figure 7.2

Sample distribution, sampling distribution of the mean, population distribution. The sample distribution on the left represents most of the information that we know at this point in the text. We have one sample mean, $M = 19.41$ lifts of the medicine ball. We could compute other statistics on the sample. The middle of this figure shows a number line with the sample mean on it. Theoretically, we could take all possible samples like the one shown on the left, compute means on every sample, and arrange those means into a sampling distribution of M; 19.41 would be one of those sample means. On the right, we see a number line with 21 on it, the hypothesized value of the population mean. The distribution of scores that would be averaged to obtain the population mean is called the population distribution; the distribution is missing because we rarely know the shape or variance of a population of scores.

the view of the big picture. We are trying to generalize from the sample to the population. We pull a sample from the population and study the data by creating graphs and computing statistics. Then we want to look back at the population and make generalizations about the population. Yet we know that different samples will give us different data, leading to different values for statistics like the sample mean. We need to quantify the sampling variability by use of a sampling distribution, which we will use like a bridge to travel from the sample back to the population in the process of making inferences. Next we will learn more details about the sampling distribution of M.

Central Limit Theorem

Mathematical statisticians have studied the sample mean and can tell us quite a bit about its sampling distribution. What we are covering next is specific to the sample mean and does not apply to other statistics (median, standard deviation, etc.). Under certain conditions we can know the shape of the sampling distribution of the sample mean, as well as its average value and typical spread. A gift from mathematical statisticians, the *Central Limit Theorem* says that with a large enough sample size and independent observations, the sample mean will have a sampling distribution that follows a normal distribution. Further, we could average together the pile of sample means from repeated samples—the mean of the means, if you will. The Central Limit Theorem says that the mean of the means will equal the mean of the population from which we sampled. We also could take the pile of sample means and compute a measure of spread. But we don't have to—the Central Limit Theorem says the variance of the sampling distribution of M will equal the population variance divided by the sample size, or σ^2/N.

Why is the Central Limit Theorem a gift? We want to cross the bridge from the sample to the population so that we can generalize our sample results to the population. The bridge needs to quantify how much variation we might expect across repeated samples. But we do not know what the bridge looks like or where it is located on the number line. The sampling distribution is that bridge. We want to avoid having to create a sampling distribution—but we need it to learn something about the likelihood of sample means like ours. The mathematical statisticians arrived and saved us from having to create a sampling distribution! They said, "There is a missing piece of your puzzle, and you don't know the shape or numerical characteristics of that puzzle piece. Well, you don't need to create or build that puzzle piece (sampling distribution of the mean). We can tell you what it looks like: a particular theoretical distribution defined by a mathematical formula. Use it instead of trying to create or build the sampling distribution of the mean!"

So the Central Limit Theorem gives us three crucial pieces of information about the sampling distribution of M: its shape, its mean, and its variance. The shape of the sampling distribution of M is normal. As you know, there are infinitely many normal distributions, each located in a different place on the number line with different amounts of spread. So we need to know the location and spread for this normal distribution—that is, its mean and variance. Luckily, the Central Limit Theorem tells us about the mean and variance of the sampling distribution of M. Its mean (average of the repeated sample means) equals the mean of whatever population would have been repeatedly sampled. For our pregnant women lifting the medicine ball, we guessed that the population mean was 21 lifts. If that is the population from which we actually sampled, then the sampling distribution for the sample mean would be centered on 21. As for the spread, the Central Limit Theorem tells us that the variance of the mean's sampling distribution is equal to σ^2/N. For the pregnancy/exercise example, we have not ventured a guess about a possible numeric value for σ^2 at this point, so for now we will not compute the variance of the sampling distribution for the sample mean number of lifts.

Let's look again at the idea of a sampling distribution being a bridge between a sample and a population, using what we learned from the Central Limit Theorem. The theorem says the mean's sampling distribution is normal. Every normal distribution is defined by two bits of information: a mean and a variance. Prior research may give us an idea of the population mean, which will be the mean of M's sampling distribution. If we have some idea of the numeric value of the population variance, then we can divide σ^2 by our sample size and get a measure of how spread out the repeated sample means would be in the sampling distribution. With independent scores and a large enough N, the missing bridge between the sample and the population will materialize and appear as a normal distribution, if we know the population mean and population variance.

In our example of pregnant women, do we have independent scores? Technically, mathematical statisticians expect random sampling to produce the independent observations, but applied researchers may cough, look at the ground, and then whisper to each other, "We don't have a random sample, but I think our convenience sample is good enough, don't you? After all, we measured each person without influence from anyone else." Some statisticians would disagree with those researchers' judgment about their convenience sample, but the fact is that people whisper about such limitations. Do we have a sufficiently large sample size, $N = 61$? Yes, we do. Technically, the Central Limit Theorem talks about what happens to the sampling distribution of M as the sample size approaches infinity. The mathematical researchers have tried to figure out what happens to the sampling distribution of the sample mean in reasonably sized samples. Research has been conducted using computer simulations, checking the shape and other characteristics of M's sampling distribution under various conditions. The research shows that as long as the population does not have large clumps of outliers (e.g., a subgroup of pregnant women who are Olympic athletes able to lift the medicine ball 100 times in a minute), the shape of the sampling distribution of the sample mean generally approximates a normal distribution if the sample size is 25 or larger. If the sample of 61 pregnant women came from a population with a mean number of lifts equal to $\mu = 21$ and a population standard deviation equal to $\sigma = 7.81$ (a number chosen solely for convenience in this example), then the sampling distribution of the sample mean may look somewhat like Figure 7.3. This distribution could be added to the middle section of Figure 7.2. Note that we centered this sampling distribution of M at 21, because 21 is our hypothesized value of the population mean. And we know that the sampling distribution of M has an average that equals the population mean, μ.

We already know that our single sample mean, 19.41, is slightly lower than the hypothesized population mean, 21—but how much lower? Later in this chapter, we will introduce a statistic that will tell us whether 19.41 is close to or markedly far away from the hypothesized $\mu = 21$.

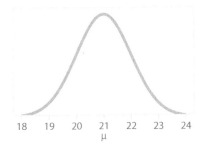

18   19   20   21   22   23   24
                      μ

Figure 7.3

Sampling distribution of the mean number of lifts. Suppose we repeatedly drew samples of $N = 61$ women from a population with a mean number of lifts = 21, and the population had a standard deviation of 7.81 (a number chosen for our convenience). Each sample's mean number of lifts could be computed, and the means could be arranged into a distribution. The Central Limit Theorem says it will look like this figure.

SCENARIO 7-A

7-5. Why do we need the Central Limit Theorem? McDowell, Fryar, Ogden, and Flegal (2008) reported on the heights of thousands of American men in their 30s. Suppose these men have a population mean height of 69.4 in. and that the population variance is 9.6. We obtain a random sample of 55 men and measure their heights. 7-6. If we are interested in the sample mean height, what would the Central Limit Theorem tell us about our mean's sampling distribution?

Suggested Answers

7-5. The Central Limit Theorem tells us about the sampling distribution for the sample mean, M. It tells us about this sampling distribution's shape (normal, if the population of heights does not have large clumps of outliers—some very short men or very tall men), average ($\mu = 69.4$), and spread (σ^2/N). This information will help us in the process of generalizing from the sample to the population. 7-6. The Central Limit Theorem would say that our sampling distribution for the mean height of 55 men would be approximately normal in shape because the sample size is large enough. Further, the theorem says that all the repeated sample means from the same population would average together to equal the population mean, 69.4 in., and the variance of all the repeated sample means would equal $\sigma^2/N = 9.6/55 = 0.17454545 \approx 0.17$.

Notice that the Central Limit Theorem does not say anything about the required shape of the population distribution. With a large enough sample size, the shape of the population distribution is mostly irrelevant. To get an intuitive glimpse at how the Central Limit Theorem can lead us to a normal distribution to replace the unobtainable sampling distribution of the mean, follow us through this contrived example. Consider a very small sample size, $N = 2$, and a population of 12 numbers. Here is our entire population:

$$1, 1, 2, 2, 3, 3, 4, 4, 5, 5, 6, 6$$

What does this population look like? Figure 7.4 shows this population distribution. This small population has 12 numbers, with six possible values that each occur twice. The shape of the distribution can be described as uniform or rectangular. We could draw a random sample of two numbers, $N = 2$, from this population and average them together. Then we could repeat the process for all possible samples of $N = 2$. Then we could create a very small sampling distribution of the mean. What will the sampling distribution of the mean look like if we draw all possible samples of two observations each from this population?

Table 7.1 contains the sums that are possible when drawing two numbers from this population. The second column contains the number of ways that each sum (i.e., numerator of M) could be found.

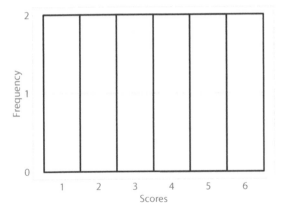

Figure 7.4

Population of 12 scores (two occurrences of six numbers). Imagine a limited population. Here are all the scores: 1, 1, 2, 2, 3, 3, 4, 4, 5, 5, 6, 6. This figure shows a histogram of the population distribution.

Table 7.1 All Possible Sums of Two Scores Taken from Our Population of 12 Numbers

Sum	Number of Ways of Obtaining the Sum
2	1
3	2
4	3
5	4
6	5
7	6
8	5
9	4
10	3
11	2
12	1

There is only one way to get a sum of 2 from our population: by randomly selecting both scores of 1 from the population. There are two ways of getting a sum of 3: by randomly selecting a 1 for the first score and a 2 for the second score—or drawing a 2 for the first score and a 1 for the second score. To get a sum of 4, there are three ways: (a) a 1 for the first score and a 3 for the second score, (b) a 3 for the first score and a 1 for the second score, and (c) both occurrences of 2. And so forth. To compute the mean, we take each sum and divide by the sample size, $N = 2$. So the first mean is $M = 1$, which can be found only one way: by sampling both scores of 1 and dividing their sum by 2. If the sum is 3, then the mean would be $3/2 = 1.5$, and this mean could occur twice. If the sum is 4, then the mean would be $4/2 = 2$, and this mean could occur three times. Table 7.2 shows all the means and the number of ways each one could occur.

So we would have a pile of 36 means ranging from 1 to 6. The extreme means equaled 1.0 and 6.0, and there was only one way to get each of those means. But there are six ways to get a mean of 3.5, with the following samples: (1 and 6), (6 and 1), (2 and 5), (5 and 2), (3 and 4), and (4 and 3). Remembering that we sampled from a uniform distribution, let's look at the sampling distribution of the sample mean based on all possible samples of $N = 2$ from our limited population (Figure 7.5).

It did not matter that we were sampling from a rectangular distribution. Even with this tiny sample size, the mean's distribution is heading in the direction

Table 7.2 All Possible Means for Samples of Two
Scores from Our Population

Mean	Number of Ways of Obtaining the Mean
1.0	1
1.5	2
2.0	3
2.5	4
3.0	5
3.5	6
4.0	5
4.5	4
5.0	3
5.5	2
6.0	1

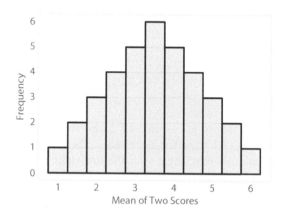

Figure 7.5

Sampling distribution of the mean for all possible samples of two scores each from our limited population. There is only one way to get a mean = 1 when we sample from the limited population of 12 scores (1, 1, 2, 2, 3, 3, 4, 4, 5, 5, 6, 6). There is only one way to get a mean = 6. But there are many ways of getting a mean = 3.5. Even with $N = 2$ scores in each sample and only 12 scores in the population, the Central Limit Theorem already is trying to push the sampling distribution in the direction of a normal curve.

7. Sampling Distributions and Estimation

of a normal distribution. What was the population mean? It was $\mu = 3.5$. You can verify by averaging the 12 numbers in the population. The Central Limit Theorem states that the mean of this sampling distribution (i.e., the average of the 36 possible averages) would be 3.5—and it is. As the sample size goes up, there are many more ways to combine sampled numbers to get all the possible means, and the histogram smoothes out and becomes less chunky, eventually approximating a normal curve. (In case you did not recognize the population, it is based on outcomes from rolling two dice.)

The best way to become more familiar with sampling distributions is to work through some online demonstrations. One of the best demonstrations, which can be found by searching for "sampling distribution demonstration," is on a website affiliated with Rice University (http://tinyurl.com/mwdmw52). The demonstration shows three distributions. The first distribution shows a population. You choose a normal, uniform, skewed, or custom population. To draw a custom distribution, you click on different places above the number line to draw a population. When you click the "Animated" button next to the distribution labeled Sample Data, scores will drop out of the population and form a sample distribution, and the mean of the sample will drop to the third number line. When you click Animated again, another sample drops from the population onto the sample distribution, and another mean drops to the third number line. In this way you can build a sampling distribution of the mean. As a shortcut, you can click the "5" button to get five sample means, or 10,000 to get that many sample means, then examine the shape and mean of the sampling distribution. Look at the mean of the sampling distribution and compare it with the mean of the population from which it came; can you confirm the Central Limit Theorem's statement? You can change the sample size to see how that affects the sampling distribution of the mean, and then you can change to a different statistic's sampling distribution. We highly recommend spending some time with this kind of demonstration.

Unbiased Estimators

Statistics have different characteristics as point estimates of population parameters. We want our point estimates to be good—but how do we define "good"? There are many characteristics that could be considered good, and we would like to mention one good characteristic. As we said earlier, the Central Limit Theorem tells us about sampling distributions of only one statistic: the sample mean. Sampling distributions can help us explain a term that we have mentioned previously. As you know, the sample mean estimates the population mean. Further, the sample mean's sampling distribution has a mean (the average of the averages) that equals the population mean—the parameter that the sample mean estimates. Any particular sample mean probably will not equal the population mean, but *on average*, the sample mean hits the target: the average of the averages equals the parameter being estimated. This fact is describing a characteristic of the mean as an estimator: it is *unbiased*. A statistic is unbiased if the mean of its sampling distribution equals the parameter estimated by the statistic.

We have seen the term *unbiased*, but without a specific definition. The unbiased variance is a statistic used as a point estimate of the population variance. If this statistic is unbiased, then the mean of its sampling distribution should be the population variance. And that is the case—if we took all possible samples of the same size from the same population and computed the unbiased variance on all of the samples, we would end up with a pile of unbiased variances. If we averaged them together, we would get the population variance, which is what the unbiased variance statistic estimates. Not all statistics are unbiased; the standard deviation, *SD*, which is the square root of the unbiased variance, is used to estimate the population standard deviation, σ, but *SD* is a biased statistic. If we took all possible samples of the same size from the same population and computed *SD* on every sample, we would have a pile of standard deviations. If we averaged them together, the average of the values of *SD* would *not* equal the population standard deviation. With very large sample sizes, however, the average of all possible standard deviations would be very close to the population standard deviation, so we do not worry about the fact that it is a biased statistic.

In case you were uneasy about the idea of a point estimate having variability across samples, the concept of unbiased estimation should provide some reassurance. Even though our particular sample value for a point estimate probably will not equal the parameter being estimated, we can take comfort in knowing that if the statistic is unbiased, the average of its sampling distribution will equal the parameter.

Check Your Understanding

7-7. Suppose we want to compute the proportion of patients in a sample who have diabetes, and we are using this statistic to estimate the proportion of patients in the population with diabetes. If we had unlimited resources and time, how would we demonstrate that the sample proportion is an unbiased point estimate of the population proportion?

Suggested Answers

7-7. First, we would take all possible samples of our chosen N from our identified population. On each sample we would compute the sample proportion of patients with diabetes. For example, suppose our chosen sample size was 80, and the first sample contained 12 patients with diabetes. The first sample proportion would be 12/80 = 0.15. We would repeat this procedure of computing the proportion for all possible samples of N = 80. Once we had the pile of sample proportions, we would average them. Assuming we already know the proportion of our population with diabetes, we would determine whether the average of our pile of sample proportions equaled the population proportion; if so, then we would have demonstrated that the sample proportion is an unbiased point estimate of the population proportion.

Standardizing the Sample Mean

We have come through a lot of theoretical material so far in this chapter. We talked about sampling variability as an important fact about statistics that must be kept in mind whenever we want to generalize from a sample to a population. We introduced sampling distributions, which quantify the sampling variability in statistics. We celebrated the Central Limit Theorem for rescuing us from the trouble of creating a sampling distribution for the sample mean. We talked about statistics as point estimates of parameters, such as the sample mean as a point estimate of the population mean. And we introduced the notion of sampling distributions being used to compute probabilities to assist the process of making inferences from a sample to a population. We will continue that train of thought.

We first saw normal distributions in Chapter 4, when we talked about z scores. You may recall our visit to fantasy land, where we pretended to know the population mean and standard deviation for the Pittsburgh Sleep Quality Index (PSQI), where a higher score indicated more trouble sleeping. We pretended that for a population of healthy adults, the population mean, μ, was 6 and the population standard deviation, σ, equaled 3. Then we asked how a patient with fibromyalgia who had a PSQI score = 17 compared with the population mean. From there we computed a z score for this patient:

$$z \text{ score} = \frac{\text{something minus its mean}}{\text{its standard deviation}}$$

$$= \frac{17-6}{3} \approx 3.67$$

But we did not know the shape of the population, and the developer of the PSQI says he is sure that the population is not normal. So we could not compute the proportion of the population with PSQI scores that were lower than our patient's score of 17. We could have computed that proportion if the population were normal because transforming PSQI scores into z scores would not change the shape of the distribution. Because a set of z scores has a mean = 0 and a standard deviation = 1, we could have used a standard normal distribution (Table A.1 in the back of the book) to find the proportion of the population with PSQI scores lower than our patient's score of 17. But we couldn't because the population wasn't normal.

In this chapter, however, we are talking about statistics, not individual scores. If we had a sample of PSQI scores, we might ask whether the sample mean for sleep quality differed from some value of a population mean. Luckily, we do have a sample of PSQI scores, provided to us by Wang et al. (2010). As discussed in Chapter 4, this study examined the effect of tai chi on patients with fibromyalgia, and the researchers measured sleep quality as one outcome variable. Before the intervention began, sleep quality (PSQI) scores were collected on all participants ($N = 66$). We might expect that the patients with fibromyalgia would suffer from sleep problems because of the pain associated with their condition. In other words, perhaps these patients differ from a population of healthy patients in terms of their mean PSQI. Buysse et al. (2008) studied healthy patients and measured their PSQI; we will springboard from that study back into fantasy land, where we find that healthy adults have a population mean PSQI = 6 and a population standard deviation = 3. The sample mean of the baseline PSQI scores for these 66 patients with fibromyalgia was 13.7 (which you may verify by downloading the data via http://desheastats.com). Taking into account sampling variability, how does this sample mean compare with the hypothesized population mean PSQI = 6?

Before we show you how we might answer this question, let's address a possible concern. It may bother you that we are talking about making a comparison between a sample of people with a chronic condition and a healthy population. But who knows? Maybe our sample is similar to the healthy population, at least in terms of sleep quality. If our sample is similar to that population, then the sample mean should be close to the population mean. "Close" is a relative term, however. We need a standardized way of measuring the distance between the sample mean and the population.

To compare the sample mean to the population mean, we are returning to the concept of a z score: (something minus its mean) divided by its standard deviation. Let's look at how we can apply this concept in a new way:

- Instead of "something" referring to a score, "something" could refer to the sample mean, M.
- "*its* mean" would be the mean of M's sampling distribution—the average of all possible sample means for this sample size and population.
- "*its* standard deviation" would be the standard deviation of M's sampling distribution.

We know from the Central Limit Theorem that the "mean of the means" equals the population mean, μ. Further, the theorem also says the variance of M's sampling distribution would be the population variance divided by the sample size, or σ^2/N. Remember from Chapter 2 that we can get from a variance to a standard deviation easily: we take the square root of the variance. So "its standard deviation" would be the square root of σ^2/N.

What we are describing is not a z "score," but a z *test statistic*: the difference between the sample mean and the population mean, divided by the square root of σ^2/N. The gap between M and μ is computed, then we divide by a kind of standard deviation, which serves as a standardized measure of distance between the sample mean and population mean. The denominator of the z test statistic has a special name. The standard deviation of M's sampling distribution is called the *standard error of the mean*. It is a measure of spread in the sampling distribution of the sample mean. This is fantastic! We can use the standard error of the mean as a kind of yardstick for saying something about the distance between the sample mean and a hypothesized value of the population mean.

But wait—is it a problem that we do not know how the sleep quality scores are distributed in the population? Chapter 4 said that if we take a group of scores and convert them to z scores, the z scores will have a distribution that looks the same as the original scores. But the z test statistic is not concerned with the relative location of an individual score; it is looking at the position of the sample mean, relative to the population mean. And the Central Limit Theorem said that with a large enough sample size (ours is $N = 66$), the sampling distribution of the sample mean will approximate a normal curve, regardless of the shape of the population of scores. The news keeps getting better, doesn't it? We can compute a z test statistic for the sample mean for Wang et al.'s 66 patients with fibromyalgia and consider whether it is close to a hypothesized population mean for healthy people's PSQI scores. And the new z test statistic's distribution will be normal, the same shape as the distribution of our M.

We will wait until Chapter 8 to complete the decision-making process about whether we think people with fibromyalgia probably are similar to or different from healthy people in terms of sleep quality. But we can complete the computation of the z test statistic, sometimes called simply the z test. (Please note that this is not the only inferential statistic that is called z. Unfortunately for students, statisticians sometimes reuse the same symbols for different statistics.) The sample mean for sleep quality is 13.7, where a higher number means more sleep problems, and the hypothesized population mean is 6. So the numerator of the z test ("something minus its mean") is $M - \mu = 13.7 - 6 = 7.7$. The denominator of the z test ("its standard deviation") is the standard error of the mean, or the square root of σ^2/N. We can rewrite this denominator as follows:

$$\text{Standard error of the mean} = \sqrt{\frac{\sigma^2}{N}}$$

$$= \frac{\sqrt{\sigma^2}}{\sqrt{N}}$$

$$= \frac{\sigma}{\sqrt{N}}$$

When we visited fantasy land in Chapter 4, we said the PSQI scores for the population of healthy people had a standard deviation $= \sigma = 3$. With a sample size of $N = 66$, this means the denominator of the z test statistic is

$$\text{Standard error of the mean} = \frac{3}{\sqrt{66}}$$

$$= \frac{3}{8.1240384}$$

$$= 0.3692745$$

We are not going to round that denominator because we are in the middle of computing a statistic. Dividing the numerator by the denominator completes the calculation of the z test statistic:

$$z \text{ test statistic} = \frac{7.7}{0.3692745}$$

$$= 20.851699$$

$$\approx 20.85$$

How do we interpret a z test statistic $= 20.85$? This result tells us that more than 20 standard errors of the mean (our standardized yardstick for distance) will fit in the gap between the sample mean and the population mean. If you tried to look up a value of z greater than 4 in Table A.1 in the back of the book, you will not find it; this distance between M and μ is huge. So do you believe that in terms of sleep quality, our sample of people with fibromyalgia came from a population of healthy people? This kind of generalization from the sample to the hypothesized population will be completed in Chapter 8. But we could not have gotten this far without the Central Limit Theorem, which rescued us from having to create a sampling distribution of the mean and allowed us to use our hypothesized population mean and population standard deviation to compute the z test statistic. In Chapter 8, we will talk about how to compare our z test statistic to certain noteworthy values of z from Table A.1 for the purpose of hypothesis testing.

Check Your Understanding

SCENARIO 7-B

Suppose we have been reading various reports about the birth weight of full-term babies in affluent Western cultures like the United States. We speculate that the population mean birth weight for these full-term infants is about 3,400 g (almost 7.5 lb) and that the population standard deviation is about 375 g (about 13 oz). Returning to the Price et al. (2012) study of pregnancy and exercise, we want to compare the mean birth weight of babies whose mothers exercised during pregnancy (data available via http://desheastats.com). The researchers reported that all but one mother in the active group ($N = 31$) delivered a full-term baby. 7-8. To compare the sample mean, $M = 3,329.71$ g, with our best guess of the population mean birth weight for full-term babies, compute the z test statistic. Explain the meaning of the number you computed. 7-9. Suppose we realized we should not have included the weight of the preterm baby, so we recompute the mean without it. Now we have $M = 3,376.43$ g and $N = 30$. Recalculate the z test statistic and explain its meaning. 7-10. Based on what the Central Limit Theorem tells us, does it matter whether birth weight is normally distributed in the population?

(*Continued*)

Suggested Answers

7-8. *The numerator of the z test statistic is* $M - \mu = 3329.71 - 3400 = -70.29$. *The denominator of the z test statistic is* $\dfrac{\sigma}{\sqrt{N}} = \dfrac{375}{\sqrt{31}} = \dfrac{375}{5.5677644} = 67.351988$. *By dividing the numerator by the denominator, we get* $-70.29/67.351988 = -1.0436218 \approx -1.04$. *Our sample mean is slightly more than one standard error below the assumed population mean.* 7-9. *The numerator of this second z test is* $M - \mu = 3376.43 - 3400 = -23.57$. *The denominator of this second z test is* $\dfrac{\sigma}{\sqrt{N}} = \dfrac{375}{\sqrt{30}} = \dfrac{375}{5.4772256} = 68.46532$. *So* $z = -23.57/68.46532 = -0.3442619 \approx -0.34$. *The revised sample mean, computed only on the full-term babies' birth weights, is about one-third of a standard error below the assumed population mean for full-term babies in this culture.* 7-10. *The Central Limit Theorem says that if our sample size is large enough, the mean's sampling distribution will approximate a normal distribution with an average of all possible sample means* $= \mu$ *and a standard deviation* $= \dfrac{\sigma}{\sqrt{N}}$. *The Central Limit Theorem does not require a specific shape for the population distribution of birth weights (although some studies indicate it is approximately normal). Our sample size was 31 for the first z test statistic and 30 for the second z test statistic, and both of these sample sizes were large enough to ensure that the sampling distribution for each mean would be approximately normal, as long as we were sampling from a population without large clumps of outliers (e.g., a number of low-birth weight babies or high-birth weight babies).*

A point estimate like the sample mean is not the only way to estimate a population mean. Researchers also sometimes use another way of estimating parameters, which we will discuss next.

Interval Estimation

Sometimes analogies help us to explain statistical concepts. We would like to start this section of the chapter with an analogy. Both of your authors live in Oklahoma, one of the most common locations in the world for tornadoes. If you asked us, "Do you think a tornado will touch down in Oklahoma on May 18 next year?" we probably would not feel confident in making such a statement. But if you asked us, "Do you think a tornado will touch down in Oklahoma between April 27 and June 8 next year?" we would feel quite confident about saying yes. We know from repeatedly observing springtime weather in this state, tornadoes are more likely to occur in spring and early summer, compared with other times of year.

Specifying the date of May 18 is analogous to a point estimate, like a single value of a statistic being used as an estimate of a parameter. The date range of April 27–June 8 is analogous to an *interval estimate*, a pair of numbers that contain a range of values that is more likely to contain the true value of the parameter being estimated. We may have little faith that a single number like the sample mean is a good, accurate estimate of a population mean because the statistic contains sampling variability. But we can perform *interval estimation*, an approach that quantifies the sampling variability by specifying a range of values in the estimation of a parameter. We may be wrong in our assertion about the dates from April 27 to June 8, but based on what we know about spring weather in Oklahoma, we know there were tornadoes between those dates over many years.

Interval estimation can be performed for most descriptive statistics. Let's turn to two more analogies: darts and horseshoes. In a game of darts, the center of the board is called the bull's-eye. It is a fixed spot that is analogous to a parameter. A player throws a dart toward the bull's-eye. The dart is like a statistic; if we are estimating the population mean, the dart would be a sample mean. If we aim at the bull's-eye every time and throw many darts, they will not all land in the same spot; there will be variability. The spread in the darts is analogous to the spread we could expect in the sample mean when computed on many different samples. So throwing a dart at a dart board's bull's-eye is like computing a point estimate of a parameter.

Now let's consider a game of horseshoes, in which the fixed spot is a stake driven in the ground (that's the parameter). When we toss a horseshoe at the stake, there

Tornado, May 31, 2013. This photo by Christopher Morrow was taken from the 24th floor of the Oklahoma Tower in downtown Oklahoma City. (Used with permission.)

7. Sampling Distributions and Estimation

is a gap between the two points on the end of the horseshoe, and we are trying to encircle the stake or get the horseshoe as close to the stake as possible. Will any particular horseshoe hit the mark? Maybe—or maybe not. But at least we are not trying to hit a target with one single point like the tip of a dart. Tossing a horseshoe at a stake is like computing an interval estimate of a parameter. We could draw repeated samples from a population, compute a statistic on each sample, and then compute an interval estimate based on each statistic. Some of the interval estimates will contain the parameter being estimated, and some will not, just as some horseshoes will encircle the stake and some will not (and for some years, the period of April 27–June 8 will have tornadoes in central Oklahoma, whereas for a few other years, there won't be tornadoes). We interpret an interval estimate based on the percentage of times across repeated samples that we would expect to capture the parameter, like the percentage of horseshoes that resulted in a point for the player.

As we said, one toss of a horseshoe may or may not result in a point for the player. Suppose a player scored on 95% of his tosses. We would have 95% confidence across time in this player's ability to score. Suppose we are computing the sample mean height of 64 American men. We could compute an interval estimate based on our sample mean. Yet we know that because of sampling variability, a different sample of 64 men probably will produce a different sample mean— with a different associated interval estimate. But we could count how often the intervals bracket a parameter. When we specify a percentage of intervals across repeated samples that successfully capture the parameter being estimated, the interval estimate is called a *confidence interval*. A 95% confidence interval is an interval estimate that could be expected to contain the true value of the parameter for 95% of the repeated samples. As we saw in sampling distributions, we actually do not perform the repeated samples. We are relying on research by math experts to inform our understanding of interval estimation.

Our analogies of darts and horseshoes break down in two ways. First, the researcher computing an interval estimate often cannot see the bull's-eye on the dartboard or the stake in the ground; that is, the researcher rarely can obtain populations and must rely on educated guesses about the numeric values of parameters. Second, researchers rarely can rely on a known value for a parameter that measures spread in the population, so usually they must use a statistic measuring spread to compute a confidence interval. As a result, different samples would have different widths of intervals; that would be like different widths of horseshoes being thrown at the stake.

Let's take the comparison of point estimation and interval estimation a bit further. We computed a single sample mean for the number of times that the women at 12–14 weeks of pregnancy could lift the medicine ball in 1 min (Price et al., 2012). We know that across repeated samples of the same size from the same population, we would get different sample means for the number of lifts, and these means could be arranged in a sampling distribution. When we compute a confidence interval based on our single sample mean, we must recognize that *different* confidence intervals could be computed, one for every sample mean that we could draw from the same population. If we performed the repeated sampling required to create a sampling distribution for *M*, we could end up with a pile of sample means. A slightly different confidence interval also could be computed using each sample of data. So we could end up with a pile of confidence intervals. Each sample mean is trying to estimate the population mean, and some could be expected to come closer than others. Similarly, all of those possible confidence intervals would be interval estimates of μ. Some of those intervals would contain the population mean, and others would not. A 95% confidence interval would mean that we could expect 95% of those confidence intervals based on the repeated sample means to bracket the true mean of the population from which we drew the samples.

To help you understand the meaning of confidence intervals, we recommend searching online for "confidence interval applet." Rice University hosts a demonstration showing repeated confidence intervals to estimate μ, and Utah State University has a demonstration of confidence intervals estimating a population proportion. These demonstrations will illustrate how the percentage of confidence comes from repeated sampling and the rate at which the repeated confidence intervals capture the true value of the parameter being estimated.

Check Your Understanding

7-11. Explain the meaning of a 90% confidence interval for estimating a population mean. 7-12. What percentage of 90% confidence intervals could we expect to fail to bracket the true population mean?

Suggested Answers

7-11. A 90% confidence interval can be interpreted as the percentage of intervals like ours that would contain the true mean of the population being sampled. 7-12. We would expect 10% of confidence intervals like ours to fail to contain the true population mean. Those intervals would be like the horseshoes that did not encircle the stake.

Calculating a Confidence Interval Estimate of μ

For almost every point estimate (descriptive statistic), there is a corresponding confidence interval estimating the same parameter as the point estimate. Confidence intervals also can be computed in conjunction with many inferential statistics. That means there are many ways of computing confidence intervals. In this section, we are going to demonstrate one way of computing a confidence interval, relying on hypothesized values of a population mean and population standard deviation. We have been avoiding formulas as much as possible in this book, but we think it is important to show some math to help you to understand a term that you most likely have heard in connection with reports about opinion polling.

Let's return to the example of sleep quality scores recorded for the patients with fibromyalgia at the beginning of the tai chi study by Wang et al. (2010).

We computed a sample mean = 13.7, where a higher number means more sleep problems. We visited fantasy land and picked up the idea that the population of healthy people had a mean sleep quality (PSQI) score = 6, and that the population of PSQI scores had a standard deviation = 3. These numbers led us to compute a standard error of the mean = 0.3692745 and a z test statistic = 20.85. If the sample mean were close to the population mean, then the z test would be close to zero, meaning we could fit hardly any standard errors in the gap between M and μ. That's not what we found; we can fit more than 20 standard errors in the gap. Question: How can we take the point estimate of M = 13.7 and use the notion of sampling variability to compute a confidence interval as an interval estimate of μ?

The answer requires some explanation of the math involved, and please remember that the example that you will see here is only one of many kinds of confidence intervals that will be discussed in the book. As we think about the z test statistic, we were measuring the gap between the sample mean and the population mean. We found out how many standard errors of the mean fit in that gap. Now we are going to use the standard error of the mean to define one number that is below the sample mean and a second number above the sample mean. Those two numbers will define a confidence interval. Instead of having a single number, M, as an estimate of μ, we will have a range of values bounded by two numbers, each a certain distance from M.

To get the confidence interval, let's begin by asking, "When comparing the sleep quality of people with fibromyalgia to a population of healthy people's sleep quality, how far apart would M and μ have to be in order to say that the difference was statistically noteworthy?" Let's use what we know about the Central Limit Theorem to find an answer. Because Wang et al. (2010) had a sample size of 66, the Central Limit Theorem tells us that the sampling distribution for the sample mean would look like a normal distribution, and the distribution of the z test statistic will have the same shape as the distribution of the sample mean. Figure 7.6 shows a distribution for our z test statistic.

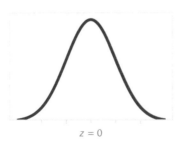

$z = 0$

Figure 7.6

Distribution of the z test statistic. The Central Limit Theorem tells us what the sampling distribution of the mean looks like. And we know that z scores have the same shape as the original variable. When we compute a z test statistic for the sample mean, the distribution of our z test statistic benefits from the Central Limit Theorem, and then we know what the sampling distribution of our z will look like: a standard normal distribution.

If the sample mean equaled the hypothesized population mean, there would be no distance between them, and the z test statistic would equal 0. The bigger the gap between M and μ, the farther that the z test statistic gets from 0 (either in the positive or negative direction). Let's say we are interested in the two values of z that will enclose 90% of the values in the distribution in Figure 7.6. (The most commonly reported confidence interval is 95%, but we are showing an example of 90% just so that you know it is possible to have different levels of confidence.) Table A.1 in the back of the book shows tail areas or middle areas (between zero and a value of z) for one side of the distribution. If we want the middle 90% of scores, we could look for a value of z that cuts off 5% of one tail area; by symmetry we would know that the negative value of that z would cut off 5% of the lower tail area. By looking in Table A.1, we find these values are $z = 1.645$ and $z = -1.645$. These two values of z enclose 90% of z values in the distribution. But we wanted to know the *distance* between the sample mean and the population mean that would get us to one of these values of z. We want to know the numerator of z, or the distance that we must travel away from the hypothesized value of the parameter to reach one of these endpoint values. What we are looking for is called a *margin of error*, a measure of spread that is used to define a confidence interval.

To find the margin of error will require some mathematical gymnastics. To help you follow the math, remember that any time we multiply, divide, add, or subtract on one side of an equation, the same thing must be done to the other side of the equals sign so that the two sides of the equation remain equal. For example, consider,

$$\frac{3}{4} = 0.75$$

If we multiply the left side of the equation by 4, then we must multiply the right side of the equation by 4 to keep the two sides equal:

$$\frac{3}{4} \times 4 = 0.75 \times 4$$

$$\frac{3 \times 4}{4} = 3$$

$$3 = 3$$

On the left side, the 4 is multiplied against the numerator, 3. Then we have 4 in the numerator and 4 in the denominator, and 4/4 = 1. So the left side of the equation simplified to 3, and, of course, 0.75 times 4 = 3.

We found in Table A.1 that $z = 1.645$ and $z = -1.645$ bracket 90% of the z values in the standard normal distribution, and we want to find out the distance between a sample mean and the hypothesized $\mu = 6$ that could result in one of these two values of the z test statistic. We recall that the z test's denominator is the standard error of the mean, so let's use SE as the abbreviation for our standard error of the mean. The z test formula is

$$z = \frac{M - \mu}{SE}$$

$$= \frac{\text{some distance}}{SE}$$

We already computed SE to be 0.3692745. And we are trying to find the distance in the numerator that would result in a critical value of 1.645. So let's replace z with 1.645 and SE with a numeric value:

$$1.645 = \frac{\text{some distance}}{0.3692745}$$

Just as we multiplied 3/4 by the number 4 to get rid of the denominator in the previous math example, now we can get rid of SE in the denominator on the right by multiplying both sides of this equation by 0.3692745:

$$0.3692745 \times 1.645 = \frac{\text{some distance}}{0.3692745} \times 0.3692745$$

$$0.6074566 = \text{some distance}$$

"Some distance" is the margin of error for this example. It is the distance that we would have to observe between the sample mean and the population mean to get $z = 1.645$ or $z = -1.645$, which encompass 90% of the z tests in a standard normal distribution.

Now that we know the margin of error, we are going to apply it to our observed sample mean for the sleep quality scores for the 66 patients with fibromyalgia. We computed $M = 13.7$. Suppose we are not sampling from a population of healthy people, but instead a population of patients with fibromyalgia who differ in sleep quality from healthy people, which seems much more likely. If we took another sample of 66 patients with fibromyalgia and computed their mean sleep quality, we could expect to get a different sample mean and not 13.7. Taking into account the expected variation across samples, we can compute a 90% confidence interval using our sample mean. The confidence interval will provide a range of values as an interval estimate of the population mean sleep quality for patients with fibromyalgia. The lower limit of the confidence interval is

$$\text{Lower limit} = M - \text{margin of error}$$
$$= 13.7 - 0.6074566$$
$$= 13.092543$$
$$\approx 13.09$$

We find the upper limit of the confidence interval by adding the margin of error to the sample mean:

$$\text{Upper limit} = M + \text{margin of error}$$
$$= 13.7 + 0.6074566$$
$$= 14.307457$$
$$\approx 14.31$$

The 90% confidence interval that was computed using our sample mean, 13.7, is the interval [13.09, 14.31]. This interval is estimating the population mean sleep quality for patients with fibromyalgia. How do we interpret this interval? We must do so with the concept of repeated sampling in mind. We can say that 90% of confidence intervals computed on data from repeated samples of the same size from the same population from which our sample mean was drawn will contain the true population mean PSQI for patients with fibromyalgia. Does our interval contain the true population mean? There is no way of knowing; either the interval does or does not contain the true μ. But 90% of intervals like ours will encompass the true population mean. So what can we say? We can say that we are 90% confident that this interval, 13.09 to 14.31, includes the true population mean of sleep quality scores for patients with fibromyalgia.

Let's do another example. In Chapter 4, we talked about a National Health Statistics Report by McDowell et al. (2008) in which one variable being measured was the height of American men in their 30s. Suppose the report leads us to speculate that American men have a mean height of 69.4 in. Further, based on the thousands of men who have been measured, a reasonable guess for the population standard deviation of these men's heights is 3.1 in. Imagine that we draw a sample of 64 American men and measure their height, and we compute $M = 70$. This point estimate of the population mean is only one value, and every time we sample 64 different American men, we probably will get a different sample mean—yet we would want to compare our single sample mean to the hypothesized population mean = 69.4.

Let's add what we know about sampling variability to compute a 95% confidence interval associated with our sample mean. If we look at Table A.1 to find values of z that enclose the middle 95% of the standard normal distribution, we find $z = 1.96$ and $z = -1.96$. To get the margin of error, we need to find some distance between the sample mean and the population mean that would produce a z test $= 1.96$:

$$1.96 = \frac{\text{some distance}}{SE}$$

So we need to know the standard error of the mean. The Central Limit Theorem tells us that it is $\frac{\sigma}{\sqrt{N}}$. We said the population standard deviation for the heights of American men in their 30s is 3.1 in., and we have a sample size = 64. For this example, $SE = \frac{3.1}{\sqrt{64}} = \frac{3.1}{8} = 0.3875$. So to find our margin of error, we solve the following equation for "some distance":

$$1.96 = \frac{\text{some distance}}{0.3875}$$

$$0.3875 \times 1.96 = \frac{\text{some distance}}{0.3875} \times 0.3875$$

$$0.7595 = \text{some distance}$$

$$= \text{margin of error}$$

Now we compute the lower limit on the 95% confidence interval by subtracting the margin of error from our sample mean:

$$M - \text{margin of error} = 70 - 0.7595$$

$$= 69.2405$$

The upper limit on the confidence interval is the sample mean plus the margin of error:

$$M + \text{margin of error} = 70 + 0.7595$$

$$= 70.7595$$

We can say that our 95% confidence interval based on our sample of 64 men's heights is [69.24, 70.76], and 95% of confidence intervals computed like ours will contain the true population mean for the heights of American men in their 30s. Notice that we said that our best guess of the population mean was 69.4, which is a value within this confidence interval. So our best-guess value for the population mean height (69.4) is a number contained within the interval [69.24, 70.76], representing plausible values for mean height of American men in their 30s, based on our sample mean and the likely sampling variability associated with it.

Check Your Understanding

SCENARIO 7-B, Continued

We speculated that the population mean birth weight for full-term infants in the United States is about 3,400 g and that the population standard deviation is about 375 g (about 13 oz). The sample mean birth weight of 3,376.43 g was computed for 30 full-term babies whose mothers exercised during pregnancy (Price et al., 2012). 7-13. Compute the standard error of the mean for this example. 7-14. To compute a 95% confidence interval based on our sample mean, use $z = 1.96$ to find the margin of error. 7-15. Compute the 95% confidence interval based on our sample mean. 7-16. Interpret the meaning of this interval estimate.

Suggested Answers

7-13. *The standard error is* $\dfrac{\sigma}{\sqrt{N}} = \dfrac{375}{\sqrt{30}} = \dfrac{375}{5.4772256} = 68.46532$. *7-14. The margin of error for the 95% confidence interval is* $1.96 \times 68.46532 = 134.19203$. *7-15. The lower limit is the sample mean minus the margin of error* $= 3376.43 - 134.19203 = 3242.248 \approx 3242.24$. *The upper limit is the sample mean plus the margin of error* $= 3376.43 + 134.19203 = 3510.622 \approx 3510.62$. *Thus, the 95% confidence interval for the sample mean is* [3242.24, 3510.62], *which is an interval estimate of the population mean of birth weight for full-term American babies based on* $N = 30$ *babies born to women who*

(Continued)

were physically active during pregnancy. Notice that the interval contains 3,400 g, which is what we speculated to be the population mean birth weight for full-term babies. 7-16. Our confidence interval tells us a range of plausible values for the sample mean, and if we repeatedly drew random samples from the same population, we could expect 95% of the confidence intervals like ours to bracket the true population mean birth weight of full-term babies with mothers who were active during pregnancy.

What's Next

This chapter had a lot of information to digest—and much of the content was quite abstract. We introduced the crucial concept of sampling variability, which led to the definition of sampling distributions. It would be a good bet that a random sample would not produce a sample mean that perfectly equaled the mean of whatever population was sampled, just as most people throwing a dart will not hit the bull's-eye. With repeated sampling, we could start to see a pattern of sample means. The Central Limit Theorem says our repeated sample means will pile up around the population mean, like darts clustering around the bull's-eye. We used the information from the Central Limit Theorem to compute a z test statistic. Even if we compute a confidence interval to quantify the sampling variability, we may obtain an interval that does not bracket the actual population mean. But with repeated sampling, a large percentage of confidence intervals computed like our interval will bracket the true mean of the population being sampled, like horseshoes clustering around a stake.

Chapter 8 will build on the information covered in this chapter, and we will formalize the process of deciding whether we think a sample mean differs from a population mean. This decision will rely on probability. Now that we know about sampling variability, we will talk about how likely it would be to obtain a sample mean that is at least as far as ours from a hypothesized population mean. We also can give a deeper interpretation of a confidence interval as an estimate of μ. A similar process later will be followed to make other decisions, such as whether two variables share a noteworthy correlation or whether a treatment group's mean differs markedly from a control group's mean.

Exercises

7-17. What is sampling variability? 7-18. Why should researchers care about sampling variability? 7-19. What does a sample distribution contain? 7-20. What does a population distribution contain? 7-21. What does a sampling distribution contain? 7-22. We obtain samples because we cannot get populations. We use the sample data to compute _____, which are estimates

(*Continued*)

of _____. Yet statistics always have uncertainty connected with them because of _____. 7-23. Use the unbiased variance statistic as an example to describe a sampling distribution. 7-24. The Central Limit Theorem is an example of information from mathematical statisticians who have rescued us from the work of creating a particular sampling distribution. What does the Central Limit Theorem say? 7-25. Has the Central Limit Theorem rescued us from all sampling distributions? 7-26. Is the sampling distribution of the median when $N = 74$ the same as the sampling distribution of the median when $N = 75$? Explain why or why not.

SCENARIO 7-C

(Based on Price et al., 2012.) The researchers studying exercise during pregnancy randomly assigned the participants to either remaining sedentary or participating in a supervised exercise program. On five occasions, participants were asked to cover 3.2 km as quickly as they comfortably could, either by walking or running. The researchers computed power scores for each participant by taking the product of the participant's weight and distance covered, divided by time. In this way, the researchers took into account the participants' weight gain during pregnancy. A higher power score meant that the participant showed greater aerobic fitness. The first occasion of measurement occurred when participants were 12–14 weeks pregnant, before the intervention began. A participant with a lower power score may have taken more time to cover the distance, compared with someone who was the same weight. Suppose we visit the researchers in fantasy land, where they tell us that the average power score for sedentary women early in pregnancy is 1100 watts, and that the population of power scores has a standard deviation = 200.

7-27. What kind of research is this, and how do you know? 7-28. What kind of variable is power score? 7-29. What kind of variable is activity during pregnancy? 7-30. Suppose we were working on this study of 62 pregnant women and we were sampling from the population described by the researchers in fantasy land. Based on the Central Limit Theorem, what three facts could we assert about the sampling distribution of the sample mean? (You may need to perform one calculation as part of your answer.) 7-31. The mean power score at Time 1 (12–14 weeks into pregnancy) equaled 1083.05 and $SD = 195.27$. Compute the z test statistic to compare this sample mean with the population mean that we were told in fantasy land. 7-32. Interpret the meaning of the numeric value of your z test statistic. 7-33. Suppose we think this sample may have come from a population of sedentary pregnant women who had a mean power score = 1100. Based on the numeric value of your z test statistic, does 1083.05 seem like a plausible mean for a sample of women from that population? 7-34. Let's say we want to compute a 95% confidence interval for our sample mean. The values of z that enclose 95% of a standard normal distribution are −1.96 and +1.96. Multiply 1.96 by the standard error of the mean to

(Continued)

get the margin of error. 7-35. Use the margin of error to find the lower and upper limits of the 95% confidence interval. 7-36. Explain the meaning of the confidence interval you computed.

SCENARIO 7-D

We read in Chapter 4 about a longitudinal healthy study in the United States in which weight measurements were taken on children at various ages (McDowell et al., 2008). Based on information in the report, we estimate that the population mean weight of 1-year-old boys is 25.5 lb, and that these boys' weights have a standard deviation = 3.7 lb. Suppose we are reviewing medical charts of children receiving health care at a clinic that serves lower-income families, and we make a list of the weights of the last 78 boys examined within a month of their first birthday. We compute $M = 24.3$ lb and $SD = 4.4$ lb. 7-37. Why can we use a standard normal distribution to answer questions about this scenario? 7-38. We know that $z = 1.645$ cuts off the top 5% of scores in a standard normal distribution. Use this value of z to compute the margin of error for a 90% confidence interval. 7-39. Compute and interpret the 90% confidence interval. 7-40. Suppose we show our results to one of the most experienced pediatricians in the clinic, and she says, "It looks to me as if these boys are similar to the rest of the population." Do you agree? Why or why not?

References

Buysse, D. J., Hall, M. L., Strollo, P. J., Kamarck, T. W., Owens, J., Lee, L., ... Matthews, K. A. (2008). Relationships between the Pittsburgh Sleep Quality Index (PSQI), Epworth Sleepiness Scale (ESS), and clinical/polysomnographic measures in a community sample. *Journal of Clinical Sleep Medicine, 4*, 563–571. Retrieved from http://www.aasmnet.org/jcsm/

McDowell, M. A., Fryar, C. D., Ogden, C. L., & Flegal, K. M. (2008, October 22). Anthropometric reference data for children and adults: United States, 2003–2006. *National Health Statistics Reports, 10*, 6–16. Retrieved from www.cdc.gov/nchs/data/nhsr/nhsr010.pdf

Price, B. B., Amini, S. B., & Kappeler, K. (2012). Exercise in pregnancy: Effect on fitness and obstetric outcomes—A randomized trial. *Medicine & Science in Sports & Exercise, 44*, 2263–2269. doi:10.1249/MSS.0b013e318267ad67

Wang, C., Schmid, C. H., Rones, R., Kalish, R., Yinh, J., Goldenberg, D. L., ... McAlindon, T. (2010). A randomized trial of tai chi for fibromyalgia. *New England Journal of Medicine, 363*, 743–754. doi:10.1056/NEJMoa0912611

Hypothesis Testing and Interval Estimation

Introduction

Imagine that you are a physician in Norman, Oklahoma, in 2002. You have noticed an increase in the number of patients you suspect of having hepatitis C. Your patients lack the usual risk factors for hepatitis C (illicit drug use, piercings or tattoos received in unclean conditions, previous clotting problems, history of dialysis, etc.). As you puzzle over the mystery, you realize that these patients have one thing in common: they received intravenous pain medication at a certain pain management clinic.

If nothing is out of the ordinary, how many new cases of hepatitis C could be expected in a year in Norman, Oklahoma? This question is hard to answer. Hepatitis C is not easy to diagnose and often has no symptoms for newly infected patients. Sometimes the symptoms are mistakenly attributed to the flu, so patients may have hepatitis C for a while before it is diagnosed. A study by Armstrong et al. (2006) estimated the overall prevalence of hepatitis C during 1999–2002 to be 1.6% (95% confidence interval: [1.3%, 1.9%]). If these rates could be applied to Norman, a city of about 100,000 residents in 2002, we might guess that fewer than 2,000 Norman residents at that time were living with hepatitis C. But this number would include people who had been diagnosed in earlier years. What about cases that were newly diagnosed in 2002? The closest period for which we could find rates was 2006–2007; the Centers for Disease Control and Prevention (CDC) reported that six American cities had annual rates for new infections ranging from 25 to 108 cases per 100,000 people.

An outbreak of hepatitis C really did happen in Norman, Oklahoma. As part of the investigation, 795 patients of the pain management clinic were tested. Comstock et al. (2004) said 86 of those patients (10.8%) tested positive for hepatitis C. If nothing were amiss and Norman resembled the six cities studied by the CDC, we might expect the entire city to have as many as 100 or so new diagnoses that year. If so, does it seem typical or does it seem unusual for one clinic to have 86 patients diagnosed with this disease? If nothing were wrong, a large number of new hepatitis C cases in one clinic would seem unlikely. If it was an unlikely event to have many new cases in one clinic, we would doubt the idea that nothing was wrong. In fact, Comstock et al. said a certified registered nurse anesthetist at the pain management clinic had reused needles between patients, a serious problem that two other nurses in the clinic had noticed and reported to supervisors.

The heartbreak of the people in Norman who were infected with hepatitis C cannot be communicated with statistics. Yet some of these patients want the story to be told as a cautionary tale for future health-care professionals so that other people do not suffer the lifelong consequences of similar mistakes. Statistics can help researchers to find answers to questions such as, "If the usual rate is this number, how likely is it to obtain a rate at least as high as this year's rate?" In other words, if nothing is wrong, what are the chances of seeing an infection rate at least this high?

This chapter builds on what we have covered in Chapters 6 and 7. We will use what we learned about probability, sampling distributions, and estimation to test a hypothesis. This chapter provides a framework for decision making that is used repeatedly throughout the rest of the book.

Testable Guesses

Hypothesis was defined in Chapter 7 as a testable guess. To help clarify what we mean by a testable guess, let's talk about what is *not* testable. We may believe that everyone should try to remain physically active throughout their lives because research has demonstrated that physical activity has many benefits, both physically and mentally. But we cannot test the following idea: "A good life requires physical activity." For one thing, who can judge whether a life is good? We would have to define what we mean by a "good life." Personally, we would be unwilling to tell people who are paralyzed that they cannot have a good life. So the notion that "a good life requires physical activity" is not a testable idea.

How could we translate the idea of the importance of physical activity into a hypothesis that can be tested? We could begin by being more specific, objective, and nonjudgmental. We might ask what variables are affected by a person's level of physical activity. A great deal of research has gone into this question. We could consider heart disease, but that is a broad term for many cardiovascular conditions. So let's be specific: high blood pressure is related to the risk of stroke, so health-care professionals recommend that patients control their blood pressure. Studies are being conducted constantly in many countries to follow large representative samples for the development of conditions such as high blood pressure. If we had access to the data from one of these longitudinal studies, we could determine retrospectively whether those who had high blood pressure had exercised less than those who did not have high blood pressure. In this way we could test a hypothesis that there was a relationship between exercise level and a diagnosis of high blood pressure. This hypothesis is more specific and objective than the assertion about a good life requiring physical activity.

Most research on high blood pressure, or hypertension, is even more specific. Recent research has investigated the effectiveness of controlling blood pressure by using a telemonitoring device. Patients used a device that took their blood pressure and automatically transmitted the readings to a secure website, where the readings were monitored by a pharmacist (Margolis et al., 2013). The investigators wanted to know whether the intervention would result in patients having lower blood pressure, compared with patients in a control group. These researchers' hypothesis was a testable guess because it was specific, involved objective measures, and was not judgmental, unlike the "good life" statement.

If *hypothesis* is defined as a testable guess, what does *hypothesis testing* mean? This phrase refers to a process of setting up two competing statements (hypotheses, which is the plural of hypothesis) that describe two possible realities. The process then involves using probability to decide whether the results of a study are typical or unusual for one of those two possible realities. If the results are unlikely for one of those realities, we may conclude that its competing statement of the opposite reality is more likely to be true. Because our decisions are based on probabilities, we cannot *prove* which reality is true. We only say what is more likely or less likely to be true. (Do not worry if this paragraph was hard to follow. This chapter will explain it in detail.)

8-1. An old expression is, "An apple a day keeps the doctor away." Explain whether this expression is a testable guess. If not, suggest how it could be rewritten to become a hypothesis.

Suggested Answers

8-1. As written, the statement is not a testable guess because it is not specific. It could be reworked into a hypothesis if we defined what was meant by "keeps the doctor away." Does the phrase mean that the person never gets sick? We might define "keeps the doctor away" as a person misses no more than two days of work per year. So our hypothesis could be stated as follows: "People who eat one apple every day will miss no more than two days from work per year, on average." Whatever definition we use for "keeps the doctor away," it would need to be objective, specific, and nonjudgmental.

The Rat Shipment Story

Many researchers perform studies using laboratory rats in the early stages of researching treatments that eventually could inform our understanding of human conditions such as obesity, diabetes, and high blood pressure. These researchers do not get their rats by catching them in alleys or fields; they buy them from companies that keep extensive records of the rats' genetic histories. Otherwise, all sorts of uncontrolled extraneous variables associated with the rats could interfere with the experiments. We would like to tell you a story that is based in reality, with some details taken from fantasy land for teaching purposes. This story is used to illustrate many concepts in this chapter. If you do not understand everything in this story, be patient. After the next Check Your Understanding section, we will go into further details about the story.

Dr. Sarah Bellum conducts animal experiments related to obesity. She uses brown Norway rats, which she purchases from a company in Massachusetts. Dr. Bellum orders 600 rats for a new study. On a winter's day, Dr. Bellum is notified that a delivery company has arrived with the rat shipment. She sends her graduate assistant, Ray D. Ology, to meet the delivery truck at a loading dock. Ray signs for the shipment and brings the rats to the lab. As Ray is unloading the rats, he thinks, "Something is not right. These rats don't act like our usual rats." He observes them for a while and concludes that they seemed sluggish, compared with the usual alert, frisky rats. Ray asks his mentor to take a look at the rats. After scrutinizing the rats' behavior, Dr. Bellum says, "I agree, something is wrong with the rats." The two researchers discuss whether the rats may have been exposed to the winter weather or an illness. Dr. Bellum is reluctant to use these rats in her research because they may continue to behave differently from her usual, predictable Norway rats. Because of her limited research budget, she needs the company to replace the rats for free. She needs evidence to persuade the company that there really is something wrong with the rats.

Dr. Bellum has large databases of information about healthy rats that were measured at the beginning of her studies before any interventions were introduced. She always tests rats in a 138-cm straight-alley maze. The test involves placing a rat at one end of the maze and a food reward at the other end. Sensors time the rat from the moment it leaves the starting position to the moment it reaches the food. From her years of research with brown Norway rats, Dr. Bellum knows that healthy rats take an average of 33 seconds to make the trip from the starting gate to the food. Further, the running times for the thousands of rats in her prior research had a standard deviation of 19 seconds. These two numbers are parameters. The population mean, μ, is 33 seconds, and the population standard deviation, σ, is 19 seconds. If there is nothing wrong with the rat shipment, then the new rats should perform like the rats from past studies.

Dr. Bellum doesn't want to test all 600 rats in the shipment. She and Ray decide to draw a random sample of 25 rats from the shipment, measure their times to complete the maze, then compare the sample's mean running time with the known population mean. The general question will be, "Does our sample seem to come from a population of healthy rats?" Following procedures to ensure the ethical treatment of animals in research, Ray randomly samples $N = 25$ rats and measures each rat in the maze. He cleans the maze between rats so that one rat doesn't follow the scent of a previous rat's path. He reports to Dr. Bellum that the sample mean running time is 44.4 seconds. So far they have confirmed their initial prediction: that the rats are sluggish and take more time to complete the task than healthy rats. But is 44.4 seconds close to or far away from 33 seconds? Dr. Bellum computes the z test statistic for the sample mean and finds $z = 3.0$, which means three standard deviations (or standard errors of the mean) can fit in the gap between the sample mean and the population mean.

If Dr. Bellum's sample came from a population of healthy rats, how likely is it to obtain a sample mean at least this extreme? Because the Central Limit Theorem said the sampling distribution of the mean is normally distributed, Dr. Bellum looks at a standard normal table like Table A.1 in the back of the book to find the probability of a z test statistic equal to or greater than 3.0. This is the same as the probability of obtaining a mean equal to or greater than 44.4. She finds a probability = .0013. If Dr. Bellum's sample came from a population of healthy Norway rats with a population mean = 33 seconds, the probability is .0013 for obtaining a sample mean of 44.4 seconds or greater. Dr. Bellum concludes that if she and Ray sampled from a population of healthy rats, the probability of obtaining a sample mean of 44.4 seconds or greater is very small—so small that she rejects the idea that the rats are healthy. For more evidence Dr. Bellum computes a confidence interval, with the prediction that the interval will be higher on the number line than $\mu = 33$ because she expects the new rats to take more time to complete the simple maze, relative to the usual rats. She computes a confidence interval that does not contain the population mean, so Dr. Bellum doubts that she has sampled from a population with $\mu = 33$. She contacts the company that shipped the 600 rats, presents her evidence, and persuades the company to replace the rat shipment with healthy rats.

The rest of this chapter formalizes the concepts presented in the rat shipment story. The steps of hypothesis testing presented in this chapter are used with different test statistics through the rest of the book. We must say that the

z test statistic for the sample mean is almost never used in real research because it requires knowledge of a population mean and population standard deviation. Those details ($\mu = 33$ and $\sigma = 19$) came from fantasy land. But the z test statistic is a great way to introduce hypothesis testing, and we refer to this story of the rat shipment repeatedly throughout the chapter.

Check Your Understanding

8-2. In telling the rat shipment story, we wrote, "If Dr. Bellum's sample came from a population of healthy rats, how likely is it to obtain a sample mean at least this extreme?" Based on the mean maze completion time for the random sample, Dr. Bellum concluded that the company should pay for another shipment of rats. Did she reach her conclusion based on a small probability or a big probability? In other words, how did probability support her conclusion?

Suggested Answer

8-2. *The notion being tested was that the sample came from a population of healthy rats, known to have a population mean = 33 seconds. Dr. Bellum wanted to know how likely it was to obtain a sample mean of 44.4 seconds or greater, given the sample came from a population of healthy rats. If this notion of healthy rats is true, then a sample mean of 44.4 seconds or greater should be a typical or highly likely result. Instead, she found that obtaining a sample mean of 44.4 seconds or greater with a sample of N = 25 rats from the healthy population would be extremely unlikely; the probability is .0013. A small probability led Dr. Bellum to conclude that it was unlikely that the notion of healthy rats was true. Based on the small probability, she rejected the idea that the sample came from a healthy population.*

Overview of Hypothesis Testing

To provide some structure for presenting the numerous concepts in this chapter, we have made the following list of steps that researchers may take to test a hypothesis using the z test statistic. This overview is intended to help you to see where we are going in this chapter. It introduces a few terms quite briefly. Don't worry, we will explain everything in detail later.

- Encounter a problem or research question. Ray encountered the problem of the new rats seeming to be more sluggish than the usual healthy Norway rats. He and Dr. Bellum wanted to know if they would take longer than the usual rats to complete the maze. In more complex studies, researchers may have a long list of questions that they wish to answer, and they may read many published studies to refine their research questions.

8. Hypothesis Testing and Interval Estimation

- Consider what is known from the past. From years of research Dr. Bellum knew the mean maze completion time for healthy Norway rats, and she knew the population standard deviation of the maze completion times.
- Formulate an idea about what may be true. Dr. Bellum and Ray believed the new rats would take longer on average to complete the simple maze than the usual healthy rats would take. We will formulate this prediction in a statement called the alternative hypothesis.
- Write a hypothesis that will be the opposite of what the researcher believes to be true. Dr. Bellum and Ray did not believe that the new rats came from the population of healthy rats. But they knew facts about the maze performance of healthy rats, so they tested the idea that the sample came from a population of healthy rats. The statement that is opposite of the alternative hypothesis is called the null hypothesis.
- Decide what will be considered a small probability for the results. When Dr. Bellum and Ray were thinking about possible results for the rats, they had to consider the likelihood of various outcomes compared with the known performance of healthy rats. If a result was to be considered unusual, then it would be improbable for healthy rats. How improbable? We will talk about a standard for a small probability known as the significance level.
- Collect data, and then compute the sample mean, the z test statistic, and a certain probability. The statistics will serve as evidence as Dr. Bellum and Ray try to determine whether what they believe (that the new rats are sluggish and perhaps sick) is more likely to be true—or the opposite is probably true (that the new rats are similar to healthy rats).
- If results were predicted in a certain direction, check whether the prediction was correct. Dr. Bellum and Ray had to make sure that the sample of rats had a mean maze completion time that actually was greater than the mean for healthy rats. If their sample had zipped through the maze more quickly than the usual rats, then their prediction would have been wrong, and the evidence would not have supported the idea that the rats were sluggish.
- Compare the computed probability with what is considered a "small" probability. The computed probability will be called the p value, and it will be compared with the significance level.
- Test the null hypothesis. This step may lead researchers to keep the null hypothesis as the most probable statement of reality for their research question. Or this step may lead them to throw away the null hypothesis in favor of a reality described in the alternative hypothesis. This hypothesis test depends on the comparison of the p value and the significance level. (There is a second way to do the hypothesis test, which we will explain later.)
- Draw conclusions. Dr. Bellum and Ray found evidence that made them toss aside the null hypothesis and conclude that the rats were sluggish and probably unhealthy. Therefore, they asked the company to replace the shipment.

In the next section, we begin filling in the details for each of these main steps of hypothesis testing. If you start getting lost in the details, perhaps looking back at this overview will help you to see the big picture.

Two Competing Statements About What May Be True

The rat shipment example skimmed over some details of hypothesis testing and interval estimation, and we will expand on those details through the rest of the chapter. Ray knew quite a bit about healthy rats when he went to the loading dock. When he was unpacking the new shipment, he noticed that the rats were acting in a way that did not match his understanding of healthy rats; he suspected that something was wrong with the rats. So far we are describing the first two steps in our overview of hypothesis testing: encountering a problem and considering what is known from the past. The next two steps are to formulate an idea about what the researcher believes to be true and to write a hypothesis that will be opposite to what the researcher believes to be true. Restating the beginning of the story, here are the two competing ideas being considered:

- The new rats are sluggish and unhealthy.
- The new rats are alert and healthy, similar to the usual rats.

The first idea is that these rats are more sluggish than healthy rats, indicating that maybe they are sick. The second idea is that nothing unusual is going on—the new rats are like the healthy rats encountered in the past. The researchers formalized these competing ideas in a way that allowed objective, specific measurement of the rats. They chose to measure the rats in the 138-cm straight-alley maze because years of research with healthy Norway rats provided them with the knowledge of the population mean running time and population standard deviation. The first idea, which reflects what the researchers believe to be true, can be restated as follows:

> Our sample comes from a population of Norway rats that has a population mean running time greater than 33 seconds.

In other words, the new rats are expected to differ from healthy rats in a specific way: they will take longer to complete the simple maze. This idea is the *alternative hypothesis*, a statement that reflects what the researchers believe. To compare the new rats with what is known from the past involves the second idea, which basically says nothing different is going on with this new sample. This idea could be restated as follows:

> Our sample comes from a population of Norway rats that has a population mean running time of 33 seconds (or perhaps less than 33 seconds).

The statement of what we do not believe to be true reflects a possible reality where nothing different from prior research is going on. This statement is an example of a *null hypothesis*, the idea to be tested. The null hypothesis and alternative hypothesis are opposites of each other. Dr. Bellum and Ray think the rats seem too slow—that is what they *do* believe to be true. But they know a lot more about healthy rats than unhealthy rats. They know how long healthy rats take on average to complete the simple maze. So the researchers will test this idea of the new rats being similar to the usual healthy rats. In other words, they are setting up a null hypothesis that they hope to discredit with evidence.

The concepts of null and alternative hypotheses can be difficult for students to grasp. Why is the null hypothesis tested when it is the idea we do not believe to be true? Let's use an example. Suppose we have lost a set of car keys in our home. At first we look in the usual places where we tend to leave the keys. Having no success, we must become more systematic in our search. We start with the kitchen: we do not think we left the keys in the kitchen. The alternative hypothesis, therefore, is that the keys are not in the kitchen; this is a statement of what we believe. The null hypothesis is a statement opposite to the alternative hypothesis: the keys are in the kitchen. How can we eliminate the kitchen as the location of the keys unless we look there? That is what we do—we test the null hypothesis (the keys are in the kitchen) by looking in the kitchen. The lack of keys in the kitchen is evidence against the null hypothesis. So we throw out the null hypothesis and conclude the keys are elsewhere. This decision about the null hypothesis does not tell us where the keys are, but we have shortened the list of possible places. Now we need a new null hypothesis. We do not think we left the keys in a bedroom, so the new null hypothesis would be, "The keys are in a bedroom." By testing a series of null hypotheses, we can rule out each location until we eventually find the keys in the dining room.

Let's consider another example of testing a null hypothesis. Suppose we are talking about professional basketball players, and we make the claim that all professional basketball players have been at least 6 ft tall (about 1.8 m). You think we are wrong. You think there have been some professional basketball players shorter than 6 ft tall; that is your alternative hypothesis. The null hypothesis, which you do not believe is true, says all professional basketball players have been at least 6 ft tall. To try to settle the argument, we access an online list of professional basketball players, and we show you a list of a thousand players, all of whom were at least 6 ft tall. Has our list persuaded you that all professional basketball players have been at least 6 ft tall? No. In response to our argument, how many players do you need to name to have evidence that contradicts the null hypothesis? Only one. (For instance, Spud Webb was 5 ft 7 in., or 1.7 m.)

The examples of the lost keys and heights of basketball players both allow us to find definitive answers: the keys eventually are found, and we find a basketball player who was known to be shorter than 6 ft. But in statistics, we cannot know for sure what is true. We make decisions about a null hypothesis based on probability. Before we go into the details about using probability, we will explain how to write the null and alternative hypotheses in a more exact way that corresponds to the z test statistic, which we will use to test the null hypothesis.

Check Your Understanding

SCENARIO 8-A

Suppose we think people who eat an apple every day will be absent from work because of sickness fewer than four times a year on average. 8-3. Is the preceding statement reflective of a null hypothesis or an alternative hypothesis? 8-4. Write an opposite statement and identify what kind of hypothesis it would be.

Suggested Answers

8-3. The statement shows what we believe, so it would be an alternative hypothesis. 8-4. The opposite statement would be a null hypothesis. Our null hypothesis may be stated as: People who eat an apple every day will have an average of four or more sickness-related absences from work in a year.

Writing Statistical Hypotheses

We have tried to avoid symbols and formulas as much as possible in this book, but hypothesis testing requires some symbols, which also tend to show up in journal articles. We need to be quite specific about the hypotheses that correspond to inferential statistics such as the z test statistic. The exact hypotheses that correspond to the inferential statistic chosen for a given hypothesis testing situation are called *statistical hypotheses*, which usually are written with symbols. When we write hypotheses, we tend to start with what we believe or expect to be true, based on our understanding of the research topic or situation. We find it easier to write the alternative hypothesis first. Second, we write a null hypothesis that is a statement opposite of the alternative hypothesis.

So let's start with the alternative hypothesis. The symbol for the alternative hypothesis varies across statisticians; we will use H_1 (your instructor may prefer H_A, another commonly used symbol). We are using the number one as a subscript to indicate something is present; the alternative hypothesis can reflect our belief that there is some effect, relationship, or difference to be detected. Dr. Bellum and Ray thought something was wrong with the new rats and expected that they would take longer than healthy rats to complete the maze. We can summarize that idea with the following use of symbols:

$$H_1: \mu > 33 \text{ seconds}$$

The symbol "H_1:" is saying, "This is the alternative hypothesis." So the entire symbolic expression can be translated as

This is the alternative hypothesis: Our sample comes from a population with a mean straight-alley maze running time that is greater than 33 seconds.

The symbol μ is summarizing a lot of words: "Our sample comes from a population with a mean running time...." Although we rely on sample statistics for the hypothesis test, the alternative hypothesis reflects the idea that we want to generalize back to the population. The above translation of the statistical alternative hypothesis (H_1: $\mu > 33$ seconds) encompasses that process of sampling from the population to which we want to generalize our conclusions.

Let's turn to the null hypothesis. There are different ways of writing the symbol that means, "This is the null hypothesis." We will use the symbol H_0 followed by a colon. The subscript is the number zero. Some people pronounce this symbol "H.O." or "H-sub-oh," but we tend to say "H-naught," using the British term for zero or nothing (because, as you know, the null hypothesis is essentially saying, "nothing is going on" or "nothing is different from what we have known in the past"). For the rat shipment example, if nothing is going on with the rats, then they are similar to the thousands of previously studied healthy rats, who had a mean running time of 33 seconds for the straight-alley maze. We can summarize that idea with the following use of symbols:

$$H_0: \mu = 33 \text{ seconds}$$

The symbolic statement above can be translated into the following statement:

This is the null hypothesis: Our sample comes from a population with a mean straight-alley maze running time of 33 seconds.

As a statement opposite to what we believe, however, we may need to modify this statistical null hypothesis because Dr. Bellum and Ray believed the new rats would take *longer* than 33 seconds to complete the simple maze. If the null hypothesis is going to be truly opposite to the alternative hypothesis, then H_0 should include maze completion times that are equal to or less than 33 seconds. There are mathematical reasons for writing the null hypothesis as we have presented it above (H_0: $\mu = 33$), and your instructor may prefer that way. We prefer to write a null hypothesis for this situation that includes $\mu = 33$ as well as all outcomes that are less than 33. Therefore, we would write the null hypothesis as

$$H_0: \mu \leq 33 \text{ seconds}$$

The symbolic statement above can be translated into words as follows:

This is the null hypothesis: Our sample comes from a population with a mean straight-alley maze running time of 33 seconds or less.

Sometimes journal articles will use symbols for *research hypotheses*, which are statements predicting certain outcomes in a study. Research hypotheses tend to correspond to alternative hypotheses, and most studies have multiple research hypotheses, often involving different outcome variables. Dr. Bellum's research hypothesis might have been stated as, "A random sample of rats from the shipment will complete the maze task in a mean running time that is substantially greater than 33 seconds." Research hypotheses sometimes are denoted in journal articles with the capital letter H, followed by a numbered subscript. Researchers who use this sort of numbering on research hypotheses—H_1, H_2, H_3, and so on—will organize their written results so that they can easily refer to the research question that was being investigated by a particular statistical analysis. This numbering system can help the reader of a journal article to understand the results being presented.

Check Your Understanding

SCENARIO 8-B

We are studying the emergence of diabetes in American men in their 30s, and we have collected initial data on 36 participants. These men seemed particularly tall to us, and we decide to practice what we are learning about hypothesis testing. McDowell, Fryar, Ogden, and Flegal (2008) reported the mean height of American men in this age group to be 69.4 in. 8-5. Write the statistical alternative hypothesis using the provided population mean. 8-6. Translate the statistical alternative hypothesis into a sentence, as we did in this chapter. 8-7. Write the statistical null hypothesis. 8-8. Translate the statistical null hypothesis into a sentence.

Suggested Answers

8-5. H_1: $\mu > 69.4$. 8-6. *Here is the alternative hypothesis: Our sample of men comes from a population with a mean height that is greater than 69.4 in. 8-7. H_0: $\mu \leq 69.4$. 8-8. Here is the null hypothesis: Our sample of men comes from a population with a mean height that is less than or equal to 69.4 in.*

Directional and Nondirectional Alternative Hypotheses

When we introduced the rat shipment story, we said Ray noticed that something was wrong with the rats, which did not seem to be acting like the usual alert, frisky rats. We could have stopped there. Instead, we presented the additional detail that he thought the rats seemed sluggish. This detail suggests that the rats may not be as fast-moving as the usual healthy rats. Researchers often do not predict a direction on the number line for their results. In fact, they often want to leave open the possibility of detecting noteworthy results that are extreme in either direction on the number line. A *nondirectional alternative hypothesis* is more general—it does not predict whether an outcome will be greater than some value or less than some value. Instead, a nondirectional alternative hypothesis predicts an outcome will differ from some value. An alternative hypothesis that does predict an outcome in a specific direction on the number line is called a *directional alternative hypothesis*. Dr. Bellum and Ray had a directional alternative hypothesis because they expected that the rats would take longer on average than the usual, healthy rats to complete the maze:

$$H_1: \mu > 33 \text{ seconds}$$

Notice that the directional sign ">" is pointing toward the upper end of the number line. It reflects the prediction of a mean to the right of 33 on the number line. *We always look at the alternative hypothesis to check which direction, if any, is predicted for the results.* But suppose that Dr. Bellum and Ray had not noticed sluggishness in the rats, just that something was wrong with them. They could not articulate what exactly was wrong with the rats, but they agreed that the rats seemed different from the usual, healthy rats. So they wanted to look for evidence that the new rats *differed from* the usual rats in terms of their average maze completion time. The researchers could have written the following nondirectional alternative hypothesis:

$$H_1: \mu \neq 33 \text{ seconds}$$

The "not equal to" symbol, $\neq$, is not pointing one direction or the other on the number line. This alternative hypothesis says that the rats come from a population where the maze completion time is different from 33 seconds. Now the researchers are not predicting a mean completion time greater than 33 seconds or less than 33 seconds; they are leaving open the possibility of results in either direction.

Using a nondirectional alternative hypothesis requires an opposite statement for the null hypothesis. If Dr. Bellum and Ray had not predicted a direction for their results and had written, "$H_1: \mu \neq 33$ seconds," they would have used the following null hypothesis:

$$H_0: \mu = 33 \text{ seconds}$$

Translating this symbolic statement into words, we may write, "Here is the null hypothesis: Our sample of rats comes from a population with a mean maze completion time equal to 33 seconds." Does this mean that the only evidence in support of the null hypothesis would be a sample mean exactly equal to 33 seconds? No, a sample mean may be a small distance from μ as a result of sampling variability. A test statistic is used to determine whether a difference between the sample mean and the population mean is noteworthy.

A quick note about directional signs: a lot of people accidentally mix up the meaning of the symbols "<" and ">." Think of reading the symbols for "less than" or "greater than" in the way you are reading this sentence, from left to right.

Starting with <, the small pointy side corresponds to "less." You read the small pointy side of the symbol first, from left to right, so you will say "less than." If you read the symbol > from left to right, then first you encounter the open side, which means "greater." When you encounter the bigger side of the symbol first, "greater than" is what you will say.

How do we use the statistical null and alternative hypotheses? They define specifically what we are studying with any given test statistic, describe the predicted direction, and help us to interpret the observed results. The following paragraphs step through three brief examples. Let's start with the nondirectional alternative hypothesis. If Dr. Bellum and Ray think something is wrong with the rats and if they do not make a directional prediction for the results, the alternative hypothesis would be H_1: $\mu \neq 33$. Dr. Bellum and Ray obtained a sample mean $= 44.4$ seconds. Obviously, $M = 44.4$ seconds does not equal 33, so this sample mean appears to be evidence supporting the alternative hypothesis. But the researchers must determine statistically whether 44.4 seconds is an arbitrary, small distance from 33 seconds or a large, noteworthy distance from 33 seconds. We will explain shortly how we make that determination.

Now let's consider a second example of judging $M = 44.4$ seconds. Suppose Dr. Bellum and Ray have a directional alternative hypothesis that says H_1: $\mu > 33$ seconds. This hypothesis predicts that the sample came from a population with a mean maze completion time greater than 33 seconds. After measuring the

"Marginalized-GDP Per Capita" (6'x8'), by Gary Simpson, used with permission. Thin copper rods measure the per-capita gross domestic product (GDP) and are cement embedded. The artist writes, "Each country is shown individually with one inch equaling $667. They are placed randomly, perhaps unlike our planet."

8. Hypothesis Testing and Interval Estimation

rats and finding $M = 44.4$ seconds, they can tell that 44.4 is, in fact, greater than 33 seconds, which would be evidence in support of the alternative hypothesis and not in support of the null hypothesis, which said H_0: $\mu \leq 33$ seconds. But again, a determination must be made about whether the distance between 44.4 and 33 is inconsequential or noteworthy.

Just to be complete in our explanation, let's consider a third example. Continuing with the directional alternative hypothesis predicting the rats would take longer on average to complete the maze than the usual healthy rats, let's say Dr. Bellum and Ray get a sample mean of 26 seconds. Dr. Bellum and Ray had written H_1: $\mu > 33$ seconds. Does $M = 26$ seem to support the alternative hypothesis? No, because the "greater than" sign is like an arrow head pointing toward the right side of the number line, predicting a result that is to the right of 33 seconds. But 26 is less than 33 seconds; that is, 26 is to the left of 33 seconds on the number line. $M = 26$ seconds would appear to support the null hypothesis, which said H_0: $\mu \leq 33$ seconds ("Here is the null hypothesis: Our sample comes from a population where the mean maze completion time is less than or equal to 33 seconds"). Thus, the alternative hypothesis is absolutely crucial in our interpretation of results. When you use a directional alternative hypothesis, you should be especially careful and double-check every time to make sure that the results came out in the predicted direction.

Choosing a directional alternative hypothesis limits researchers. Even if $M = 26$ is an extraordinarily fast time for completing the maze, it could not be detected as statistically noteworthy because the alternative hypothesis was H_1: $\mu > 33$ seconds. That is because $M = 26$ would agree with the null hypothesis that said $\mu \leq 33$. So when using a directional alternative hypothesis, researchers must be quite certain that the only possible noteworthy outcome would be in the predicted direction.

But we are getting ahead of ourselves by looking at a sample mean. One important detail that we need to make explicit: null and alternative hypotheses are written in advance of collecting data. Remember the steps of scientific methods discussed in Chapter 1: we encounter a problem or research question, then we make predictions and define measures, long before we collect data. If we specified the hypotheses *after* we saw the data, then it would be like choosing which team to support in a sporting event *after* the final score was announced.

Another detail that is decided in advance of collecting the data has to do with choosing a standard for what would be considered improbably extreme results. The next section will explain the details of this step in hypothesis testing: deciding what will be considered a small probability for the results.

Check Your Understanding

SCENARIO 8-A, Continued

This scenario concerns a sample of 36 American men in their 30s who seemed relatively tall, compared with the population mean of 69.4 in.

(Continued)

for American men in their 30s. We said the alternative hypothesis is $H_1: \mu > 69.4$ in. 8-9. Which hypothesis would a sample mean of $M = 68.8$ appear to support? Explain your answer.

Suggested Answer

8-9. *The alternative hypothesis said $\mu > 69.4$ in., which means the null hypothesis would be $H_0: \mu \leq 69.4$. The alternative hypothesis is predicting a sample mean that is somewhere to the right of 69.4 on the number line. But an observed $M = 68.8$ would be to the left of 69.4 on the number line, so this sample mean would appear to be evidence in support of the null hypothesis.*

Choosing a Small Probability as a Standard

Let's continue expanding on the details of hypothesis testing with the story of the rat shipment. For now, we will use the directional alternative hypothesis that said $H_1: \mu > 33$. Later we will go through the example again using the nondirectional alternative hypothesis; we have pedagogical reasons for wanting to start with the directional case. This directional alternative hypothesis suggests a possible reality in which our sample comes from a population with a mean maze completion time greater than 33 seconds—that is, the new rats will take longer on average than the usual, alert, and healthy rats to complete the straight-alley maze. This alternative hypothesis is paired with a null hypothesis that says $H_0: \mu \leq 33$. We do not believe that our sample comes from a population with a mean maze completion time less than or equal to 33 seconds, but we are going to test this idea. (Similarly, we did not believe we left the keys in the kitchen, but we had to go look there to eliminate it as a possibility.)

How much greater than 33 seconds would a sample mean, M, have to be in order for us to declare that it was substantially greater than this hypothesized μ? This question must be answered within the context of the null hypothesis and will require some knowledge of sampling distributions (Chapter 7). *Given that* the null hypothesis is true ($H_0: \mu \leq 33$), we would be sampling from a population of rats with running times that averaged 33 seconds or less. We need to go back to the concept of the sampling distribution of M and consider the variation that could be expected when sampling from a population with $\mu \leq 33$. We do not know the shape of the distribution of maze completion times. We do know, however, that if we repeatedly sample from the population of healthy rats and compute their mean running time for all possible samples, we can graph all those sample means—and the sample means will be normally distributed. Further, we know the average of all possible sample means will equal the population mean, so the sampling distribution of M is centered on 33. We could compute a z test statistic for each sample mean and arrange the z test statistics in a distribution. The distribution would look just like the one for the sample means: a normal distribution. The z test statistics will be centered on zero because any set of z scores has a mean $= 0$ (as we saw in Chapter 4). The population mean, $\mu = 33$, corresponds to a z test statistic $= 0$, as shown in Figure 8.1. The distribution is drawn as if the null hypothesis is true.

Notice that Figure 8.1 has two number lines: one for the maze completion time and one for the z test statistic. The population mean, 33, corresponds to $z = 0$; that is, if the sample mean equals the population mean, the z test statistic will be zero. If we get a sample mean that is far away from $\mu = 33$, then the z test statistic will be either a large positive number or a small negative number. The alternative hypothesis predicted a larger sample mean than 33, so noteworthy results would be unusually large positive z test statistics.

We have not answered the question posed above: *how much* greater than 33 seconds would a sample mean, M, have to be in order for us to declare that it was substantially greater than this hypothesized μ? We would need a sample mean that would be *extremely unusual* to obtain if the null hypothesis is true. This section of the text is about choosing a small probability to serve as a standard for unlikelihood under the null hypothesis. That small probability is called the *significance level*, a probability chosen in advance of data collection to serve as a standard for how unlikely the results must be to declare that the results are evidence against the null hypothesis. The significance level has a symbol called alpha, which is the lower-case Greek letter α. Traditionally, researchers use only certain numeric values for α, usually .01 or .05. What would these numbers mean? Let's use $\alpha = .05$. (The preceding mathematical expression with the symbol for

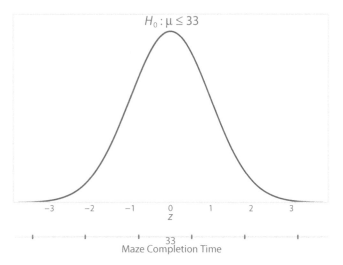

Figure 8.1

Distribution of z test statistic. This distribution of the z test statistic is drawn as if the null hypothesis is true. The number line for z can be compared with the number line for the maze completion time. If the sample mean equals the hypothesized population mean, then the z test statistic will equal zero.

alpha is how the authors of journal articles commonly tell their readers about the chosen significance level.) This significance level of $\alpha = .05$ means that if the null hypothesis is true, then a result at least as extreme as ours would be found 5 times out of 100 by chance alone when sampling from a population described in the null hypothesis.

The previous sentence contains a big concept to grasp. The standard normal distribution in Figure 8.1 represents what the sampling distribution of M would give us, after converting each repeated sample mean into a z test statistic—but only if the null hypothesis is true. In the alternative hypothesis, Dr. Bellum and Ray predicted a larger sample mean than 33, so an unusual outcome in our predicted direction would be a positive z score. The researchers' H_1 points to the right, so the unusual or rare results that would be evidence against the null hypothesis can be found only in the right tail. Please do not gloss over this detail! The alternative hypothesis is crucial for interpreting the results of our test statistic.

How far to the right on the number line of z test statistics do we need to travel before we run into a value of z that would indicate a substantial difference between M and μ? We would need to go to the point where a value of z cuts off an area equal to $\alpha = .05$ in the upper tail. Let's look in Table A.1 for the standard normal distribution and find a tail area $= .05$. You should be able to find $z = 1.645$. If the null hypothesis is true, then 5 times out of 100 by chance alone we could get a sample mean that would produce a z test statistic $= 1.645$ or greater (because many z values are beyond 1.645 and also would be unusually large values of z). Adding to the earlier figure, Figure 8.2 now includes 1.645 and α.

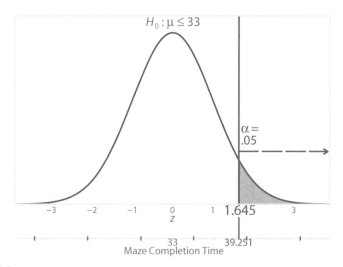

Figure 8.2

Distribution of z test statistic, with α in one tail. When the alternative hypothesis predicts that the mean will be in the upper tail, we place α in the upper tail. The critical value of $z = 1.645$ cuts off 5% of the distribution in the upper tail.

8. Hypothesis Testing and Interval Estimation

We could work backward from $z = 1.645$ by plugging in $\mu = 33$, $\sigma = 19$, and $\sqrt{N} = \sqrt{25} = 5$, and we could figure out that $M = 39.251$ would result in $z = 1.645$. But you do not have to worry about the mathematical gymnastics of figuring out this value of M. We suggested the idea of a sample mean that was unusual simply to motivate this discussion of the significance level. If we want to find an unusual or rare z test statistic, it will have to be at least as extreme as the value $z = 1.645$, which cuts off an area $= .05$ in the upper tail. The red tail area in Figure 8.2 corresponds to $\alpha = .05$. The numeric value of z that cuts off α is called a *critical value*. Here, the critical value equals 1.645. It is critical to our decision about whether the sample mean differs substantially from the population mean. The critical value is a point on the horizontal number line for the z test statistic. A vertical line drawn through the critical value defines the red tail area equal to α. Look back at Figure 8.2 to fully understand the meaning of that last sentence. Remember the analogy of the bleeding armadillo cake in Chapter 4? The critical value is like the location where a knife slices through the end of the cake, cutting off 5% of the cake, if $\alpha = .05$.

Critical values and α go together. They define each other. Where is the critical value? It is the point on the number line for the test statistic where we could draw a vertical line, and the tail area being cut off would equal α. Where is α? In this example, it is a tail area with a vertical line forming its boundary on one side. Where is the vertical line drawn? The line is drawn vertically through the critical value. We cannot separate critical values and α. The difference between a critical value and α is that a critical value exists only on the number line for the test statistic, whereas α is a probability (or area). Please reexamine Figure 8.2 to persuade yourself that critical values and α are inseparable.

Later in the book, we explain why some researchers may wish to use $\alpha = .01$ as a standard of smallness, whereas other researchers may choose $\alpha = .05$. Now that we have chosen $\alpha = .05$ as a standard for what is a small probability, we can compute the z test statistic for Dr. Bellum and Ray—and compute a certain probability that is connected with the test statistic. We can make decisions about the null hypothesis in one of two ways: we can compare an observed test statistic with a critical value or we can compare our "certain probability" with α. Both ways will lead us to the same conclusion, and in actual research, we usually use the way involving a comparison of two probabilities. The next section explains the "certain probability" that Dr. Bellum and Ray will compute.

Check Your Understanding

SCENARIO 8-A, Continued

This scenario pertains to the sample of 36 American men who seemed relatively tall, relative to a population mean $= 69.4$. The alternative hypothesis is H_1: $\mu > 69.4$ in. 8-10. If we have a z critical value $= 2.325$, what α must

(Continued)

we have chosen? 8-11. Instead of the α that you answered in Question 8-10, suppose we chose α = .025. What would be the z critical value?

Suggested Answers

8-10. Alpha would be .01, which can be found by looking up the critical value in Table A.1. This z = 2.325 is between two values of z: z = 2.32, which cuts off a tail area of .0102, and z = 2.33, which cuts off a tail area of .0099. 8-11. To find the z critical value using α = .025, we look for the tail area of .025 in Table A.1. Looking through the third column for the tail area, we find this amount of area is cut off by z = 1.96, which would be the critical value.

Compute the Test Statistic and a Certain Probability

If you check the outline of the steps of hypothesis testing, you will see that we are in the middle of the step that says, "Collect data on the rats, and compute the sample mean, the z test statistic, and a certain probability." After Dr. Bellum and Ray wrote their null and alternative hypotheses and chose their significance level, α =. 05, they collected data and computed some statistics: the sample mean, the z test statistic, and this certain probability. This certain probability, which is linked to the z test statistic, will be compared with the chosen significance level.

Dr. Bellum directs Ray to measure 25 rats in the straight-alley maze, and he computes a sample mean, $M = 44.4$. Dr. Bellum knows from years of research that healthy rats have maze completion times with a population average of $\mu = 33$ and a population standard deviation of $\sigma = 19$. She and Ray now have all the information needed to compute the observed z test statistic. Remember, the z test statistic follows the pattern of "(something minus its mean) divided by its standard deviation." The "something" is the sample mean, M. According to the Central Limit Theorem, "its mean" is the mean of M's sampling distribution: the population mean, μ. And "its standard deviation" is the standard error of the mean, $\sigma/\sqrt{N}$, which gives us the following formula for the z test statistic:

$$z = \frac{M - \mu}{\sigma/\sqrt{N}}$$

Dr. Bellum and Ray put their numeric values into the formula:

$$\frac{44.4 - 33}{19/\sqrt{25}}$$

We recommend writing down each computation that is performed. The numerator is equal to

$$M - \mu = 44.4 - 33$$
$$= 11.4$$

The denominator is equal to

$$\frac{\sigma}{\sqrt{N}} = \frac{19}{\sqrt{25}}$$
$$= \frac{19}{5}$$
$$= 3.8$$

Putting the numerator over the denominator, we find

$$\frac{11.4}{3.8} = 3.0$$

The z test statistic $= 3.0$ is a positive number because the sample mean is greater than the population mean ($44.4 > 33$). This is an important observation to make because our alternative hypothesis predicted a direction for the outcome. We had written H_1: $\mu > 33$. The alternative hypothesis' directional sign, $>$, was pointing toward the upper tail, meaning we predicted the sample mean would be to the right of the population mean on the number line. So far, so good—we have found evidence supporting that prediction. (Note that checking on the direction of the results was another step in our overview of hypothesis testing.) Now let's continue to interpret the observed z. The z test statistic $= 3.0$ means that three standard errors of the mean will fit in the gap between $M = 44.4$ and $\mu = 33$. But that is not enough of a conclusion. We are in the middle of a hypothesis testing example, and we need to know something about how likely we are to obtain a z test statistic at least this extreme if the null hypothesis is true.

Table A.1 tells us the proportion of the z values found in different sections of a standard normal distribution. We can find the area between $z = 0$ and a computed value of z, or we can find the area above a computed value of z. As a thought exercise, let's consider the implication of having a sample mean close to the population mean. If the sample mean is very close to the population mean, then a large portion of the distribution will be as extreme as M or more extreme than M. We can link back to the idea of sampling distributions; many of the repeated sample means would be piled up around the population mean. If Dr. Bellum and Ray had obtained $M = 33.5$, many sample means would be this distance or farther from $\mu = 33$, making our results highly likely or typical if the null hypothesis is true. But Dr. Bellum and Ray obtained $M = 44.4$. A greater distance between M and μ (such as the distance between 44.4 and 33) would occur less frequently, if we truly are sampling from a population with $\mu = 33$. Very few sample means would be this distance or farther from μ, making it less likely that our sample came from a population of healthy rats.

Back to Dr. Bellum and Ray, who computed a z test statistic $= 3.0$. We need to know the probability of obtaining a sample mean at least three standard errors greater than the population mean; in other words, we need a tail area. By consulting Table A.1, we find what Dr. Bellum and Ray found: a probability $= .0013$. This is the probability of obtaining a sample mean of 44.4 seconds or greater, given that we have sampled from a population with $\mu = 33$. This "certain probability" is called a *p value*, which is a probability of observing a test statistic at least as extreme as the one computed on our sample data, given that the null hypothesis is true.

To interpret the meaning of the p value, let's remember what we covered in Chapter 7. We could create a sampling distribution of the mean for the maze completion times, obtaining all possible samples of the same size from the same population, computing the mean time for each sample, and arranging the means in a distribution—but we do not have to. The Central Limit Theorem tells us what that distribution would look like. This p value $= .0013$ says that out of all those thousands of repeated means, the proportion of them equal to 44.4 or greater is $p = .0013$, if the null hypothesis is true. Another way of saying it is that 13 out of every 10,000 sample means would be at least as large as 44.4 seconds, if we are sampling from the population described by the null hypothesis. Adding to Figure 8.2, Figure 8.3 shows the observed z test statistic and its p value. A tiny tail area of $p = .0013$ is shown in blue, with the observed $z = 3.0$ cutting off that blue area. This blue area is only the tip of the right tail. As indicated by the red arrow, α is the whole tail area beyond the critical value, including the tiny blue area.

Just as a critical value and α were inseparable, the observed z test statistic and the p value are inseparable. We cannot compute $p = .0013$ without drawing a vertical line through a certain point on the horizontal number line. That point is the observed test statistic, $z = 3.0$. In the analogy of the bleeding armadillo cake, the observed z test statistic is a place where a knife could be used to slice off a very small piece of the cake (the blue area). The critical value, $z = 1.645$, is the location where we could slice of the *entire* tail area, including the tiny blue area.

Is $p = .0013$ a small probability? Let's consult our standard for smallness. We chose a significance level, $\alpha = .05$, meaning that a small probability would be .05 or anything smaller. Our p value, .0013, is indeed smaller than .05. What does that tell us? Our chosen significance level was our way of saying, "If the null hypothesis is true, then any result with a probability as small as .05 or smaller will be considered to be improbable, leading us to doubt that the null hypothesis is true." A test statistic of $z = 3.0$ or more extreme would occur by chance alone 13 times out of every 10,000 sample means if the null hypothesis ($\mu \leq 33$) is true. The computed p value was even smaller than the standard we had set for smallness ($\alpha = .05$), so we would consider our sample mean to be extremely unusual if the null hypothesis were true. It is so unusual, in fact, that we will *reject the null hypothesis*, a decision to conclude that the evidence contradicts the reality described in the null hypothesis. Rejecting the null hypothesis allows us to

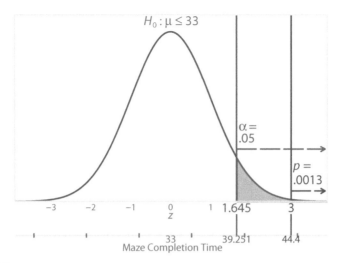

$H_0 : \mu \le 33$

$\alpha = .05$

$p = .0013$

1.645

−3   −2   −1   0   1   1.645   3
z

33   39.251   44.4
Maze Completion Time

Figure 8.3

Distribution of z test statistic, with α and p in one tail. This figure is just like Figure 8.2, except now we have added three details: the observed sample mean, M = 44.4 seconds; the corresponding z test statistic = 3; and the p value = .0013, shown as the tiny blue area in the upper tail of the distribution.

draw a conclusion that the results are *statistically significant*. We do not use the term *significant* unless a null hypothesis has been rejected. We also do not say "highly significant" or "more significant" because the decision about the null hypothesis is binary. Either we reject H_0 and conclude the results are statistically significant or we do not reject H_0 and conclude that the results are not statistically significant.

If the *p* value had been bigger than .05, we would have reached a different decision: to *retain* the null hypothesis, or *fail to reject* the null hypothesis. To retain the null hypothesis means we have found results that would be probable or typical if the null hypothesis is true, so we cannot rule out the null hypothesis as a possible description of reality. When we fail to reject the null hypothesis, we conclude that the results are *not* statistically significant. Let's look at an example of retaining the null hypothesis. Instead of a mean maze completion time of M = 44.4, suppose we had observed a sample mean of M = 33.5, which would result in a z test statistic that equals 0.1315789. (Test yourself—compute this value of z, using μ = 33, σ = 19, and N = 25.). This z test statistic, which we are rounding to 0.13 for simplicity, would cut off an area of p = .4483 in the upper tail. (Can you find this area in Table A.1?) Figure 8.4 illustrates this observed test statistic and p value.

The small tail section in red still represents α = .05, the area defined by the critical value, 1.645. In the analogy of the bleeding armadillo cake, we would slice the cake at the point where z = 1.645 and cut off 5% of the cake. The observed test statistic, z = 0.13, is close to z = 0, and a vertical line has been drawn through

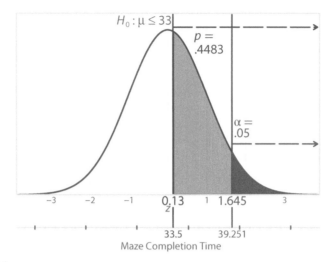

$H_0 : \mu \leq 33$

$p = .4483$

$\alpha = .05$

| -3 | -2 | -1 | 0.13 | 1 | 1.645 | 3 |

z

33.5   39.251

Maze Completion Time

Figure 8.4

Distribution of z test statistic, showing a large p value. If the rats took 33.5 seconds on average to complete the maze, the z test statistic would equal approximately 0.13. This value of z cuts off a large area in the upper tail: $p = .4483$. The p value is all of the area to the right of the vertical line drawn through $z = 0.13$.

$z = 0.13$. In the cake analogy, slicing the cake at $z = 0.13$ results in almost half of the cake being cut off, including the red tail area. All of the area to the right of that vertical line at $z = 0.13$ represents the p value. The blue area, including the red area, equals $p = .4483$. When we slice through the distribution where $z = 0.13$, we cut off 44.83% of the distribution. This p value says that if the null hypothesis is true and we are sampling from the population of healthy rats, more than 4 times out of 10 we would get a sample mean at least as extreme as ours. That is a large probability, compared with $\alpha = .05$, so we would have to retain the null hypothesis. The researcher would conclude that the evidence supports the null hypothesis that $\mu = 33$. We can say that $M = 33.5$ and $\mu = 33$ are statistically indistinguishable, or that the difference between them is not significant.

We are picky about the terminology used in hypothesis testing. Notice that both actions—rejecting and retaining—are associated with the null hypothesis. We do not take any action on the alternative hypothesis. It just sits there as a possible alternative description of reality. We suggested only a few terms associated with the actions taken on the null hypothesis: *rejecting* is one action and the other action is called either *retaining* or *failing to reject* the null hypothesis. If we fail to reject the null hypothesis, we must hang onto the null hypothesis, even though we had predicted an outcome described in the alternative hypothesis. We never embrace the null hypothesis as definitely true because we had good reasons for thinking it was not the best description of reality. Therefore, we do not want you to use the word *accept* in connection with a null hypothesis. You may say *retain the null hypothesis* or *fail to reject the null hypothesis*, just as juries will issue decisions of *not guilty*. Prosecutors may have reasons for thinking someone committed a crime, but the evidence may be insufficient to find the defendant guilty, and a jury may reach the decision of *not guilty*. In reality, the defendant may or may not be innocent. A verdict of not guilty is not the same thing as innocence, and failing to reject a null hypothesis is not the same thing as accepting it as truth.

Returning to the rat shipment example with $M = 44.4$ seconds, we want to remind you of an important step in our hypothesis test: we checked to make sure that the sample mean for the new rats was in the predicted direction. The alternative hypothesis said $H_1: \mu > 33$, and the sample mean, 44.4 seconds, is in the predicted direction because 44.4 is greater than 33. It may seem to you as if we are saying 44.4 is μ—but that is not the case. Remember, in this statistical alternative hypothesis, μ is translated as saying, "Our sample comes from a population where ...." When we check whether the results turned out as predicted, we are asking, "Now that we have $M = 44.4$, does this result support the idea that we sampled from a population with a mean > 33?" Any sample mean greater than 33 would lead us to say yes, so we have confirmed that the results are in the predicted direction.

It is possible to get a small p value and still retain the null hypothesis. Suppose the rats were not sluggish on the day that Ray took them into the lab for testing. Maybe they were tired from traveling and after a good night's sleep, they zipped through the maze in a mean time = 23.5 seconds. This sample mean would give us a test statistic that equals $z = -2.5$. (Test yourself again—calculate this number, using $\mu = 33$, $\sigma = 19$, and $N = 25$.) If we use Table A.1 and look up a tail area for this value of z, we would find a p value = .0062. (Surely you are a skeptical reader and are turning to Table A.1 right this second to check our work!) Is .0062 a small probability? Well, .0062 is less than α, which we chose to be .05. But if we get in a hurry and simply look at whether the p value is less than or equal to α, we can lose sight of the fact that we predicted *slower* rats than the usual healthy rats. Our prediction was that the rats take more time to complete the maze. In fact, this well-rested sample of rats was *faster* on average and took less time to navigate the maze. We cannot reject the null hypothesis because the mean was in the wrong direction. We recommend that students always draw a picture to accompany a hypothesis test.

Figure 8.5 shows the outcome for the extremely quick rats. The picture is drawn as if the null hypothesis is true, and the location of α in the upper tail is dictated by the alternative hypothesis, which said $H_1: \mu > 33$. The red upper tail area represents $\alpha = .05$, which is cut off by the critical value of $z = 1.645$. We draw a vertical line through the observed test statistic, $z = -2.5$, which is less than zero on the horizontal number line. The observed test statistic cuts off the blue area in the lower tail; that area is $p = .0062$. But only the upper tail represents the outcomes that the alternative hypothesis predicted as the unusual or rare events, if the null hypothesis is true. Our results run contrary to the alternative hypothesis. Because the sample mean was in the wrong tail, we must retain the null hypothesis and conclude that our new rats are not significantly slower than the usual rats. Based on the p value, can we say, "Oops—well, there is something wrong with these rats, but just in a way that was different from our prediction"? Unfortunately, no. We chose to make a directional prediction and missed an opportunity to detect

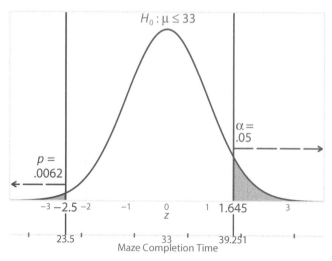

Figure 8.5

Distribution of z test statistic, when results were not in the predicted direction. If the sample of rats took 23.5 seconds on average to complete the maze, the z test statistic would equal –2.5. This value of z cuts off a small p value (.0062). But we must remember our prediction: that the rats would take longer than 33 seconds to complete the maze. Therefore, we must retain the null hypothesis.

a result as statistically significant. As ethical researchers, we cannot change our minds about the hypotheses after we see the data.

This section contained some of the most crucial information for understanding hypothesis testing. You may need to read it more than once. We have gone through all the steps in our overview of hypothesis testing. Next we will summarize the rules for testing a null hypothesis in two ways: (1) by comparing a p value to the significance level and (2) by comparing an observed test statistic to a critical value. We also illustrate the rules when using a nondirectional alternative hypothesis.

Check Your Understanding

SCENARIO 8-A, Continued

Continuing with our example of American men in their 30s, we had an alternative hypothesis that said μ > 69.4 in. Suppose we are told that the population standard deviation = 3.1. We choose α = .05. Then we measure the 36 men's heights and compute a sample mean = 70 in. 8-12. Does our sample mean appear to support the null hypothesis or the alternative hypothesis? 8-13. Compute the z test statistic. 8-14. Use Table A.1 to find a p value for our observed z test statistic. 8-15. Should we reject or retain the null hypothesis, and why? 8-16. What does this decision on the null hypothesis mean about this sample of men?

(Continued)

8-12. H_1 predicts that our sample comes from a population with a mean greater than 69.4 in. In fact, $M = 70$ is greater than 69.4, so at first glance it appears the sample supports the alternative hypothesis. But we have to ask: is 70 close to 69.4, or is it relatively far away, given the amount of sampling variability we might expect to see in the sampling distribution of M? 8-13. Our z test statistic's numerator is $M - \mu = 70 - 69.4 = 0.6$. The z test statistic's denominator is $\sigma/\sqrt{N} = 3.1/\sqrt{36}$. If we do the math, we find the denominator is $3.1/6 = 0.5166667$ (on our calculator). Do not round this figure! To get the final answer for z, we divide the unrounded denominator into 0.6, and we get $1.1612903 \approx 1.16$. (We round only the final answer.) 8-14. Table A.1 shows a tail area of .1230 for this value of the test statistic. 8-15. Although the sample mean was in the predicted direction and was greater than the population mean, the p value, .1230, is greater than $\alpha = .05$. Therefore, we must retain or fail to reject the null hypothesis. 8-16. We would conclude that it is fairly likely that our sample comes from a population of American men with a mean height of 69.4 in. Another way to state the result is that our sample mean is not significantly different from $\mu = 69.4$ in.

Decision Rules When H_1 Predicts a Direction

Hypothesis testing involves *decision rules*, the requirements for taking the action to either reject or retain the null hypothesis. The decision rules can focus on the comparison of probabilities (p and α), or they can focus on the comparison of test statistics (the observed z test statistic and a critical value). Our explanation of hypothesis testing mostly has emphasized the comparison of a p value with our standard for a small probability: the significance level, α. We have mentioned the possibility of reaching the same decision about the null hypothesis by comparing an observed z test statistic with a critical value. As we talk about decision rules, we will reinforce the idea that these two approaches—the comparison of probabilities and the comparison of test statistics—are two sides of the same coin. This section focuses on the decision rules when the alternative hypothesis predicts a direction for the results. The section on Decision Rules When H_1 Is Nondirectional covers the decision rules when no prediction is made about the direction of the results.

We already have shown how to test the null hypothesis by comparing p and α in the case of a directional alternative hypothesis. But to be clear, let's state the decision rule. For situations in which we have a directional alternative hypothesis, we have a one-tailed p value decision rule. All of α is in the predicted direction, and the p value is in one tail, so we call it a one-tailed p value. This hypothesis test is called a *one-tailed test*. The one-tailed p value decision rule is

If the observed test statistic is in the tail predicted by H_1

AND

if the observed p value in the predicted tail is less than or equal to α,

then reject the null hypothesis.

Otherwise, retain the null hypothesis.

We are comparing two probabilities (p and α). Both parts of the one-tailed p value decision rule must be true: the results must be in agreement with the direction predicted by the alternative hypothesis, and the p value must be small enough. Another way to check whether the z test statistic is in the predicted direction is to see whether the sample mean corresponds to the prediction in the alternative hypothesis. In the rat shipment example, we said H_1: $\mu > 33$. This statement is saying that our sample comes from a population with a mean greater than 33. If the researchers computed $M = 44.4$, then this sample mean appears to be in line with that statement.

This one-tailed p value decision rule has a counterpart that involves the comparison of the observed z test statistic with a critical value. This counterpart is called the one-tailed critical value decision rule, which says

If the observed test statistic is in the tail predicted by H_1,

AND

if the observed test statistic is equal to or more extreme than the critical value

reject the null hypothesis.

Otherwise, retain the null hypothesis.

We are comparing two values of the z test statistic: the observed z and the critical value of z, both of which can be found as points on the horizontal number line. The only way that the observed z test statistic can be more extreme than the only critical value in a one-tailed test is by occurring in the predicted direction. So our decision rule actually is redundant when it says "if the observed test statistic is in the tail predicted by H_1." We just want to emphasize the need to check the direction of the results.

Let's take another look at an earlier figure, reproduced here as Figure 8.6. The observed z test statistic for the sample of new rats was 3.0. The z critical value was 1.645. Because 3.0 is more extreme than 1.645, the one-tailed critical value decision rule tells us to reject the null hypothesis. In this example, the only way to find significance is when the observed test statistic is in the predicted tail because that is where the critical value is. The observed z test statistic is in the upper tail, meaning the sample mean is larger than the population mean, as predicted by H_1. The fact that the observed z is more extreme than the critical value allows us to reject the null hypothesis.

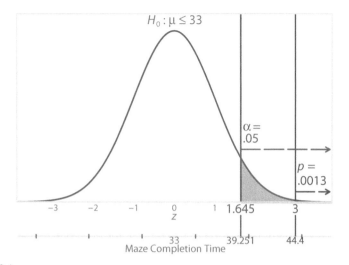

$H_0 : \mu \leq 33$

$\alpha = .05$

$p = .0013$

Maze Completion Time

Figure 8.6

Distribution of z test statistic, with α and p shown as tail areas. The alternative hypothesis said $\mu > 33$. We always rely on H_1 to decide where to put α. The directional sign, $>$, pointed toward the upper tail, and that is where α is located (the red shaded area).

The critical value is on the right-hand side of the horizontal number line because of the link between the critical value and α. The critical value cuts off α, an area that is in the upper tail because Dr. Bellum and Ray's alternative hypothesis said $H_1: \mu > 33$. The directional sign points toward the upper tail because the researchers predicted that the new rats would take longer on average than the population mean of 33 seconds for the usual healthy rats to complete the simple maze. Study Figure 8.6 and persuade yourself that the one-tailed p value decision rule and the one-tailed critical value decision rule are leading you to the same conclusion about the rats. Notice that the critical value decision rule compares two values of the test statistic, whereas the p value decision rule compares two probabilities. With the critical value decision rule, we are comparing an observed z with a critical value for z. With the p value decision rule, we are comparing an obtained probability, the p value, with α, our standard for what is a small probability.

When the alternative hypothesis predicts results in the lower tail, the one-tailed decision rules are the same as when H_1 predicted results in the upper tail. For the one-tailed p value decision rule, we would check to make sure the observed test statistic was in the lower tail as predicted. If the observed test statistic was *not* in the lower tail, then we would stop right there and retain the null hypothesis; there would be no need to compare p with α. If the observed test statistic did appear in the lower tail, then we would compare p with α. With the one-tailed critical value decision rule, we would check whether the observed test statistic was equal to or more extreme than the critical value in the lower tail.

SCENARIO 8-B

Suppose we have a sample of 78 boys who are 1 year old, and we suspect they may be underweight on average. To compare their mean weight with the population mean for boys of that age, we compute the z test statistic. We find a p value = .0052. 8-17. Can we conclude that these boys are significantly underweight?

Suggested Answer

8-17. We cannot draw a conclusion until we check whether the results were in the predicted direction. The scenario says we suspect the boys may be underweight—that is, we think our sample comes from a population with a mean less than the (unstated) known population mean for 1-year-old boys. This implies H_1: $\mu <$ ___ (some unstated value). Unless we find out that the z test statistic is a negative number (indicating that the sample mean is less than the population mean) or we see the numeric values of the sample mean and the unstated but previously known μ, we do not have enough information to test the null hypothesis.

Decision Rules When H_1 Is Nondirectional

Let's consider the p value and critical value decision rules for testing a nondirectional alternative hypothesis. Returning to the situation where Ray thought something was wrong with the rats, let's stop there and not say anything about the rats seeming sluggish. He and Dr. Bellum discuss the unusual behavior of the new rats. Suppose they choose an alternative hypothesis that does not predict whether the rats will go faster or slower on average, compared with the known population mean for healthy rats; they simply think the new rats will differ from the usual rats on the mean maze completion time. If there is something wrong with the new rats and they differ significantly from healthy rats in either direction—by running too fast or too slow on average—then the researchers want to return the rats and get a new shipment. The alternative hypothesis therefore will not contain a directional sign pointing one way or the other. It will say H_1: $\mu \neq 33$. The researchers will test the null hypothesis that says H_0: $\mu = 33$.

Let's draw a picture (Figure 8.7). Our sampling distribution still looks like a normal distribution. The distribution is centered on zero, a numeric value that would be obtained if the sample mean equaled $\mu = 33$. Note the null hypothesis written above the distribution. It is there because the picture is still drawn as if the null hypothesis is true, and we still are looking for improbable results if H_0 is true. Now that we have a nondirectional alternative hypothesis, we want to detect if the rats take significantly longer on average or if the rats take significantly less time on average to complete the maze. We are looking for rare events in either

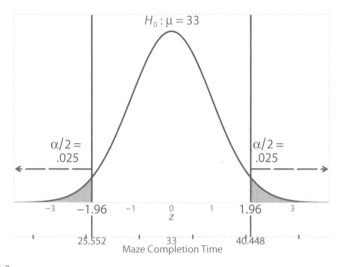

Figure 8.7

Distribution of z test statistic when no direction is predicted. Remember that we always look at the alternative hypothesis to decide where to put α. Now that we have H_1: $\mu \neq 33$, there is not a directional symbol pointing in one direction. So we must split up α, putting half of it in each tail of the distribution.

direction, so we will perform a two-tailed test. A *two-tailed test* has two critical values, one in the left tail and one in the right tail of the distribution.

We will continue to use a significance level of .05 as our standard for a small probability. But α is a *total* probability defining unusual or unlikely events, given the null hypothesis is true. So we will split up α, putting half of .05 in the upper tail and the other half of .05 in the lower tail. Each tail will have a critical value. The critical value in the lower tail will cut off .025, and the critical value in the upper tail will cut off .025. We cannot use 1.645 anymore as a critical value because it cuts off .05 in one tail. Look in Table A.1 for a tail area of .025. You should find that $z = 1.96$ cuts off an area of .025 in the upper tail. By symmetry we know that $z = -1.96$ cuts off an area of .025 in the lower tail. These two numbers, $z = -1.96$ and $z = +1.96$, are our critical values, as shown in Figure 8.7.

When we covered one-tailed tests, we talked about the *p* value decision rule first, followed by the critical value decision rule. For two-tailed tests we will start with the critical value decision rule:

If the observed test statistic is equal to or more extreme than either

critical value,

reject the null hypothesis.

Otherwise, retain the null hypothesis.

With a two-tailed test we do not have to check whether our results turned out in a predicted direction because there is no such prediction. We want to detect a significant result no matter whether the new rats are running too fast or too slow on average, compared with the population mean for healthy rats. To practice hypothesis testing using the critical value decision rule for a two-tailed test, let's use the three values of the z test statistic already mentioned in this chapter. Then we will choose a fourth value of the test statistic to lead us into a discussion of the two-tailed p value decision rule.

First, let's test the null hypothesis using $z = 3.0$, which is what Dr. Bellum and Ray computed for the rats that ran the maze in a mean time of 44.4 seconds. Reread the critical value decision rule for a two-tailed test, and then ask yourself: is the observed test statistic, $z = 3.0$, more extreme than one of the critical values (-1.96 or $+1.96$)? Yes, so we can reject the null hypothesis. Figure 8.8 has added the observed z test statistic $= 3.0$ to Figure 8.7.

The observed $z = 3.0$ is farther up the number line to the right of the critical value 1.96. Our decision to reject the null hypothesis means that the new rats have a mean running time that is significantly different from the population mean, $\mu = 33$. We could take the interpretation a step further. Now that we have rejected the null hypothesis using the two-tailed critical value decision rule, we can look at the sample mean, $M = 44.4$, and say it is significantly greater than $\mu = 33$. (Your instructor may say we are wrong to interpret the result of a two-tailed test by talking about the direction of the results. We think that *after* the significant difference is detected, the direction can be discussed. But it is a matter of opinion.)

Now let's use a second value for the z test statistic to test the same H_0: $\mu = 33$. A previous example described remarkably quick rats who completed the simple maze in $M = 23.5$ seconds. In that example, we computed an observed z test statistic

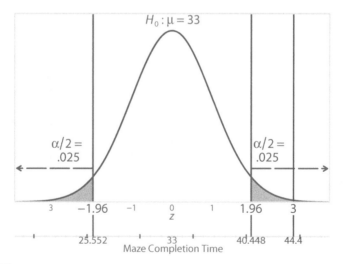

Figure 8.8

A nondirectional H_1 and two-tailed critical value decision rule. When we do not predict a direction, $\alpha = .05$ is divided between the two tails. We have a critical value in each direction on the number line (-1.96 and $+1.96$). This figure shows the observed z test statistic $= 3$, which is more extreme than one of the critical values.

8. Hypothesis Testing and Interval Estimation

= −2.5. Is this observed z more extreme than a critical value? Yes. Because −2.5 is more extreme than critical value of −1.96, we reject the null hypothesis. (Notice that −2.5 is a smaller number than −1.96. An observed test statistic that is extreme in the lower tail will be a number that is smaller than a critical value. That is why our critical value decision rules say "equal to or more *extreme* than" instead of "equal to or *greater* than." As we see here, it is possible to have an observed test statistic that is *less* than a critical value in the lower tail.) Rejecting the null hypothesis means our sample of new rats had a mean maze completion time that was significantly different from the population mean, $\mu = 33$. Now that we have rejected the null hypothesis, we can look at the sample mean, $M = 23.5$, and conclude that the new rats ran the maze in a significantly shorter average amount of time than the usual healthy rats.

Let's use the two-tailed critical value decision rule with a third value for the observed test statistic. An earlier example said a sample of new rats took an average of 33.5 seconds to complete the maze, which resulted in an observed z test statistic = 0.1315789. Figure 8.9 shows this result, rounded to $z = 0.13$.

Is the observed z more extreme than a critical value? No, it is not. This $z =$ 0.13 is close to zero and is between the critical values. It is not more extreme than a critical value in either tail; therefore, we must retain the null hypothesis. This decision means our new rats have a mean maze completion time that is not significantly different from the mean for healthy rats ($\mu = 33$).

Real research is more likely to use p value decision rules instead of critical value decision rules, and we have one more decision rule to cover: the two-tailed p value decision rule. But this kind of p value requires some explanation. One more example using the critical value decision rule will help us to transition to an explanation of the two-tailed p. Suppose we have the same nondirectional

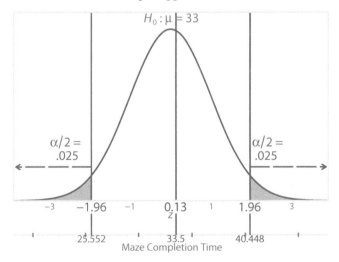

Figure 8.9

Critical value decision rule for a two-tailed test when results are not extreme. If the sample of rats took 33.5 seconds to complete the maze, the z test statistic would be approximately 0.13, which is not more extreme than either critical value.

alternative hypothesis and the same critical values as the previous example. But in this new example, the questionable rats take an average of 39.99 seconds to complete the simple maze. We compute an observed z test statistic = 1.84 (rounded). (Check to make sure you can get the same answer, using $\mu = 33$, $\sigma = 19$, and $N = 25$.) Before addressing the p value decision rule for a two-tailed test, let's use the two-tailed critical value decision rule. Is the observed z test statistic (1.84) equal to or more extreme than a critical value? No. 1.84 is between the two critical values (−1.96 and +1.96) and does not go beyond either critical value. Our decision, therefore, is to retain the null hypothesis and conclude that the new rats come from a population with a mean maze completion time that is not significantly different from $\mu = 33$, the population mean for healthy rats.

This example may not seem much different from the one that preceded it, but we chose our numbers carefully. As you know, an observed z test cuts off an area called the p value. Look in Table A.1 and find a tail area for $z = 1.84$. You should find a probability = .0329. Are you concerned that we retained H_0 using the critical value decision rule, yet this probability is less than .05? Previously, when a probability associated with the observed test statistic was less than or equal to α, we rejected a null hypothesis. Here is the catch: now we have a two-tailed significance level, $\alpha = .05$, but *the probability that you just looked up is a one-tailed probability*. We need to compute a *two-tailed p value* because the alternative hypothesis is nondirectional.

You may think, "But wait! We have an observed test statistic in only one location on the number line because we have only one sample mean. How can we have a two-tailed p value?" You're right, we found $M = 39.99$, and then we computed $z = 1.84$. But we did not know in advance of running the study which way the results would go. We set up the hypothesis test with half of α in each tail. We need a two-tailed p value to compare with the two-tailed α. Figure 8.10 shows the two critical values (−1.96 and +1.96), which cut off half of α as an area in each tail. The figure also shows our observed z test statistic, $z = 1.84$, which cuts off a probability = .0329 in the upper tail. Finally, the figure shows $z = -1.84$, which is what we might have gotten if the sample mean had been equally extreme in the opposite direction from what we obtained. This other value of z cuts off a probability = .0329 in the lower tail. The two-tailed p value is .0658 (i.e., .0329 + .0329). The two-tailed p value is compared with a two-tailed significance level.

Remember, the arrows help us to understand how much area is associated with $p/2$ and $\alpha/2$ in each tail. The red arrows go with the red shaded areas for $\alpha/2$ in each tail. The blue arrows remind us that in this graph, the areas for $p/2$ in each tail include both the blue and red shaded areas. (If the p value were smaller than α, as in Figure 8.6, the red arrow would remind us that the entire tail area was included in α.) Now we can state the two-tailed p value decision rule for testing our nondirectional alternative hypothesis:

If the two-tailed p is less than or equal to the two-tailed α,

reject the null hypothesis.

Otherwise, retain the null hypothesis.

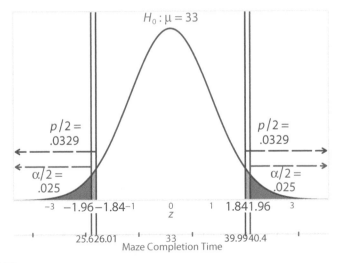

$H_0 : \mu = 33$

$p/2 = .0329$

$p/2 = .0329$

$\alpha/2 = .025$

$\alpha/2 = .025$

-3 -1.96 -1.84 -1 0 1 1.84 1.96 3
z

25.6 26.01 33 39.99 40.4
Maze Completion Time

Figure 8.10

Two-tailed p values. If the rats took an average of 39.99 seconds to complete the maze, the observed z test statistic would equal 1.84. This value cuts off an area = .0329 in the upper tail. But we need a *two-tailed p value* to compare to the two-tailed α. The two-tailed p is found by looking for the area cut off by $z = -1.84$; because of the symmetry of a normal distribution, we know the area will be .0329. The two-tailed p value will be .0658 (i.e., .0329 + .0329). By comparing the total p with the total α, we decide to retain the null hypothesis, the same decision reached with the critical value decision rule.

Let's use this two-tailed p value decision rule to test our null hypothesis for our situation in which $M = 39.99$ and $z = 1.84$. The two-tailed p is .0658. Is this probability smaller than or equal to $\alpha = .05$? No, it is not. We must retain the null hypothesis and conclude that our sample mean for the rats' maze completion time did not differ significantly from the population mean for healthy rats. This is the same decision that was reached with the critical value decision rule for a two-tailed test.

When we use test statistics, we need to know the conditions under which they will provide valid results. These conditions for the z test statistic will be discussed in the next section.

Check Your Understanding

SCENARIO 8-B, Continued

Instead of expecting the boys to be underweight, we simply want to check whether they differ on average from the population mean of 25.5 lb. This mean comes from a longitudinal study of thousands of American children (McDowell et al., 2008). 8-18. Write the alternative hypothesis in symbols

(Continued)

and words. 8-19. We weigh 78 boys who are 1 year old, compute their mean weight, and calculate the z test statistic. Then we find a p value = .0052. Using $\alpha = .05$, do we have enough information to test our revised null hypothesis?

Suggested Answer

8-18. We think the boys' mean will differ from the known population mean, so the alternative hypothesis is H_1: $\mu \neq 25.5$ lb. Because we have a nondirectional alternative hypothesis, we can use the two-tailed p value decision rule and ask, "Is p less than or equal to α?" Because .0052 is less than .05, we reject the null hypothesis and conclude that this sample of 1-year-old boys differs significantly on average from the population mean weight of 1-year-old boys. We do not have enough information here to judge whether this sample is overweight or underweight.

Assumptions

The use of the z test statistic and the computation of accurate p values require that certain conditions be met. These conditions are called *assumptions*, which are statements about the data or the population that allow us to know the distribution of the test statistic and to compute p values. Just think, if we did not know that the z test statistic had a sampling distribution that followed a normal curve, then we could not obtain p values and we could not test hypotheses. When we introduced the Central Limit Theorem, we described it as a gift from mathematical statisticians that freed us from having to create sampling distributions for M. We made a swap: we used a theoretical distribution instead of a sampling distribution for M. We did not want to go through the arduous task of creating a sampling distribution for the z test statistic. We were able to set aside that unobtainable sampling distribution and use a standard normal distribution instead. What allowed us to make that swap? Assumptions. To be certain that the sampling distribution of the z test statistic will look like the standard normal distribution, assumptions must be met. If the sampling distribution *does not* look like the standard normal distribution, then our p values will not be accurate.

The z test statistic has two assumptions:

- Scores are independent of each other.
- Scores in the population are normally distributed.

If scores are independent, they do not have any connection with each other. That makes sense. Think of independence as being similar to an exam requirement for students to do their own work. When we have achieved the condition as stated in the assumption, we say the assumption has been *met*. When we have not achieved the condition as stated in the assumption, we say the assumption has been *violated*. Suppose we are running a study and we discover that we have three sets of twins in the treatment group. We would have good reason to worry that we had violated the independence assumption because within a set of twins, there are two people who are a lot alike. (It is possible to study twins, but not in a one-group study that would use the z test statistic.)

How do we make sure that the independence assumption is met? Typically, random sampling from the population would assure that we obtained independent scores from the participants. The rat shipment example used random sampling for this reason and for the purpose of assuring they could generalize their results to the entire shipment. Otherwise the research assistant Ray may have tested only the slowest rats who were easiest to get out of their cages. Yet many studies use convenience samples. Does that mean the researchers have violated the independence assumption? We don't know. Many researchers make the judgment that their convenience sample is sufficiently similar to the sample that they would have obtained through random sampling from a population. There is no way to know whether a convenience sample provides data similar to what would be obtained from a random sample. Usually researchers trade the risk of a biased sample for the ability to complete the study at all. That is because random sampling sometimes is impossible. Imagine trying to draw a random sample from among all children with type 1 diabetes in a given culture. Researchers often have to conduct their studies in only one or a few locations, so random sampling throughout a country would be impossible. Further, researchers must provide ethical, safe treatment of participants, especially children, taking into account their willingness to participate and the parents' informed decision to let the children participate. Obviously people cannot be forced to participate in research, even if they were randomly sampled. Seeking volunteers to participate may be the only way to obtain a sample.

Even if a study has random sampling, careful experimental methods must be followed to assure that the scores remain independent. Let's consider a situation where participants have been allowed to influence each other. Suppose we are running a study in which we ask patients who are considering hip replacement surgery to rate their typical pain during 10 different activities—walking, climbing stairs, sleeping, and so on. The data are collected in the waiting room of a surgeon's office. Then we find out that during one busy morning of data collection, eight patients had talked to each other about the survey. One of them said, "I'm going to answer these questions based on my *worst* typical pain." After thinking about this statement, several other patients changed their answers. Would the scores of those patients be independent of each other? Maybe not.

Now let's consider the normality assumption. How do we make sure that it is met? The good news here is that we know the sample mean, M, is normally distributed as long as the population of scores is not unusual in shape. Again, we are making use of the Central Limit Theorem. Because M is normally distributed, the z test statistic is also normally distributed. That is enough for us to say that the normality assumption for the z test statistic is met. So even though we do not technically meet the assumption that the population of scores is normal, the end result is that the z test statistic is normally distributed, the same as if we had a normally distributed population of scores.

Every test statistic has assumptions, and the assumptions are not the same for all statistics. We can think of assumptions as being like the owner's manual for a test statistic. It seems that every device comes with an owner's manual, but there are different rules for different devices. The consequences of breaking the rules are not equally serious, either. The manual for a handheld hair dryer might have rules such as, "Do not use this blow dryer while taking a bath or shower," "This blow dryer should not be used by children," and so forth. A person can be electrocuted by using a hair dryer while showering, so clearly this rule must be taken seriously. On the other hand, a 14-year-old is a child, yet who is going to keep a teenager from using a blow dryer? Violating some assumptions is like blow-drying one's hair in the shower, and violating other assumptions is like letting a teenager use the blow dryer. With the assumptions for the z test statistic, violating the independence assumption is like the electrocution case, and violating the normality assumption is like the teenager using the blow dryer. Violation of the independence assumption can destroy the validity of the z test statistic—that is, the z test statistic's sampling distribution would not match the standard normal distribution. As a result, the p value may not be trustworthy. As this book progresses, you will learn more test statistics. With each new test statistic you should think about its assumptions and whether a violation of an assumption is like electrocution or like a teenager using the hair dryer. We will talk about assumptions in greater detail in Chapter 11, when we describe statistical research that shows there are situations in which one assumption of a certain test statistic can be violated, yet we can still get valid p values (i.e., the test statistic's sampling distribution still matches the shape of a theoretical distribution).

We introduced the z test statistic in Chapter 7, where we also first discussed interval estimation. In this chapter we have shown how to test a null hypothesis using the z test statistic. Next we will describe hypothesis testing using a confidence interval for the population mean, μ.

Check Your Understanding

8-20. What are assumptions? 8-21. Why do we need to know about assumptions?

Suggested Answers

8-20. Assumptions are statements about the population or the data that describe the conditions that allow a valid use of a test statistic and accurate computation of a p value for hypothesis testing. 8-21. For the p value to be accurate, the sampling distribution of the test statistic needs to match the shape of a theoretical distribution. The z test statistic has a sampling distribution that will match the shape of the standard normal distribution if the z test statistic's assumptions are met.

8. Hypothesis Testing and Interval Estimation

Testing Hypotheses with Confidence Intervals: Nondirectional H_1

We introduced interval estimation in Chapter 7. As we said, an interval estimate is called a confidence interval when we specify a percentage of intervals (which could be computed on repeated samples) that would contain the parameter being estimated. We do not perform the repeated sampling; we rely on knowledge about sampling distributions to allow us to say what would happen if we were to perform repeated samples. Now we will use confidence intervals to test hypotheses. Why did we lead you through the rest of this long chapter if we can reach the same decisions about hypotheses by using confidence intervals? We find that students understand confidence intervals better after they learn the above material, which remains pervasive in the scientific literature. An advantage of also reporting confidence intervals in journal articles is that the readers can see how precise the estimate is. Consider our example in Chapter 7 when we talked about our confidence in predicting that a tornado would strike our area between two dates. If we tell you in December that we are 95% confident that a tornado will touch down near our city during the second week of May, that would be a fairly impressive prediction. But if we said we are 95% confident that a tornado would hit our area next year—a much wider window of time—then you may be less impressed. Similarly, interval estimates provide more information for readers of journal articles about the variables being measured. Those familiar with the variables being measured can judge how wide or narrow the interval estimates seem to be.

So far in this chapter we have used the sample mean, M, as a point estimate of the population mean, μ. Then we computed the z test statistic to compare M with a hypothesized value of $\mu = 33$. Now we will compute some examples of interval estimates of M, and we will use the confidence intervals to test various null hypotheses. Again we will subject you to the never-ending rat shipment example, which we dearly love as a teaching tool. For simplicity we will begin with the nondirectional case, in which Dr. Bellum and Ray said something was wrong with the rats, but they did not know if they would go faster or they would go slower through the simple maze. The alternative hypothesis said

$$H_1: \mu \neq 33 \text{ seconds}$$

This symbolic statement says our sample comes from a population where the mean maze completion time is different from 33 seconds. This approach will allow us to detect a significant difference between M and μ if the rats take a long time to complete the maze or if they finish the maze very quickly.

We know that this nondirectional alternative hypothesis will require us to split α and put half of α in each tail of the standard normal distribution, which gave us the critical values and p values for hypothesis testing using the z test statistic. Let's take a look at this distribution again. Figure 8.11 reproduces Figure 8.7 for the rat shipment example.

Five percent of the possible values of z test statistics are represented on the number line as being equal to or beyond the two critical values (-1.96 and $+1.96$),

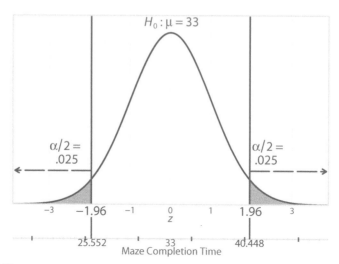

$H_0 : \mu = 33$

$\alpha/2 = .025$

$\alpha/2 = .025$

| -3 | -1.96 | -1 | 0 z | 1 | 1.96 | 3 |

25.552   33   40.448

Maze Completion Time

Figure 8.11

Figure 8.11

Distribution of z test statistic when no direction is predicted. The alternative hypothesis said H_1: $\mu \neq 33$. With no direction being predicted, α is divided between the two tails.

and 95% of the possible values of z test statistics are represented between the critical values. A two-tailed test with $\alpha = .05$ will correspond to a 95% confidence interval. If we are sampling from the population described in the null hypothesis, which is a population similar to healthy rats, then there are many numeric values around $\mu = 33$ that would be typical values for M. With repeated samples we would not expect to get the same sample mean every time because of sampling variability. As you will recall from Chapter 7, this idea of repeated samples is crucial for interpreting confidence intervals. We are quantifying the sampling variability inherent in any estimate of a population parameter.

To find an interval estimate of the population mean, μ, we need to know the margin of error or the distance that we would have to go between the sample mean and the population mean to reach a significant difference. We get that distance from a calculation involving a critical value for the z test statistic. As shown in Figure 8.11, one of these critical values is 1.96; we will use the positive value for simplicity. Reviewing what we covered in Chapter 7: this critical value can be defined as the margin of error divided by the standard error of the mean, SE:

$$1.96 = \frac{\text{margin of error}}{SE}$$

To isolate the margin of error on one side of the equation, we need to multiply both sides of the equation by the standard error of the mean:

$$SE \times 1.96 = \left(\frac{\text{margin of error}}{SE} \right) \times SE$$

$$= \frac{\text{margin of error} \times SE}{SE}$$

8. Hypothesis Testing and Interval Estimation

Now the right side of the equation has *SE* times the margin of error, divided by *SE*. The *SE* in the numerator divided by the *SE* in the denominator equals 1, so we are left with the margin of error. The margin of error for this example equals 1.96 times the standard error of the mean.

Let's compute the margin of error for the rat shipment example. The standard error of the mean equals the population standard deviation divided by the square root of the sample size, or $\sigma/\sqrt{N}$. In this example, Dr. Bellum knew from years of research that the standard deviation of healthy rats' maze completion times was 19 seconds, and she and Ray used a sample of 25 rats. So $SE = 19/\sqrt{25} = 19/5 = 3.8$. Now we can compute the margin of error:

$$\text{margin of error} = 3.8 \times 1.96$$
$$= 7.448$$

This is the distance that we would have to travel from $\mu = 33$ in either direction to reach a sample mean that is significantly different from 33. How much sampling variability might we expect to be associated with a sample mean? Dr. Bellum and Ray obtained $M = 44.4$, which appears to be quite a precise number, but we know that if they drew a different sample of 25 rats, most likely they would obtain a different sample mean. As we saw in Chapter 7, the margin of error gives us a range of values to estimate the population mean, instead of the point estimate (i.e., the sample mean). The lower limit of the confidence interval is

$$\text{Lower limit} = M - \text{margin of error}$$
$$= 44.4 - 7.448$$
$$= 36.952$$

The upper limit of the confidence interval is

$$\text{Upper limit} = M + \text{margin of error}$$
$$= 44.4 + 7.448$$
$$= 51.448$$

The 95% confidence interval as an estimate of the population mean is [36.952, 51.848]. Reporting this kind of interval estimate gives readers of research articles more information about how precise the estimate of the population mean is. If we could know for a fact that we are sampling from a population with a mean of 33 seconds, then our sample mean probably would be one of the values that would be somewhere near 33. But if it turns out that we actually are sampling from a population with a mean that is different from 33 seconds, then our sample mean probably would be close to some other value of μ. Taking into account the sampling variability inherent in our sample estimates of μ, does 33 belong within the range of plausible values for the mean of the population from which our sample was drawn?

Figure 8.12

A point estimate and 95% confidence interval, both estimating μ. The purple circle represents the population mean, hypothesized to equal 33. The purple diamond shows the sample mean, 44.4 seconds to complete the maze. The orange vertical lines show the lower and upper limits of the 95% confidence interval. The shaded area represents the entire range of values in the interval. Both the interval and the point estimate, M, were estimates of μ.

To answer this question, we ask whether our interval contains μ = 33. Figure 8.12 shows the population mean as the purple circle, the sample mean as the purple diamond, and the lower and upper limits of the interval estimate as orange lines. The entire interval estimate is the shaded area from the lower limit to the upper limit.

The number line in Figure 8.12 corresponds to the lower number line shown in many of the previous graphs, such as Figure 8.11. The sample mean is in the middle of the confidence interval, which quantifies the sampling variability in the estimation of the population mean. To test a null hypothesis that says μ = 33, we will ask: does the 95% confidence interval contain the population mean? No, it does not. Therefore, we reject the null hypothesis that said μ = 33, and we will conclude that 95% of the time that we conducted repeated samples from the same population that gave us the sample with M = 44.4, we would obtain a range of values that did not include 33. It is unlikely that we have sampled from a population with a mean of 33 seconds. The following is the confidence interval decision rule for testing a null hypothesis about μ when we have not predicted a direction for the results:

If the 95% confidence interval for estimating the population mean

does not include the hypothesized value for μ,

then reject the null hypothesis.

Otherwise, retain the null hypothesis.

In some disciplines, researchers report only the 95% confidence interval and the conclusion about the hypothesis, leaving out the p value entirely. In trying to obtain a new shipment of rats, Dr. Bellum may write to the supplier, "We found that a sample of N = 25 rats took significantly longer to complete a simple maze, compared with the usual population mean of 33 seconds, M = 44.4 (95% CI: 36.952, 51.848)." If we have tested the null hypothesis using the p value decision rule, must we also compute a 95% confidence interval and test the null hypothesis again? No, we do not have to do so. But it is a good idea to report the confidence interval because it allows readers to judge the precision of our point estimates.

8. Hypothesis Testing and Interval Estimation

Continuing with our example of American men in their 30s, suppose we had an alternative hypothesis that said $\mu \neq 69.4$ in. We had a population standard deviation $= 3.1$ in., a sample size of $N = 36$, and a sample mean $= 70$ in. 8-22. Compute the margin of error. 8-23. Compute the 95% confidence interval. 8-24. Determine whether the sample mean is significantly different from the population mean.

Suggested Answers

8-22. The margin of error will be the product of 1.96 and the standard error of the mean, SE. We know $SE = \sigma/\sqrt{N}$, so our $SE = 3.1/\sqrt{36} = 3.1/6 = 0.5166667$. Now we multiply this number by 1.96, and we get 1.0126667. 8-23. The lower limit of the 95% confidence interval is $M - $ margin of error $= 70 - 1.0126667 = 68.987333 \approx 68.987$. The upper limit is $M +$ margin of error $= 70 + 1.0126667 = 71.012667 \approx 71.013$. So the 95% CI is [68.987, 71.013]. 8-24. The null hypothesis would say $\mu = 69.4$ in. This value is contained within the confidence interval. Because the interval brackets the hypothesized mean, our sample mean is within sampling variability of the hypothesized mean, and $M = 70$ is not significantly different from $\mu = 69.4$.

Testing Hypotheses with Confidence Intervals: Directional H_1

Confidence intervals almost always are used with nondirectional alternative hypotheses. That makes sense—we have a range of values in both directions from the sample mean, reflecting the sampling variability that goes along with any estimate of a parameter. Things get trickier when a directional alternative hypothesis is involved. In the original story of the rat shipment example, Dr. Bellum and Ray thought the rats seemed sluggish, as if they would take longer than the usual rats to complete the simple maze. The alternative hypothesis said

$$H_1: \mu > 33 \text{ seconds}$$

Using $\alpha = .05$, we would look at the directional alternative hypothesis, notice the directional sign pointing toward the upper tail, and put α in the upper tail, as shown in Figure 8.13.

Five percent of the z test statistic values will be equal to or beyond the critical value of 1.645. The rest of the values of the z test statistic will be less than 1.645. Question: *How do we compute a confidence interval when there is no limit on the values in one tail?* That is the same question we had when we started thinking about this section of the book, and we are going to tell you what seems to be the most common approach to the problem. We need to use the critical value of 1.645 to find the margin of error—that is, the distance that the sample mean

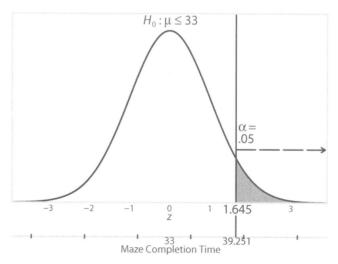

$H_0 : \mu \leq 33$

$\alpha = .05$

−3   −2   −1   0   1   1.645   3

z

33   39.251

Maze Completion Time

Figure 8.13

Distribution of z test statistic when results are predicted to be in the upper tail. When the alternative hypothesis said H_1: $\mu > 33$, all of α was placed in the upper tail.

would have to be from the population mean to find statistical significance. If we compute the confidence *interval* with two values, one on either side of the mean, then we would have a 90% confidence interval. It would be like having $\alpha = .05$ in the upper tail and $\alpha = .05$ in the lower tail. But when we have a directional alternative hypothesis, we will care about an extreme result in only one direction, so we will have to pay close attention to the location of the interval in relation to the location of the hypothesized μ.

Let's continue with the rat shipment example. To find the margin of error, we can use our previously computed standard error of the mean, 3.8, and multiply it by 1.645, the critical value when we have $\alpha = .05$ in the upper tail. In other words,

$$\text{Margin of error} = 3.8 \times 1.645$$
$$= 6.251$$

The sample mean was $M = 44.4$. The lower limit of the 90% confidence interval is

$$\text{Lower limit} = M - \text{margin of error}$$
$$= 44.4 - 6.251$$
$$= 38.149$$

The upper limit of the 90% confidence interval is

$$\text{Upper limit} = M + \text{margin of error}$$
$$= 44.4 + 6.251$$
$$= 50.651$$

8. Hypothesis Testing and Interval Estimation

μ = 33   M = 44.4

38.149   50.651

Figure 8.14

A point estimate and 90% confidence interval, both estimating μ. The purple circle still represents μ, hypothesized to be 33, and the purple diamond still represents the sample mean. Now a 90% confidence interval has been computed. The orange lines represent the lower and upper limits of the interval, and the shaded orange area shows the entire range of values in the interval estimate.

These results indicate that the 90% confidence interval would be [38.149, 50.651]. This interval is shifted up the number line, above the hypothesized population μ = 33. Our sample mean, 44.4, is significantly greater than the population mean because 33 is not contained within the interval. We can conclude that these rats probably did not come from a population of healthy rats. It is more likely that they came from the population of sluggish, unhealthy rats.

You may be concerned that we want our significance level, α, to be .05, but we are computing a 90% confidence interval. But what is most important is the location of the confidence interval relative to the hypothesized value of μ. Remember, the alternative hypothesis said μ > 33. The fact that the sample mean turned out to be greater than 33 (and the confidence interval is above 33 on the number line) shows us that the results were in the predicted direction. Because the lower limit of the confidence interval is higher than μ = 33, we know that the sample mean is significantly greater than the population mean.

Figure 8.14 illustrates this 90% confidence interval. Again, the population mean is indicated by the purple circle in the middle of the number line, similar to its location in the middle of the lower number line in Figure 8.13. The sample mean again is shown with the purple diamond. The lower and upper limits of the confidence interval are shown as orange lines, and the entire interval is shaded between those lines. The upper limit of this confidence interval really does not matter. We predicted that the sample came from a population with a higher mean than 33, and the lower limit is greater than the population mean—and that is the difference allowing us to say the results are significant. With any directional alternative hypothesis, it is crucial to pay attention to the relative location of the confidence interval and the population mean. We cannot get into a lazy habit of simply saying, "Oh, the interval does not contain the parameter, so there is statistical significance—yippee!" Let's illustrate the erroneous conclusion that would be reached if we got lazy.

Earlier in the chapter, we were talking about this same alternative hypothesis (H_1: μ > 33), and we used a different sample mean. What if the rats were tired from traveling, and after resting overnight they zipped through the maze in a mean time = 23.5 seconds. Our margin of error is still 6.251 (the product of 1.645 and the standard error of the mean, 3.8). Let's compute the 90% confidence interval using M = 23.5. The lower limit will be

$$\text{Lower limit} = M - \text{margin of error}$$
$$= 23.5 - 6.251$$
$$= 17.249$$

The upper limit will be

$$\text{Upper limit} = M + \text{margin of error}$$
$$= 23.5 + 6.251$$
$$= 29.751$$

The interval, ranging from 17.249 to 29.751, does not include the population mean $= 33$. *But look back at the alternative hypothesis.* It predicted a *greater* mean maze completion time ($H_1: \mu > 33$). We must retain the null hypothesis because the sample mean was not in the predicted direction. Therefore, when we have a directional alternative hypothesis, the confidence interval's decision rule for testing the null hypothesis about μ is

If the results are in the predicted direction

AND

if the 90% confidence interval for estimating the population mean

does not include the hypothesized value for μ,

then reject the null hypothesis.

Otherwise, retain the null hypothesis.

Check Your Understanding

SCENARIO 8-B, Continued

We are concerned about some 1-year-old boys in foster care who are suspected of being underweight. The sample consists of 31 children. Researchers from FantasyLand Studies tell us that on average, 1-year-old boys in the United States weigh 25.5 lb (about 11.6 kg) with a standard deviation of 4.1 lb. 8-25. Write the alternative hypothesis, using symbols. Then write the null hypothesis. 8-26. Why would we want to compute a 90% confidence interval? 8-27. Suppose the children were weighed during routine well-child visits, and we find a sample mean $= 26.8$ lb. What conclusion can you draw?

(Continued)

Suggested Answers

8-25. The alternative hypothesis would be $\mu < 25.5$ lb. The null hypothesis to be tested would be $\mu \geq 25.5$ lb. 8-26. We would use the 90% confidence interval because it corresponds to the situation where 5% of results are beyond one critical value. Our focus will be on whether the results are in the predicted direction. 8-27. When we have predicted a direction for our results, we can take the first step in testing the null hypothesis without computing an inferential statistic. The alternative hypothesis said $\mu < 25.5$ lb, meaning we think our sample came from a population with a mean that is lower on the number line than 25.5 lb. But the sample mean is 26.8 lb, which is higher on the number line than 25.5. So we must retain the null hypothesis and conclude that our sample came from a population with a mean of 25.5 lb or greater. We do not need to compute the 90% confidence interval or the z test statistic when the results are not in the predicted direction.

When reporting the results of the confidence interval, we would have to specify that a directional prediction was made about μ, leading to a 90% confidence interval. Then we would say whether the results were in the predicted direction and whether the difference between the sample mean and the hypothesized μ was significant.

What's Next

When we reject a null hypothesis, are we *certain* that our sample comes from a population that is described in the alternative hypothesis? No. Hypothesis testing relies on probability. We reject a null hypothesis when the results are *unlikely* to occur if the null hypothesis were true. Unlikely results under the null hypothesis make us doubt H_0 and reject it. But maybe we do have a sample that comes from the population described in the null hypothesis, and the sample happened to perform differently from that population on the day of the study. With the rat shipment example, when Dr. Bellum and Ray computed a sample mean maze completion time of 44.4 seconds and ended up finding statistical significance, they concluded that the rats must have been sick because they took significantly longer than the usual, healthy rats to complete the maze. *They might have been wrong.* Maybe they happened to randomly sample the laziest rats in the shipment. So it is possible to make mistakes: we might reject the null hypothesis when we should have retained it, and we might retain the null hypothesis when we should have rejected it. The problem is that we never know whether we have made a correct decision or a mistake in hypothesis testing because we cannot know what is truly happening in the population. Ensuring a decent probability of making correct decisions and reducing the chances of errors are the topics of Chapter 9.

8-28. Which hypothesis contains a statement of what we do not believe to be true in the population? Where do we state what we do believe?

8-29. Explain what it means to have a hypothesis that says $H_0: \mu = 60$.

8-30. Write the alternative hypothesis that would accompany a null hypothesis that says $H_0: \mu \leq 45$.

8-31. If we are using $H_0: \mu \leq 45$, have chosen $\alpha = .05$, and plan to run a study that will use a z test statistic, where would α be placed in a standard normal distribution? How is a critical value connected with α in this case?

8-32. Explain how the formula for the z test statistic corresponds to the verbal definition of any z, and then explain where each piece of the formula comes from.

8-33. Using a significance level of .05, find the critical value or values for each of the following examples. For each example, sketch a standard normal distribution and add the critical value(s) and α to the sketch.

    a. $H_0: \mu \leq 18$
    b. $H_0: \mu = 18$
    c. $H_1: \mu < 18$
    d. $H_1: \mu \neq 18$

8-34. Find the one-tailed p value for each of the following z test statistics. For each example, sketch a standard normal distribution and add the z test statistics and one-tailed p values to the sketch.

    a. z test statistic = 0.5
    b. z test statistic = 2.5
    c. z test statistic = −0.5
    d. z test statistic = 0.05

8-35. Find the two-tailed p value for each of the following z test statistics.

    a. z test statistic = −0.5
    b. z test statistic = 2.5
    c. z test statistic = 0.5
    d. z test statistic = 0.05

8-36. Pretend that each of your answers from Question 8-34 reflects the result of a study that was designed to test a null hypothesis that says $H_0: \mu \leq 45$. If your significance level is .05, use each answer to test this null hypothesis.

8-37. Pretend that each of your answers from Question 8-35 reflects the result of a study that was designed to test a null hypothesis that says $H_0: \mu = 45$. Use each answer to test this null hypothesis.

(*Continued*)

8-38. Suppose we are told that someone has computed a z test statistic that has a p value = .006. Using α = .05, can we test a null hypothesis even if we have not been told its details? If yes, then do so. If not, then explain the details that you need to know.

SCENARIO 8-C

(Inspired by Cserjesi et al., 2012. Details of this scenario may differ from the actual research.) Dutch researchers were studying children who had been born moderately preterm, which they defined as 32–36 weeks' gestational age. Specifically, they were looking at IQ scores for these children at age 7 years. Suppose we are collaborating with the Dutch researchers, who tell us that these children grew up in fairly rich areas of the Netherlands. As a result, we guess that these children had many advantages to help them overcome any developmental obstacle resulting from having been born moderately preterm. Therefore, we suspect that the sample comes from a population of children with higher average IQ than typical children who were born pre-term. We take a trip to fantasy land, where we are told that many years of research have produced the following known information: the population mean for typical 7-year-olds who were preterm at birth is 95, with a standard deviation = 20. 8-39. What kind of research is this study (observational/descriptive, quasi-experimental, experimental)? 8-40. What kind of variable is IQ of 7-year-old children? 8-41. Explain whether the Dutch researchers are able to say whether being born moderately preterm causes lower IQ at age 7. Connect your answer with the concept of internal validity. 8-42. Using symbols, write the statistical alternative hypothesis. 8-43. Translate the statistical alternative hypothesis into sentences. 8-44. Using symbols, write the statistical null hypothesis. 8-45. Translate the statistical null hypothesis into sentences. 8-46. Suppose we are told the Dutch researchers have tested 248 children who were moderately preterm. These children have a mean IQ = 101.2. Compute the z test statistic. 8-47. Draw a picture of a standard normal distribution, determine where to put the significance level, and, using α = .05, look up the critical value(s). 8-48. Find the p value for our z test statistic. 8-49. Using the appropriate p value decision rule, test the null hypothesis. Then using the appropriate critical value decision rule, test the null hypothesis. (Your decision on the null hypothesis should be the same using either decision rule.) 8-50. What does your decision mean about these children's preterm births and their mean IQ? 8-51. Compute a 90% confidence interval for μ for this scenario. 8-52. Explain the meaning of your 90% confidence interval, using the variable names. 8-53. Test the null hypothesis, using the confidence interval, and explain how you reached the decision.

(Continued)

SCENARIO 8-D

(Inspired by Macdonald, Hardcastle, Jugdaohsingh, Reid, & Powell, 2005. Details of this scenario may differ from the actual research.) We are planning to collect data and run a z test statistic on data from a sample of postmenopausal women who are long-time moderate beer drinkers. We think the dietary silicon that they have consumed in the beer could have given them greater bone mineral density (BMD) than postmenopausal women who do not drink alcohol. But then again, if they drink beer regularly, perhaps they take worse care of their health in general, leading to lower BMD. We want to know whether these women's average BMD differs from that of the average postmenopausal woman. BMD is measured in units of grams per centimeters squared (g/cm^2). We find a researcher who has conducted bone scans on thousands of nondrinking postmenopausal women over the years, and she says, "When measured at the hip bone, these women in general have a mean BMD equal to 0.88 g/cm^2, with a standard deviation = 0.6." 8-54. In words, write the alternative hypothesis, using the variable names. Then write the statistical alternative hypothesis in symbols. 8-55. In words, write the null hypothesis, using the variable names. Then write the statistical null hypothesis in symbols. 8-56. Using $\alpha = .05$, draw a picture of a standard normal distribution, determine where to put the significance level, and look up the critical value(s). 8-57. We run our study and collect hip-bone BMD measures on 36 postmenopausal women who are long-time moderate beer drinkers. We find the following results: sample mean = 1.03, sample median = 0.99, and SD = 2.1. Compute the z test statistic. 8-58. Find the p value for our z test statistic. 8-59. Using the appropriate p value decision rule, test the null hypothesis. Then using the appropriate critical value decision rule, test the null hypothesis. (Your decision on the null hypothesis should be the same using either decision rule.) 8-60. Explain the meaning of your decision as it related to these beer-drinking women. 8-61. Why would a 95% confidence interval for μ be appropriate for this scenario? 8-62. Compute the 95% confidence interval for μ. 8-63. Explain the meaning of your 95% confidence interval, using the variable names. 8-64. Test the null hypothesis, using the confidence interval, and explain how you reached the decision.

References

Armstrong, G. L., Wasley, A., Simard, E. P., McQuillan, G. M., Kuhnert, W. L., & Alter, M. J. (2006). The prevalence of hepatitis C virus infection in the United States, 1999 through 2002. *Annals of Internal Medicine, 144,* 705–714. doi:10.7326/0003-4819-144-10-200605160-00004

Comstock, R. D., Mallonee, S., Fox, J. L., Moolenaar, R. L., Vogt, T. M., Perz, J. F., … Crutcher, J. M. (2004). A large nosocomial outbreak of hepatitis C and hepatitis B among patients receiving pain remediation treatments. *Infection Control and Hospital Epidemiology, 25,* 576–583. doi:10.1086/502442

Cserjesi, R., Van Braeckel, K. N. J. A., Butcher, P. R., Kerstjens, J. M., Reijneveld, S. A., Bouma, A., ... Bos, A. F. (2012). Functioning of 7-year-old children born at 32 to 35 weeks' gestational age. *Pediatrics, 130,* e838–e846. doi:10.1542 /peds.2011-2079

Macdonald, H., Hardcastle, A., Jugdaohsingh, R., Reid, D., & Powell, J. (2005). Dietary silicon intake is associated with bone mineral density in premeno-pausal women and postmenopausal women taking HRT. *Journal of Bone Mineral Research, 20,* S393.

Margolis, K. L., Asche, S. E., Bergdall, A. R., Dehmer, S. P., Groen, S. E., Kadrmas, H. M., ... Trower, N. K. (2013). Effect of home blood pressure telemonitoring and pharmacist management on blood pressure con-trol: A cluster randomized clinical trial. *Journal of the American Medical Association, 310,* 46–58. doi:10.1001/jama.2013.6549

McDowell, M. A., Fryar, C. D., Ogden, C. L., & Flegal, K. M. (2008, October 22). Anthropometric reference data for children and adults: United States, 2003–2006. *National Health Statistics Reports, 10,* 6–16. Retrieved from www.cdc.gov/nchs/data/nhsr/nhsr010.pdf

9

Types of Errors and Power

Introduction

Suppose we are health-care providers treating a middle-aged woman named Eileen Dover, who has chronic low back pain and arthritis. Eileen asks whether she should take glucosamine to help with her back problem. Glucosamine is found in healthy cartilage, and supplements that come from shellfish have been promoted as a possible way to help people with arthritis. Fortunately we have read about recent research involving glucosamine (Wilkens, Scheel, Grundnes, Hellum, & Storheim, 2010). The paper reported that a randomized controlled trial found no significant effect of glucosamine on ratings of pain, disability, or quality of life. Based on this journal article, we tell Eileen that we do not recommend glucosamine for low back pain.

What if the journal article's conclusions were wrong? By "wrong," we do not mean the results were misinterpreted by the researchers, nor do we mean that the results were reported incorrectly by mistake. The researchers may have done everything correctly in the study and reported the sample results accurately—but the results actually might not reflect the reality in the population. Decisions in hypothesis testing are based on probabilities. There is never a guarantee that the sample results truly reflect what is happening in the population. So a decision to reject the null hypothesis might be incorrect, yet *we would not know that it was wrong*. A decision to retain the null hypothesis also can be made incorrectly because we cannot know what is happening in the population. For these reasons, we can never say "prove" in connection with a hypothesis.

These statements may be nerve wracking to you. We can imagine students reading this book and saying, "You just spent all these pages explaining about hypothesis testing and drawing conclusions, and now you tell me that even if I do everything right, *my decisions may be wrong?!*" Well … yes, that is what we are saying. But we will try to minimize the likelihood of errors in hypothesis testing, and at the same time try to improve our chances of making correct decisions. These topics are covered in this chapter. As you read the chapter, think about the potential effect of errors in hypothesis testing on science as a whole and the important role of replication across multiple studies.

Possible Errors in Hypothesis Testing

Whenever we test a null hypothesis, we have two possible decisions: to reject the null hypothesis or to retain the null hypothesis. When we reject the null hypothesis, there are two possible realities in the population:

- We are correct to reject the null hypothesis because it actually is false in the population.
- We are wrong to reject the null hypothesis because it actually is true in the population.

Similarly, when we retain the null hypothesis, there are two possible realities in the population:

- We are correct to retain the null hypothesis because it actually is true in the population.
- We are wrong to retain the null hypothesis because it actually is false in the population.

Scientists use special names for the two errors that we have described. Rejecting the null hypothesis when it is actually true in the population is called a *Type I error*. Retaining the null hypothesis when it is actually false in the population is called a *Type II error*. Let's consider the meaning of these possible errors in terms of the glucosamine study. After randomly assigning participants to the treatment group or control group, the researchers compared the groups on a number of variables so that they could report whether the groups actually seemed to be equivalent at the beginning of the study. The researchers computed a number of inferential statistics and reported that the two groups were not significantly different on gender, smoking status, body mass index (BMI), duration of low back pain, and so on. In other words, the two groups had relatively the same numbers of men versus women, as well as smokers versus nonsmokers, and the two groups were equivalent on their average BMI and their mean duration of low back pain. (Later in the book, we will cover some test statistics that allow us to compare two groups. So far, we have covered only one test statistic, the z test statistic, which is computed on scores from one sample.)

The researchers might have been wrong in their statements about these four characteristics at baseline. The treatment group and the control group could have represented two populations that actually differed in terms of one or more of these variables: gender, smoking status, BMI, and duration of low back pain. What kind of error might the researchers have made? They may have made a Type II error on any of the hypothesis tests where they retained the null hypothesis. No one can know for sure whether the decision to retain the null hypothesis was correct.

There was one variable on which the researchers said the two groups differed at baseline. It was a measure of five dimensions of quality of life, using visual analog scales. This kind of scale presents the respondent with a statement or question, such as, "How do you feel today?" The statement or question is presented next to a line of a certain length. Each end of the line is anchored by opposite ideas, such as *best imaginable health state* and *worst imaginable health state*. Respondents place a mark somewhere on the line, which acts like a thermometer showing where they are on the continuum. The researcher uses a ruler to measure the distance from one end of the line and apply a number to the respondents' answers. The glucosamine researchers reported that the two groups differed significantly in their mean responses to this measure of quality of life. What kind of error might the researchers have made? They may have made a Type I error because "differed significantly" implies that they rejected a null hypothesis, and a Type I error occurs when we reject the null hypothesis in situations where the null hypothesis is true in the population. No one can know for sure whether the decision to reject the null hypothesis was correct. If the two groups came from populations in which people truly are not different in terms of their average quality of life, then the researchers made a Type I error. On the other hand, the two groups may represent populations that actually *are different* in the population, and in that case the decision to reject the null hypothesis would be correct. We cannot know whether the decision to reject the null hypothesis is correct because we cannot know what is true in the population.

Let's summarize the combinations of the decisions (reject or retain H_0) and the possible realities (H_0 is true or H_0 is false). We will use the term *True State of the World* to describe the two possible realities in the population. We cannot know the True State of the World because we cannot obtain the population. With any given hypothesis test, the True State of the World may be that the null hypothesis is true, or the True State of the World may be that the null hypothesis is false. Table 9.1 shows a 2 × 2 table (i.e., a table consisting of two rows and two

Table 9.1 Two Possible Realities, Two Possible Decisions

		True State of the World	
		H_0 is True	H_0 is False
Decision on a Given Hypothesis Test	Reject H_0	This combination is a Type I error	This combination is a correct decision
	Retain H_0	This combination is a correct decision	This combination is a Type II error

columns) that appears in most introductory statistics textbooks. The table represents four possible combinations based on two elements: a decision on a given hypothesis test and a possible True State of the World.

Scientists in general agree on the names *Type I error* and *Type II error*, and statistics teachers love to test students on this material, so it is a good idea to memorize and understand the information in Table 9.1. At the beginning of the chapter, we asked you to keep in mind the effect of errors on science as a whole. If a study results in a Type I error, what is the larger effect? If a statistically significant result is found, but it actually is a Type I error, then the result does not reflect a real phenomenon in the population. Remember, we never know whether a significant result represents what really is happening in the population or is an aberration based on one study. Through replication of studies, researchers can determine whether a real phenomenon has been observed. What about Type II errors? A nonsignificant result could be real—the phenomenon does not exist in the population. Or it could be a Type II error, meaning the phenomenon does exist, but a study failed to find evidence of the phenomenon. If a nonsignificant result has been found, it may not even be reported in the scientific literature, where significant results often seem to be preferred. In that case, others may try to examine the same phenomenon. Replication again can help researchers to determine whether the phenomenon exists or not.

Check Your Understanding

9-1. Returning to the rat shipment example in Chapter 8: Dr. Sarah Bellum and Ray D. Ology found that the sample of 25 rats took significantly longer to complete the simple maze, compared with the known mean for the population of normal, healthy Norway rats. What kind of error might they have made in their hypothesis test?

Suggested Answer

9-1. This decision could have been a Type I error, which occurs when we reject the null hypothesis when actually the null hypothesis is true in the population. This question says "significantly," which means a null hypothesis was rejected. The only kind of error possible when we reject a null hypothesis is a Type I error.

It may be alarming to you to think that we could be making a mistake and drawing erroneous conclusions whenever we perform a hypothesis test. It is possible because we depend on probability to test hypotheses. The next section will begin our discussion of the probabilities of errors and correct decisions, including the ways that researchers try to limit the chances of errors and at the same time try to increase the chances of correct decisions.

Probability of a Type I Error

It may surprise you to learn that we already have talked about the probability of committing a Type I error in a hypothesis test. Let's take a look at an earlier figure, reproduced here as Figure 9.1.

Figure 9.1 shows a standard normal distribution reflecting a reality in which the null hypothesis is true. To refresh your memory: we used this figure in the rat shipment example, where the null hypothesis said the population mean maze completion time was less than or equal to 33 seconds. We *believed* that we were sampling from a population of rats that would take longer than 33 seconds on average to complete the maze, so we had written an alternative hypothesis that said $H_1: \mu > 33$. This alternative hypothesis predicted a sample mean that would be higher on the number line than 33, corresponding to a positive z test statistic, so we put all of α in the upper tail of the standard normal distribution. But this distribution is drawn as if the null hypothesis is true because we knew some facts about healthy rats' maze completion times. By putting α in the upper tail, we

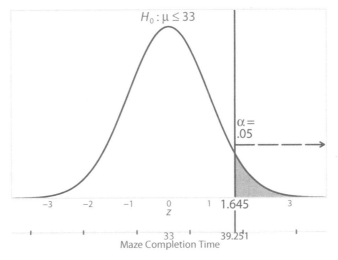

Figure 9.1

Distribution of z test statistic when results are predicted in the upper tail. When Dr. Bellum and Ray predicted that the rats came from a population with a mean maze completion time greater than 33, they put $\alpha = .05$ in the upper tail of the standard normal distribution. If they reject the null hypothesis, could they be making an error?

specified what would be *unusual* results under a true null hypothesis. Through repeated sampling from the population of healthy rats, we could find a variety of different sample means. By choosing $\alpha = .05$, we were saying, "Given that the null hypothesis is true, a result at least this extreme could be found through repeated sampling 5 times out of 100 or even less frequently by chance alone. Such a result will be deemed evidence against the null hypothesis."

If the observed z test statistic occurred to the left of the vertical line for the critical value in Figure 9.1, we would retain the null hypothesis. If the observed z test statistic occurred to the right of the critical value in Figure 9.1, we would reject the null hypothesis. But any outcome along the horizontal number line in Figure 9.1 is possible *even when the null hypothesis is true*. In other words, the True State of the World may be that the null hypothesis is true—and we *still* could get an extreme mean running time. An extreme mean may be less likely if the null hypothesis is true, but it is possible.

To elaborate, consider that Dr. Bellum and Ray ran the rats through the maze and found a sample mean maze completion time of 44.4 seconds, which resulted in a z test statistic $= 3.0$. This value of the z test statistic would be in the upper tail of this distribution, which is drawn as if the null hypothesis is true. The sample mean of 44.4 is a possible result *even if* the True State of the World is that our sample came from a population where the mean time was 33 seconds or less. We previously defined the significance level, α, as a small probability chosen in advance of data collection to serve as a standard for how unlikely the results must be to declare that the results are evidence against the null hypothesis. In our current context, α is the probability of finding a statistically significant result when the True State of the World says the null hypothesis is true. We now have a new way of defining α, the significance level: α is the probability of a Type I error that was set in advance of the study. This is good news because we, the researchers, get to choose the significance level. We make the choice after thinking about the consequences of a Type I error for our research situation. If the consequences are more severe, we may decide to reduce the chance of a Type I error by choosing a smaller α, like .01.

Let's return to the glucosamine example. We said the researchers found a significant difference in means for the treatment and control groups on a measure of five dimensions of quality of life, using a visual analog scale. So the researchers rejected a null hypothesis that said the population means for the two groups were equal, concluding instead that the groups differed on average quality of life at baseline. *The researchers might be wrong*. It is possible that the null hypothesis was rejected at the same time that the population means were equal in the True State of the World. Before analyzing the data, the researchers chose a significance level of .05. What does this number mean in terms of the probability of a Type I error? They were weighing the cost of finding statistical significance incorrectly—that is, the cost of rejecting the null hypothesis when the True State of the World reflected the null hypothesis. With repeated sampling from the same population, the choice of $\alpha = .05$ means that 5 times out of 100 by chance alone, they could expect to reject the null hypothesis when in fact the null hypothesis was true in the population. What is the cost of a Type I error in this instance? If the samples actually came from populations where the means for quality of life were equal, then the cost was almost nonexistent for these researchers because they *wanted* the groups to be equivalent at the beginning of the study.

A Type I error can be more costly in other hypothesis tests. What if the glucosamine researchers had reported a significant difference in mean quality of life at the *end* of their study? That is not what the study actually found. But let's pretend for a moment that they did find a significant difference. In that case, they might have recommended that people use glucosamine supplements for low back pain. If the True State of the World says there is no difference in mean quality of life for people taking glucosamine supplements versus people taking a placebo, then the researchers would have been making a Type I error. In that case, the cost would be high: they would recommend a treatment that in fact was useless. When choosing a significance level, therefore, researchers must weigh the cost of a Type I error. When the stakes are higher, a smaller significance level may be needed.

A quick aside: how did we know the glucosamine researchers chose $\alpha = .05$? The paper by Wilkens et al. (2010) mentioned "a 2-sided significance level of .05" (p. 47). We will discuss the context for this statement later in this chapter.

Check Your Understanding

9-2. When Dr. Bellum and Ray tested the null hypothesis that said $\mu \leq 33$, they chose $\alpha = .05$ and observed a sample mean of 44.4 seconds and a z test statistic $= 3.0$. The one-tailed p value for the test statistic equaled .0013. When they planned their study, what were they willing to risk as the probability of a Type I error?

Suggested Answer

9-2. The significance level, $\alpha = .05$, was the probability of a Type I error that Dr. Bellum and Ray chose in advance. If we could know that the True State of the World was that the rats are healthy, rejecting the null hypothesis would be a Type I error. The other details in this question, such as the p value, are distractors that aren't needed to answer the question.

Probability of Correctly Retaining the Null Hypothesis

The introduction to this chapter said we want to try to improve our chances of making correct decisions in hypothesis testing. One correct decision would be to retain the null hypothesis when the True State of the World says the null hypothesis is true in the population. If the null hypothesis actually is true, we do not want to throw it out. When we are planning a study, how can we increase the probability of this correct decision? Such a probability is directly connected with the significance level. Let's look at Figure 9.2.

To the right of the vertical line that goes through the critical value of $z = 1.645$, we find an area associated with $\alpha = .05$. Now we know that .05 also is the chosen

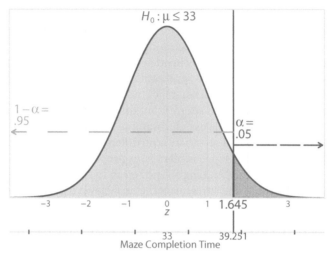

$H_0 : \mu \le 33$

$1 - \alpha = .95$

$\alpha = .05$

−3   −2   −1   0   1 1.645   3
Z

33   39.251
Maze Completion Time

Figure 9.2

Dividing the null distribution into two parts. This distribution is drawn as if the null hypothesis is true. The red tail area in this distribution is α, the probability of a Type I error. The rest of the distribution, shown in gray, is another probability, which is equal to $1 - \alpha = .95$. This probability is associated with retaining the null hypothesis when H_0 is true.

probability of a Type I error. It is the probability associated with rejecting the null hypothesis when the True State of the World says the null hypothesis is true in the population. The rest of the distribution, located to the left of the vertical line through the critical value, has an area $= 1 - \alpha$. This larger area, shown in gray, is the probability of retaining the null hypothesis when the True State of the World says H_0 is true in the population. So when $\alpha = .05$, the probability of correctly retaining the null hypothesis is $1 - .05 = .95$.

By choosing a smaller α, we get a larger probability of correctly retaining the null hypothesis. That should make sense—a small α means we are setting a stringent standard for finding significance, so if it is harder to reject the null hypothesis, then it should be easier to retain the null hypothesis. Take a look at Figure 9.3, showing $\alpha = .01$.

Compared with Figure 9.2, where $\alpha = .05$, we now have Figure 9.3, where $\alpha = .01$. As a result, the critical value (2.33) is farther out in the upper tail. Instead of the observed z test statistic having to reach or exceed a critical value of 1.645, which was the case when $\alpha = .05$, now the z test statistic will have to reach or exceed a critical value of 2.33 because $\alpha = .01$. To get a larger observed z test statistic requires a bigger distance between the sample mean and the population mean. When designing a study with $\alpha = .01$, we know that through repeated samples from the population described in the null hypothesis, we would get z test statistics equal to or more extreme than the critical value once out of every 100 samples. If the True State of the World is described by the null hypothesis, then rejecting H_0 would be a Type I

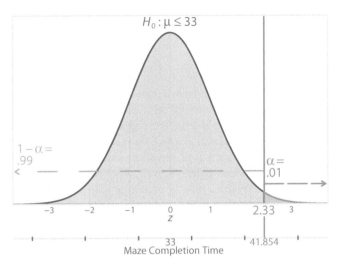

$H_0 : \mu \leq 33$

$1 - \alpha = .99$

$\alpha = .01$

-3   -2   -1   0   1   2.33   3
Z

33   41.854
Maze Completion Time

Figure 9.3

The probability of a Type I error and the probability of a correct decision. If $\alpha = .01$, the probability of a Type I error is small; here, it is shown as the small orange tail area. As a result of choosing a smaller α, we have a greater probability of correctly retaining the null hypothesis, shown as the gray area.

error. Again, we cannot know the True State of the World. If an incorrect rejection of H_0 is costly, then a smaller α should be chosen. If we repeatedly sampled from the same population when the null hypothesis is true, 99% of the time we would make the correct decision to retain the null hypothesis. In the next section, we will link the probability of a Type I error with confidence intervals.

Check Your Understanding

SCENARIO 9-A

Suppose we are looking at a U.S. governmental report containing several years of health statistics about large representative samples. The report includes information on the resting heart rates of children who were 6–8 years old. Suppose the report says that 20 years ago, the mean for children (ages 6–8 years) was 82 beats per minute. We think about the current problem of childhood obesity, and we wonder whether American children today might have a different average resting heart rate than children 20 years ago. We want to compare a large sample of children today with a previously reported mean for resting heart rate. We want to do what we can to avoid the mistake of finding a significant difference when actually no difference exists in the population. 9-3. What kind of error are we trying to avoid? 9-4. How can we reduce the chances of making that error?

(*Continued*)

Suggested Answer

9-3. We think our sample comes from a population with a mean resting heart rate that differs from the previously reported population mean of 82; that would be our alternative hypothesis. The implied null hypothesis is that our sample comes from a population in which the mean resting heart rate is equal to 82. We are trying to avoid a Type I error: rejecting the null hypothesis (i.e., concluding the sample mean differs from 82) when the null hypothesis is true (i.e., the sample actually came from a population with a mean of 82). 9-4. We can reduce the probability of a Type I error by making the significance level very small, such as $\alpha = .01$, or even smaller.

Type I Errors and Confidence Intervals

In Chapter 8, we computed a confidence interval for the rat shipment example. Instead of relying on a point estimate, the sample mean, to estimate the population mean, we computed an interval estimate of μ. The 95% confidence interval was used to test the null hypothesis that said the sample of rats came from a population with a mean maze completion time of 33 seconds (i.e., H_0: $\mu = 33$). This time we were not predicting the direction of the results. Quantifying the sampling variability to be expected across repeated samples, the 95% confidence interval was [36.952, 51.848]. Because the interval did not contain $\mu = 33$, we concluded that the value of 33 was not within the range of plausible values for the mean of the population that we sampled. We therefore rejected the null hypothesis. Our confidence interval could be interpreted as follows: we could expect that 95% of confidence intervals to bracket the true mean of the population that we sampled, if we were to sample repeatedly and compute an interval estimate using the data from each sample.

How do confidence intervals relate to Type I errors? We rejected the null hypothesis because the interval of [36.952, 51.848] did not bracket $\mu = 33$. But it is possible that we actually did sample from a population with a mean maze completion time of 33 seconds, and we just happened to sample some rats that took longer to complete the maze. If we actually did sample from the population with a mean time of 33 seconds, our rejection of the null hypothesis would have been a Type I error. The probability of a Type I error when using a confidence interval is the same probability that we saw previously: the significance level, α. A 95% confidence interval is linked to $\alpha = .05$ through the critical value that was used to compute the margin of error; this computation was explained in Chapters 7 and 8.

Suppose Dr. Bellum and Ray, the researchers who received the rat shipment, actually preferred to use $\alpha = .01$ instead of $\alpha = .05$ so that their probability of a Type I error would be smaller. What effect does the smaller α have on the confidence interval? Let's work through a numeric example. For simplicity, we will continue with the nondirectional alternative hypothesis, H_1: $\mu \neq 33$, so half of α would be in each tail, and the critical values would be +2.575 and −2.575. The margin of error would be the standard error of the mean (*SE*) times a critical value, 2.575. As shown in Chapter 8, the *SE* was 3.8, so the margin of error would be

$$\text{Margin of error} = 2.575 \times SE$$
$$= 2.575 \times 3.8$$
$$= 9.785$$

To find the lower limit of the confidence interval, we subtract the margin of error from the sample mean, $M = 44.4$:

$$\text{Lower limit} = 44.4 - 9.785$$
$$= 34.615$$

To find the upper limit of the confidence interval, we add the margin of error to the sample mean:

$$\text{Upper limit} = 44.4 + 9.785$$
$$= 54.185$$

Thus, the 99% confidence interval would be [34.615, 54.185]. This interval does not bracket $\mu = 33$, so we would reject the null hypothesis and decide that our rats probably came from a population that had a mean maze completion time that was different from 33 seconds. In this situation, we might conclude that we have a sample of sick rats.

Are we correct in this decision to reject H_0? There is no way of knowing because we cannot obtain the population from which our rats were sampled and discover the true population mean. We reduced the chances of a Type I error, however, by using the smaller α, .01. The 99% confidence interval means that 99% of the intervals computed on repeated samples from the same population that we sampled would bracket the true population mean, whatever that value might be. And it is unlikely that we would get a confidence interval like [34.615, 54.185] if 33 is the mean of the population that we sampled—but it is possible! If 33 actually is the mean of the population that gave us the sample of rats, then the decision to reject the null hypothesis is a Type I error. Now let's consider the probability of correctly retaining the null hypothesis. As shown previously, this probability is $1 - \alpha$. For a 99% confidence interval, the probability of the correct decision to retain H_0 would be .99, if the interval contains the hypothesized value of μ. Again, we cannot know whether any decision on a hypothesis test is correct or not because we cannot know the True State of the World in the population.

Notice that the smaller α, .01, corresponded to a greater confidence level, 99%, and a wider interval. The 95% confidence interval was [36.952, 51.848], and the 99% confidence interval was wider: [34.615, 54.185]. There is less precision in the interval estimate of μ when we use a smaller α. You may recall our analogy about tornado prediction in Chapter 7. We were not very confident about predicting a tornado in Oklahoma on a specific day next year. But if you gave us a range of dates (April 27 to June 8) and asked us how confident we were that a tornado would touch down in Oklahoma during that range, our confidence in saying yes would increase. The less precise interval gave us greater confidence, but an interval that is extremely wide is not very informative. (If we wanted to be 100% confident that the interval contained the true value of the parameter, then all intervals would be minus infinity to positive infinity, which is uninformative at best.) Now that we have talked about the probability of a Type I error and the probability of correctly retaining the null hypothesis, we will move on to the probability of a Type II error and the probability of correctly rejecting the null hypothesis.

Check Your Understanding

9-5. If researchers computed a 90% confidence interval for a situation in which they had written a nondirectional alternative hypothesis, what would be the probability of a Type I error? 9-6. What would the probability .90 represent?

Suggested Answers

9.5. *The probability of a Type I error would be .10. The researchers must have had very little concern about the probability of rejecting the null hypothesis when it was true; a significance level of $\alpha = .10$ corresponds to a 90% confidence interval in the case of a nondirectional alternative hypothesis. 9-6. The value .90 in this situation would represent the probability of correctly retaining the null hypothesis when it was true.*

Probability of a Type II Error and Power

The probability of a Type I error is straightforward: it is our chosen significance level, α. Controlling the probability of a Type I error at some small value is direct: we choose a small α. As a result of the choice of α, we also were able to decide that the probability of correctly retaining the null hypothesis would be $1 - \alpha$. The probability of a Type II error and how to control it at some small value will take considerably more explanation. As we have said, a Type II error occurs when the True State of the World says that the null hypothesis is false, yet we have retained the null hypothesis. How can that happen? Let's return to the rat shipment example, but we will change one number to help us to explain Type II errors. Dr. Bellum and Ray thought the rats seemed sickly and slow moving, so their alternative hypothesis said $\mu > 33$. That is, they thought their sample of rats came from a population where the mean maze completion time was greater than 33 seconds, the known mean for healthy rats. Using $\alpha = .05$, the researchers would have a critical value of 1.645, as shown in Figure 9.2, which was drawn as if the null hypothesis (H_0: $\mu \leq 33$) was true.

As we saw in Chapter 8, Dr. Bellum also knew that healthy rats had a population standard deviation of 19 seconds, and Ray drew a random sample of $N = 25$ rats from the shipment. But suppose the sample mean maze completion time

     9. Types of Errors and Power

turned out to be 37.94 seconds, which is a different value for M than the ones we used as examples in Chapter 8. Let's compute the z test statistic:

$$z = \frac{M - \mu}{\sigma / \sqrt{N}}$$

$$= \frac{37.94 - 33}{(19 / \sqrt{25})}$$

$$= \frac{4.94}{(19 / 5)}$$

$$= \frac{4.94}{3.8}$$

$$= 1.3$$

We can see in Figure 9.2 that an observed z test statistic $= 1.3$ would not be more extreme than the critical value, which was 1.645, so we would retain the null hypothesis. *But what if we were wrong?* What if there exists an alternative distribution that is shifted upward on the number line, with a population mean that is greater than 33? And what if our sample came from that population? This alternative distribution would represent *one* possible reality if the True State of the World agrees with the alternative hypothesis (H_1: $\mu > 33$).

Figure 9.4 shows two distributions. The distribution on the left is the usual standard normal distribution, which reflects the reality described by the null hypothesis. The left-hand distribution is centered on $z = 0$, corresponding to the population mean, $\mu = 33$. The second normal distribution is shifted to the right on the number line. The curve on the right represents *one possible location* for an alternative distribution that would describe a different True State of the World. This right-hand curve represents one possibility if, in fact, the researchers sampled from a population where the alternative hypothesis was true (H_1: $\mu > 33$). The vertical line is still drawn through the critical value for a one-tailed z test statistic when $\alpha = .05$.

In our latest example, the observed z test statistic, 1.3, was not more extreme than the critical value, 1.645, so we retained the null hypothesis. Remember, 1.645 cuts the distribution for the null hypothesis into two pieces: α and $1 - \alpha$ (here, .05 and .95). But we have no way of knowing whether retaining H_0 was a correct decision or a Type II error. Look at the point where $z = 1.3$ is located on the number line in Figure 9.4. There is overlap between the two distributions. It is as if we are looking at two clear mountains, and we are trying to figure out whether a person (represented by $z = 1.3$ on the number line) is standing at the base of the left-hand mountain or the base of the right-hand mountain. Either mountain could represent the True State of the World. We do not know whether the null hypothesis or the alternative hypothesis is true in the population. We test the null hypothesis based on the distribution that was drawn as if the null hypothesis is true. It could be that our sample actually came from a population with a mean maze completion time that is slightly higher than 33, yet the difference between $M = 37.94$ and $\mu = 33$ was not large enough to be detected as statistically significant.

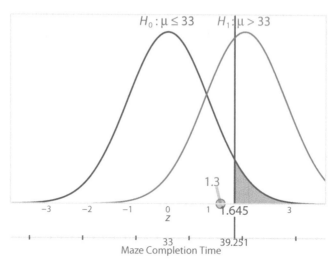

$H_0 : \mu \leq 33$ $H_1 : \mu > 33$

1.3

-3   -2   -1   0   1   1.645   3
Z

33   39.251
Maze Completion Time

Figure 9.4

One possible location of an alternative distribution if H_1: $\mu > 33$ is true. The observed z test statistic in this example equals 1.3, which is not more extreme than the z critical value = 1.645, so we would retain the null hypothesis. If the True State of the World is that the alternative hypothesis is true, we still do not know exactly where the alternative distribution is located because "$\mu > 33$" is a general statement in H_1. The right-hand curve shows one possible location of the alternative distribution.

The True State of the World could be that the alternative hypothesis is true and the sample actually came from a population with a mean maze completion time greater than 33 seconds. There are many possible locations for distributions that would agree with the alternative hypothesis. The right-hand distribution in Figure 9.4 represents only one of the possible locations for the alternative distribution. Let's pretend for a moment that the True State of the World is that the researchers sampled from a population with a mean maze completion time that was greater than 33 seconds—that is, the alternative hypothesis is true. Yet the researchers who computed a z test statistic = 1.3 retained the null hypothesis because the observed test statistic was not more extreme than the critical value. The researchers concluded that the evidence supported the reality described by H_0. If H_1 is actually true in the population, yet we have retained H_0, we have made a Type II error—retaining the null hypothesis when we should have rejected it.

Check Your Understanding

9-7. In the study of glucosamine for patients with low back pain, the researchers found no significant difference in the means for the treatment and control groups on their ratings of lower back pain while resting. What kind of error might the researchers have made?

(Continued)

Suggested Answer

9-7. If the researchers found no significant difference, then they must have retained a null hypothesis. The only possible error in hypothesis testing when retaining the null hypothesis is to do so incorrectly: a Type II error.

If we could know exactly where the alternative distribution is located on the number line, we could look at how the critical value is cutting the alternative distribution into two parts. Figure 9.5 shows the two distributions again, but now the alternative distribution is partly shaded in green; we will explain why shortly. The null distribution again shows a one-tailed test with the critical value in the upper tail. We reject the null hypothesis if the observed test statistic is to the right of the critical value in Figure 9.5, and we retain the null hypothesis if the observed test statistic is to the left of the critical value in this example. Focusing on the right-hand distribution, we can see there is an area associated with retaining the null hypothesis; it is the green shaded area to the left of the critical value. This rather large green area represents the probability of a Type II error.

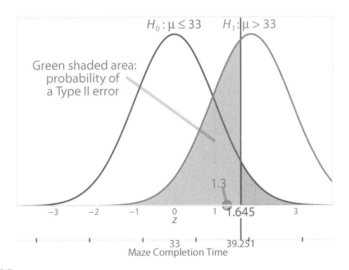

Figure 9.5

One possible location of an alternative distribution and the probability of a Type II error. This graph shows one possible location for an alternative distribution if H_1 is true in the population. Adding to the previous figure, we now focus on the green shaded area. This green area is to the left of the critical value, $z = 1.645$—but the green area is part of the alternative distribution. The observed test statistic, $z = 1.3$, is not more extreme than the critical value, so we would retain H_0. *But what if we are wrong* and the alternative hypothesis is true? Then we have made a Type II error, and its probability is the green shaded area.

Let's return to the analogy of the two clear mountains and the person (shown as $z = 1.3$ on the number line in Figure 9.5) standing at the base of one of those mountains. We can never really know whether the person is standing at the base of the left-hand mountain or the base of the right-hand mountain. We make the decision to say, "The person is standing on the left-hand mountain," because the person is standing to the left of the vertical line (that is, the observed test statistic is not more extreme than the critical value). If we are right and the null hypothesis is true in the population, we made a correct decision. If we are wrong and the null hypothesis is false in the population, we made a Type II error.

There are many possible locations for the alternative distribution if the alternative hypothesis ($\mu > 33$) is true. Let's take a look at another possibility. Compared with the last graph, Figure 9.6 shows a bigger difference in the distributions. That is, the alternative distribution now has shifted farther to the right on the number line.

Focus on the right-hand distribution, representing one possible location for an alternative distribution. The green tail area on the left side of the alternative distribution is the probability of a Type II error, which would occur if we retain the null hypothesis when in fact H_0 is false in the population. In Figure 9.6, the probability of a Type II error is depicted as being small.

We definitely want to minimize our chances of making wrong decisions in hypothesis testing, so how do we make that probability small? Unlike the probability of a Type I error, which we choose directly when selecting our significance level, α, the probability of a Type II error is determined indirectly. Focus for a moment on the

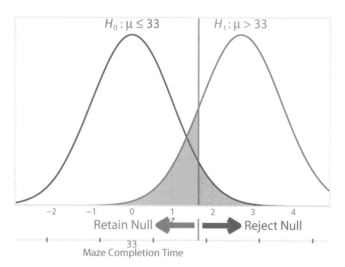

Figure 9.6

Another possible location of an alternative distribution. Compared with the last two figures, this graph shows a bigger difference between the null distribution and the alternative distribution. The gray vertical line is still the critical value, and the size of α does not change; the probability of a Type I error, shown as the red shaded area, remains .05. But the probability of a Type II error, shown as the green shaded area, is smaller than it appeared in Figure 9.5.

9. Types of Errors and Power

right-hand distribution in Figure 9.6, specifically the larger part that appears to the right of the vertical line representing the critical value. That larger part is the probability of rejecting the null hypothesis when the null hypothesis is false—that is, when the True State of the World is in agreement with the reality described by the alternative hypothesis. Rejecting the null hypothesis when it actually is false would be a correct decision. The probability of this correct decision has a special name: *power*.

In statistics, when we talk about power, we always are referring to a probability (think of the two p's going together: *power* is a *probability*). We want to have a very good probability of rejecting the null hypothesis when we should, which is when the null hypothesis is false. If the probability of correctly rejecting the null hypothesis is big, then by default, the probability of a Type II error is small, as shown in Figure 9.6. Suppose we told you that in a given hypothesis test, .90 is the probability of rejecting the null hypothesis when it is false. This statement means that power = .90 and the probability of a Type II error = .10. The probability of a Type II error has a symbol associated with it: the lower-case Greek letter beta, β. In Figure 9.6 the right-hand distribution has a green tail area to the left of the vertical line for the critical value; that small tail is β. In the left-hand distribution, representing the null hypothesis being true, there also is a small tail area, which corresponds to the probability of rejecting the null hypothesis when the null hypothesis is true. The decision to reject a true null hypothesis is a Type I error, with a probability equal to α. So the small tail area in each distribution in Figure 9.6 represents the probability of an error. The small tail area shown in red in the null distribution is α, and the small tail area shown in green in the alternative distribution is β. The bigger part of the null distribution is $1 - \alpha$, the probability of correctly retaining the null hypothesis when it is true. The bigger part of the alternative distribution is $1 - \beta$, the power, the probability of correctly rejecting the null hypothesis when it is false (i.e., the alternative hypothesis is true).

Notice that the two distributions in Figure 9.6 each have been cut into two pieces, resulting in four probabilities that could be discussed. Those four probabilities correspond to the four possible outcomes in the 2 × 2 table that we presented earlier. Let's modify the table to include these four probabilities (Table 9.2).

See if you can link the four pieces of the distributions in Figure 9.6 with the probabilities listed in the four combinations in Table 9.2.

Table 9.2 Two Possible Realities, Two Possible Decisions, Four Probabilities

		True State of the World	
		H_0 is true	H_0 is false
Decision on a Given Hypothesis Test	Reject H_0	This combination is a Type I error, which occurs with a probability $= \alpha$	This combination is a correct decision, which occurs with a probability $= 1 - \beta$
	Retain H_0	This combination is a correct decision, which occurs with a probability $= 1 - \alpha$	This combination is a Type II error, which occurs with a probability $= \beta$

We said the probability of a Type II error is decided indirectly by getting a large probability for power, which is the probability of rejecting the null hypothesis when the True State of the World says the null hypothesis is false. That is because both of these probabilities are related to the True State of the World being a true alternative hypothesis:

- If we reject the null hypothesis when it is false (i.e., the alternative hypothesis is true), we make a correct decision that has a probability that equals power, or $1 - \beta$.
- If we retain the null hypothesis when it is false (i.e., the alternative hypothesis is true), we make a Type II error, which has a probability of β.

Suppose power $= .80$. That is the same thing as saying $1 - \beta = .80$. So the probability of a Type II error is $\beta = .20$. Now suppose power $= 1 - \beta = .90$. That means the probability of a Type II error is $\beta = .10$. By setting up a study to have more power, we will have a smaller probability of a Type II error. If the cost of a Type II error is high, then we would want to make sure that power is large. For example, suppose we are testing a new nonaddictive pain medication, and we have a general null hypothesis that says, "This drug does not work." If the null hypothesis is true, we do not want to reject it and recommend the production of an ineffective medicine, which would be a Type I error. If the null hypothesis is false, we do not want to retain it, missing the opportunity to provide pain relief to people who are suffering but who cannot take narcotic medications. Both of these errors come at a cost. If missing the opportunity to find statistical significance would be costly, then we make the probability of that Type II error small by increasing power.

So, how does a researcher set up a study to ensure the probability known as power is within an acceptable range? That is a question with a complex answer, which we cannot cover completely in this book. Power is a big topic in statistics, and no introductory statistics course can teach you everything about the topic. In the next section, we will go over the factors that influence power and help you to understand how power is discussed in journal articles.

Check Your Understanding

9-8. Earlier in the book, we talked about a study of exercise during pregnancy (Price, Amini, & Kappeler, 2012). Women with a history of being sedentary were randomly assigned early in pregnancy to two groups. The researchers manipulated their activity level. One group was told to remain sedentary, and the other group engaged in regular physical activity. The journal article says, "... 30 subjects per group were adequate to detect a 10% difference in a 3.2-km (2-mile) walk time, with a significant difference at the 0.05 level and a power of 0.80" (p. 2265). This statement implies a certain probability of a Type II error. What is it?

(Continued)

Suggested Answer

9-8. *The probability of a Type II error in this case would be .20. Here is how we can find the answer: the article says power is equal to .80, which is the probability of rejecting the null hypothesis when the True State of the World says the null hypothesis is false (i.e., the alternative hypothesis is true). We can visualize this probability as the larger part of an alternative distribution, somewhat like the one in Figure 9.6. This larger part represents power, which equals $1 - \beta = .80$. In other words, the total area under the right-hand curve is 1, and if we cut off the smaller tail area, β, we are left with .80. So $\beta = .20$. This probability is associated with a Type II error, which is retaining the null hypothesis when the alternative hypothesis is true in the population.*

Factors Influencing Power: Effect Size

Let's talk generally about power, also called statistical power. We know it is a probability. If the True State of the World says the null hypothesis is false, then a correct decision would be to reject H_0, a decision with a probability called power. We want to have a good chance of making that correct decision. If we have too little power in a study, then we will not be able to detect clinically interesting differences or relationships. If we have too much power in a study, then tiny, arbitrary differences or correlations could be detected as statistically significant, and that is not in the best interest of science. So we want enough power for our test statistics to be significant when they encounter the smallest differences or weakest relationships that have reached the threshold of being clinically noteworthy. The judgment about what is clinically noteworthy is not a statistical issue; it depends on the expertise of researchers within the applied area of study.

The last Check Your Understanding question gave the following quote from the Price et al. (2012) study of exercise during pregnancy: "… 30 subjects per group were adequate to detect a 10% difference in a 3.2-km (2-mile) walk time, with a significant difference at the 0.05 level and a power of 0.80" (p. 2265). The 10% difference in the (average) time that the two groups took to walk 3.2 km appears to be what the authors considered to be clinically noteworthy. A smaller difference would not be noteworthy, in their expert determination. If a 10% difference is clinically and statistically significant, then a larger difference also would be noteworthy, both clinically and statistically. That is why we specify the *smallest* difference or relationship that would be noteworthy in a practical sense. The difference in means is an example of an *effect size* (although an effect size statistic, not covered here, would standardize this difference by dividing it by a kind of standard deviation). There are many ways to define effect size. We can say it is the magnitude of the impact of an independent variable on a dependent variable, or we can say it is the strength of an observed relationship. The pregnancy/exercise researchers reported that for a hypothesis test using a significance level of .05, 30 participants per group would give them power = .80 to detect the specified difference in means. This level of power, .80, is a probability of rejecting the null hypothesis when it is false. Across repeated samples from a population in which a 10% difference in mean walk times exists, the researchers could expect to detect a statistically significant difference in means 80% of the time, using $\alpha = .05$ and two groups of 30 people each. Because power = $1 - \beta$, the probability of a Type II error would be $\beta = .20$ in this case. In this example, we could define the 10% difference as the effect size.

How is effect size related to power? Let's think about the effect of a low-dose aspirin on pain, compared with the effect of a prescription pain medication. Suppose a person is suffering from back pain. Which pill do you think will have a bigger effect on the person's pain? We would hope that the prescription pain medication would be more effective than aspirin, resulting in a bigger reduction in pain than aspirin would provide. Which pill's effect would be easier to detect? It should be easier to "see" the bigger effect, which came from the prescription pain medication. This is true in statistics, too. A larger effect size is easier for the test statistics to "see," leading to a greater probability of a statistically significant result. In other words, as effect size increases, power tends to increase. If researchers are interested in detecting a small effect size because it is clinically important, they will need more power.

Small effect sizes can be important. Rosnow and Rosenthal (1989) describe a study in which thousands of people were randomly assigned to taking either an aspirin or a placebo every day. The study involved the risk of a heart attack, which is a rare event. That is why huge sample sizes were used: if researchers are looking for something that occurs rarely, they will need to track a lot of people. The study found that 104 out of 10,933 people who took an aspirin daily had a heart attack (<1%), compared with 189 out of 10,845 people taking the placebo (1.7%). That would appear to be a small difference in the two groups. But Rosnow and Rosenthal (1989) looked at the data more closely. Out of the 104 aspirin takers who had heart attacks, 5 of the patients died (5%). By comparison, out of the 189 people in the placebo group who had heart attacks, 18 people died (10%). So people who took an aspirin a day were half as likely to die if they had a heart attack, compared with people in the control group. Sometimes researchers get caught up in easy-to-follow rules for what might be considered a small, medium, or large effect size. It is better to think about the effect on patients and what would be clinically noteworthy.

Factors Influencing Power: Sample Size

We sometimes talk about the sensitivity of test statistics. This is different from the topic of sensitivity and specificity discussed in Chapter 6; we are now using the term *sensitive* in an informal way. If a test statistic is sensitive, then we are more likely to find a statistically significant result. That is the same thing as saying we will have more power. Researchers have several ways that they can manage how sensitive their inferential statistics are. The factor that is most easily changed by researchers is sample size, and it has a huge effect on power. As N goes up,

9. Types of Errors and Power

power generally goes up. Another way of saying the same thing is that as sample sizes get larger, test statistics tend to become more sensitive.

The most common question that researchers bring to a statistician is, "How many subjects should I have in my study?" Sometimes researchers talk about calculating power, but in fact researchers calculate the sample size that will give them the amount of power that they want. The quotation from the article by Price et al. (2012) in the last Check Your Understanding question is exactly how researchers tend to report the details of sample size calculations. These researchers decided that they wanted a decent probability of finding significance, but they did not want to set up the study so that statistical significance was guaranteed. That is, they did not want their test statistics to be so sensitive that some small, arbitrary difference or relationship would be detected as statistically noteworthy. So they specified power = .80, which means they were willing to have a probability of a Type II error = .20. They chose α = .05, which was the probability of a Type I error, and they identified the smallest difference in means that would be clinically important. Then, once these details were decided, the researchers performed some calculations to tell them how large their samples should be to achieve power = .80.

The details of calculating sample sizes to achieve a targeted power are beyond the scope of this book, but you can tell from this discussion that there is a constellation of details being taken into account when sample sizes are calculated. These details include the significance level, the targeted level of power (which also determines the probability of a Type II error), and the smallest effect size that would be clinically important.

Factors Influencing Power: Directional Alternative Hypotheses

We already said that larger sample sizes bring more statistical power to a study. Another factor that affects power is related to the alternative hypothesis. The choice of a directional hypothesis can make it easier for researchers to find statistical significance, with one important caution: power goes up only if the results turn out in the predicted direction. We will illustrate this concept using two figures from Chapter 8, reproduced here as Figures 9.7 and 9.8. Figure 9.7 shows the null distribution and critical values for the situation in which Dr. Bellum and Ray did not know whether the rats would complete the maze in a shorter amount of time or a longer amount of time. Their alternative hypothesis did not predict that the results would go in a particular direction, so they had to split up the significance level and put half of α in each tail.

To find a statistically significant result, the observed z test statistic had to exceed one of the critical values (-1.96 or $+1.96$). Now let's look at Figure 9.8, showing the situation in which Dr. Bellum and Ray predicted that the sample came from a population with a longer mean maze completion time than 33 seconds. When a directional alternative is used, all of α is placed in one tail, as shown in Figure 9.8. The critical value in this case is 1.645. Think about the difference between M and μ that would be necessary to reach a critical value. Would a bigger difference be required for finding significance with a critical value of 1.645 (Figure 9.8) or with critical values of -1.96 or $+1.96$ (Figure 9.7)? Answer: The nondirectional case would require a

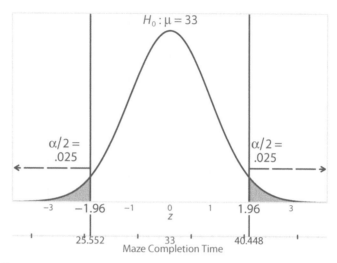

$H_0 : \mu = 33$

$\alpha/2 = .025$

$\alpha/2 = .025$

-3   −1.96   −1   0   1   1.96   3
Z

25.552   33   40.448
Maze Completion Time

Figure 9.7

Distribution of z test statistic when no direction is predicted. This standard normal distribution is drawn as if the null hypothesis is true and the researchers sampled from a population of rats with a mean completion time equal to 33 seconds. They thought something was wrong with the rats and did not predict a direction for the results, so they split up α and put half of it in each tail.

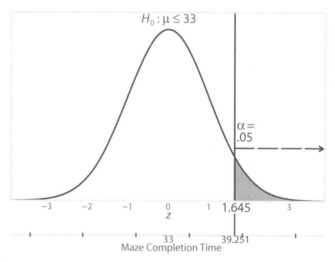

$H_0 : \mu \leq 33$

$\alpha = .05$

-3   -2   -1   0   1   1.645   3
Z

33   39.251
Maze Completion Time

Figure 9.8

Distribution of z test statistic when results are predicted in the upper tail. As usual, the standard normal distribution is drawn as if the null hypothesis is true, but now Dr. Bellum and Ray have predicted that the rats came from a population with a mean maze completion time greater than 33 seconds. Because the alternative hypothesis predicted results in the upper tail, we put α there.

bigger difference between M and μ. So it would be harder to reject the null hypothesis in the nondirectional case than the directional case, meaning there would be a lower probability of finding significance (i.e., less power) in the nondirectional case. If it is easier to find significance with the one-tailed test shown in Figure 9.8, then the directional case has more power. But remember our big exception: we would have more power *only* if the direction of the results was predicted correctly. Consider the directional case in Figure 9.8 again. Suppose the sample mean for the 25 rats turned out to be $M = 22$ seconds. Are the results in the predicted direction? No, because $M = 22$ is not evidence that would support the idea expressed in H_1: $\mu > 33$; the z test statistic would be negative because its numerator would be $M - \mu$ $= 22 - 33 = -11$. The observed z test statistic would be in the lower tail, so we would retain the null hypothesis. The alternative hypothesis predicted the wrong direction for the results, so we had *zero power*—no probability of rejecting the null hypothesis. In sum, a directional hypothesis provides more power than a nondirectional hypothesis, but only if the direction is predicted correctly. If the results turn out in the opposite direction from the prediction, the cost is high: power is zero. This is why many statisticians say that researchers should think carefully about how certain they are that their results will turn out in the predicted direction.

Check Your Understanding

SCENARIO 9-B

The results of the glucosamine study were reported in Wilkens et al. (2010). These researchers used the Roland Morris Disability Questionnaire (RMDQ), which often is used to gather self-reported data on low back pain. The authors of the journal article wrote, "A change in score of 3 points on the RMDQ is considered the lowest level of clinical importance to be used for sample size calculations in trials. We estimated that 250 patients should be enrolled based on a clinically important difference of 3 with 80% power, a 2-sided significance level of .05, and adding 20% for possible dropouts" (p. 47). 9-9. What is the probability of a Type I error that the researchers were willing to have? 9-10. What is the probability of a Type II error that the researchers were willing to have? 9-11. Suppose we are planning a similar study, but we think a clinically important difference would be 8 points on the RMDQ. All else being the same, would we need a larger or smaller N than these researchers to achieve the same level of power?

Suggested Answers

9-9. The researchers set their significance level at .05, which is the probability of a Type I error that they were willing to have. 9-10. If power = $1 - \beta = .80$, then the probability of a Type II error would be $\beta = .20$. 9-11. It would be easier for us to detect a larger effect size as statistically significant. To maintain the same level of power as these researchers, we would use a smaller sample than they did.

Factors Influencing Power: Significance Level

We use a lot of examples with $\alpha = .05$, but sometimes researchers need to be stricter about controlling the probability of a Type I error because the cost of finding significance incorrectly is high. What happens to power if researchers choose a smaller α, such as $\alpha = .01$? Let's compare two distributions, both representing directional predictions, but one with $\alpha = .05$ and another with $\alpha = .01$ (Figure 9.9).

The left-hand distribution is similar to Figure 9.8, with an area of $\alpha = .05$ being cut off by a critical value of 1.645. The right-hand distribution shows a critical value of 2.33 cutting off an area of $\alpha = .01$. Thinking about the numerator of the z test statistic, $M - \mu$, will it be easier to reach a critical value of 1.645 or 2.33? The difference between M and μ would have to be bigger if the z test statistic is to reach the critical value 2.33 shown in the right-hand distribution. So it would be easier to find statistical significance with a larger significance level, compared with finding a significant result with a smaller α. All else being the same, as α goes up, power goes up. And as α goes down, like from .05 to .01, power goes down. Be very careful here not to confuse α going down with the critical value's direction of change. For an upper-tailed test, switching from $\alpha = .05$ to the smaller $\alpha = .01$ means we switch from a critical value $= 1.645$ to the larger critical value $= 2.33$, as shown in Figure 9.9. If we were doing a lower-tailed test, then the more extreme critical value would be a smaller number that is farther out in the left tail.

We also can see the effect of α in the picture of the two overlapping distributions, representing the null and alternative hypotheses. Figure 9.10 is similar to Figure 9.6, except now we have $\alpha = .01$. Alpha $= .01$ is shown in Figure 9.10 as the small right-hand tail of the left-hand distribution, representing the True State of the World with the null hypothesis being true. The vertical line now goes through a critical value of 2.33, cutting off α. Compared with Figure 9.6, the vertical line has shifted to the right a small distance (from $z = 1.645$ to 2.33). As a result, the larger portion of the alternative distribution in Figure 9.10 is smaller than

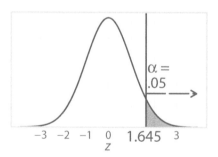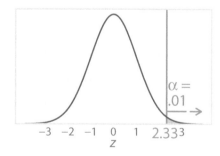

Figure 9.9

Comparing distributions when $\alpha = .05$ and when $\alpha = .01$. The left-hand distribution shows the probability of a Type I error $= .05$ because $\alpha = .05$. The right-hand distribution shows the probability of a Type I error $= .01$ because $\alpha = .01$. Compare the two critical values. The z test statistic's numerator is $M - \mu$. All else being the same, we would need a bigger difference between the sample mean and the population mean to reach the critical value when $\alpha = .01$.

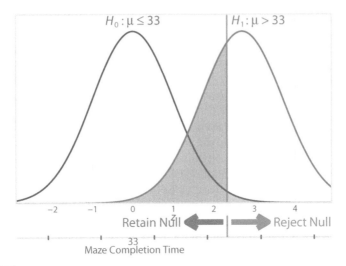

$H_0 : \mu \leq 33$ $H_1 : \mu > 33$

Retain Null   Reject Null

33
Maze Completion Time

Figure 9.10

One possible location of an alternative distribution if H_1: $\mu > 33$ is true and $\alpha = .01$. We never know exactly where the alternative distribution is located. Compared with the situation shown in Figure 9.6, we have changed one thing: now $\alpha = .01$ (the red shaded area). What happened to the probability of a Type II error (the green shaded area)? What happened to power?

it was in Figure 9.6. This larger portion of the right-hand distribution is power. So power, the probability of finding statistical significance, goes up as the area representing α increases, and power goes down as α gets smaller. (For practice, compare Figure 9.10 to Figure 9.6, and see if you can identify how the change from $\alpha = .05$ to $\alpha = .01$ affected the probability of a Type II error.)

Check Your Understanding

9.12. In the glucosamine study, Wilkens et al. (2010) said, "... 250 patients should be enrolled based on a clinically important difference of 3 with 80% power, a 2-sided significance level of .05, and adding 20% for possible drop-outs" (p. 47). What would happen to power if the researchers lowered α or changed to a one-tailed test?

Suggested Answer

9-12. If the researchers lowered the significance level to .01, power would be reduced. If they changed to a directional alternative hypothesis, they would use a one-sided significance level. If they correctly predict the direction of the results, they would have more power than the situation in the actual research, where they had a nondirectional alternative hypothesis. But if they are wrong in predicting the direction, they will have no power (i.e., they will have no probability of detecting a statistically significant effect).

Factors Influencing Power: Variability

Back in the dark ages before most of you were born, radios had knobs: one for volume, one for tuning. Some radios also had buttons for jumping between specific radio frequencies. Alternatively, we could use the tuning knob to scan through the radio frequencies. Between stations there was noise, a sound sort of like a steady rain falling on pavement, often called static. As we turned the tuning knob, we could come across a far-away radio station, and a signal could be heard through the noise. If the signal was strong enough, the music or talk would become clearer, and there would be less noise. Scientists sometimes talk about a signal-to-noise ratio, using a number to measure the amount of signal and a number to measure the amount of noise. The number for the signal is divided by the number for the noise to give us a signal-to-noise ratio. If there is more signal than noise, then the ratio is a number greater than 1. If there is less signal than noise, then the ratio is a number less than 1.

Statisticians like the analogy of signal and noise. The signal is like an effect, such as a difference in means. A bigger effect in research is like a stronger signal; a bigger difference in means would correspond to a larger effect size. The noise is like random variability. We could redefine the z test statistic as a signal-to-noise ratio (although for now we will ignore the situation in which the z test statistic is negative). Let's use the rat shipment example again. If the alternative hypothesis is true and our sample comes from a population in which rats take longer than 33 seconds on average to complete the maze, then a big difference between M and μ would be the signal. But rats vary from each other in the population in terms of their running times, and across repeated samples we also will have variation in the sample mean. The Central Limit Theorem told us how much variation we could expect in the sample mean: M's sampling distribution has a standard deviation called the SE, which is $\sigma/\sqrt{N}$. That is the noise. Is there more signal than noise in the rat shipment example? If we find a significant result, we could answer yes.

How is variability or noise related to statistical power? There are two ways to make the signal-to-noise ratio big: we can increase the signal, or we can decrease the noise. When we have a treatment that leads to a bigger effect (like the earlier example of a prescription pain medication for back pain), we are increasing the signal by using that treatment, rather than a treatment that has a smaller effect (like aspirin). With the z test statistic, the denominator represents the noise. There are two ways that we can make the denominator smaller; increasing the sample size is one way. Let's see a numeric example, using the SE, the denominator of the z test statistic. Our rat shipment example used $\sigma = 19$ and $N = 25$, so the denominator of the z test statistic was

$$SE = \frac{\sigma}{\sqrt{N}}$$

$$= \frac{19}{\sqrt{25}}$$

$$= \frac{19}{5}$$

$$= 3.8$$

What if we had used $N = 100$? Let's look at the effect on the SE:

$$SE = \frac{\sigma}{\sqrt{N}}$$

$$= \frac{19}{\sqrt{100}}$$

$$= \frac{19}{10}$$

$$= 1.9$$

With $SE = 1.9$, we could fit twice as many standard errors of the mean between M and μ, making the z test statistic bigger and more likely to be significant.

A second way to make the denominator of the z test statistic smaller would be to make the population standard deviation smaller, and that is not so easy to do. Generally speaking, variability can be reduced by controlling extraneous variables. For example, suppose Dr. Bellum's years of maze studies with Norway rats had been conducted by graduate students working in a laboratory near a busy hallway. When classes were in session, the hallway was fairly quiet, and the rats would take their time to explore the maze. When classes let out, the volume of ambient sound went up in the laboratory. The foot traffic and talking in the hallway seemed to disturb the rats, and they tended to run through the maze more quickly during noisy times. Overall, there was a lot of variability in the rats' maze completion times. Suppose Dr. Bellum obtains funding for a new lab, located in the basement of a quiet building where classes are not held. Now the rats behave much more alike, and the variability in their running times goes down. Perhaps Dr. Bellum will find after a few years that healthy Norway rats have a smaller population standard deviation than she originally thought. If we were computing the z test statistic, the SE would become smaller if the population standard deviation, σ, is smaller. Instead of $\sigma = 19$, suppose Dr. Bellum told us that she has tested thousands of rats in the new, quieter laboratory, and now she thinks the population standard deviation is 16. If we were to recompute the SE, we would find $\sigma/\sqrt{N} = 16/\sqrt{25} = 16/5 = 3.2$, instead of our original 3.8. Slightly more standard errors of the mean would fit in the gap between M and μ, making the z test statistic bigger and more likely to be statistically significant. So reducing extraneous variability will increase power because it will make it easier for the statistics to detect the signal in the presence of the inevitable noise.

One other factor influences the calculation of a sample size for achieving a certain level of power: the test statistic that the researchers plan to use. So far, we have presented only one inferential statistic, the z test statistic. The calculations of sample size to achieve a certain level of power will depend on which test statistic will be used. Be sure that you do not latch onto some number as being a sufficient sample size for every research situation.

9.13. In the glucosamine study, Wilkens et al. (2010) described the inclusion and exclusion criteria for participants. Among the inclusion criteria, the researchers accepted patients if they were 25 years or older, had non-specific chronic back pain below the 12th rib for at least 6 months, and had a certain score or higher on a measure of self-reported low back pain. Participants were excluded if, among other things, they had certain diagnoses (pregnancy, spinal fracture, etc.) and prior use of glucosamine. In terms of signal and noise, what effect would the inclusion and exclusion criteria probably have on the power in this study?

Suggested Answer

9-13. If the researchers let any adult into the study, there would be much more extraneous variability. For example, if people with spinal fractures were admitted to the study, their experience of back pain probably would differ considerably from similar people who did not have spinal fractures. The inclusion and exclusion criteria defining the sample would reduce some of this extraneous variability (noise), allowing the researchers a better chance of detecting any effect of glucosamine (signal). Thus, the researchers would have a better chance of finding statistical significance, meaning that the power would be greater.

Factors Influencing Power: Relation to Confidence Intervals

Some of the same factors that influence power also will have an effect on confidence intervals. We already talked in Chapter 8 about the effect of one-tailed tests and their corresponding confidence intervals; if the results are not in the predicted direction, we have no probability of rejecting the null hypothesis (that is, we have no power). We also have talked about the effect of the significance level, α. Compared with the situation when $\alpha = .05$ and a 95% confidence interval is computed, a smaller α like .01 corresponds to a wider confidence interval, specifically a 99% confidence interval. A wider confidence interval means the interval is more likely to bracket the hypothesized parameter, which would lead to a decision to retain the null hypothesis. So a smaller α corresponds to a wider interval and a lower probability of finding significance—that is, a lower power. The rest of this section will talk about the other factors that influence power and how they are related to confidence intervals.

Larger effect sizes are easier to detect, so we have a greater probability of finding statistical significance; that is, we have more power. A large effect size will impact the location of a confidence interval on the number line. Let's compare two situations, both involving the rat shipment example, where the null hypothesis said the population mean maze completion time was 33 seconds:

Situation 1: We will use Dr. Bellum and Ray's sample mean of 44.4 seconds.
Situation 2: We will pretend we had a different sample mean: 68 seconds.

Which situation would represent a larger effect size? The bigger difference between M and μ in Situation 2 would represent a larger effect, which should be easier to detect as statistically significant. That means the larger effect provides more power. Now let's think about a 95% confidence interval in these two situations. In both cases we would reject the null hypothesis if the 95% confidence interval does not bracket $\mu = 33$. All else being equal, the confidence interval in Situation 2 would be farther away from $\mu = 33$, compared with the confidence interval in Situation 1. The 95% confidence interval in Situation 1 was [36.952, 51.848]. Only one fact changes in Situation 2: the location of the sample mean. As a result, the only difference between the confidence intervals will be their locations. The 95% confidence intervals have the same width in both situations, but Situation 2 reflects a larger effect—that is, the rats are taking more than twice as long as healthy, normal rats to complete the simple maze. As a result of the larger effect, the confidence interval in Situation 2 is farther to the right on the number line from $\mu = 33$. Now the confidence interval is [60.552, 75.448].

Figure 9.11 illustrates these two situations for using confidence intervals to test a nondirectional null hypothesis that says $\mu = 33$. This figure displays the location of the population mean from the null hypothesis, $\mu = 33$, as a purple circle. If the two sample means, 44.4 and 68, are representative of two alternative population distributions, then the sample with $M = 68$ probably comes from a population that is shifted farther to the right on the number line, compared with the population of healthy rats. The greater the effect, the more likely we are to find statistical significance. That is, we will have more power and a lower probability of a Type II error (retaining H_0 when in fact the alternative hypothesis is true).

The rat shipment example was nonexperimental because Dr. Bellum and Ray did not randomly assign rats to groups or manipulate an independent variable. So the effect size in that case was largely out of their control. We chose the two situations with different sample means to make a point about the effect size having an influence on the location of the confidence interval on the number line.

As we have said, the factor that is most easily changed by researchers is sample size, which has the greatest impact on power. How does N affect confidence intervals?

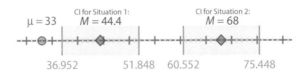

Figure 9.11

Comparing confidence intervals for two situations. Keeping everything else the same, we computed two 95% confidence intervals. In Situation 1, we used $M = 44.4$ seconds from the rat shipment example. In Situation 2, we pretended we had a higher mean: $M = 68$ seconds. Situation 2 represents a case with a larger effect size; the rats were much slower on average than the population mean $= 33$. Compared with the confidence interval for Situation 1, the 95% confidence interval for Situation 2 is farther to the right on the number line from $\mu = 33$, the value in the null hypothesis.

Let's answer that question by doing a numeric example. We already saw the 95% confidence interval of [36.952, 51.848] in Situation 1. This is an interval estimate of the mean of the population from which the questionable rats were sampled. The point estimate of that population mean was the sample mean, $M = 44.4$. The margin of error told us the location of the lower and upper limits of the confidence interval. We computed this margin of error in Chapter 8 by taking a critical value, 1.96, and multiplying it by the SE, $\sigma/\sqrt{N}$. (This appearance of N should be a clue about where we are going with this example.) We were told that $\sigma = 19$ and $N = 25$. That is how we got 3.8 for the SE and a margin of error of 7.448. (As a skeptical reader, you should check our math and make sure you understand where we got these numbers.) Let's change only one fact: what if Ray had randomly sampled 100 rats from the shipment? What is the effect on the confidence interval when N goes up?

We already showed in the section on variability that the SE gets smaller as the sample size gets larger. With $N = 100$, we computed the SE to be 1.9. Let's compute the margin of error:

$$\begin{aligned}
\text{Margin of error} &= \text{critical value} \times SE \\
&= 1.96 \times 1.9 \\
&= 3.724
\end{aligned}$$

Now let's find the limits of the confidence interval:

$$\begin{aligned}
\text{Lower limit} &= M - \text{margin of error} \\
&= 44.4 - 3.724 \\
&= 40.676
\end{aligned}$$

$$\begin{aligned}
\text{Upper limit} &= M + \text{margin of error} \\
&= 44.4 + 3.724 \\
&= 48.124
\end{aligned}$$

When $N = 100$, the 95% confidence interval is [40.676, 48.124]. Let's compare this confidence interval with the one that we computed earlier with $N = 25$: [36.952, 51.848]. The larger sample size leads to a narrower confidence interval. If the sample size continued to increase, the confidence interval would become even shorter. We can imagine ever increasing sample sizes. Eventually with a huge sample size, we would obtain the entire population, and the interval would disappear into a point: the population mean.

So we like narrow confidence intervals as estimates of parameters. But as tools in hypothesis testing, huge sample sizes can give us such a narrow confidence interval that any random difference from a hypothesized value of a parameter (like $\mu = 33$) would be detected as statistically significant. Larger samples mean more power and more narrow confidence intervals, but researchers must plan how much power they want to attain so that an appropriate sample size can be chosen. Then tiny, random differences or relationships will not be detected as statistically significant. Instead, the researcher will be able to target clinically or practically noteworthy effects to detect as statistically significant.

We have seen that variability influences power, and variability also affects confidence intervals. We saw that we could make the *SE* smaller by increasing the sample size or decreasing σ. We mainly decrease variability by controlling extraneous variables and otherwise reducing the noise so that the signal can be easier to detect. If we reduce the variability and keep everything else constant, then power goes up. With less variability, confidence intervals become narrower, giving the same result as larger *N*: higher power. So the interval is more focused around the parameter, and we are more likely to find significance.

Check Your Understanding

SCENARIO 9-B, Continued

In the glucosamine study, the RMDQ is a measure of pain-related disability specifically associated with back problems. Wilkens et al. (2010) said, "A change in score of 3 points on the RMDQ is considered the lowest level of clinical importance to be used for sample size calculations in trials. We estimated that 250 patients should be enrolled based on a clinically important difference of 3 with 80% power, a 2-sided significance level of .05, and adding 20% for possible dropouts" (p. 47). 9–14. Name two details in the researchers' plans that could be changed to increase power and shorten a confidence interval.

Suggested Answer

9-14. A larger sample size would increase the power and shorten the confidence interval. A larger α (say, .10) would increase the power and make the interval estimate more narrow, but at the cost of a higher probability of a Type I error.

What's Next

Out of all the factors influencing power, sample size is the factor that is most easily changed by researchers. If scientists wanted to be guaranteed to find statistical significance, they could use huge sample sizes, which would make their test statistics extremely sensitive to small, arbitrary differences or relationships. Thankfully, it is becoming widespread practice to report how sample sizes are chosen. Science is advanced only when noteworthy effects are reported as statistically significant. It would be unhelpful if arbitrarily large samples make the statistics so sensitive that any randomly small difference or relationship could be detected as statistically significant. We quoted a couple of journal articles in this chapter so that you could see how researchers report the information about the probability of errors and power. As a skeptical reader of research, you will need to be able to discern the probability of different kinds of errors and correct decisions, recognizing that we really do not know whether our decisions in hypothesis testing are correct. They are based on probability, not proof. And one study is not sufficient evidence that a phenomenon exists in the population.

We liked providing the quotations from the journal articles because they gave you a realistic view of research. In contrast, the z test statistic is quite unrealistic. Researchers hardly ever know both the population mean and population standard deviation. In Chapter 10, we will introduce you to another test statistic and a new way of computing a confidence interval for the mean that you would be more likely to find in a journal article.

Exercises

SCENARIO 9-C

Mammography is a common screening for breast cancer. In the United States, about 20% of breast cancers are missed by mammograms, according to the National Cancer Institute. 9-15. If a radiologist reading a mammogram uses a null hypothesis of "there is no breast cancer," what kind of error would occur if the radiologist says the patient has breast cancer, when in fact she or he does not? 9-16. Based on the figure 20% from the previous description, we could state that the probability is .20 for a certain error in hypothesis testing for mammograms as screening tools for breast cancer. What kind of error is associated with that probability? 9-17. What would be the cost of a Type I error in this scenario? 9-18. What would be the cost of a Type II error in this scenario? 9-19. Describe two correct decisions that could be made in a screening mammogram. Then compute the probability of one of those correct decisions, using information given in the scenario.

SCENARIO 9-D

(Based on Fayers, Morris, & Dolman, 2010.) Many patients avoid treatment that involves injections because they fear the pain of the needle sticks. In recent years, researchers have looked for inexpensive ways to reduce the pain without affecting the treatment that requires the injections. Researchers were investigating whether a vibrating device could be used to disrupt pain signals. The study involved surgery on the eyelids, which required patients to undergo local anesthesia. Immediately before one eyelid received an injection of anesthesia, the vibrating device was placed on the patient's forehead and moved in a small circle; this was the treatment condition. Before the other eyelid was injected with anesthesia, the device was turned off and held on the same spot, but not moved in a circle; this was the control condition. The order of conditions was randomized. Patients rated their pain after each injection, using a scale of 0 (*no pain*) to 10 (*worst pain imaginable*). The researchers wrote, "… we decided that on a scale of 0 to 10, a difference in pain score of 1.5 could be considered clinically significant. The power calculation for a paired t test with 90% power and $P = .05$ demonstrated a sample size of 80 patients to be more than sufficient" (p. 1455). The paired t test is an inferential statistic that you will learn in Chapter 11. It compares two means that are linked in a pairwise fashion. In this

(Continued)

example, the same person is giving ratings of pain for the injection in his or her left and right eyelids, and the researchers compared the mean pain ratings for each eye. 9-20. What concept covered in this chapter is related to the "difference in pain score of 1.5"? 9-21. Based on the above quotation from the journal article, what significance level did the researchers choose? 9-22. What is the meaning of the researchers' significance level in terms of the probability of an error in hypothesis testing? 9-23. The probability of a correct decision can be calculated using the significance level. Which correct decision is it? Compute its probability. 9-24. Explain the meaning of "90% power." Why would your textbook authors quibble with the wording of that phrase? 9-25. The probability of another error in hypothesis testing can be computed using the value for power. Which error is it? Compute its probability.

SCENARIO 9-E

(Inspired by Macdonald, Hardcastle, Jugdaohsingh, Reid, & Powell, 2005. Details of this scenario may differ from the actual research.) Scenario 8-D in Chapter 8 described a study in which we pretended we would collect data on postmenopausal women who were long-time moderate beer drinkers. Dietary silicon in the beer could have given them greater average bone mineral density (BMD) than postmenopausal women who do not drink alcohol. Alternatively, the regular beer drinking could be an indication that the women were less concerned with their health, which could mean they take other risks—being sedentary and having a poor diet, for example. Those risks could lower their BMD, which is measured in units of grams per centimeters squared (g/cm^2). Suppose a researcher with years of experience in measuring BMD said the population mean BMD for nondrinking postmenopausal women was 0.88 g/cm^2, with a standard deviation = 0.6. The alternative hypothesis was that our sample came from a population with a mean BMD that was different from 0.88 g/cm^2. Let's say we collected the BMD data on a sample of women who were long-time beer drinkers. Results showed that $M = 1.03$ for $N = 36$ women, resulting in a z test statistic = 1.5 and a 95% confidence interval of [0.834, 1.226]. 9-26. What kind of research is this—nonexperimental, quasi-experimental, or experimental—and how do you know? 9-27. What kind of variable is BMD? 9-28. What kind of variable is physical activity level? 9-29. Use the confidence interval to test the null hypothesis. 9-30. What kind of error could we have made in the hypothesis test in the previous question? 9-31. If we had run the study with $N = 72$ women instead of $N = 36$ women, what would have been the effect on power? 9-32. If we had used $N = 72$ instead of $N = 36$, what would have been the effect on the probability of a Type II error? 9-33. If we had used $N = 72$ instead of $N = 36$, what would have been the effect on the confidence interval? 9-34. What would have been the effect on power if we had used $\alpha = .01$ instead of $\alpha = .05$? 9-35. Why should the scenario have described what the research team considered to be a clinically important difference from $\mu = 0.88$? And in what context would such a description appear? 9-36. Why is it important for researchers to choose their sample sizes carefully?

References

Fayers, T., Morris, D. S., & Dolman, P. J. (2010). Vibration-assisted anesthesia in eyelid surgery. *Ophthalmology, 117*, 1453–1457. doi:10.1016/j.ophtha.2009.11.025

Macdonald, H., Hardcastle, A., Jugdaohsingh, R., Reid, D., & Powell, J. (2005). Dietary silicon intake is associated with bone mineral density in premenopausal women and postmenopausal women taking HRT. *Journal of Bone Mineral Research, 20*, S393.

Price, B. B., Amini, S. B., & Kappeler, K. (2012). Exercise in pregnancy: Effect on fitness and obstetric outcomes – a randomized trial. *Medicine & Science in Sports & Exercise, 44*, 2263–2269. doi:10.1249/MSS.0b013e318267ad67

Rosnow, R. L., & Rosenthal, R. (1989). Statistical procedures and the justification of knowledge in psychological science. *American Psychologist, 44*, 1276–1284. doi:10.1037/0003-066X.44.10.1276

Wilkens, P., Scheel, I. B., Grundnes, O., Hellum, C., & Storheim, K. (2010). Effect of glucosamine on pain-related disability in patients with chronic low back pain and degenerative lumbar osteoarthritis: A randomized controlled trial. *Journal of the American Medical Association, 304*, 45–52. doi:10.1001/jama.2010.893

10

One-Sample Tests and Estimates

Introduction

So far you have learned one test statistic and a confidence interval. The *z* test statistic was a great way to introduce you to hypothesis testing because we could use it in an uncomplicated example, such as our story of the rats that seemed different from the usual healthy rats (in the version of the story with no direction predicted). Unfortunately, we doubt you ever will find this *z* test statistic in a scientific journal article. It is an unrealistic statistic for two reasons. First, we had to know the numeric values of both the population mean and population standard deviation, which in reality we rarely can know. Second, we are limited to a simple hypothesis. For example, the rat shipment example had a null hypothesis that said the sample came from a population with a mean equal to 33 seconds. Research hypotheses usually are more nuanced, leading to more complex ways of setting up studies and therefore more complex statistical analyses.

Researchers occasionally do compare a sample mean to a population mean, particularly when they want to compare *M* to some standard. Can we test a hypothesis about a single population mean or compute an interval estimate for μ in situations where we do not know the population standard deviation or variance? Yes, and the purpose of this chapter is to explain a statistic and confidence interval for this situation. The good news is that we will follow the decision rules that you already have learned for hypothesis testing. In fact, many of the steps in hypothesis testing will be the same for the remaining test statistics in this book, so when we introduce a new inferential statistic, we will concentrate on what is different. We also will introduce a graph that incorporates a confidence interval in estimating a population mean.

One-Sample *t* Test

You may be preparing for a career that has a reputation for requiring long work hours, which can affect the quality and quantity of sleep. Let's take a look at a journal article that investigates the sleep quality of medical students. Brick, Seely, and Palermo (2010) explored the effect of many variables (exercise, caffeine intake, etc.) on medical students' sleep. One criterion variable in the study was the Pittsburgh Sleep Quality Index (PSQI), which we have mentioned in other chapters. This measure produces higher scores for people with more sleep problems (i.e., lower sleep quality). Among other predictions in the study, the researchers wrote, "In this sample, we hypothesized that medical students would report worse sleep quality in comparison to published normative samples of healthy, young adults" (p. 114).

In Chapter 4, we defined norming as a process of gathering scores and assessing the numerical results for a large reference group. We said the mean and standard deviation of the reference group often are called norms. When the article on medical students' sleep quality says "in comparison to published normative samples," the authors are saying they will compare the mean sleep quality for the sample of medical students to the mean of a reference group that is being treated like a population. In this case, the population consists of healthy, young college students with a mean PSQI of 5.6 (Carney, Edinger, Meyer, Lindman, and Istre, 2006). Let's translate the prediction of Brick et al. (2010) into an alternative hypothesis. They predicted that on average, medical students would have worse sleep quality (higher mean PSQI) than other healthy adults, whose mean was 5.6. Our alternative hypothesis could be written as follows:

$$H_1: \mu > 5.6$$

This statistical hypothesis can be translated into the following statement:

Our sample comes from a population with a mean PSQI greater than 5.6, where a higher PSQI score means worse sleep quality.

Our null hypothesis would be an opposite statement:

$$H_0: \mu \leq 5.6$$

In words, the null hypothesis can be translated as follows:

Our sample comes from a population with a mean PSQI less than or equal to 5.6.

The null and alternative hypotheses are identical to the ones we would have written when we were using the z test statistic. But Brick et al. (2010) computed a slightly different test statistic, one that does not require knowledge of a population standard deviation. The test statistic that they used was called the *one-sample t test*. Similar to the z test statistic, the one-sample t test is used to compare a sample mean with a population mean. The difference is that instead of having a population standard deviation as part of the denominator, the one-sample t test uses the sample standard deviation, as part of the denominator. For that reason, we say that the one-sample t test follows the verbal definition of "(something minus its mean) divided by its *estimated* standard deviation." The "something" is the sample mean, M. "Its mean" is the population mean, μ. And "its estimated standard deviation" is $SD/\sqrt{N}$. So the formula for the one-sample t test is

$$\text{one-sample } t = \frac{M - \mu}{SD / \sqrt{N}}$$

We use the one-sample t test in situations similar to those in which we would use the z test statistic: we have one sample, and we are interested in comparing the sample mean to a population mean. The difference is that we choose the one-sample t test when we do not know the population standard deviation, σ, or the population variance, σ^2. The only difference between the formulas for the one-sample t test and the z test statistic is the use of SD in the denominator of the one-sample t. And by the way, we always say "one-sample" in front of "t test" because there are *lots* of t test statistics, some of which you will read about in this book. Saying "one-sample" helps to clarify which t test statistic is being used. The one-sample t test's numeric value also is interpreted in a similar way to the z test statistic: it is the number of estimated standard errors that fit in the gap between M and μ.

The present example shows a directional alternative hypothesis, but it is possible to have a nondirectional alternative hypothesis when using the one-sample t test, if we do not have theory or prior research to inform a directional prediction. The rest of the process of hypothesis testing will look quite familiar. We will determine whether the results are in the predicted direction (if we made such a prediction). Then we will compare an observed one-sample t test with a critical value, or we will compare a p value with a chosen significance level, α. The decision rules about rejecting or retaining the null hypothesis will be the same as the ones in Chapter 8.

To continue our present example, the researchers predicted that medical students would have a higher mean PSQI (indicating worse sleep quality) than other healthy, young adults. In Chapter 8, we said the one-tailed p value decision rule was

If the observed test statistic is in the predicted tail

AND

if the observed p value in the predicted tail is less than or equal to α,

then reject the null hypothesis.

Otherwise, retain the null hypothesis.

This rule applies with our current example, except now the observed test statistic is the one-sample t test, and the p value comes from a different distribution, not the standard normal distribution. So here we have seen two details of hypothesis testing that are different: the one-sample t statistic and the distribution from which we get our critical value and p value. Next we will explain the distribution that we need to use when testing our null hypothesis about the mean sleep quality of medical students.

Check Your Understanding

SCENARIO 10-A

Normal human body temperature is said to be 98.6°F. Let's pretend we are medical researchers, and we have some reason to suspect that this traditional number is not really the average body temperature for healthy adults. We obtain a data set that contains the body temperatures of military enlistees who were measured on their first day of basic training. We exclude enlistees whose records indicated any illness in the week before or the week after the first day of basic training, leaving a sample of $N = 245$. Is the average body temperature for this sample significantly different from 98.6? 10-1. Write the alternative hypothesis in symbols and words. 10-2. Why would the one-sample t test be appropriate for this scenario?

Suggested Answers

10-1. The alternative hypothesis would be $H_1: \mu \neq 98.6$. In words, the alternative hypothesis says the sample comes from a population with a mean body temperature that is not equal to 98.6. 10-2. The one-sample t test would be appropriate because the study has one sample; we are interested in comparing a sample mean to a population mean; a population mean (98.6) is known; but a population standard deviation is unknown.

Distribution for Critical Values and p Values

To clarify why we need a different distribution for testing hypotheses with the one-sample t test, let's review what we learned in Chapters 7 and 8 about sampling variability. We know that one particular sample of sleep quality scores will give us a sample mean to compare with a known value for a population mean. But then we remember the idea of sampling variability. If we were to draw different samples from the same population, we could expect to get different numeric values of the sample mean. If we could draw all possible samples of the same size from the same population and compute the mean sleep quality on every sample, we could create a sampling distribution of the mean. The Central Limit Theorem told us that the mean's sampling distribution would be shaped like a normal curve. *(Look! Flashing lights and authors jumping up and down to indicate the importance of the following two sentences!)* We also know that the z test statistics computed using all those possible sample means would have a distribution that looks like the mean's distribution. That is why we could look at a standard normal distribution for our critical values and p values for hypothesis testing.

But in our present example of the medical students' mean sleep quality, we have switched to the one-sample t test. The idea of sampling variability is still important. If we had different samples of medical students, we still could obtain a variety of sample means for sleep quality. The Central Limit Theorem still would tell us that the sample mean's sampling distribution would be normal. We could compute the one-sample t test for each of those possible sample means. But all possible one-sample t tests would *not* form a normal distribution. Why not? Look back in the previous paragraph where we used flashing lights and jumped up and down: the z test statistic was distributed the same as the original distribution. But the one-sample t test is slightly different from the z test statistic, so its distribution also will differ. The one-sample t test requires the use of an *estimate* of the sampling variability of M because we do not know σ or σ^2. That is, the one-sample t test uses $SD/\sqrt{N}$ as an *estimated* standard error of the mean, not the exact standard error of the mean described in the Central Limit Theorem.

The use of this estimated standard error of the mean brings in more sampling variability associated with a statistic, SD, being used in the denominator of the test statistic. Let this idea of "more sampling variability" soak in for a moment. The numerator of the one-sample t statistic has M in it, so the one-sample t has some sampling variability because of M—the z test statistic had that same amount of sampling variability. But the one-sample t also has SD in its denominator, and SD is a sample statistic with its own sampling variability. Remember that the z test statistic had σ in its denominator, and σ was a known constant, not a statistic with sampling variability. So using SD adds more sampling variability to the one-sample t that was not present in the z test statistic. As a result, the one-sample t tests that could be computed on all those possible sample means will have a distribution that is slightly more spread out than a standard normal distribution. A distribution of all possible one-sample t tests is called a t distribution. A t distribution looks a lot like a standard normal distribution, but not quite as tall in the middle and a little thicker in the tails, affecting the probabilities determined by those crucial tail areas. Figure 10.1 shows the standard normal distribution and a *t* distribution.

We had to be careful in the previous paragraph, where we said "a *t* distribution" because there actually are many possible *t* distributions. The sample size affects the spread of a *t* distribution. With an extremely large sample size, a *t* distribution becomes almost indistinguishable from a standard normal distribution. With smaller sample sizes, the *t* distributions differ from a standard normal distribution, although most people glancing at a graph of a *t* distribution would say it was a normal distribution. But the differences are important in those tail areas, which give us critical values and *p* values. The exact shape of any particular *t* distribution is specified by something called *degrees of freedom*. In practical terms, degrees of freedom are needed so that we can know which *t* distribution to use to find critical values and *p* values. The exact definition of degrees of freedom is more complex, and we do not think you need it in a beginning statistics course. To help you to gain a conceptual understanding of degrees of freedom (abbreviated as *df*), we need to take a brief detour from our sleep quality example.

Let's pretend for a moment that we had a sample of five numbers written on a piece of paper, and we knew that the sample mean equaled 3. But we dropped a piece of pizza (messy side facing down, of course) on the paper, and now we can read only four of the numbers:

$$3, 5, 4, 4$$

If we already knew that the five numbers had a mean of 3, then what is the fifth number? Well, the mean is the sum of the scores divided by the number of scores. Five numbers that have an average of 3 must add up to 15 (i.e., $M = 15/5 = 3$). Let's add up these four numbers:

$$3 + 5 + 4 + 4 = 16$$

Our fifth number *must* have been a negative 1. We can check: $3 + 5 + 4 + 4 - 1 = 15$, so the mean of these five numbers is 3.

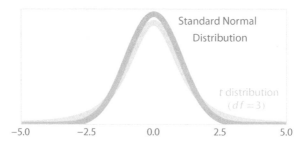

Figure 10.1

Comparing a standard normal distribution and a *t* distribution. The standard normal distribution in this figure is a little taller in the middle, compared with a *t* distribution. This *t* distribution has thicker tails than the standard normal distribution.

10. One-Sample Tests and Estimates

Let's do another example using the same $M = 3$ and $N = 5$. Now let's say the four legible numbers are 1, 5, 3, and 4. What is the fifth number? The sum of the five numbers must be 15 if $M = 3$, and these four numbers add up to 13. Our fifth number must be 2. Let's check:

$$1+5+3+4+2=15$$

and

$$M=15/5=3$$

The sample mean is an estimate of the population mean. Four of the five numbers being used to compute a mean are *free to vary*. We know the value of M, so once those four numbers have been found, the fifth number is *not free to vary*.

Why did the concept of degrees of freedom come up now? The one-sample t test's use of SD instead of σ is responsible for the introduction of df. As it turns out, the formula for SD contains the sample mean as an estimate of the population mean. And as we have seen above, all but one of the numbers that go into computing M are free to vary. Degrees of freedom matter to us because they define the different t distributions used to find critical values and p values.

As an aside, we now have another way of defining the unbiased variance. As you will recall, SD is the square root of the unbiased variance. The unbiased variance has a numerator that takes the difference between each score and the sample mean, then squares the differences and adds them up. In Chapter 2, this numerator was called the sum of squared differences from the mean. The denominator of the unbiased variance is $N - 1$, which also happens to be its degrees of freedom. So we can define the unbiased variance as a ratio of the sum of squared differences from the mean, divided by df.

Here is the formula for the degrees of freedom for the one-sample t test:

$$df = N - 1$$

Other test statistics have different formulas for df. Next we will describe the connection between df and the distributions used in hypothesis testing with the one-sample t test.

Check Your Understanding

SCENARIO 10-A, Continued

We have a sample size of 245 military enlistees, and we plan to use the one-sample t test to compare the sample mean with the norm of 98.6° F. 10-3. How many degrees of freedom do we have?

Suggested Answer

10-3. We have $df = N - 1 = 245 - 1 = 244$.

Critical Values for the One-Sample *t* Test

Why did we start talking about *df*? We were explaining how the one-sample *t* test differed from the *z* test statistic and why the standard normal distribution does not give us critical values for the one-sample *t* test. Table B.1 gives us values of *t* statistics similar to the one-sample *t* test, whereas Table A.1 showed values in the standard normal distribution. Table A.1 showed many values of *z* along the positive end of the number line and areas associated with those values—both middle areas and tail areas. Table A.1 was used when we computed *z* scores as descriptive statistics for measuring relative location of a score within a sample or a population. Table A.1 was used again when we computed *z* test statistics for hypothesis testing.

Table B.1 is quite different. It represents many *t* distributions, not just one distribution like Table A.1. Each *t* distribution is defined by its degrees of freedom, so each line of Table B.1 contains critical values for a different *t* distribution. Another major difference between the tables is that Table A.1 allowed us to find a value of *z*, then look for an area under the curve. In contrast, Table B.1 tells us a few different areas (values of α) and, based on the *df* for a given situation, we find values of *t* statistics to use as critical values. Statistical software also uses *t* distributions to compute *p* values, but our table is limited to critical values associated with values of α. Figure 10.2 shows an excerpt of Table B.1. To use Table B.1, we will follow these steps:

- Choose the significance level, α.
- Determine whether we have a one-tailed or a two-tailed test.
- Compute *df*.
- Use that value of *df* to find a row in Table B.1.
- And find a critical value or values on that row in the table.

Then we can test a null hypothesis using the critical value (for a one-tailed test) or critical values (for a two-tailed test). In actual research, critical values are rarely used. Instead, researchers rely on *p* value decision rules. Consider how long Table B.1 would have to be if it contained *p* values instead of a few critical values. For one particular value of *df*, there would have to be a table like Table A.1, showing a large number of possible values of the one-sample *t* test and the associated areas for different portions of the *t* distribution for that *df*. Then for every new value of *df*, there would be a separate table representing a different *t* distribution. We would need a separate table for every value of *df* that we would choose to include. Table B.1 alone would be as long as this book! That does not even include the possibility of *df* being a fractional number, which is possible for some kinds of *t* statistics. So in real research, we let statistical software compute the *p* values, and for the one-sample *t* test, the software finds *p* values in *t* distributions defined by *df*. Whenever we give you a *p* value for the rest of the test statistics in this book, it will have come from statistical software.

Let's examine Table B.1. It has a separate row for many different values of *df*. Each row represents a different *t* distribution. The first column lists the *df* for all the whole numbers from 1 to 30, then there are lines for *df* = 35, 40, 45, 50, 55, and 60.

Table B.1 Critical Values for *t* Distributions

	Total α for a Two-Tailed Test					
	.20	.10	.05	.02	.01	.001
	Total α for a One-Tailed Test					
df	.10	.05	.025	.01	.005	.0005
1	3.078	6.314	12.706	31.821	63.657	636.619
2	1.886	2.920	4.303	6.965	9.925	31.599
3	1.638	2.353	3.182	4.541	5.841	12.924
4	1.533	2.132	2.776	3.747	4.604	8.610
5	1.476	2.015	2.571	3.365	4.032	6.869
6	1.440	1.943	2.447	3.143	3.707	5.959
7	1.415	1.895	2.365	2.998	3.499	5.408
8	1.397	1.860	2.306	2.896	3.355	5.041
9	1.383	1.833	2.262	2.821	3.250	4.781
10	1.372	1.812	2.228	2.764	3.169	4.587
11	1.363	1.796	2.201	2.718	3.106	4.437
12	1.356	1.782	2.179	2.681	3.055	4.318
13	1.350	1.771	2.160	2.650	3.012	4.221
14	1.345	1.761	2.145	2.624	2.977	4.140
15	1.341	1.753	2.131	2.602	2.947	4.073
16	1.337	1.746	2.120	2.583	2.921	4.015
17	1.333	1.740	2.110	2.567	2.898	3.965
18	1.330	1.734	2.101	2.552	2.878	3.922
19	1.328	1.729	2.093	2.539	2.861	3.883
20	1.325	1.725	2.086	2.528	2.845	3.850
21	1.323	1.721	2.080	2.518	2.831	3.819
22	1.321	1.717	2.074	2.508	2.819	3.792
23	1.319	1.714	2.069	2.500	2.807	3.768
24	1.318	1.711	2.064	2.492	2.797	3.745
25	1.316	1.708	2.060	2.485	2.787	3.725
26	1.315	1.706	2.056	2.479	2.779	3.707
27	1.314	1.703	2.052	2.473	2.771	3.690
28	1.313	1.701	2.048	2.467	2.763	3.674
29	1.311	1.699	2.045	2.462	2.756	3.659
30	1.310	1.697	2.042	2.457	2.750	3.646
35	1.306	1.690	2.030	2.438	2.724	3.591
40	1.303	1.684	2.021	2.423	2.704	3.551
45	1.301	1.679	2.014	2.412	2.690	3.520
50	1.299	1.676	2.009	2.403	2.678	3.496
55	1.297	1.673	2.004	2.396	2.668	3.476
60	1.296	1.671	2.000	2.390	2.660	3.460
70	1.294	1.667	1.994	2.381	2.648	3.435

Figure 10.2

Excerpt from Table B.1, which appears in the back of the book. This table provides critical values for the one-sample *t* test and other *t* statistics.

Then the lines go by tens (70, 80, 90), then the table skips to 120. The last line shows 100,000. Each line gives critical values for the *df* of that line and the different columns. Clearly we will never do a study with *df* = 100,000, but that line provides what should be some familiar numbers. As the sample size gets enormous, the *t* distribution becomes practically indistinguishable from the standard normal distribution, and you can see some familiar critical values, such as 1.645 and 1.96. What if Table B.1 does not contain exactly the same *df* that we need for a particular situation? In real research, we hardly ever use tables in hypothesis testing, but we will show you examples where Table B.1 does not contain a line for the value of *df* that we need. When that happens, we look for the closest *smaller* value for *df*. We cannot give ourselves more degrees of freedom because it could increase the likelihood of a Type I error, so we use the critical value for a close but smaller *df*.

The column headings for Table B.1 require some explanation. Across the top of the page there is a set of column labels with a heading that says, "Total α for a Two-Tailed Test." Immediately below that heading is a line giving different values of α: .20, .10, .05, .02, .01, and .001. Below those numbers there is another set of column labels with a heading that says, "Total α for a One-Tailed Test," followed by a line with different values of α: .10, .05, .025, .01, .005, and .0005. When we use Table B.1, we must pay careful attention to whether we have a one-tailed or a two-tailed test, which will determine which set of column labels we will use. We are saving space by having two sets of column labels. To illustrate, suppose α = .20 for a two-tailed test. That means half of α is in each tail of a *t* distribution. A positive critical value would cut off α/2 = .10 in the upper tail, and a negative critical value would cut off α/2 = .10 in the lower tail. If we have a two-tailed test with α = .20, then we would go down the first column of critical values. But this column can be used in another situation. What if we had a total α = .10 for a *one-tailed* test? That would be the same thing as using one of the critical values in the case of a two-tailed test with a total α = .20.

Now let's talk about the row labels, which appear in the first column, labeled *df*. As we have said, the degrees of freedom determine the exact appearance of a *t* distribution, so every line represents a different *t* distribution. By choosing the correct row (for *df*) and the appropriate column label (one- or two-tailed test and the total α), we can find critical values for the correct *t* distribution for our test statistic and α.

Let's look at an example in Table B.1. Look at the row labeled *df* = 30. This row represents a particular *t* distribution, the exact shape of which is defined by *df* = 30. Let's pretend we have a two-tailed test with α = .05. We look at the top of Table B.1 and look at the set of labels for a two-tailed test (totally ignoring the labels for a one-tailed test). Go down the middle column (α = .05 for a two-tailed test) until you reach the row for *df* = 30. You should find the number 2.042. This is a *t* critical value—but remember, we said we were doing a two-tailed test. The *t* distributions are symmetrical, like the standard normal distribution, so we would use +2.042 and −2.042 as the critical values. Figure 10.3 shows the *t* distribution with *df* = 30. The *t* critical values of +2.042 and −2.042 are shown on the horizontal number line, and vertical lines have been drawn through those values.

The tail areas reflect the total $\alpha = .05$, so half of that area (.025) is cut off in the upper tail and the other half (.025) is cut off in the lower tail.

Now let's consider another situation, but we are going to keep using Figure 10.3. What if we had $df = 30$ for a one-sample t test, except now we had a directional alternative hypothesis and $\alpha = .025$? In that case, we would perform a one-tailed test, and all of $\alpha = .025$ would be in one tail. Let's pretend that the alternative hypothesis came from a situation similar to the sleep quality of medical students, where H_1: $\mu > 5.6$, except now we have only 31 students in our present sample. All of $\alpha = .025$ would go in the upper tail. Look at Figure 10.3 again. What would be the critical value if we had $df = 30$? The critical value would be $+2.042$, which cuts off a tail area corresponding to $\alpha = .025$. So we could cover up the lower tail in Figure 10.3 and proceed with the hypothesis test. Now let's see if we can find this critical value in Table B.1. We will still use the row for $df = 30$, but now we need to look at the column heading for the total α for a one-tailed test when $\alpha = .025$. This column label is directly below the two-tailed test's column heading for $\alpha = .05$ and leads us to the critical value of $t = 2.042$. So a critical value of $t = 2.042$ could be used with a one-tailed test for $\alpha = .025$, or it could be used along with -2.042 as the two critical values for a two-tailed test when $\alpha = .05$ (with half of that area in each tail).

Look for the critical value decision rules in Chapter 8. You will notice that these rules did not mention the name of the test statistic. Instead of saying "the z test statistic," we intentionally said "the test statistic." Those decision rules apply to the one-sample t test. If an observed one-sample t is more extreme than a critical value, then we will reject the null hypothesis; otherwise, we will retain the null hypothesis. Chapter 8's p value decision rules also apply directly to the one-sample t test. For example, with a one-tailed test, if the results are in the predicted direction *and* the one-tailed p value is less than or equal to α, then we reject the null hypothesis. The meaning of the p value remains the same. The p value is computed based on a distribution drawn as if the null hypothesis is true. For a one-sample t test, a t distribution for a particular value of df represents all possible values of the one-sample t test computed on repeated sample means from the same population. If the null hypothesis is true, the probability of obtaining a one-sample t test at least as extreme as ours by chance alone is the p value.

Figure 10.3

A t distribution with $df = 30$. By looking in Table B.1 in the back of the book, we can find critical values for t distributions, which are defined by their degrees of freedom. This t distribution has $df = 30$. If $\alpha = .05$ and the alternative hypothesis does not predict a direction, we would have a two-tailed test with the two critical values shown.

If the p value is bigger than the significance level, then we would conclude that our observed one-sample t test is a typical result for the situation where the null hypothesis is true, so we would retain H_0. If the p value is less than or equal to α, then it is small enough for us to conclude that getting a one-sample t test at least as extreme as ours is unlikely under the null hypothesis, so we reject H_0.

Hypothesis testing usually proceeds exactly as it did in Chapter 8, with the only differences being the names of the test statistics and the distributions where we get the critical values or (via statistical software) p values.

Check Your Understanding

SCENARIO 10-A, Continued

The medical researchers who think the average human body temperature differs from the norm of 98.6 planned to use the one-sample t test, which we found in a previous question to have $df = 244$. 10-4. Using $\alpha = .01$, find the critical values in Table B.1.

Suggested Answers

10-4. Table B.1 does not have a row for $df = 244$. We do not have infinite degrees of freedom, so we will use the table's next smaller value of $df = 120$. We have a nondirectional alternative hypothesis, so we will use the column labels for the total α for a two-tailed test. Going down the column labeled .01, we find a value of 2.617. So our critical values are +2.617 and −2.617.

Completing the Sleep Quality Example

Let's complete our sleep quality example from Brick et al. (2010) using a significance level of .05. The researchers computed a global score for sleep quality using the PSQI, which has higher numbers for worse sleep quality. The sample mean was 6.37 ($SD = 2.57$). The norm for healthy adults was 5.6, which came from the work of Carney et al. (2006). Our alternative hypothesis for this study said $H_1: \mu > 5.6$, predicting that the medical students on average would have worse sleep quality than other young adults.

Because we have a directional alternative hypothesis, we must check whether the results turned out in the predicted direction. Is the sample mean for the medical students greater than the norm for healthy adults? Yes, because 6.37 is greater than $\mu = 5.6$. We also could check whether the one-sample t test is positive, but we have not computed it yet. Is the difference between 6.37 and 5.6 statistically significant? Remember, our null hypothesis says that our sample comes from a population in which the mean sleep quality is 5.6 or less, with smaller numbers meaning better sleep quality. We do not believe that is the case, but pretending momentarily that the null hypothesis is true, how likely is it to obtain a sample mean of 6.37 or greater? To answer this question, we need a p value. Brick et al. (2010) reported a one-sample $t = 5.13$ and a one-tailed $p < .001$. This p value is less than $\alpha = .05$, so we would reject the null hypothesis. These authors concluded that "the medical students had significantly worse self-reported sleep quality" compared with the norm for young adults (p. 116).

We would like for you to complete the computation of these researchers' one-sample $t = 5.13$. We already have stated their sample mean, the population mean, and SD, so all we need is the sample size. The article stated that the total number of respondents to the researchers' survey was $N = 314$. But when we used this sample size in the computation of the one-sample t, we got a different result from the researchers' reported one-sample $t = 5.13$. Why? It is common in survey research for respondents to skip questions. When we are using a scale such as the PSQI, the lack of a response to an item generally means the researchers cannot compute a total score for that respondent. We suspect that PSQI scores were calculated only for those who answered all the questions. By working backward and solving for N in the formula for the one-sample t, we found that using $N = 293$ would produce a one-sample $t = 5.13$, as reported by the researchers. Please do not skip the next Check Your Understanding question!

Check Your Understanding

10-5. Let's see if you can compute the one-sample t test from the study by Brick et al. (2010) and test the null hypothesis. The alternative hypothesis said H_1: $\mu > 5.6$. The researchers reported a mean global PSQI of $M = 6.37$ ($SD = 2.57$), and we think 293 respondents' scores were used to compute those two descriptive statistics. Compute the one-sample t test and df, then use Table B.1 to test the null hypothesis using the one-tailed critical value decision rule and $\alpha = .05$.

Suggested Answers

10-5. *The numerator of the one-sample t test is $M - \mu = 6.37 - 5.6 = 0.77$. The denominator is the estimated standard error of the mean = $SD/\sqrt{N} = 2.57/\sqrt{293}$ = 2.57/17.117243 = 0.150141. So the one-sample $t = 0.77/0.150141 = 5.1285124 \approx$ 5.13, as reported in the journal article. The $df = N - 1 = 293 - 1 = 292$. Table B.1 has a line for $df = 120$ and $df = 100,000$. We cannot give ourselves more degrees of freedom than $df = 292$, so we will use the critical value for $df = 120$. We consult the column labels and go down the column for a one-tailed test's total $\alpha = .05$. On the row for $df = 120$, we find a critical value of 1.658. The results are in the predicted direction because 6.37 is greater than 5.6, as predicted in the directional alternative hypothesis. Our observed one-sample t test, 5.13, is more extreme than 1.658, so we reject the null hypothesis and conclude that the medical students had significantly worse sleep quality than the norm for healthy, young adults.*

Assumptions

As we saw in Chapter 8, the computation of accurate p values depends on certain conditions or assumptions being met. We could get accurate p values without assumptions if we went through the process of obtaining all possible samples of the same size from the same population—that is, if we created sampling distributions. But to avoid that hassle, we want to take advantage of the knowledge of mathematical statisticians, who give us valuable information, such as the Central Limit Theorem. Then we can use a theoretical distribution to find critical values and p values. To trade in the sampling distribution and use a theoretical distribution in its place, assumptions must be made.

The one-sample t test has two assumptions, and they are the same as the assumptions for the z test statistic:

- The scores are independent of each other.
- The scores in the population are normally distributed.

We usually cannot know whether the scores on a particular outcome variable are normally distributed in the population. But the Central Limit Theorem saves us from that concern and tells us that the sample mean, which is central to the one-sample t test, will be normally distributed. Independence of observations typically is assumed to come from some form of random sampling from a population. But researchers usually make a judgment that the scores will be independent as long as participants are unrelated individuals who provided the scores without influencing each other.

Next we explain a confidence interval for estimating the population mean. It is similar to the one we covered in Chapter 8, except this one will rely on the one-sample t test's critical values.

Check Your Understanding

SCENARIO 10-A, Continued

The medical researchers who suspected that normal human body temperature was different from 98.6°F collected data from 245 military enlistees, who had a mean = 98.4 and $SD = 14.2$. We already found in Question 10-4 that the critical values are +2.617 and −2.617 for the two-tailed one-sample t test ($\alpha = .01$). 10-6. Compute the one-sample t test. Test the null hypothesis, and explain the meaning of the decision.

Suggested Answers

10-6. The numerator is $M - \mu = 98.4 - 98.6 = -0.2$. The denominator is $SD/\sqrt{N} = 14.2/\sqrt{245} = 14.2/15.652476 = 0.9072047$. So the one-sample t test $= -0.2/0.9072047 = -0.2204574 \approx -0.22$. So the sample mean is about one-fifth of an estimated standard error below the population mean. Because the observed test statistic, −0.22, is not more extreme than a critical value, we will retain the null hypothesis and conclude that our sample's mean body temperature is not statistically significantly different from the norm of 98.6.

10. One-Sample Tests and Estimates

Confidence Interval for μ Using One-Sample *t* Critical Value

The sample mean can be deceiving as an estimate of the population mean because sometimes we forget that it contains sampling variability. After all, when we see $M = 6.37$ for the average sleep quality score for the medical students in the study by Brick et al. (2010), we see only one number, which seems precise. But if we were to repeatedly draw the same size of sample from the same population that provided the sample of medical students and then compute the mean sleep quality repeatedly, we would get different values of M. Interval estimation provides a way to quantify the sampling variability. A 95% confidence interval as an estimate of μ tells us that 95% of the confidence intervals like ours will contain the true mean of whatever population we are sampling. One particular confidence interval may or may not contain the true mean of the population being sampled. Maybe we are sampling from a population described in a null hypothesis, or maybe we are sampling from some other population. But we do know that 95% of such intervals include the true value of μ, whatever it is.

When we introduced the interval estimation of the population mean, the calculations included a critical value from the standard normal distribution. But that was when we were using the z test statistic, which required knowledge of a numeric value of the population standard deviation or variance. Now we have switched to the one-sample t test, which allows us to use SD instead of σ. We have to change our way of calculating the confidence interval too.

We will not review the logic behind the computation of the margin of error; you can read it again in Chapter 8. We pick up on the idea of using a margin of error to compute an interval estimate of μ that reflects the sampling variability inherent in a point estimate of a parameter. Chapter 8 showed the margin of error as the product of a critical value and the standard error of the mean. In this chapter, we have been using the *estimated* standard error of the mean, $SD/\sqrt{N}$. Let's continue with the example of the PSQI scores for medical students, where a higher number means worse sleep quality. For simplicity, let's pretend that the researchers thought the medical students would *differ* from the norm of 5.6 for average sleep quality of young adults, corresponding to a nondirectional alternative hypothesis:

$$H_1: \mu \neq 5.6$$

The corresponding null hypothesis would be

$$H_0: \mu = 5.6$$

This null hypothesis could be translated as follows:

Our sample comes from a population in which the mean PSQI equals 5.6.

The researchers reported $M = 6.37$ ($SD = 2.57$), and we said these statistics were computed on data from 293 medical students. So the estimated standard error of the mean is

$$\text{estimated standard error of the mean} = \frac{SD}{\sqrt{N}}$$

$$= \frac{2.57}{\sqrt{293}}$$

$$= \frac{2.57}{17.117243}$$

$$= 0.150141$$

To get the margin of error, we need to multiply the estimated standard error of the mean and a critical value. To find a critical value, we look at Table B.1 and find the column for a total $\alpha = .05$ for a two-tailed test. We will have to use the critical value for $df = 120$ because the table does not contain a listing for $df = 292$, and the next smaller value is $df = 120$. The table shows a critical value of 1.98. Now we can compute a margin of error:

$$\text{margin of error} = \text{estimated standard error of the mean} \times \text{critical value}$$

$$= 0.150141 \times 1.98$$

$$= 0.2972792$$

So the margin of error is approximately 0.3, but we cannot round yet, as we are in the middle of a computation. The lower limit of the interval estimate is the sample mean minus the margin of error, and the upper limit is M plus the margin of error. The lower limit is

$$\text{Lower limit} = M - \text{margin of error}$$

$$= 6.37 - 0.2972792$$

$$= 6.0727208 \approx 6.07$$

The upper limit is

$$\text{Upper limit} = M + \text{margin of error}$$

$$= 6.37 + 0.2972792$$

$$= 6.6672792 \approx 6.67$$

The 95% confidence interval is [6.07, 6.67]. Now we have a range of values estimating the mean of the population that gave us the sample of medical students, and we have a sense for the sampling variability that could be expected across multiple samples. To test the null hypothesis that said our sample came from a population with a mean PSQI equal to 5.6, we can use the same decision rule that appeared in Chapter 8, when we computed a confidence interval using a critical value for the z test statistic. We check whether the interval contains the population mean, 5.6. If so, we retain the null hypothesis. If not, we reject the null hypothesis. Because the interval [6.07, 6.67] does not bracket $\mu = 5.6$, we can conclude that our sample mean for medical students differs significantly from the norm for sleep quality of young adults. The interval is higher on the number line than $\mu = 5.6$, so the medical students had significantly worse sleep quality than young adults in general.

You may wonder about the use of an imprecise critical value in this computation of a confidence interval. The confidence interval actually is not drastically affected, especially with larger sample sizes. In real research, we almost always compute statistics using specialized software; SAS, SPSS, EpiData, NCSS, and Stata are some of the software packages used by health sciences researchers. These software packages would use mathematical formulas to find the exact critical value for $df = 292$ to compute this confidence interval. For example, SPSS says the critical values would be $t = \pm 1.96812140700915$. If you use the positive t critical value to compute the margin of error in the above example, you will get results that round to the same confidence interval that we computed using a critical value based on $df = 120$. (For practice with confidence intervals, give it a try!)

Check Your Understanding

SCENARIO 10-A, Continued

The researchers who thought normal human body temperature differed from 98.6°F used the critical values +2.617 and −2.617 and performed a two-tailed using the one-sample t test and $\alpha = .01$. Their sample mean was 98.4, and their estimated standard error of the mean was 0.9072047. 10-7. Calculate the 99% confidence interval for μ. 10-8. Test the null hypothesis using the confidence interval, and explain the meaning of the decision. 10-9. What would happen to the confidence interval if we used a 95% level of confidence?

Suggested Answers

10-7. The margin of error is a critical value times the estimated standard error of the mean. In this case, the margin of error $= 2.617 \times 0.9072047 = 2.3741547$. The lower limit of the confidence interval is $98.4 - 2.3741547 = 96.025845 \approx 96.03$, and the upper limit is $98.4 + 2.3741547 = 100.77415 \approx 100.77$. So the 99% confidence interval is [96.03, 100.77]. 10-8. The interval brackets 98.6, so the sample mean is not statistically significantly different from the usual norm. The value 98.6 is one of the plausible values for the population mean. 10-9. If we had used a 95% level of confidence and all other details were unchanged, the interval estimate would be more narrow. (For practice, you could look up a critical value for a two-tailed test using $\alpha = .05$ and compute the interval for yourself.)

The researchers in the study of medical students had a directional alternative hypothesis, predicting higher mean PSQI (worse sleep quality) for the medical students, compared with the norm for healthy adults. We need a confidence interval that would correspond to the one-tailed test. Now all of α would be in one tail. As you will recall from Chapter 8, researchers typically report a 90% confidence interval when they have a directional alternative hypothesis and α =.05. The focus will be on only one side of the confidence interval, however, corresponding to the directional prediction.

We confirm that the sample mean (6.37) is, in fact, greater than the norm (5.6), so the results are in the predicted direction. To compute the margin of error, we would need to find a critical value when $\alpha = .05$ in one tail. Look in Table B.1. Find the column for a total $\alpha = .05$ for a one-tailed test. Go down to the row for $df = 120$, the closest smaller value than our actual $df = 292$. We find a critical value of 1.658, and we compute

$$\text{Margin of error} = \text{estimated standard error of the mean} \times \text{critical value}$$
$$= 0.150141 \times 1.658$$
$$= 0.2489338$$

The limits on the 90% confidence interval are

$$\text{Lower limit} = M - 0.2489338$$
$$= 6.37 - 0.2489338$$
$$= 6.1210662 \approx 6.12$$

$$\text{Upper limit} = M + 0.2489338$$
$$= 6.37 + 0.2489338$$
$$= 6.6189338 \approx 6.62$$

We established that the sample mean is greater than the norm of 5.6, and now we are checking whether the difference is statistically noteworthy. We use the same decision rule that we saw in Chapter 8: if the interval does not contain the population mean, then the sample mean and the population mean differ significantly; if the interval does contain μ, then there is no significant difference. Here, we are most interested in the lower limit of the confidence interval. The prediction was that the interval estimate would be higher on the number line than the hypothesized value of μ. The lower limit of the confidence interval is 6.12, which is greater than 5.6, so the interval does not bracket the population mean. We can reject the null hypothesis and conclude that the medical students had significantly worse sleep quality on average, compared with the norm for healthy young adults.

10. One-Sample Tests and Estimates

Graphing Confidence Intervals and Sample Means

Researchers sometimes display confidence intervals in graphs. Let's return to the tai chi study by Wang et al. (2010), which was discussed earlier in the book, to see an example of graphing confidence intervals. The researchers investigated whether tai chi, a meditative practice involving gentle movements, would be helpful for people with fibromyalgia. The Fibromyalgia Impact Questionnaire (FIQ) measures how much this complex condition has affected patients' well-being in the last week. Higher FIQ scores indicate greater difficulty with every-day activities. The researchers randomly assigned patients to one of two groups. Participants in the treatment group attended two 1-hour tai chi classes per week for 12 weeks. Those in the control group attended twice-weekly 40-minute discussion sessions related to fibromyalgia, with each session followed by 20 minutes of stretching. The researchers measured the participants at the beginning of the study, after 12 weeks, and after 24 weeks (i.e., 12 weeks after the supervised activities ended).

Can we compare the mean FIQ scores for the two groups in Week 24? We have not covered the inferential statistic that would allow us to say whether the two means differed significantly, but let's take a look at a graph of the two means. We will clump together the scores of the participants in the control group and graph their mean, and we will clump together the scores of those in the tai chi treatment group and graph their mean. Figure 10.4 shows these two means, computed on the data that Wang et al. (2010) graciously provided; you may download the data via http://desheastats.com.

In Week 24, the FIQ mean for the control group was 57.8 ($SD = 17.9$) and the FIQ mean for the treatment group was 34.3 ($SD = 20.5$). The heights of the bars in Figure 10.4 represent these two means. But Figure 10.4 is different from the graphs shown in Chapter 3. The short horizontal lines connected by the vertical lines represent the 95% confidence intervals for each group. These lines are called *error bars* because they represent the margin of error for each group's estimation of the population mean. Each group's margin of error is calculated separately, using that group's *SD*. The bottom horizontal lines on the error bars represent the lower limits of the confidence intervals and the top horizontal lines on the error bars represent the upper limits of the confidence intervals. A bar graph with error bars provides a quick view of how much variability could be expected in each mean. An end-of-chapter exercise provides you with an opportunity to calculate the confidence intervals for these two groups.

A common mistake in interpreting a graph like Figure 10.4 involves the comparison of the two confidence intervals. Sometimes researchers want to say whether one mean differs from the other, based on the relative location of the two error bars. Making that comparison of confidence intervals would be a mistake. Each of those error bars is based on a critical value from the one-sample *t* test. But the one-sample *t* test by definition is not designed to compare two means.

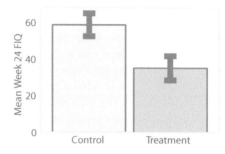

Figure 10.4

Displaying confidence intervals for two group means. The heights of the bars in this graph represent the two means for the Fibromyalgia Impact Questionnaire scores for the treatment group and the control group in Week 24 of the study by Wang et al. (2010). The "I"-shaped bars represent confidence intervals. Sometimes this kind of graph is misinterpreted. (Data from "A randomized trial of tai chi for fibromyalgia," by C. Wang et al., 2010, *The New England Journal of Medicine, 363*, 743–754.)

In Chapter 11, we introduce *t* statistics that allow us to compare two means, and we learn about confidence intervals that are used when we want to test a null hypothesis about the *difference* in two means.

What's Next

So far in this book, we have covered two test statistics that have one kind of hypothesis. Both the *z* test statistic and the one-sample *t* test are used to test null hypotheses that involve a known population mean. Both statistics have been computed on data from a single sample. But the *z* test statistic and the one-sample *t* test are not the only one-sample test statistics. We could compare a single proportion to a hypothesized proportion. For example, suppose we work at a hospital where it seems as if a lot more baby boys than girls are born. We could compare the proportion of boys born in our hospital to some hypothesized population value, such as .5 (if we thought half of the babies should be boys). Notice that a proportion is different from a mean that is computed on continuous quantitative data, such as the sleep quality scores in the study of medical students. Gender of babies is a categorical variable, and Chapter 14 describes a confidence interval for estimating one population proportion.

There are other one-sample test statistics in addition to the ones we have covered. A widely used one-sample inferential statistic tests whether two variables have a significant linear relationship. We talked about Pearson's correlation coefficient in Chapter 5 as a descriptive statistic measuring bivariate correlation. That same statistic is used as an inferential statistic too. We cover hypothesis testing with that statistic in Chapter 13. For now, we continue our focus on sample means. In Chapter 11, we expand your repertoire of statistical tricks to include some two-sample test statistics. These procedures will give you the ability to compare the means of a treatment group and a control group, making it the most realistic test statistic covered so far.

Exercises

SCENARIO 10-B

(Inspired by Brick et al., 2010. Details of this scenario may differ from the actual research.) The researchers who studied medical students' sleep quality by using the PSQI, a well-researched scale for measuring how much trouble people report with sleeping. Higher numbers on the PSQI mean more sleep trouble and worse sleep quality. The team's research question was, "Do medical students have higher mean PSQI scores than the known mean for healthy adults?" All medical students at a university in the northwestern United States received an email invitation to participate in the research. 10-10. What kind of research is this? 10-11. What kind of variable is PSQI? 10-12. What kind of variable is the number of years of medical school completed? 10-13. What can we say about the internal validity of this study? 10-14. What can we say about the external validity of this study?

SCENARIO 10-C

(Based on Wang et al., 2010. Details of this scenario may differ from the actual research.) Part of the study of tai chi for patients with fibromyalgia considered the effect on sleep quality, measured by the PSQI. As we said earlier, higher scores on the PSQI indicate more trouble sleeping and worse sleep quality. People with fibromyalgia tend to have musculoskeletal pain that interferes with daily life, including sleep. Suppose we are analyzing the data from Wang et al. (2010), available for download via http://desheastats.com. We want to know whether the patients in the tai chi study started out with significantly worse sleep quality than adults in general who are about the same age. We find an article by Buysse et al. (2008), which studied an adult community sample. These adults were diverse in race/ethnicity, with an average age of 59.5 years. The community sample recruited by Buysse et al. seems comparable to the sample in the study by Wang et al., aside from the diagnosis of fibromyalgia, so we decide to use the mean PSQI from Buysse et al. as a norm. Buysse et al. reported a mean PSQI = 6.3. 10-15. Write the alternative hypothesis in symbols and words. 10-16. Write the null hypothesis in symbols and words. 10-17. If you have access to statistical software, graph the data for the baseline PSQI scores for all 66 participants in the Wang et al. study. Among other graphs, create a histogram and a boxplot. 10-18. What advantage does each of these graphs provide? 10-19. For students with access to statistical software, compute some descriptive statistics and see if they match the following results: M = 13.7, median = 14, SD = 3.39, skewness = −0.15. 10-20. Explain the meaning of the descriptive statistics in Question 10-19. 10-21. Compute the df for the one-sample t test. 10-22. For students without access to statistical software, look up the critical value for this test, using Table B.1 and α = .05. 10-23.

(*Continued*)

Compute the one-sample *t* test. 10-24. For students without access to statistical software, test the null hypothesis using the critical value decision rule. For students with access to statistical software, test the null hypothesis using the *p* value decision rule. 10-25. Explain the meaning of your decision in Question 10-24, using the variable names.

SCENARIO 10-C, Continued

In this chapter's section "Graphing Confidence Intervals and Sample Means," we talked about graphing a confidence interval. The bar graph in Figure 10.4 included error bars to represent confidence intervals. The graph showed the means for the treatment and control groups, with the dependent variable being the FIQ. Higher scores on the FIQ indicate greater impairment of the patient's daily functioning because of the fibromyalgia. Earlier we gave the following means and *SD*s for the FIQ scores, measured in Week 24: for the control group, $M = 57.8$ ($SD = 17.9$), and for the treatment group, $M = 34.3$ ($SD = 20.5$). Each group had 33 participants. Now we are going to compute an interval estimate of μ for each group. 10-26. Compute *df* for each group's one-sample *t* test. 10-27. Look up a critical value to use in computing a 95% confidence interval. 10-28. Compute the margin of error for estimating the population mean for treatment. 10-29. Compute the margin of error for estimating the population mean for control. 10-30. Why is there a different margin of error for each group? 10-31. Compute the 95% confidence interval for the treatment group. 10-32. Compute the 95% confidence interval for the control group. 10-33. Explain the meaning of the confidence interval for the treatment group. 10-34. Explain the meaning of the confidence interval for the control group. 10-35. How do your results compare with Figure 10.4?

References

Brick, C. A., Seely, D. L., & Palermo, T. M. (2010). Association between sleep hygiene and sleep quality in medical students. *Behavioral Sleep Medicine, 8,* 113–121. doi:10.1080/15402001003622925

Buysse, D. J., Hall, M. L., Strollo, P. J., Kamarck, T. W., Owens, J., Lee, L., & Matthews, K. A. (2008). Relationships between the Pittsburgh Sleep Quality Index (PSQI), Epworth Sleepiness Scale (ESS), and clinical/polysomnographic measures in a community sample. *Journal of Clinical Sleep Medicine, 4,* 563–571.

Carney, C. E., Edinger, J. D., Meyer, B., Lindman, L., & Istre, T. (2006). Daily activities and sleep quality in college students. *Chronobiology International, 23,* 623–637. doi:10.1080/07420520600650695

Wang, C., Schmid, C. H., Rones, R., Kalish, R., Yinh, J., Goldenberg, D. L.,... McAlindon, T. (2010). A randomized trial of tai chi for fibromyalgia. *The New England Journal of Medicine, 363,* 743–754. doi:10.1056/NEJMoa0912611

11

Two-Sample Tests and Estimates

Introduction

So far, our focus in hypothesis testing and interval estimation has been on one population mean. We told the story of Dr. Sarah Bellum and Ray D. Ology, who suspected that something was wrong with a shipment of laboratory rats. They compared a sample mean for maze completion time with a known value of μ from years of research with healthy rats. In addition to computing M as the point estimate of the population mean, the researchers computed an interval estimate of μ. The z test statistic and the confidence interval relying on a z critical value required knowledge of the numeric values of the population mean, μ, and the population standard deviation, σ. In Chapter 10, we added one twist: what if we do not know σ? The solution was to switch to the one-sample t test. Instead of using σ in the denominator, the one-sample t test uses SD, a sample estimate of σ. We used the one-sample t test to compare the mean sleep quality of medical students with the norm for healthy adults. We also found a confidence interval for the mean sleep quality, this time relying on a t critical value in the computation.

The problem with all of these statistics is their limited usefulness. We used the rat shipment example because we could not find any research articles in the health sciences that reported the results of a z test statistic. We searched for hours to find journal articles that reported the use of a one-sample t test. Both the z test statistic and the one-sample t test require knowledge of a population mean or norm. Here is the good news: the statistics in this chapter are more realistic because they do not require knowledge about numeric values of any parameters. This chapter will describe statistics for comparing two means. We will begin with a statistic that has a strong connection with the formula for a one-sample t test. This new statistic can be used with one sample measured on the same variable at two different occasions in time; the statistic also can be used with two samples that have a specific kind of connection between them. Then we will talk about statistics that can be used to compare the means of two independent groups, like a treatment group and a control group.

Pairs of Scores and the Paired t Test

We will begin our introduction to the next test statistic with a long, informal description of one situation in which it can be used. This description will make it easier for us to explain many details of the statistic.

Dentists, dermatologists, and many other medical professionals give injections to patients to numb an area before a procedure. The shot itself can be painful, making some patients avoid treatment. Researchers have looked for ways that they can reduce the pain of needle sticks. Studies have shown that vibration applied near the injection site can distract the pain signals to the brain, so patients experience less pain. Pain researchers have developed vibrating devices for use during injections or blood draws (e.g., see http://buzzy4shots.com). Pain researchers know that people tend to vary quite a bit in their experience of injections. Some people tend to be unbothered by shots, while others are quite sensitive to needle pain. In statistical terms we would say the patient's sensitivity to needle pain is an important extraneous variable that needs to be controlled. We could randomly assign participants to groups, then manipulate an independent variable: the vibration (present or absent). But there is another way that we could control this extraneous variable. We could record the same participant's pain ratings twice: once after a shot with vibration and a second time after a shot without vibration. If the same person is being measured twice, then in both vibration conditions (present and absent) we would have someone with the same degree of needle sensitivity.

Let's look at an example involving your authors. Your first author has try-panophobia, a fear of needles. She has gotten better over the years, although she still gets nervous about having blood drawn. Suppose she has to get two shots. The first shot is given without a vibratory distraction, and on a scale from 0 (*no pain*) to 10 (*worst pain imaginable*), she rates her pain as 8. The second shot is given while a vibrating device is placed near the injection to distract the pain signals, and she rates her pain as 6. Now suppose your second author does not have needle phobia and he too has to get two shots. The first shot is given without the vibratory distraction, and he rates his pain as 3. The second shot is given with a vibratory distraction, and he rates his pain as 1.

We can see that the first author gave higher ratings of pain (8 and 6) than the second author did (3 and 1). To compare the pain ratings with and without the vibration, we can compute the difference between the two pain ratings. Each person's two scores exist as a pair, and when we subtract one score from the other within the pair, we have created a *difference score*. Looking at the distance between each author's pair of scores means we no longer are focusing on how high or low either individual's scores may be. In this way we control the extraneous variable of needle sensitivity. For each author, let's compute the following difference score, which we will symbolize with the letter d:

$$d = \text{pain rating with vibration} - \text{pain rating without vibration}$$

For the first author, the difference score is $d = 6 - 8 = -2$. That is, the pain rating for the shot with vibration is two points lower than the pain rating for the shot without vibration. For the second author, the difference score is $d = 1 - 3 = -2$. So his pain rating for the shot with vibration also is two points lower than his pain rating for the shot without vibration. (We could reverse the order of the subtraction; we will come back that idea soon.)

Both authors reported lower ratings for shots with vibration versus shots without vibration. Focusing on the difference scores keeps us from being distracted by the fact that the two authors differed from each other in their experiences of needle pain. What matters to us is the effect of the vibration, not whether the first author is a wimp and the second author is stoic, at least when it comes to shots. By having pairs of scores on the same outcome variable, it is as if we are cloning each participant and controlling many of the extraneous variables associated with each person. The first author is almost exactly the same person during both occasions of measurement, remaining needle-phobic and having the same demographic characteristics during both shots. The second author also is almost exactly the same person on both occasions. That is why we sometimes say the participants act as their own controls in this kind of study; each person is like an individual version of a control group with the same characteristics as the person in the treatment group.

There is another advantage of looking at difference scores. Computing a difference score for each participant reduces our focus from two pain ratings per participant to one difference score per participant. The new statistic will perform the same calculations as a one-sample t test, but now the data are the difference scores. Let's restate that to make sure you understand. If we computed a set of difference scores for a sample, then we set aside the original pain ratings and looked only at those difference scores, we would have one set of d's—and we could use the one-sample t test formula to analyze those d's. We will not call the test statistic a one-sample t test, though. We will call it the *paired t test*. Statistical software generally does not produce output showing the computation of the difference scores when it performs the paired t test. But we will explain how the paired t test relies almost entirely on the formula you learned for the one-sample t test. (The connection between these statistics is so strong that we even have seen journal articles specifying that the authors used a one-sample t test to analyze difference scores; e.g., see Sowell et al., 2004.)

Let's go back to our example of the two shots, one with a vibratory distraction near the injection and the other without vibration. The experience during the first shot could affect the person's experience during the second shot. What if the second shot was reported as less painful, simply because the person knew what to expect after the first shot—regardless of any effect of the vibration? This is an example of an *order effect*, which is a special kind of extraneous variable associated with the chronological order in which conditions are presented to participants and the influence of that order on the outcome variable. In our example, suppose we always present the two conditions in this order: "shot without vibration" followed by "shot with vibration." In that case, we could never know whether the results are attributable to the vibration (present/absent) or the fact that the conditions were presented in this order. We can control for the order effect by randomly assigning half of the participants to getting the vibration first, whereas the remaining participants get the vibration second.

It is important to compute the difference score with the same order of subtraction for all participants, regardless of which condition came first in time, so that the difference scores are comparable across the members of the sample. Suppose you are a third person in our example with the authors, and you receive the shots in the opposite order from the authors' shots. You receive the first shot with vibration and rate your pain as a 5. You receive the second shot without vibration and rate your pain as a 6. But when we defined the difference score, we did not say "d = second rating minus first rating." We said,

$$d = \text{pain rating with vibration} - \text{pain rating without vibration}$$

We still need to compute the difference in your two pain ratings in the same order that we did the subtraction for the authors' ratings ("with" minus "without") so that your difference score is comparable to ours. For your scores, the difference would be $5 - 6 = -1$. So your pain ratings showed the same pattern as the authors' ratings: the pain rating was lower for the shot with vibration than without vibration.

By doing the subtraction in the same order for all participants, we would be able to tell when some people had a different pattern of results or no difference in the two pain ratings. Suppose we ask two more people to participate. Sam and Ella agree to get two shots. Sam gets the shot with vibration first and reports a pain rating of 3. He gets the second shot without vibration, and he again rates his pain as 3. So the difference in his two ratings is $3 - 3 = 0$. This difference is easy to interpret: the zero means there is no difference in his two ratings. Ella's first shot is without vibration, and she rates her pain as 4. Her second shot is with vibration, and she rates her pain as 7. Remember, we defined the difference score as

$$d = \text{pain rating with vibration} - \text{pain rating without vibration}$$

Ella's difference score is $d = 7 - 4 = 3$. That is, the rating with vibration was higher than the rating without vibration, so the use of vibration was linked with an increase in her pain rating. Sometimes results are counterintuitive, and researchers need to remain open to unexpected findings like this one.

Must we compute the difference scores as "with" minus "without"? No, we could compute them as "without" minus "with," as long as the same direction of subtraction is used for all participants. Participants are randomized to the order of the conditions, not the order of subtraction in the difference scores. We will be able to use the paired t test to answer the question, "On average, what is the effect of the vibration on the pain ratings?"

Check Your Understanding

SCENARIO 11-A

Fayers, Morris, and Dolman (2010) conducted a study of pairs of pain ratings from patients who underwent injections of anesthesia (pain-blocking medicine) before surgery on their upper eyelids. Vibration was applied to the forehead between the eyes before an injection in one eyelid but not the other. After each injection, the patients rated their pain from 0 (*no pain*) to 10 (*worst pain imaginable*). 11-1. Name one advantage that these researchers gained in collecting data from the same participants under two conditions.

Suggested Answer

11-1. One advantage is that many extraneous variables associated with the participant were controlled. Each person was like a perfect duplicate, measured under two conditions. Potential extraneous variables that would be controlled include age, gender, tendency to be sensitive to needle pain, health conditions that may influence the experience of pain, and so forth.

Two Other Ways of Getting Pairs of Scores

The new statistic that we have introduced goes by different names in the research literature. We are calling it the paired t test because this is the term we have seen most frequently in the health sciences literature. The name also reminds us that we have pairs of scores as the focus of the analysis. The statistic also is called the dependent-samples t test, the matched-pairs t test, the t test for related samples,

Student's *t* test for paired samples, and other names. Perhaps you have noticed that this chapter is called "Two-Sample Tests and Estimates," but the example of vibration and pain ratings had *one* sample of participants. You can think of this example as having two samples of *scores*: pain ratings when vibration is used and pain ratings when vibration is not used. But this statistic also can be used with two samples of participants that have a pairwise connection.

Let's look at how two samples can give us pairs of scores for analysis with a paired *t* test. Studies have examined the loss of bone mass in people who have been immobilized by spinal cord injuries (e.g., Bauman, Spungen, Wang, Pierson, & Schwartz, 1999). How much of the loss is the result of the spinal cord injury and not the loss that may be experienced with normal aging? Let's imagine trying to control most of the extraneous variables that could influence bone mineral density and coming up with an appropriate comparison group for the people with spinal cord injuries. For each person with a spinal cord injury, we would need to identify someone who had the same demographics (such as age, sex, and race/ethnicity) and similar general health history (such as smoking history and family history of osteoporosis). Suppose we found someone without a spinal cord injury who could act as a match for the person with the spinal cord injury and who would be willing to participate in our study. These two people would be treated as a researcher-matched pair, and we could compute a difference score using their scores on a measure of bone mineral density. Then for every additional person with a spinal cord injury who had agreed to participate in our research, we would have to find a match. In the end, we would have two samples: (1) people with spinal cord injuries and (2) people without spinal cord injuries, with each person in one group matched with someone in the other group.

Creating pairs in this artificial way controls the extraneous variables that the researcher used to match the pairs (the demographic variables, the health history variables). Researcher-created matches sometimes are used in case-control studies. One common mistake is that researchers often forget to take the matching into account in the analysis of the data. Bloom, Schisterman, and Hediger (2007) looked into this common error in studies focusing on polycystic ovary syndrome. They found that 10 out of 11 studies with researcher-created pairs failed to take the matching into account in the data analysis. To help you understand why that is a problem, let's imagine running the study of needle pain, except now we throw all of the "no vibration" pain ratings together and we throw all of the "vibration" pain ratings together, without keeping the ratings in pairs. Now the first author's "no vibration" pain rating of 8 is in the same pile of scores with the second author's "no vibration" pain rating of 3, and we have lost the connections to the pain ratings of 6 and 1 in the "vibration" pile. Now there is more noise interfering with our ability to detect the signal (i.e., the effect of vibration), plus we will have violated an independence assumption by treating pairs of numbers as independent scores. As you read journal articles that mention matching, try to figure out whether the researchers remembered to take the pairs into account in the data analysis.

Another way of obtaining pairs of scores is to use naturally occurring pairs. Instead of having to create the pairs ourselves, we could study people who already are paired. Let's go back to our example of people with spinal cord injuries and bone loss. Bauman et al. (1999) conducted a small study on this topic, but their participants were monozygotic (identical) twins, one with a spinal cord injury and one without. These researchers were able to control many extraneous variables because the twins have the same DNA, the same early family history and upbringing, the same demographics, and so forth. Among other analyses in the study, a paired t test compared the means for bone mineral density for two groups: the people with spinal cord injuries and their unaffected twins. Because of the link between the two people within each pair, difference scores had to be analyzed, giving the researchers the advantage of controlling many extraneous variables.

In sum, there are three ways that we can get pairs of scores for the paired t test: one group measured twice, pairs that are artificially created by the researcher to control some chosen extraneous variables, or naturally occurring pairs like twins. Next we will go back to our example of the needle sticks with and without vibration to illustrate what is meant by paired means and how they are connected with difference scores.

Fun Fact Associated with Paired Means

In our example of the needle sticks with and without vibration, we can compute the mean pain rating for each condition. But we must remember that these are *paired means*, or averages computed on scores that are linked in a pairwise manner. The analysis will have to keep those two ratings paired together. Before we explain the hypotheses for the paired t test, we need to talk some more about the data. As we said before, the computations in the paired t test are performed on one set of difference scores. The direction of subtraction can be performed either way, as long as (1) the same direction of subtraction is used for all pairs of scores, and (2) the differences are computed within each pair of scores (*not* some random difference between any score in one group and any score in the other group). Let's look back at the five participants in the previous example so that we can illustrate a fun fact about the math of difference scores and means. You may feel as if we are wandering a bit, but follow us for a while—we have a point to make. Table 11.1 lists the participants, their scores, and two ways of computing the difference scores.

If we compute the difference scores putting the pain ratings without vibration first, we have three positive numbers, a zero, and a negative number. If we swap the order of the subtraction, we get three negative numbers, a zero, and a positive number. Now we are going to illustrate the fun fact:

- Compute the mean of the "Without Vibration" column, then compute the mean of the "With Vibration" column. (No, really—get out a calculator and do this math. It will mean more if you do it.)
- Next, compute the difference in these two means by taking the "Without Vibration" mean minus the "With Vibration" mean.

You should get 4.8 for the mean of the pain ratings in the "Without Vibration" column and 4.4 for the mean of the "With Vibration" column. For the difference in these two means ("without" minus "with"), you should get $4.8 - 4.4 = 0.4$.

Table 11.1 Pain Ratings and Difference Scores

| Name | Pain Ratings | | Difference Scores | |
	Without Vibration	With Vibration	Without Minus With	With Minus Without
Lise	8	6	2	−2
Larry	3	1	2	−2
You	6	5	1	−1
Sam	3	3	0	0
Ella	4	7	−3	3

Now compute the mean of the third column of numbers, which are the difference scores for "Without Minus With." You should get 0.4 again. So this is the fun fact: the difference in the two means (also called the *mean difference*) equals what you will get by averaging the difference scores.

What good is this fun fact? We said this chapter would cover statistics for comparing *two* means. This fun fact gives us justification for using the paired *t* test, which analyzes *one* set of difference scores and has a formula that looks a lot like the formula for the one-sample *t* test. Even if we have only one sample of participants, as in our small example in Table 11.1, we can compare two means: (1) the average pain rating for the shot with vibration and (2) the average pain rating for the shot without vibration. But we can compute one set of difference scores. If we had two samples, like the researchers who compared the bone mineral density of people with spinal cord injuries and their unaffected twins, we have two means but we also could compute one set of difference scores. The fact that the difference in means is the same as the mean of the difference scores also will result in more ways to write hypotheses. Before we present the formula for the paired *t* test, we need to describe the hypotheses associated with this statistic.

Check Your Understanding

SCENARIO 11-A, Continued

Fayers et al. (2010) conducted the study of vibration for patients receiving injections of anesthesia (pain-blocking medicine) before upper-eyelid surgery. They gave one shot without vibration and the other shot with vibration applied to the forehead between the eyes. After each injection, the patients rated their pain from 0 (*no pain*) to 10 (*worst pain imaginable*). The researchers reported a mean pain rating of 3.3 for the injections when vibration was applied and a mean pain rating of 4.5 for the injection when the vibratory device was turned off. 11-2. Compute the mean of the difference scores.

Suggested Answer

11-2. *The mean of the difference scores is the same thing as the difference in the two means. Subtracting one mean from the other, the answer could be either 4.5 − 3.3 = 1.2, or it could be 3.3 − 4.5 = −1.2.*

Paired *t* Hypotheses When Direction Is Not Predicted

Hypothesis testing with the test statistics in this chapter will be similar to other hypothesis tests we have performed:

- We can have a directional alternative hypothesis, or we can write the alternative hypothesis to be nondirectional.
- The null hypothesis will be a statement opposite to the alternative hypothesis.
- After computing the test statistic, we will check whether the results are in a predicted direction (if H_1 is directional), and we will test the null hypothesis one of two ways: (1) by comparing an observed test statistic to a critical value or (2) by comparing a p value with the significance level, α.
- The critical value decision rules and the p value decision rules will be the same as the ones we used with the one-sample t test.

The paired t test differs markedly from the one-sample t test in the writing of the hypotheses. There are many ways that the hypotheses may be written. Let's start with the simplest way to write the alternative hypothesis, using a nondirectional example. Suppose we think vibration will make a difference in the mean pain ratings, but we are not sure whether the vibration will relieve pain or increase pain. Another way to state this prediction is as follows:

Our sample comes from a population in which the mean of the pain ratings for a shot with vibration will differ from

the same people's mean of the pain ratings for a shot without vibration.

We could write this nondirectional alternative hypothesis as follows:

$$H_1: \mu_{\text{vibration}} \neq \mu_{\text{no vibration}}$$

These subscripts are differentiating the two conditions: a shot with vibration and a shot without vibration. They are *not* the averages of vibrations, but the averages for pain ratings. So the subscripts can be confusing to some students. Just recognize that averages are computed for two conditions, and both of the means are calculated on pain ratings. The subscripts could be changed to the words "with" and "without." Or we could write ourselves a note saying, "Condition A will involve vibration, and Condition B will not have vibration," then we could use the subscripts A and B. We also could swap the two μ's, putting $\mu_{\text{no vibration}}$ first in the statement. In any case, the statement is saying the mean pain ratings for the two conditions are not equal; that is, there will be a difference in the two means. The null hypothesis will be an opposite statement:

$$H_0: \mu_{\text{vibration}} = \mu_{\text{no vibration}}$$

This null hypothesis is saying the following:

> Our sample comes from a population in which the mean pain
> rating after an injection with vibration equaled
> the same people's mean pain rating after an injection without vibration.

There are two other ways that we could write the nondirectional alternative hypothesis. Understanding these other ways will help you to understand the formula for the paired t test. If two means are not equal, then that is the same thing as saying, "There is a difference." A mathematical way of saying the same thing is, "The difference is not equal to zero," a statement that may sound awkward to you. Here is how we can write this H_1:

$$H_1: \mu_{\text{vibration}} - \mu_{\text{no vibration}} \neq 0$$

This expression is saying that if we take the mean of the pain ratings collected after the shot with vibration and we subtract the same people's mean of the pain ratings after the shot without vibration, we will get a difference that is not zero. In other words, there will be some difference in the two means. The subtraction could be performed the opposite way: $\mu_{\text{no vibration}} - \mu_{\text{vibration}}$. As long as the order of subtraction is consistent across participants and we understand the meaning of the direction of subtraction, it does not matter which way the subtraction is performed. We are saying that the two means will not be equal, so there will be some difference between them—either a positive difference or a negative difference. This second way of stating H_1 can be translated as follows:

> Our sample comes from a population where the difference in the mean pain ratings
> from the same people studied in the two conditions
> (with and without vibration) is not zero.

Corresponding to this second way of writing H_1 is the following null hypothesis:

$$H_0: \mu_{\text{vibration}} - \mu_{\text{no vibration}} = 0$$

This null hypothesis can be translated as follows:

> Our sample comes from a population in which
> the mean pain rating for a shot with vibration
> minus the mean pain rating for same people
> receiving a shot without vibration equals zero.

That is, the difference in means is zero, which is the same thing as saying there is no difference in the means.

11. Two-Sample Tests and Estimates

The first two ways of writing the nondirectional alternative hypothesis linked back to the idea of computing two means. The third way to write this alternative hypothesis goes back to the difference scores. Remember our fun fact: the mean of the difference scores is equal to the difference in the two means. Thus we can write the nondirectional alternative hypothesis to say that in the population, the mean of the difference scores is not equal to zero, or

$$H_1: \mu_d \neq 0$$

The symbol μ_d is the population mean of the difference scores, with the subscript d representing the difference scores. In words, we might say the following:

Our sample comes from a population
where the mean of the difference scores for the pain ratings
collected from the same people under the two conditions is not zero.

This H_1 goes along with the following null hypothesis:

$$H_0: \mu_d = 0$$

This null hypothesis can be translated as follows:

Our sample comes from a population in which
the mean of the difference scores equals zero.

Students often do not like this third way of writing the null hypothesis, but this is the one that will help us to link the formula for the paired t test to the formula for the one-sample t test. All three ways of writing the nondirectional alternative hypothesis are equivalent, and it does not matter which one is used. Your instructor can advise you on how she or he prefers to write these hypotheses.

Check Your Understanding

SCENARIO 11-B

Bauman et al. (1999) compared the average bone mineral density of patients immobilized by spinal cord injuries with the average bone mineral density of their identical twins. Suppose the researchers wanted to leave open the possibility of finding a difference in means in either direction. That is, if the patients had higher or lower average bone mineral density than their unaffected twins, the researchers wanted to detect the difference. 11-3. Write the alternative hypothesis.

(Continued)

11-3. The alternative hypothesis would say that the samples of patients and their twins come from populations in which the means for the two groups' bone mineral density are different or not equal to each other. The alternative hypothesis could be written in the following ways:

$$H_1: \mu_{patients} \neq \mu_{unaffected\ twins}$$

$H_1: \mu_{patients} - \mu_{unaffected\ twins} \neq 0$ (The two population means could appear in the opposite order in either of these first two ways of writing H_1.)

$H_1: \mu_d \neq 0$, where d = bone mineral density of the patient minus bone mineral density of the unaffected twin (or vice versa).

Paired *t* Hypotheses When Direction Is Predicted

It may seem as if we already presented too many ways of writing hypotheses, but we are not done. What if a directional outcome is predicted? Let's go through some examples—and again there will be multiple ways of writing the same hypothesis. The research on vibration suggests that on average, the presence of vibration should reduce the pain of needle sticks, compared with the no-vibration condition. The idea can be expressed as follows:

Our sample comes from a population in
which the mean pain rating for the shot with vibration
will be less than the same people's mean pain rating
for the shot without vibration.

This alternative hypothesis can be expressed as follows:

$$H_1: \mu_{vibration} < \mu_{no\ vibration}$$

The corresponding null hypothesis will be an opposite statement:

Our sample comes from a population in which
the mean pain rating for injections with vibration
will be greater than or equal to the same people's mean pain rating
when vibration is absent.

This null hypothesis can be written in the following abbreviated style:

$$H_0: \mu_{vibration} \geq \mu_{no\ vibration}$$

(Your instructor may prefer for you to write this null hypothesis without the "greater than" part; that is, you may be instructed to write this null hypothesis as $\mu_{vibration} = \mu_{no\ vibration}$. There are mathematical reasons to write the null hypothesis that way.)

The same idea can be conveyed by reversing the order of the two means and the directional sign. Instead of putting $\mu_{vibration}$ first, we could put $\mu_{no\ vibration}$ first. But to maintain the meaning of the prediction, the directional sign will have to be turned around too. Study the following two statements and persuade yourself that they are saying the same thing:

$$H_1: \mu_{vibration} < \mu_{no\ vibration}$$

$$H_1: \mu_{no\ vibration} > \mu_{vibration}$$

In both statements, the vibration condition is associated with the smaller mean pain rating. It is crucial that we keep tabs on these directional signs, because we will have to check whether the sample means for the two conditions came out in the predicted direction.

Another way to write the directional alternative hypothesis is similar to the second way that we wrote the nondirectional alternative hypothesis: with one population mean subtracted from the other. We are predicting lower pain ratings on average for the vibration condition, compared with the average pain rating for the no-vibration condition. So if we take the smaller mean and subtract the larger mean, we would get a negative number. We would write the following:

$$H_1: \mu_{vibration} - \mu_{no\ vibration} < 0$$

This H_1 is translated as follows:

Our sample comes from a population in which
the difference in the means for the pain ratings
(with the computation performed as "with" minus "without")
will be a negative number.

That is an awkward way of saying the same prediction that we stated above: the vibration condition will have the smaller of the two means. The null hypothesis corresponding to this H_1 would be

$$H_0: \mu_{vibration} - \mu_{no\ vibration} \geq 0$$

This null hypothesis is saying

> Our sample comes from a population in which
> the mean pain rating with vibration present
> minus the mean pain rating when vibration was absent
> is a difference greater than or equal to zero.

We could rewrite the same prediction in an alternative hypothesis with the expected-to-be-bigger mean listed first. If a bigger number has a smaller number subtracted from it, the result will be a positive number:

$$H_1: \mu_{\text{no vibration}} - \mu_{\text{vibration}} > 0$$

Notice that when we reversed the order of subtraction, the directional sign had to be reversed too. This alternative hypothesis would correspond to the following null hypothesis:

$$H_0: \mu_{\text{no vibration}} - \mu_{\text{vibration}} \leq 0$$

The third way that we wrote the nondirectional alternative hypothesis used μ_d. We can write our directional alternative hypothesis using μ_d too. The next Check Your Understanding question will cover the two ways that the hypotheses can be written using μ_d. (Hint: There are two ways, because the difference score can be computed in two ways.) Next we will show the strong connection between the formulas for the paired t test and the one-sample t test.

Check Your Understanding

Based on prior research, we believe that the pain ratings will be lower on average for the injection while vibration is applied, compared with the same people's mean pain rating when vibration is absent. For each participant we can compute a difference score, d. 11-4. Why did we say it is crucial to keep track of the order of subtraction when the difference scores are computed? 11-5. Choose one way of computing the difference scores and write the directional alternative hypothesis using the symbol μ_d. 11-6. Write the null hypothesis corresponding to H_1 in your previous answer. 11-7. Change the way of computing the difference scores and rewrite both H_1 and H_0.

(*Continued*)

11. Two-Sample Tests and Estimates

Suggested Answers

11-4. *The interpretation of the results of the paired t test will depend on under-standing what kind of difference was predicted in the alternative hypothesis and what is meant by a positive or negative difference. 11-5. We can compute the d = pain rating for the vibration condition minus the pain rating for the no-vibration condition. We are predicting lower pain ratings for shots with vibration, compared with shots without vibration. This difference score would mean the smaller number had the bigger number subtracted from it, resulting in negative difference scores. So the directional alternative hypothesis would say H_1: $\mu_d < 0$, or our sample comes from a population in which the mean of the difference scores (computed as "with" minus "without") is less than zero. 11-6. The null hypothesis would be H_0: $\mu_d \geq 0$, or our sample comes from a population in which the mean of the difference scores (computed as "with" minus "without") is greater than or equal to zero. 11-7. We change the difference score to be computed as d = pain rating for the no-vibration condi-tion minus the pain rating for the vibration condition (i.e., "without" minus "with"). We still would be predicting lower pain ratings when vibration was present during the shot, so this difference score would mean we had the bigger number minus the smaller number. So we predict the mean of the difference scores would be positive. The directional alternative hypothesis would say H_1: $\mu_d > 0$, and the null hypothesis would say H_0: $\mu_d \leq 0$.*

Formula for the Paired *t* Test

It may have seemed like a lot of trouble to go through all those versions of the null hypothesis, but it is good practice for increasing your awareness about the direc-tion of subtraction. We might have in mind an alternative hypothesis that would predict positive difference scores, yet after using statistical software to analyze the data, we might obtain output that shows results in the opposite direction. We would need to look carefully at whether the software performed the subtraction in the same way that we had in mind. The easiest way to tell is by looking at the means, because our fun fact told us the difference in means equals the mean of the difference scores. It could be that the results in the computer output actually do support our prediction, but that the subtraction was performed in the opposite direction. Or the difference scores could have been computed in the same way that we had in mind, but the participants responded in the opposite direction from our prediction, in which case we would retain the null hypothesis. So we have reasons for torturing you with multiple ways to write the same hypothesis.

The good news about creating difference scores is that we simplify the situa-tion from having two scores (in pairs) to having one difference score (per pair).

We know how to deal with one sample of scores to ask questions about an average. That is what we did with the one-sample t test, when we compared a sample mean for one set of scores to a known value of a population mean, μ. Notice that in our present example, however, we have not mentioned anything about a known numeric value for a population mean, which was a detail needed for the one-sample t test. For the paired t test, our focus is on whether there is a *difference* in the population means for the pain ratings under the two conditions—regardless of what the numeric values of those population means might be. If our null hypothesis is saying that vibration will make no difference in the pain ratings, then on average, the difference scores should be zero; this statement corresponds to the null hypothesis that said $\mu_d = 0$.

Let's review the formula for the one-sample t:

$$\text{One-sample } t = \frac{M - \mu}{SD / \sqrt{N}}$$

Now let's look at the formula for the paired t test, which we are going to simplify in a moment:

$$\text{Paired } t = \frac{M_d - \mu_d}{SD_d / \sqrt{N_d}}$$

Keeping in mind the concept of "(something minus its mean) divided by its estimated standard deviation," let's compare the two formulas:

- Each formula has a sample mean in it ("something"). For the paired t, the sample mean has a subscript of d, because it is computed on the difference scores.
- The one-sample t test used a known value of μ in the numerator, and that value came from the null hypothesis ("its mean"). The paired t test also has a population mean in it, but this μ needs a subscript of d to indicate it is the population mean of the differences. But we already said something about that value: if vibration had no effect, then the average of the difference scores will be zero. So μ_d is hypothesized to be zero in the null hypothesis, meaning we do not have to write anything but M_d in the numerator of the paired t test. Here is the simplified version of the formula for the paired t test:

$$\text{Paired } t = \frac{M_d}{SD_d / \sqrt{N_d}}$$

Let's complete our comparison of the formulas for the one-sample t test and the paired t test:

- The denominators of these two formulas ("its estimated standard deviation") are almost identical, except that the formula for the paired t test has subscripts of d to indicate that the math is being performed on difference scores.

11. Two-Sample Tests and Estimates

In sum, both formulas follow the pattern of "(something minus its mean) divided by its estimated standard deviation." For the paired t test, the "something" is the mean of the difference scores (or, equivalently, the difference in the means for the two conditions). "Its mean" comes from the null hypothesis, where the μ_d was zero. "Its estimated standard deviation" is the estimated standard error of the paired mean difference (i.e., the estimated standard error of a sampling distribution for the difference in the two paired means).

The last sentence is a good reminder that every statistic has a sampling distribution, so the paired t test does too. That sentence also needs some explanation: we could imagine drawing all possible samples of the same size from the same population, repeating the study of the pain ratings for shots with and without vibration, computing the paired t test on each sample's data, and arranging those paired t tests' numeric results in a distribution. What would the sampling distribution look like? If the null hypothesis were true and the vibration made no difference, and if two assumptions are met, then the distribution would look like a theoretical t distribution.

We just mentioned that the paired t test has two assumptions. The assumptions are the following:

- Pairs of scores (or the d's) are normally distributed in the population.
- Pairs of scores (or the d's) are independent of each other.

Clearly there is a connection within each pair of scores: either the same person is being measured twice, or we are measuring two people who have either a natural connection (like twins) or a researcher-created connection. In the case of the twins being compared on bone mineral density, one twin is not independent of the other—but that is not what the assumption of independence says. The assumption is that each pair is independent of every other pair. No set of twins has any connection to any other set of twins in the spinal cord injury study. With these two assumptions, we again can see a similarity with the one-sample t test, which assumed that the scores were normally distributed in the population and that the scores were independent of each other.

You may recall from Chapter 10 that there are many t distributions, each one slightly different in shape, depending on the degrees of freedom (df). We need to know the paired t test's df so that we can know which t distribution will provide critical values and p values in any given research scenario. Here is another detail where we can compare the one-sample t test and the paired t test. The df for the one-sample t test equals $N - 1$. Similarly, the paired t test has the following formula for df:

$$df = N_d - 1$$

We simplified the situation by going from two pain ratings to one difference score, so we have one set of d's, and the df is the number of d's minus one. This formula for df also can be stated as $N_{pairs} - 1$, because we will always have the same number of difference scores as we have pairs of scores. We can follow all the same decision rules for hypothesis testing used for the z test statistic and the one-sample t test.

Let's work through an example of a hypothesis test using the results from the study by Fayers et al. (2010), who asked patients to rate their pain for two injections of anesthesia. Patients received one injection per eyelid: one shot was given while vibration was applied between the eyebrows, and the other shot was given with the vibratory device turned off. (The article describes some interesting steps for making the no-vibration condition quite similar to the vibration condition; your instructor may wish to have you find the article and read it.) These researchers predicted that vibration would be associated with lower mean pain ratings, compared with the no-vibration condition. This alternative hypothesis could be written in any of the following ways (study these statements and make sure you can see that they all are saying the same thing):

$$H_1\text{: } \mu_{\text{vibration}} < \mu_{\text{no vibration}}$$

$$H_1\text{: } \mu_{\text{no vibration}} > \mu_{\text{vibration}}$$

$$H_1\text{: } \mu_{\text{no vibration}} - \mu_{\text{vibration}} > 0$$

$$H_1\text{: } \mu_{\text{vibration}} - \mu_{\text{no vibration}} < 0$$

$H_1\text{: } \mu_d > 0$, if $d =$ pain rating without vibration − pain rating with vibration

$H_1\text{: } \mu_d < 0$, if $d =$ pain rating with vibration − pain rating without vibration

Let's choose one of these ways of writing the alternative hypothesis:

$$H_1\text{: } \mu_{\text{no vibration}} - \mu_{\text{vibration}} > 0$$

We expect the bigger mean for the pain ratings from the "no vibration" condition, compared with the "vibration" condition, so a big number minus a small number would give us a positive difference in means (greater than zero). In addition to writing our hypotheses, we need to specify one other detail in advance of the data analysis: let's use $\alpha = .05$.

The journal article reported that the mean pain rating for the shot with vibration was 3.3 ($SD = 1.9$), and the mean pain rating for the shot without vibration was 4.5 ($SD = 2.0$). The article did not report the numeric value of the paired t test that the researchers computed. Like many health sciences publications, the journal *Ophthalmology* does not always report the numeric result for the test statistic, choosing instead to report only the name of the statistic and the computed p value. For this difference in means (3.3 vs. 4.5), the paper said $p = .0003$. By using the df and the one-tailed p value, we were able to work backward and determine the numeric value that these researchers must have computed for the paired t test. Now we can illustrate the computation of the paired t test with their results.

First, the numerator of the paired t test: our fun fact said the mean of the difference scores was the same thing as the difference in the two sample means. To be consistent with our alternative hypothesis, let's take the mean pain rating of the no-vibration condition and subtract the mean pain rating of the vibration condition:

$$4.5 - 3.3 = 1.2$$

This is the numerator of the paired t test because the difference in the means is the same as the mean of the differences, or M_d. When we gave you these means, we listed two numbers for SD—but those are the standard deviations of the pain ratings for each condition, not SD_d, which is the standard deviation of the difference scores. Based on our backward calculations, we think the researchers' SD_d was approximately 3.002. A sample of 80 participants was measured twice, so the denominator of the researchers' paired t test would be

$$\frac{SD_d}{\sqrt{N}} = \frac{3.002}{\sqrt{80}}$$

$$= \frac{3.002}{8.9442719}$$

$$= 0.3356338$$

The final step is to take the numerator and divide it by the denominator (unrounded):

$$\text{Paired } t \text{ test} = \frac{1.2}{0.3356338}$$

$$= 3.5753252$$

$$\approx 3.58$$

A paired t test ≈ 3.58 means that about 3.58 estimated standard errors of the paired mean difference fit in the gap between the two means. To test our null hypothesis, let's use the p value decision rule for a one-tailed test. First, are the results in the predicted direction? The researchers predicted a greater mean pain rating for the no-vibration condition, compared with the vibration condition. They reported a mean of 4.5 for the pain ratings for the no-vibration condition, and a mean of 3.3 for the vibration condition, so the answer is yes, the results were in the predicted direction. Second, we need to check whether the one-tailed p value is less than or equal to alpha. The researchers reported $p = .0003$, meaning that a paired t test of 3.58 cuts off an area of .0003 in the upper tail of the t distribution that is defined by

$df = N - 1 = 80 - 1 = 79$. Is $p = .0003$ less than or equal to .05, our chosen alpha? Yes, so we reject the null hypothesis, and we conclude that the mean of the pain ratings in the no-vibration condition ($M = 4.5$) was significantly higher than the same people's average pain rating in the vibration condition ($M = 3.3$). The presence of the vibration near the injection site was associated with significantly lower pain ratings on average, compared with the mean of the pain ratings for the shot without vibration.

Researchers tend to depend on p value decision rules, but to be complete in our explanation, we will conduct the same hypothesis test using the critical value decision rule. We have to check whether the results are in the predicted direction, which we have confirmed already. Then we ask if the observed paired t test is equal to or more extreme than a critical value. Our alternative hypothesis said $H_1: \mu_{\text{no vibration}} - \mu_{\text{vibration}} > 0$, meaning that if we take the subtraction in that same order, we are expecting a positive number for the paired t test. So $\alpha = .05$ will go in the upper tail of the distribution. The critical value came from Table B.1, which contains critical values for t distributions. We have $df = 79$, but the table does not list critical values for that df. We cannot give ourselves more df, so we must go to the row for a smaller number of df, which is $df = 70$. Figure 11.1 shows the theoretical t distribution defined by $df = 70$.

In Table B.1, we use the column for a significance level of .05 for a one-tailed test. The intersection of this column and the row for $df = 70$ gives us the critical value of 1.667. As usual, the critical value cuts off a tail area equal to alpha, and the observed test statistic cuts off an area equal to the p value. The results are in the predicted direction because the no-vibration condition had a higher mean

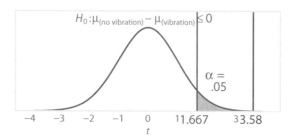

Figure 11.1

A t distribution with $df = 70$ and a directional prediction. If we expected a higher mean for the "no vibration" condition than the "vibration" condition, and if we planned to compute $M_{\text{no vibration}} - M_{\text{vibration}}$, then the paired t test would be in the upper tail. The researchers in the vibration study confirmed the prediction of a higher mean for the "no vibration" condition, and we computed an observed paired t test = 3.58. The example has $df = 79$, but because Table B.1 does not list critical values for $df = 79$, we must use the table's next-smaller $df = 70$. See if you can find the critical value of $t = 1.667$ in Table B.1.

11. Two-Sample Tests and Estimates

than the vibration condition, so next we ask, "Is the observed test statistic more extreme than a critical value?" Because our observed paired t test, 3.58, is more extreme than the critical value, 1.667, we reject the null hypothesis and conclude that the no-vibration condition resulted in a significantly higher mean pain rating than the vibration condition.

Next we will complete our discussion of paired means by looking at an interval estimate of the difference in paired population means.

Check Your Understanding

SCENARIO 11-C

Nurses and other health-care professionals cannot always use the upper arm to take blood pressure readings on patients. Some patients have intravenous catheters, injuries, obstruction in the lymphatic system causing swelling, and other problems that prevent the use of the upper arm for blood pressure readings. Schell, Morse, and Waterhouse (2010) wanted to know whether people would have the same mean blood pressure readings taken on the upper arm and the forearm. The study also involved measuring patients in two positions: (1) when the patient was lying flat (supine) and (2) when the head of the bed (HOB) was raised to a 30° angle. Instead of focusing on blood pressure, let's look at the study's results on heart rates. The researchers knew that changing the angle of the bed could affect the heart rate, and they wanted to know whether waiting 2 minutes between positions would allow the patients' heart rates to stabilize. They measured the heart rate in one position, changed the angle of the HOB, waited 2 minutes, then measured the heart rate again. They reported a paired $t = 0.15$, $p = .8787$. 11-8. What was the most likely alternative hypothesis for the heart rate? 11-9. Explain the meaning of the results.

Suggested Answers

11-8. The researchers probably thought there might be a difference in mean heart rate for the patients lying supine, compared with the same patients lying with the head raised at a 30° angle. This alternative hypothesis could be stated as H_1: $\mu_{flat} \neq \mu_{inclined}$. 11-9. Using a significance level of .05, we would retain the null hypothesis because .8787 is greater than .05. We could conclude that the participants had the same mean heart rate in the supine position as they had when the head was raised to the 30° position. It appears the 2-minute waiting time between positions was sufficient to allow the heart rate to stabilize.

Confidence Interval for the Difference in Paired Means

Our discussion of paired means has focused on the use of an inferential statistic, the paired t test. Now let's compare point estimates with interval estimates and talk about a confidence interval associated with the paired t test. We said the numerator of the paired t test could be written as a sample mean computed on difference scores and that M_d is mathematically the same as the difference in the participants' two means for the two conditions. We demonstrated this fun fact using numbers from our example of pain ratings for two injections on the same person, with one shot occurring with vibration near the injection site. The sample mean difference (i.e., $M_{vibration} - M_{no\ vibration}$) is a point estimate of a difference in two population means that are paired (i.e., $\mu_{vibration} - \mu_{no\ vibration}$). In other words, the entire expression "$M_{vibration} - M_{no\ vibration}$" is a statistic estimating this entire expression: $\mu_{vibration} - \mu_{no\ vibration}$. We can make an equivalent statement using M_d. If the difference score, d, is defined as "the pain rating for the shot with vibration minus the pain rating for the shot without vibration," then M_d is a point estimate for μ_d, the population mean of the difference scores.

We could compute an interval estimate instead of a point estimate for the population mean difference (or its equivalent, the population mean of the difference scores). The other confidence intervals that we have covered took a critical value and multiplied it by a standard error, and the result was a margin of error. Let's think about the margin of error that would give us a range of values for an interval estimate of the difference in two paired population means. This is what is being estimated: $\mu_{vibration} - \mu_{no\ vibration}$. If the null hypothesis is true and there is no effect of vibration, then the paired population means are the same; that is, their difference is zero. One particular study could give us a sample mean for the vibration condition and a sample mean for the no-vibration condition, and we could compute the point estimate: $M_{vibration} - M_{no\ vibration}$. This particular difference in sample means might not be exactly zero. What could we expect to happen for the sample mean difference across repeated samples? We are asking about the sampling distribution for the sample mean difference, M_d.

In the study by Fayers et al. (2010), the ophthalmologists asked patients to rate their pain for injections of anesthesia in each eyelid. One injection was given while vibration was applied between the eyebrows, and the other injection was given without vibration. The point estimate for $\mu_{vibration} - \mu_{no\ vibration}$ was computed as $M_{vibration} - M_{no\ vibration} = 4.5 - 3.3 = 1.2$. Other samples would produce slightly different results. To quantify how much sampling variation exists in the estimation of $\mu_{vibration} - \mu_{no\ vibration}$, we can compute a confidence interval for these paired population means. To simplify this example, we will show only the two-tailed confidence interval. The nondirectional alternative hypothesis says there is some difference in the paired population means; that is, the difference is not zero:

$$H_1: \mu_{vibration} - \mu_{no\ vibration} \neq 0$$

This H_1 corresponds to a two-tailed test, like our nondirectional confidence interval. Using the closest smaller value for the df (i.e., $df = 70$) and alpha for a two-tailed test, Table B.1 shows a critical value of 1.994. We previously computed the estimated standard error for the paired mean difference to be $SD_d/\sqrt{N} = 3.002/\sqrt{80} = 0.3356338$. The margin of error will be the product of the critical value, 1.994, and this estimated standard error:

$$\text{Margin of error} = t \text{ critical value} \times \text{estimated standard error}$$
$$= 1.994 \times 0.3356338$$
$$= 0.6692538$$

Remember, we are estimating a difference in paired population means. This margin of error, 0.6692538, is subtracted from the sample mean difference $(M_{\text{vibration}} - M_{\text{no vibration}} = 1.2)$ to give us the lower limit of the 95% confidence interval, and the margin of error is added to the sample mean difference to give us the upper limit:

$$\text{Lower limit} = 1.2 - 0.6692538$$
$$= 0.5307462 \approx 0.531$$
$$\text{Upper limit} = 1.2 + 0.0.6692538$$
$$= 1.8692538 \approx 1.869$$

Our 95% confidence interval for the difference in paired population means is [0.531, 1.869]. How do we interpret this interval? Remember that the nondirectional alternative hypothesis says there is some difference in the paired μ's, so the null hypothesis is that the difference is zero. Does this interval contain zero? No, it does not. We can conclude that there is a significant difference in the paired means. This particular interval [0.531, 1.869] may or may not contain the true mean difference in the population. But 95% of confidence intervals computed like ours will bracket the true population mean difference.

This concludes our explanation of the statistics involving paired means. The rest of the chapter will be devoted to two more t tests and confidence intervals, which can be used to compare the means of two independent groups, like a treatment group and a control group.

Check Your Understanding

SCENARIO 11-C, Continued

Schell et al. (2010) conducted the study of blood pressure taken on the upper arm and forearm, as well as when patients were lying flat versus lying with the head inclined at a 30° angle. Suppose we have conducted a similar study, except we measure all of our patients while they are lying

(Continued)

flat. We randomize half of the patients to having the upper-arm reading taken first. After taking each patient's blood pressure on the upper arm and on the forearm, we analyze the data for systolic blood pressure (the first or "top" number in a blood pressure reading). We find $M = 122$ for the upper arm and $M = 128$ for the forearm. We compute the following 95% confidence interval for the paired mean difference: [3.9, 8.4]. 11-10. Why was it a good idea to measure the same participants twice, instead of taking forearm measures on some participants and upper-arm readings on different participants? 11-11. Test a null hypothesis that says the population means for readings on the upper arm and forearm are equal. 11-12. Explain the meaning of the interval estimate.

Suggested Answers

11-10. Measuring the same participants twice controlled many extraneous variables associated with the people, most importantly whether the patients tended to have high blood pressure or low blood pressure. We are interested in the difference in the body location where the readings were taken, not whether some participants have higher blood pressure than others. 11-11. The 95% confidence interval of [3.9, 8.4] does not contain zero, so we may conclude there is a significant difference in the mean systolic blood pressure readings for the upper arm versus forearm. By examining the means, we can see that the mean systolic blood pressure was higher when taken on the forearm. 11-12. The interval [3.9, 8.4] may or may not contain the true population mean difference, but through repeated samples like ours taken from the same population, 95 out of 100 confidence intervals like ours would bracket the true mean difference.

Comparing Means of Two Independent Groups

The first author's fear of needles makes her extremely sympathetic to any readers who may share her phobia. But she figures that most needle-phobic people would not be attracted to the health sciences and required to read this book, so our next example also will describe research involving injections.

Childhood immunizations can be upsetting to both children and parents. In addition, some studies have shown that untreated pain in babies may affect the developing nervous system. Efe and Özer (2007) investigated a way of soothing infants receiving shots for routine childhood immunizations. They thought that infants' multisensory experience of being breast-fed would soothe them and provide relief from the pain of injections. Mothers were approached for participation in the study with their babies, who were 2–4 months old and receiving routine immunization for diphtheria, tetanus, and pertussis (DPT, a three-in-one shot).

The study included only full-term babies who were healthy and routinely breast-fed. The researchers randomly assigned 66 babies to either a treatment group or a control group. The intervention involved the mother reclining in a comfortable chair, cradling the mostly unclothed baby so that skin-to-skin contact was maintained, and breast-feeding the baby for 3 minutes before the injection in the baby's thigh. The mothers were instructed to encourage the babies to continue breast-feeding after the injection. The babies in the control condition were not fed; they were wrapped in a blanket with one thigh exposed for the injection, and they were placed on a padded examination table for the shot. Mothers were encouraged to talk soothingly during the shot, and they cuddled their babies afterward.

How would the researchers know whether the babies in the treatment group had a less painful experience than the babies in the control group? They followed the precedent of earlier research and measured the number of seconds that each baby cried in the first 3 minutes after the injection. (They set a limit on the maximum number of minutes because of the potential effect of an outlier—one or two babies crying for a long time could skew the results.) The researchers reported that they performed a two-tailed test, which means they did not predict a directional outcome. They wanted to know whether the mean crying time for the babies in the treatment group differed from the control group's mean crying time.

This study was an experiment. The babies were randomly assigned to two groups, each with 33 babies. Then the researchers manipulated an independent variable: whether the babies were breast-feeding or receiving the standard of care. The dependent variable was the duration of crying in the first 3 minutes after the shot. Unlike the other studies we have described in this chapter, the two groups in this study had no connection to each other, and the babies had no connection to each other, pairwise or any other way. So the researchers did not use the paired *t* test to compare the means for these two groups. They used the *independent-samples t test*, which is used to compare the means of two independent groups with equal and sufficiently large sample sizes. Like the paired *t* test, the independent-samples *t* test has many names that appear in the research literature: Student's *t* test, *t* test for unpaired samples, two-samples *t* test (which is a poor name because the paired *t* test also can involve two samples), the independent *t*, and so forth. We like the term *independent-samples t test* because it reminds us that the samples are independent of each other. As you can tell by now, it is crucial to be specific when talking about *t* tests. We tell our students always to use a *t* test's first name: one-sample, paired, independent-samples, or whatever. Otherwise, it would be as if you worked with three people named Smith (Dan Smith, Marco Smith, and Lavonne Smith) and you tell your supervisor, "Smith helped me with this project yesterday." Your supervisor would have no idea whether Dan, Marco, or Lavonne helped you.

Next we will go through the hypotheses for the independent-samples *t* test. Fortunately, they will look a lot like some of the hypotheses that you learned for the paired *t* test.

Check Your Understanding

SCENARIO 11-D

Hoffman, Meier, and Council (2002) wanted to know whether people living in urban areas and people living in rural areas differed in terms of their average chronic pain. The researchers defined chronic pain as "constant pain or pain that flares up frequently and has been experienced for at least 6 months at a time" (p. 216). Using stratified random sampling and addresses from current telephone directories, they sent out surveys to people living in urban and rural areas. Among other research questions, the study investigated whether urban dwellers with chronic pain differed in their average ratings of pain intensity when compared with the mean ratings of pain intensity from rural residents with chronic pain. 11-13. Why might the independent-samples *t* test be appropriate for this scenario?

Suggested Answers

11-13. The scenario describes two independent groups and an interest in comparing means for the ratings of pain intensity. Whether the sample sizes are equal and sufficiently large has not been verified, however. (We will talk more about sample sizes later in the chapter.)

Independent *t* Hypotheses When Direction Is Not Predicted

In the scenario of babies receiving immunizations, Efe and Özer (2007) used a two-tailed test, meaning that they did not predict which group's mean crying time would be bigger than the other. They could have written their alternative hypothesis as follows:

$$H_1: \mu_{treatment} \neq \mu_{control}$$

This expression can be translated as follows:

Our samples come from populations in which
the mean crying time for the babies who were breast-fed during the injection
differs from the mean crying time for the babies who were not
breast-fed during the shot.

Notice that when we have more than one sample, we talk about the samples (plural) coming from populations (plural). This alternative hypothesis would correspond to the following null hypothesis:

$$H_0: \mu_{treatment} = \mu_{control}$$

This null hypothesis is saying

> Our samples come from populations in which
> the mean crying time for babies who were breast-fed during the injection
> is the same as the mean crying time for babies who were not
> breast-fed during the shot.

The subscripts designate the two conditions, and the population means represent the average crying times in the populations. Notice that these hypotheses are quite similar to the ones that we wrote for the paired t test. The difference is that with the paired t test, we had to keep in mind that there was a pairwise connection between the scores, which meant the population means were related or dependent. In the vibration example, we measured one sample twice, with and without vibration, so the "with vibration" population mean was related to the "without vibration" population mean. Here, we have two different groups of babies without any pairing. The two population means in the above hypotheses are therefore independent of each other.

Like the hypotheses for the paired t test, the hypotheses for the independent-samples t test can be written as the difference in two population means. The alternative hypothesis for the baby scenario could be written as follows:

$$H_1: \mu_{treatment} - \mu_{control} \neq 0$$

This alternative hypothesis is saying

> Our samples come from populations in which
> the difference between the mean crying time for the treatment group
> and the mean crying time for the control group is not zero.

In other words, there is some difference in the population means. This alternative hypothesis would go along with the following null hypothesis:

$$H_0: \mu_{treatment} - \mu_{control} = 0$$

This null hypothesis is saying

> Our samples come from populations in which
> the difference in the mean crying times for the treatment group and the
> control group is zero.

In other words, there is no difference in the population means. When we are not predicting a direction, the order of subtraction can be turned around without affecting the meaning. In other words, if the null hypothesis is saying there is no difference in the population means, then "$\mu_{treatment} - \mu_{control} = 0$"

means the same thing as "$\mu_{control} - \mu_{treatment} = 0$." In the case of two independent samples, we do not compute difference scores because we do not have pairs of scores, so we will not write the hypotheses with μ_d, as we did in the paired-samples scenarios. Like the paired t test, the independent-samples t test can be used when researchers hypothesize a directional outcome. Next we will show how to write those hypotheses.

Check Your Understanding

SCENARIO 11-D, Continued

This scenario described research by Hoffman et al. (2002), who wanted to know whether people suffering from chronic pain differed in terms of their average ratings of pain intensity, depending on whether they lived in urban areas or rural areas. 11-14. Write the alternative hypothesis, using both words and symbols.

Suggested Answers

11-14. The use of the word "differed" does not indicate a directional prediction. In symbols, we may write H_1: $\mu_{urban} \neq \mu_{rural}$. Our alternative hypothesis is that our samples come from populations in which the mean rating of pain intensity for chronic-pain sufferers in urban areas differs from (is not equal to) the mean rating of pain intensity for chronic-pain sufferers in rural areas. The alternative hypothesis also could be written as one μ subtracted from the other, with the difference not equaling zero.

Independent t Hypotheses When Direction Is Predicted

If prior studies had provided greater support for a prediction that breast-feeding would be associated with shorter crying times after an injection, Efe and Özer (2007) could have written a directional alternative hypothesis. Future researchers who want to demonstrate the same effect with other samples of babies might write the following:

$$H_1: \mu_{treatment} < \mu_{control}$$

This directional alternative hypothesis says

Our samples come from populations in which
the mean crying time for babies breast-fed during shots
is less than the mean crying time for babies who receive the
usual care during shots.

11. Two-Sample Tests and Estimates

The corresponding null hypothesis would be

$$H_0: \mu_{\text{treatment}} \geq \mu_{\text{control}}$$

The above H_0 is saying

> Our samples come from populations in which
> the mean crying time for the babies receiving the intervention
> is greater than or equal to the mean crying time for the
> babies not receiving the intervention.

(Your instructor may prefer for you to write this null hypothesis with an equals sign only.)

As we have seen, there are other ways that these same ideas can be written. If we reverse the order of these two population means, the directional sign also will have to be turned around. Study the following two ways of writing this directional alternative hypothesis and make sure that you understand they are saying the same thing:

$$H_1: \mu_{\text{treatment}} < \mu_{\text{control}}$$

$$H_1: \mu_{\text{control}} > \mu_{\text{treatment}}$$

Both of the above ways of writing H_1 are predicting a larger mean for the control group than the treatment group. We also can write the alternative hypothesis as a difference in two population means:

$$H_1: \mu_{\text{treatment}} - \mu_{\text{control}} < 0$$

This alternative hypothesis is saying

> Our samples come from populations in which
> the difference in means (computed as the treatment mean
> minus the control mean)
> would be less than zero.

That is an awkward but equivalent way of saying that the population mean for the treatment group is smaller than the population mean of the control group. The null hypothesis corresponding to this H_1 would be

$$H_0: \mu_{\text{treatment}} - \mu_{\text{control}} \geq 0$$

Again, if we reverse the order of the means in the subtraction, the directional sign must be turned around. The directional alternative hypothesis could be written as follows:

$$H_1: \mu_{\text{control}} - \mu_{\text{treatment}} > 0$$

The corresponding null hypothesis would be

$$H_0: \mu_{control} - \mu_{treatment} \leq 0$$

None of these ways of writing the hypotheses is superior to the others. We recommend that you use the one that makes the most sense to you, and be sure to examine the sample means to determine whether the results turn out in the predicted direction.

Check Your Understanding

SCENARIO 11-E

Grant and Hofmann (2011) investigated whether health-care professionals would be more likely to wash their hands if they saw a sign that emphasized their own health or a sign that emphasized the health of others. We are running a similar study in a hospital where we know how much hand sanitizing gel tends to be used in a large number of locations around the hospital. We identify two wings, each with 30 dispensers of sanitizing gel. Each wing uses the same amount of gel every month. We decide to run a quasi-experiment. Above the dispensers in one wing, we place signs saying, "Hand hygiene prevents you from catching diseases." Above the dispensers in the other wing, we place signs saying, "Hand hygiene prevents patients from catching diseases." For each dispenser we measure how much sanitizing gel is used in a month. We predict that the wing with the signs referring to the health of patients will have a higher mean for the amount of gel that is used, compared with the wing with the signs referring to the health of the hospital workers. 11-15. What is the unit of analysis in this scenario? (That is, what is the entity providing the data? The term *unit of analysis* was introduced in Chapter 1.) 11-16. Write the alternative hypothesis in words and symbols.

Suggested Answers

11-15. The unit of analysis is the dispenser. People are not being measured individually; the dispensers are providing the data, the amount of gel used at each dispenser in a month. 11-16. We are predicting that the dispensers with the signs referring to patient health will have a higher mean amount of gel used in a month, compared with the mean amount of gel dispensed by the dispensers with the sign referring to the health of the hand-washer. Our samples come from populations in which the mean amount of gel used at dispensers with patient-referent signs will be greater than the mean amount of gel used at dispensers with self-referent signs. This hypothesis can be written in the following ways: (1) $H_1: \mu_{patient} > \mu_{self}$ (2) $H_1: \mu_{self} < \mu_{patient}$. (3) $H_1: \mu_{patient} - \mu_{self} > 0$. (4) $\mu_{self} - \mu_{patient} < 0$.

Formula for the Independent-Samples *t* Test

Remember when we said earlier in the book that we liked small and cute formulas, but not big and ugly formulas? The independent-samples *t* test has a big, ugly formula. We are going to explain the conceptual meaning of the formula, but not subject you to its computational details.

Like the other *t* tests that you have learned, the independent-samples *t* test follows the pattern of "(something minus its mean) divided by its estimated standard deviation." The "something" in the independent-samples *t* test is the difference in sample means. Efe and Özer (2007) predicted some difference in the population mean crying times for the treatment group versus the control group. Let's use the following hypotheses:

$$H_0: \mu_{\text{treatment}} - \mu_{\text{control}} = 0$$

$$H_1: \mu_{\text{treatment}} - \mu_{\text{control}} \neq 0$$

To be consistent with these hypotheses, we would compute our "something" as

$$M_{\text{treatment}} - M_{\text{control}}$$

This "something" is a point estimate of the population mean difference, $\mu_{\text{treatment}} - \mu_{\text{control}}$. Our "something" ($M_{\text{treatment}} - M_{\text{control}}$) is a statistic with a sampling distribution, and its mean is the population mean difference, $\mu_{\text{treatment}} - \mu_{\text{control}}$. What is the numeric value of $\mu_{\text{treatment}} - \mu_{\text{control}}$? The null hypothesis says it equals zero. So when we compute "something minus its mean" in the numerator of the independent-samples *t* test, we have our "something" ($M_{\text{treatment}} - M_{\text{control}}$) minus zero, leaving us with only the sample mean difference, $M_{\text{treatment}} - M_{\text{control}}$, as the numerator.

The denominator of the independent-samples *t* test is the estimated standard deviation for our "something" ($M_{\text{treatment}} - M_{\text{control}}$). That is an estimated standard error for the difference in independent means, and it is so big and ugly that we are not showing it to you. The denominator combines measures of variability for both groups, the sample sizes for both groups, and the formula for the *df*, which is

$$df = n_1 + n_2 - 2$$

This is the first time we have shown lowercase *n* with subscripts. It represents the sample size for a particular group. The entire sample size, or the total number of people in both groups put together, is *N*. But now that we have two independent groups, we need to be able to say how many people are in each group. The subscripts could be the words "treatment" and "control" instead of the numbers 1 and 2.

The study by Efe and Özer (2007) had 33 babies in each group, so they had $df = 33 + 33 - 2 = 64$. This number is used to look up critical values for the independent-samples *t* test, using the same Table B.1 that we used with the other *t* tests. As you know, Efe and Özer had a nondirectional alternative hypothesis, so we would need two critical values, one in each tail. Table B.1 does not list critical values for $df = 64$, so we would use the next-smaller df in the table, $df = 60$. For a two-tailed test and $\alpha = .05$, the critical values would be -2.0 and $+2.0$. Figure 11.2 shows the t distribution with $df = 60$ and these critical values, which cut off a total area of $\alpha = .05$, with half of alpha in each tail.

Let's complete the hypothesis test for these researchers' results, using the critical value decision rule. They reported an independent-samples t test $= 3.64$. We use the same decision rules that applied to the one-sample t test. Because we are not predicting a directional outcome, we ask whether the observed independent-samples t test is equal to or more extreme than a critical value. Yes, 3.64 is more extreme than the upper critical value, 2.0, so we reject the null hypothesis.

How do we interpret the result of rejecting the null hypothesis? We stated the null hypothesis as follows:

$$H_0: \mu_{treatment} - \mu_{control} = 0$$

We know that the independent-samples t test has a numerator containing the difference in the sample means. Here, the independent-samples t test is a positive number. Does that mean the treatment mean was bigger than the control mean? At this point, we cannot say! Even with a nondirectional alternative hypothesis, we cannot interpret the meaning of the independent-samples t test without knowing the numeric values of the sample means. All we could say at this point is that there was a significant difference in the mean crying time for the two groups. Reading Efe and Özer (2007), we found a statement about the sample means. The researchers said the mean crying duration for the breast-feeding babies was 35.85 seconds ($SD = 40.11$), and the mean crying time for the control group was 76.24 seconds ($SD = 49.61$). Now that we have found a significant difference, we can observe that the treatment mean was *smaller* than the control mean; the babies who were breast-fed during the injections cried about half as long as the babies in the control group.

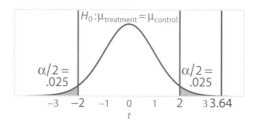

Figure 11.2

A t distribution with $df = 60$ and no prediction of direction. The study by Efe and Özer (2007) had $df = 64$, but Table B.1 does not list critical values for that value of df. Looking at the table's next smaller value of df and using $\alpha = .05$, we find the critical values shown in this figure. Can you find these values in Table B.1? The researchers reported an observed independent-samples t test $= 3.64$, shown here in blue.

11. Two-Sample Tests and Estimates

Now let's do the same hypothesis test again, using the p value decision rule, which is what the researchers probably did. Their paper stated that a significance level of .05 was chosen, and the researchers reported that the observed independent-samples t test had a two-tailed $p = .001$. So we ask: is p less than or equal to alpha? Yes, .001 is less than .05, so we reject the null hypothesis and draw the same conclusion that we reached using the critical value decision rule.

Just for fun, let's run through the hypothesis test as if we predicted the following outcome:

$$H_1: \mu_{treatment} - \mu_{control} < 0$$

This alternative hypothesis is predicting a smaller mean (shorter average crying time) for the treatment group, compared with the mean for the control group. Using the critical value decision rule when direction is predicted, we look up a critical value for a one-tailed test. Table B.1 shows that if $\alpha = .05$ for a one-tailed test and if $df = 60$ (the closest smaller value to our actual $df = 64$), the critical value would be 1.671. Wait—the alternative hypothesis above is predicting a difference that is negative, and the directional sign is pointing toward the lower tail. Should the critical value be in the lower tail? That is, should it be −1.671? It depends on which way the subtraction is performed in the numerator of the independent-samples t test. Remember, the above alternative hypothesis could be rewritten as follows:

$$H_1: \mu_{control} - \mu_{treatment} > 0$$

Which way did Efe and Özer (2007) do the subtraction? They reported that the mean for the treatment group was 35.85 seconds and the mean for the control group was 76.24 seconds, and they reported an independent-samples t test $= 3.64$. To get a positive number for the independent-samples t test, they must have taken the bigger mean and subtracted the smaller mean, or $M_{control} - M_{treatment}$. So the numerator of the independent-samples t test would have been $76.24 - 35.85 = 40.39$. (The difference in means was divided by a big, ugly estimated standard deviation to get the final answer for the independent-samples t.) This order of subtraction of means would correspond to the alternative hypothesis that is written as follows: $H_1: \mu_{control} - \mu_{treatment} > 0$. So our t distribution would have alpha in the upper tail and the critical value of 1.671, as shown in Figure 11.3.

Now we can perform the hypothesis test for our directional case. First, we need to ask: were the results in the predicted direction? Yes, we predicted that the control group would have a bigger mean than the treatment group, and a big number minus a small number would give a positive independent-samples t test. We also can see that $M_{control} = 76.24$ seconds, which is bigger than $M_{treatment} = 35.85$, as predicted. Second, we ask whether the observed test statistic was equal to or more extreme than the critical value. Yes, because 3.64 is more extreme than 1.671, so we reject the null hypothesis and conclude that the babies in the control group cried significantly longer than the babies in the treatment group. Because it was an experiment, we can say that breast-feeding most likely was responsible

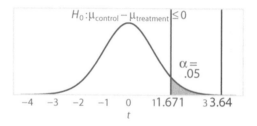

$H_0 : \mu_{control} - \mu_{treatment} \leq 0$

$\alpha = .05$

-4   -3   -2   -1   0   1.671   3.64
                    t

Figure 11.3

A t distribution with $df = 60$ and a directional prediction. Efe and Özer (2007) reported a positive independent-samples $t = 3.64$ and a smaller mean for the treatment group, as our current example predicted. To obtain a positive value for the independent-samples t test, the control group's larger mean must have come first in the numerator (i.e., $M_{control} - M_{treatment}$). This distribution shows the critical value and observed test statistic. The distribution has $df = 60$ because Table B.1 does not show critical values for the researchers' actual $df = 64$.

for causing the difference in mean crying durations. (Efe and Özer reported the positive and rounded value of the independent-samples $t = 3.64$ in the text, but a negative, unrounded result of $t = -3.637$ in a table. Whether the test statistic was positive or negative does not matter; we looked at the sample means to check which group had the longer average crying time.)

Just to be complete, let's do the hypothesis test again using the p value decision rule when a directional outcome is predicted. First, we would ask whether the results were in the predicted direction. We have confirmed that is true. Second, we ask whether the one-tailed p value is less than or equal to $\alpha = .05$. The researchers reported a two-tailed $p = .001$. Half of that p value would be in each tail in the two-tailed test—but we care only about the part of p in the upper tail. The one-tailed p value of $.001/2 = .0005$ is smaller than .05. Therefore, we can reject the null hypothesis and conclude that the mean crying duration for the control group was significantly greater than the mean crying time for the breast-fed babies. It is equally acceptable to say that the breast-fed babies had a significantly shorter mean crying duration than the babies in the control group.

We have demonstrated the independent-samples t test with a scenario from experimental research, but this statistic also can be used in observational research or quasi-experimental research. For example, we could compare means for males and females in a descriptive study. You may have noticed that we have not talked about the assumptions of the independent-samples t test. Further, you may have wondered why we said the sample sizes needed to be equal and sufficiently large to use this test statistic. Next we will go into a fair amount of detail about the assumptions and sample sizes for the independent-samples t test. We have one more inferential statistic to cover in this chapter, and it used in similar situations as the independent-samples t test. The difference is that the independent-samples t test has a weakness in certain situations, and we would direct you to use the other test statistic in those situations.

Check Your Understanding

SCENARIO 11-F

(Inspired by Stephens, Atkins, & Kingston, 2009. Details of this scenario differ from the actual research.) Pain researchers have some standardized ways of testing how long people can tolerate pain. One way of testing pain tolerance is to chill water to a temperature slightly above freezing, then time how long a person can stand to keep a hand submerged in the water. Let's say we want to research whether people who are cursing—that is, saying swear words—can tolerate pain longer than people who are told not to swear. We recruit people who say they sometimes swear, and we randomly assign them to either the "no swearing" condition or the "swearing" condition. We tell the participants that they will be timed on how long they can keep their hand in icy water. Half of the participants are told, "Please curse or swear while your hand is in the water." The other participants are told, "Please don't curse or swear while your hand is in the water." Does the swearing group tolerate the icy water for a longer average length of time than the nonswearing group? Here are our results: for the swearing group, $M = 160.2$ seconds ($SD = 90.5$), and for the nonswearing group, $M = 119.3$ seconds ($SD = 81.6$). We computed an independent-samples t test $= 1.986$, which had a one-tailed $p = .0256$. 11-17. Test the null hypothesis using $\alpha = .05$ and explain the results.

Suggested Answers

11-17. The scenario suggests that we expect the swearing participants to be able to tolerate the icy water longer than the nonswearing participants. That is, the alternative hypothesis could be written as H_1: $\mu_{swearing} > \mu_{nonswearing}$. The null hypothesis would be an opposite statement and could be written as H_0: $\mu_{swearing} \leq \mu_{nonswearing}$. To test H_0, we first must check whether the results are in the predicted direction. Yes, because the swearing group's mean of 160.2 seconds is greater than the mean of 119.3 seconds for the nonswearing group. Next, we ask whether the one-tailed p value is less than or equal to alpha. Because .0256 is less than .05, we reject the null hypothesis and conclude that the swearing condition had a significantly greater mean for the length of time that participants could keep a hand in the icy water, compared with participants in the nonswearing condition. In this study we cannot say whether the swear words helped the participants tolerate the pain longer or whether suppressing swearing in people who are used to saying curse words reduced their ability to stand the painful stimulus.

Assumptions

The other test statistics that we have covered so far had assumptions of normality and independence. The z test statistic and the one-sample t test both assume that the scores are normally distributed in the population and the scores are independent of each other. Then we said the paired t test also assumed normality and independence, but these assumptions applied to the population of difference scores. The independent-samples t test has three assumptions, and the first two assumptions will be familiar:

- Normality: the scores are normally distributed in the two populations that provided the samples.
- Independence: the scores are independent of each other.
- Equal population variances: the two populations of scores are equally spread out.

If the assumptions of normality and equal population variances are met, then the two populations will look like identical normal distributions, with the possible exception being that the two distributions could be located in different places on the number line due to different population means. If the null hypothesis is true (and the population means are equal), the two distributions would overlap entirely and look like one normal distribution.

As we said in Chapter 10, we usually cannot know whether the scores are normally distributed in the population. Now we are talking about two populations: one population for the treatment group, and one population for the control group. The normality assumption often is violated in actual research; that is, one or both populations of scores may not be normally distributed. Is that a problem? Remember why we made assumptions: we wanted to use a theoretical distribution instead of having to create a sampling distribution of our test statistic to get p values. We needed to ensure that the theoretical distribution would look just like the sampling distribution of the test statistic. Meeting the assumptions of the test statistic ensures that a theoretical distribution is a good match for the sampling distribution, so we can use the theoretical distribution instead. If the normality assumption often is violated in situations where we want to use the independent-samples t test, does that mean we cannot use a theoretical t distribution for hypothesis testing? No, we can still use the theoretical t distribution, except for situations in which we have unusual departures from normality. Except for those situations, we generally can violate the normality assumption without disturbing the shape of the independent-samples t test's sampling distribution, and it will look like the theoretical t distribution. So we can still trust the p values that we would get from a theoretical t distribution.

What about the independence assumption? In line with what we said in Chapter 8, random sampling from the two populations would assure that we obtained independent scores from the participants. But most research involves

11. Two-Sample Tests and Estimates

convenience samples, and there is no way of knowing whether convenience samples provide data similar to what would have been obtained via random sampling. If we violate this assumption, it may be disastrous. Not only will the sampling distribution of the independent-samples t test be poorly matched by a theoretical t distribution, a violation of independence probably means there is a fatal flaw in the research.

Let's consider how the independence assumption might be violated. Imagine we are running a study that uses ice packs to numb an injection site, and we recruit volunteers from a health sciences university campus. We later discover that 24 of our participants were pharmacy students who were taking a class together, and they talked to each other about the research. They decided to sign up together as volunteers. Throughout the study they talked to each other about their experiences. The connections between these participants could violate the independence assumption and ruin the study, especially if the study was supposed to be double-blinded. Suppose the participants were blinded to their experimental treatment, then compared notes with their classmates and figured out how their group differed from the other group in the study. Their shared knowledge could make participants respond similarly to their classmates, especially if psychological variables were measured. The independence assumption also would be violated if the researcher actually pairs the participants, but then uses the incorrect data analysis of an independent-samples t test.

So far, we have talked about the consequences of violating two assumptions: normality and independence. Before talking about the third assumption, let's go into a little more detail about these first two assumptions. Violating the normality assumption in the world of statistics is like encountering a cold virus in the

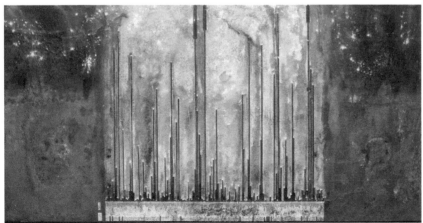

"Disparity Three-Defense Budget Per Capita" (4' × 8'), by Gary Simpson, used with permission. The artist writes, "Brass rods measure the millions of dollars spent per capita by country for defense."

world of personal health. Unless a person is immune-compromised, generally a cold virus is something that can be survived without long-term effects. We would say that most people are robust to a cold virus. In statistics, *robustness* is a term that may be used to describe the ability of the sampling distribution of a test statistic to resist the effects of a violation of an assumption. (Robustness also sometimes is used to describe a statistic's resistance to outliers, but that is not what we are talking about here.) If the test statistic is robust in the presence of a violated assumption, then the statistic's sampling distribution still would match the theoretical distribution, and we can trust the p values. If the test statistic is not robust in the presence of an assumption being violated, then the statistic's sampling distribution would look different from the theoretical distribution. Then we could not trust the theoretical distribution's p values.

Having nonnormally distributed scores in one or both populations is a violation of the normality assumption of the independent-samples t test. Is this test statistic robust when the data come from nonnormal populations? Yes, except for weirdly shaped populations. If we have reason to believe that there would be a clump of outliers in one tail of a population distribution, then the independent-samples t test may have a sampling distribution that does not look like a theoretical t distribution. So nonnormality is like a cold virus: most of the time, the independent-samples t test can survive some nonnormality in the populations.

The independence assumption is crucial. Violating the independence assumption is like being near a nuclear plant when the core melts down and a nuclear disaster occurs. No one near the site of the meltdown would survive. Similarly, when there is a violation of independence, we cannot trust the independent-samples t test's p value from the theoretical t distribution. Dependence in the data can cripple or destroy the study. Let's look at a real example. Out of respect for the researchers, we will be vague about the details. We know of a study that, among other things, compared the average body temperature of men and women. The study reported an independent-samples t test comparing the means for these two groups. We noticed that the df for the independent-samples t test did not equal $n_1 + n_2 - 2$; in fact, the df was much greater than the total N. How is that possible, when $n_1 + n_2$ should equal N? Rereading the article, we realized that the authors had taken multiple temperature readings on many participants. Some participants were measured once, but others were measured on many occasions. All of their scores were placed in the same group, as if every score were independent of the others. Then the means were computed for the two groups, males and females. Let's think about this: is your body temperature today independent of your body temperature tomorrow? No, because the same person is being measured repeatedly, and some people's body temperatures tend to run high or run low. The point is, the researchers had violated the independence assumption for the independent-samples t test, so their results were not trustworthy. So violation of the independence assumption is like a person being exposed to massive amounts of radiation: the independent-samples t test and other test statistics are not robust to violations of their independence assumptions.

Now let's talk about the third assumption, equal population variances. Violating this assumption would mean that the two populations of scores were not equally spread out. How much of a problem is this violation? The answer to that question will help us to explain why we need one more test statistic to compare the means of two independent groups. Let's consider an analogy: would you be robust if you were exposed to a measles virus? The answer: it depends. If you have had a measles immunization that is still effective, the answer probably is yes. If you are not current on your measles immunization, then the answer probably is no. Your vulnerability to the measles virus depends on whether you are current on the measles inoculation. Encountering unequal population variances is like being exposed to the measles virus. The independent-samples t test can have an inoculation that will keep it from being vulnerable to its version of the measles virus: unequal population variances. Without the inoculation, the independent-samples t test's sampling distribution will look very different from a theoretical t distribution, and the p values from the theoretical t distribution will not be trustworthy.

What is the inoculation that protects the independent-samples t test from the effects of unequal variances? The inoculation is equal sample sizes with at least 15 people per group. The study by Efe and Özer (2007) had 33 babies in each group. The sample sizes were equal, and each sample had 15 or more people. That means Efe and Özer could have sampled from populations that had unequal variances, and the independent-samples t test would have a sampling distribution that still was well-matched by the theoretical t distribution with $df = 64$. The violation of the assumption would have no effect on the trustworthiness of their test statistic's p value, just as a measles virus would not threaten the health of someone with a current measles immunization. When the two samples are equal in size and have enough people in each group (at least 15), then we do not have to worry about the equal variances assumption.

It is unusual for a study to have equal sample sizes, though. We were surprised that not even one baby (out of $N = 66$) in the Efe and Özer (2007) study had to be excluded. Researchers often try hard to have equal sample sizes, but then someone does not show up for the study, or someone arrives on the day of the study with a fever, which the study's protocol says disqualifies the person from participation. What happens if the study has unequal sample sizes, like 14 in one group and 19 in the other, or $n_1 = 33$ and $n_2 = 31$? Or what if the research involves a rare health condition and has only 10 people per group? Or what if the study has both small and unequal sample sizes, like 10 in one group and 13 in the other? In all of these cases, the independent-samples t test lacks its inoculation, and its p value may not be trustworthy.

When we have two independent groups and we want to compare their means, but we are lacking the inoculation (equal samples of at least 15 per group), we can switch to another test statistic. We call this test statistic the *Aspin-Welch-Satterthwaite t test*, and we will abbreviate it as the AWS t test. This test statistic has other names. You might see it referred to by any of the three people's names given here (Aspin's *t*, Welch's *t*, Satterthwaite's *t*, or some combination of these names). It also is called the independent *t* test for unequal variances, the two-group *t* test when equal variances are not assumed, and so forth. (We know—you wish that statisticians would get together and agree on one name per test statistic! Unfortunately, different names have become traditional in different fields and areas of the world. And there even are different statistics that share the same symbol.)

When would we use the AWS *t* test? Suppose that one baby running a fever had to be disqualified from the study of breast-feeding during injections. The researchers would have had 33 babies in one group and 32 babies in the other group. The independent-samples *t* test assumes that the samples were drawn from two populations that were equally spread out. This assumption is necessary if we are going to be able to use a theoretical *t* distribution instead of the statistic's sampling distribution to find *p* values. In the event that the two populations are not equally spread out, the independent-samples *t* test needs to be protected, and this protection comes from having equal and sufficiently large sample sizes. Without the inoculation of equal sample sizes that are large enough, the independent-samples *t* test is not robust to unequal variances, just as a person who lacks a current measles immunization is vulnerable to the measles virus. So the baby being excluded from the study would result in unequal sample sizes, leaving the statistic vulnerable to the effect of unequal population variances and making the *p* value potentially untrustworthy.

Using all the same hypotheses and decision rules, the researchers could switch from the independent-samples *t* test to the AWS *t* test. That is because the AWS *t* test has no assumption about variances; it assumes only normality and independence. Its sampling distribution cannot be affected by unequal population variances. The AWS *t* test will let the researchers test exactly the same hypotheses about two means. The formula follows the same general pattern of other *t* statistics: (something minus its mean) divided by its estimated standard deviation. Like the independent-samples *t* test, the AWS *t* test has the difference in the sample means in the numerator and an estimated standard deviation for the mean difference in the denominator. This statistic's *df* has a big, ugly formula, often resulting in a fraction, like $df = 38.4$. If the AWS *t* test is unaffected by unequal variances, why don't we use it all the time, even when sample sizes are equal and large enough? Generally the independent-samples *t* test will be slightly more sensitive to true differences in the population means when sample sizes are equal and large enough, compared with the AWS *t* test. That is, the independent-samples *t* test gives us a little more power. And speaking of power, a study might have two groups with 15 people per group, which would mean the independent-samples *t* test could be used. But 15 people per group may not provide sufficient power to detect whatever population mean difference would be clinically noteworthy. So do not latch onto any particular sample sizes as universally good.

A common mistake that students make is to say, "Oh, this study has unequal *n*'s and that means we violated the equal variances assumption." That would be like saying, "I haven't had my measles inoculation, therefore I have the measles."

No, having unequal sample sizes is similar to a person not being inoculated against the measles, meaning the person now is vulnerable. The person may or may not encounter the measles virus. Violating the equal variances assumption is the measles virus, and we usually do not know whether the population variances are equal or unequal—just as we do not know until it is too late that we have encountered a virus.

To summarize our analogy, here are several versions of a silly story that your first author tells her students:

- Lise and her dog Tate are walking down the street. They walk past a nuclear plant that suddenly melts down, explodes, and kills them both. The End! (We did not say it was a good story.) The nuclear meltdown is dependence in the data, and Lise is the independent-samples t test. Dependence in the data ruins the trustworthiness of the p value and probably the credibility of the rest of the study.
- Lise and her dog Tate are walking down the street. They bump into a new neighbor. Lise shakes the neighbor's hand before realizing the person has a cold virus. Thankfully, Lise is generally healthy. Even though she gets the sniffles a few days later, she survives and wishes she had not touched the neighbor's hand. In this analogy, the cold virus is nonnormality in the populations that provided the samples for the independent-samples t test, and Lise is the independent-samples t test. The sampling distribution of the independent-samples t test generally is unaffected by most nonnormality and will be well-matched by a theoretical t distribution, so we can trust the p value.
- Lise and her dog Tate are walking down the street. They walk through a crowd of people, not realizing that someone in the crowd is carrying the measles virus. Here is the first version of the measles story: Lise is current on her measles inoculation, despite her needle phobia, so she does not get sick after she walks near the person carrying the measles virus. The measles virus is the unequal population variances, and Lise is the independent-samples t test when the samples are equal in size and contain at least 15 people per group. The statistic's sampling distribution is unaffected by the unequal variances, so the p value is trustworthy.
- Here is the second version of the measles story: Lise is *not* current on her measles inoculation, so her health is severely threatened by the measles virus. The measles virus is still the unequal population variances, and Lise is still the independent-samples t test, but the samples are either unequal in size, or smaller than 15 per group, or both unequal in size and small. The sampling distribution for the statistic may look quite different from the theoretical t distribution, making the p value untrustworthy. This version of the story has a twist: Lise's dog Tate is the AWS t test. The dog cannot catch the human measles virus. Similarly, the AWS t test is unaffected by unequal variances. That is, the sampling distribution of the AWS t test will be well-matched by a theoretical t distribution, so the

The late film critic Roger Ebert's Law of the Economy of Characters precluded the appearance of Geordi (left) in the story that explained the robustness of the independent-samples *t* test. (Photo by Lise DeShea.)

p value for the AWS *t* test will be trustworthy. (This example also illustrates the late film critic Roger Ebert's Law of the Economy of Characters: every character is necessary to the story.)

We can summarize the assumptions of the independent-samples *t* test as shown in Table 11.2.

The independent-samples *t* test often appears on the output for statistical software with a notation that says something like, "Equal variances assumed." These words designate the *t* test as being the independent-samples *t* test. Fortunately, many statistical software packages compute the AWS *t* test at the same time as the independent-samples *t* test. The AWS *t* test often appears on the same output with a notation that says something like, "Equal variances not assumed." That phrase does not refer to the particular study. It is referring to the AWS *t* test, which does not have an assumption of equal variances.

The same output showing the independent-samples *t* test and the AWS *t* test also may show the results of a test statistic called Levene's test for equal variances. This statistic tests a null hypothesis that says the samples came from populations with equal variances. That is, the null hypothesis says, "The variance of the population that provided one sample equals the variance of the population that provided the other sample." But why use Levene's test to check whether the assumption of equal population variances has been met? Either the independent-samples *t* test will have its inoculation or it will not have its inoculation. If it does,

11. Two-Sample Tests and Estimates

Table 11.2 Robustness of the Independent-Samples *t* Test

Assumption	Are We Likely to Meet the Assumption?	If Assumption Violated, Is Independent-Samples *t* Robust?
Samples came from two normally distributed populations	No. This assumption is often violated	Usually yes. The exception: if 5%–10% of the scores are extreme in one direction, the statistic may not be robust
Samples came from two populations that were equally spread out	No. This assumption is often violated	Yes, *if* the study has equal sample sizes with at least 15 scores per group. (If the study has unequal *n*'s, small *n*'s, or both small and unequal *n*'s, use the AWS *t* test)
All scores are independent of each other	Yes, if careful experimental methods are followed	No

then we can trust its *p* value. If it does not, we will use the AWS *t* test instead. Levene's test is therefore unnecessary. Everything else about hypothesis testing with the AWS *t* test proceeds exactly like the independent-samples *t* test—the hypotheses that can be tested, the decision rules, and so forth.

The distance between two independent population means can be estimated with confidence intervals. Next we will talk about two such confidence intervals: one associated with the independent-samples *t* test, and the other linked with the AWS *t* test.

Check Your Understanding

SCENARIO 11-F, Continued

(Inspired by Stephens et al., 2011. Details of this scenario differ from the actual research.) This scenario compared the average pain tolerance of people who were told to use swear words versus the average pain tolerance of people who were asked not to swear while they kept a hand submerged in icy water. The researchers timed how long each participant kept a hand in the water, with the researchers testing only one participant at a time. Let's say someone in the nonswearing condition was tested in a room next door to a room where someone in the swearing condition was being tested. Suppose we discover that the nonswearing participant started to laugh after overhearing the person swearing in the other room. This happened with six participants in the nonswearing condition. 11-18. Is this a violation of the independence assumption?

(Continued)

11-18. The experience of the six nonswearing participants appears to be linked with the swearing they overheard in the next room, so it does appear to be a violation of the independence assumption. Further, an extraneous variable (laughter) has contaminated the independent variable. For any participants who overheard swearing, it might be a good idea to leave their data out of the study and investigate whether any other participants in the nonswearing condition overheard the swearing in the next room. Mistakes happen in research. Ethical researchers would include the information in their journal article. Readers of the article then can judge whether omitting the data from those six participants was a sufficient solution. Note that if the researchers had equal sample sizes before this incident, leaving out the data from those six participants will make the sample sizes unequal and necessitate the use of the AWS t test.

Confidence Intervals for a Difference in Independent Means

As you now know, the numerator of the independent-samples t test is the difference in the sample means, $M_{treatment} - M_{control}$; that difference in sample means is a point estimate of the difference in population means. If we did a completely new study, the new samples would produce another point estimate of the difference in population means. Continuing with the idea of treatment and control samples from repeated studies would produce a variety of mean differences. The sampling variability in the estimation of the population mean difference can be quantified with an interval estimate, similar to the ones we have seen before. We can multiply a critical value by the denominator of the observed independent-samples t test (i.e., the estimated standard error of the mean difference), and the result will be a margin of error. Placing the margin of error around the difference $M_{treatment} - M_{control}$ allows us to judge how much variation could be expected across multiple studies.

We will show one nondirectional example using the independent-samples t test and the example of soothing babies during an immunization. The null hypothesis can be written as follows:

$$H_0: \mu_{control} - \mu_{treatment} = 0$$

(The difference can be written with the subtraction performed in the opposite order, but we are trying to simplify this numeric example.) If breast-feeding had no effect on the crying time after the shot, then the difference in means would be zero—that is, no difference in the means. Efe and Özer (2007), who conducted the study of breast-feeding during immunizations, reported $M_{control} = 76.24$ and $M_{treatment} = 35.85$, so the sample mean difference was

$$M_{control} - M_{treatment}$$

$$= 76.24 - 35.85$$

$$= 40.39$$

This difference, 40.39, is the point estimate, and the interval estimate will be a range of values around this number. To find the margin of error, we need to multiply a critical value by the denominator of the independent-samples t test, which is the estimated standard error of the mean difference. The researchers did not report that number, but we worked backwards from their reported independent-samples t test and this mean difference, and we solved for the denominator, which was 11.105307. As usual, we will use the unrounded number in the computations, then round the final answers. If we used a bigger table of critical values or a computer program, we could look up an exact critical value for this scenario, which had $df = 64$, but for simplicity we will use Table B.1 and find the t critical value for $df = 60$ for a two-tailed test using $\alpha = .05$. This critical value, 2.0, was used in the nondirectional example above. The margin of error will be the product of this critical value and the estimated standard error of the mean difference:

$$\text{Margin of error} = 2.0 \times 11.105307$$

$$= 22.210613$$

The lower limit of the 95% confidence interval for the population mean difference is

$$\text{Sample mean difference} - \text{margin of error} = 40.39 - 22.210613$$

$$= 18.179387$$

$$\approx 18.18$$

The upper limit of the 95% confidence interval for the population mean difference is

$$\text{Sample mean difference} + \text{margin of error} = 40.39 + 22.210613$$

$$= 62.600613$$

$$\approx 62.60$$

How do we interpret this 95% confidence interval, [18.18, 62.60]? We can say that we are 95% confident that [18.18, 62.60] contains the true difference in the population means. This particular interval may or may not contain the true difference in the population means. Across repeated studies from the same population, 95 out of 100 studies would produce an interval that captures the true population mean difference. To use this interval for hypothesis testing, we would ask whether the interval contains zero, which is the null hypothesis' value for the difference in population means. This interval does not contain zero, so we would conclude that there is a significant difference in the means for how long the babies cried, depending on whether they were breast-fed or not. Knowing that the control group had a bigger mean than the treatment group, we can observe that breast-feeding resulted in significantly shorter crying times.

We are not going to show a numeric example of a confidence interval for the AWS t test. It would appear on the output for a statistical software program used to analyze the data. The differences are in the computation of the margin of error. It would take a t critical value based on the AWS t test's big, ugly formula for df, and it would multiply the critical value by the AWS t test's denominator. That denominator is a different computation for the estimated standard error of the mean difference than the one in the denominator of the independent-samples t test. The confidence interval would be interpreted exactly like the above example.

An important note about the confidence intervals in this chapter: these intervals are estimating the *mean difference* in the population. In Chapter 10, we showed you a graph of group means, with error bars representing the interval estimate *for one mean at a time*, not the mean difference. Such graphs often are misinterpreted. It is possible for the separate confidence intervals, each estimating a different population mean, to overlap. Some people incorrectly interpret this overlap as meaning the two means are within sampling variability of each other. In fact, the *mean difference* could be statistically significant. That is, the confidence interval for the mean difference might not bracket zero. A *difference* in two means should not be interpreted by looking at each mean's separate confidence interval.

Check Your Understanding

SCENARIO 11-F, Continued

(Inspired by Stephens et al., 2011. Details of this scenario differ from the actual research.) This was the study of pain tolerance for people who were told to swear versus people who were told not to swear. We timed how long each person was able to tolerate keeping a hand submerged in icy water. For the swearing group, $M = 160.2$ seconds ($SD = 90.5$), and for the participants who did not swear, $M = 119.3$ seconds ($SD = 81.6$). Suppose we compute a 95% confidence interval to estimate whether the mean difference is greater than zero, which is a directional prediction. The sample mean difference is computed as $M_{swearing} - M_{nonswearing}$. The lower limit of the confidence interval estimating the population mean difference is 6.55. 11-19. What does this information tell us?

(Continued)

Suggested Answers

11-19. With the mean difference being computed such that the mean for the swearing group came first in the subtraction, we would have a positive difference because the swearing group kept the hand in the icy water for a greater length of time. We need to know whether the mean difference is greater than zero. Because the lower limit of the confidence interval is greater than zero, the interval estimate is above zero. We can say that the mean pain tolerance for the swearing group is significantly greater than the mean pain tolerance for the nonswearing group. Across repeated samples, 95% of the time that we would compute a confidence interval like this one, the true difference in the population means would be captured. Chances are, the true population mean difference is greater than zero.

Limitations on Using the *t* Statistics in This Chapter

As this book progresses, we are encountering statistics that are more realistic than the first inferential statistic that you learned, the *z* test statistic. We rarely know the numeric values of population parameters, and in this chapter we have covered statistics that did not require us to know any such numbers. With the paired *t* test, we were interested in a difference in paired means. The independent-samples *t* test and the AWS *t* test both looked at the difference in two independent means. With all of these statistics, our hypotheses concerned the difference in two means, so we did not need to know where those population means were located on the number line.

Many studies in the health sciences are designed to combine the elements of independent groups (which we used with the independent-samples *t* test) and repeated measures (as we saw in one example of the paired *t* test). But when a study has both independent groups and repeated measures at the same time, these test statistics will be insufficient for analyzing the data. For example, in Chapter 1 we described a quasi-experimental study in which the researcher studied the effect of life collages on nurses' knowledge of the lives of residents in two nursing homes (Buron, 2010). The researcher could not randomly assign nurses to seeing only residents with collages or without collages, so he obtained the cooperation of two similar nursing homes. Then he manipulated an independent variable: whether collages were placed in the nursing home or not. The lack of randomization and the presence of a manipulated independent variable meant that the study was a quasi-experiment.

So far, this study may sound like a situation in which an independent-samples *t* test or an AWS *t* test might be used. There are two independent groups, and the researcher is interested in a difference in the means for the outcome variable, knowledge of the residents' lives. But the effect of the collages would take time. The nurses would not have greater knowledge of the residents at the instant that the collages were introduced; as they go through their busy work days, they might have a moment or two to read the collages and talk to the different residents about their lives. Over time, their knowledge may increase. In the other nursing home, where collages were not used, the nurses naturally may learn more about the residents over time. Change across time could be examined with a paired *t* test—if we had only one group. Here, we have two groups, and the researchers would have to measure each group on at least two occasions in time to observe change in the nurses' knowledge of the residents' lives.

The real question would be whether the change across time for the nurses exposed to the life collages would be the same as the change across time that the nurses naturally experienced in the nursing home without life collages. We are talking about a study that would have four means computed on the knowledge variable; that is, both groups of nurses would be measured before the intervention began, providing two means for the knowledge variable. Then the collages would be introduced at one of the nursing homes. At a later point in time, both groups of nurses would have their knowledge of the residents' lives measured again, providing two more means. Figure 11.4 shows the design of this study.

Each box in this schematic would represent a group of nurses at a particular point in time when knowledge of the residents (and many other variables) would be measured and a sample mean for knowledge would be computed. If knowledge were measured on more occasions in the future, then more columns would be added to the schematic. The researcher's main interest was whether the change across time for one group differed from the change across time for the other group. None of the statistics in this chapter can answer that question. We are showing you this design because it is common in the health sciences, but the test statistics to analyze the data from this design are beyond the scope of this book.

The choice of an inferential statistic depends on the researcher's hypotheses. We have met researchers who seem to fall in love with a particular way of analyzing data, and they will say, "I want to do a multiple regression study." But that is

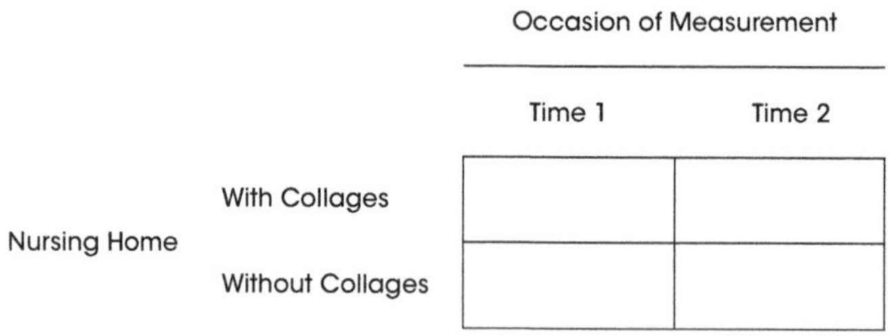

	Occasion of Measurement	
	Time 1	Time 2
With Collages		
Without Collages		

Nursing Home

Figure 11.4

Design of the nursing home study by Buron (2010). It is possible to have a study that combines independent groups (like the ones in the examples of the independent-samples *t* test) and repeated measures (one of the ways that the paired *t* test was used). This study had two independent groups and two occasions of measurement. But none of the statistics in this chapter is appropriate for analyzing the data from this study.

11. Two-Sample Tests and Estimates

like saying, "I want to use my table saw to build something." We decide what we want to build before we choose the tools that we will use to build it. We learned in Chapter 1 that we encounter a problem or question, we investigate what other researchers have learned, and we point to an unexplored area and ask, "What's over there?" Based on our understanding of prior research, we formulate hypotheses. Then those hypotheses determine which statistic should be used. Think back to the study of breast-feeding babies during injections. The researchers did not measure the crying times of the babies during two different visits. They did not have a hypothesis about whether the babies who were breast-fed during one immunization cried less when they returned at a later date for a second immunization, so the researchers did not need any statistics for repeated measures.

Figuring out which statistic to use in a given research scenario is one of the hardest tasks for students to learn. Chapter 15 will provide some instruction on how to make that decision. The exercises at the end of this chapter will give you some practice on choosing among the t tests that we have covered in this chapter.

What's Next

The independent-samples t test and the AWS t test allow researchers to compare two independent means, such as we might find with a treatment group and a control group. But what if the study needed more than two groups, such as the comparison of two different treatments and a control group? The statistics in Chapter 12 will allow us to compare the means for two or more independent groups.

Exercises

11-20. Describe three ways of obtaining pairs of scores for studies in which the paired t test might be an appropriate way to analyze the data. 11-21. Explain how to compute df for the paired t test in each of those situations.

SCENARIO 11-F, Revisited

(Inspired by Stephens et al., 2011. Details of this scenario may differ from the actual research.) In the chapter, we described a study in which we randomly assigned people to two groups, then we manipulated an independent variable: the instructions given to the participants who were asked to hold a hand in icy water as long as they could. One group was instructed to say swear words while experiencing the painful stimulus, and the other group was instructed not to swear. In truth, the researchers who conducted this study performed repeated measures; the participants were measured twice. Before the pain tolerance part of the study, everyone was asked to list five words they would say if they experienced something painful, like hitting their thumb with a hammer. The first swear word on their list became the swear word that they would say during one exposure to the icy water; one

(Continued)

person did not list a swear word and was excluded from the study. The participants also were asked to list five words that described a table. One of the table words was used in the nonswearing condition. They held a hand in room-temperature water between conditions. (The researchers also compared males and females, so they had two independent groups each measured twice, a study design that involves statistics beyond the scope of this book.) 11-22. Why might the researchers have decided to measure the same participants under both conditions, instead of randomly assigning participants to groups? 11-23. The researchers reported, "Condition order was randomized across participants." If everyone had two occasions of exposing a hand to icy water, once while swearing and once without swearing, why would it matter if the swearing condition came first for everyone?

SCENARIO 11-G

The following questions will be linked by the details of a briefly explained research situation, with the details to be revealed as needed. 11-24. Suppose we plan a study in which we are going to analyze pretest–posttest data using the paired *t* test. In our study, higher numbers on the outcome variable are better than lower numbers. We have an alternative hypothesis that said H_1: $\mu_{pretest} - \mu_{posttest} < 0$. Rewrite the alternative hypothesis in five equivalent ways. 11-25. We have determined that the pretest–posttest study requires a sample size of $N = 56$. This sample size gives the study sufficient power to detect an effect that we have identified as a minimum clinically noteworthy difference. Using $\alpha = .05$, look up the information we would need in Table B.1 to use the critical value decision rule to test the null hypothesis. 11-26. Suppose we have computed a paired *t* test for our pretest–posttest study. The result is a paired *t* = -3.87. Including your answers to the previous question, do we have enough information to test the null hypothesis? 11-27. Suppose the mean for the pretest scores was 48, and the mean of the posttest scores was 34. Test the null hypothesis. Be sure to explain how you reached your decision and what the decision means. 11-28. Let's say that we computed a 95% confidence interval to go along with the paired *t* test. Speaking in general terms, explain what such a confidence interval would estimate and how it would be interpreted. 11-29. Suppose the 95% confidence interval that we computed is [−18.9, −9.1]. Explain the meaning of this confidence interval as completely as you can.

SCENARIO 11-H

We are running a study in which we have randomly assigned participants to two groups. Participants in one group receive an injection in a room with soft music playing quietly, and the people in the other group receive an injection in an identical room with no music playing. We ask everyone to rate their pain. We want to know whether the presence of music causes any difference in the mean pain ratings for the two groups. 11-30. Name two test statistics that might be used in this scenario to compare the means,

(Continued)

explaining the circumstances in which we would use each statistic. 11-31. Write the alternative hypothesis in at least two ways, then write the null hypothesis that would correspond with each H_1. 11-32. Suppose we needed 74 participants (37 in each group) to give us enough power to detect the difference in mean pain ratings that we have identified as clinically meaningful. Further, we have chosen a significance level of .01. Compute df, then look up the information in Table B.1 that you would need to test the null hypothesis. 11-33. Why is it a good idea to have equal sample sizes? In other words, besides the effect of sample size on power, is there another reason that we would want samples this large? 11-34. Suppose your first author was a volunteer in this study, but when she found out that it involved needles, her phobia made her freak out and run screaming from the building. She refuses to rejoin the study. In terms of the statistics, what are the consequences of her departure? 11-35. We run the study with the remaining subjects, and we find the following information: $M_{music} = 4.4$ ($SD = 1.1$), $M_{no\ music} = 4.8$ ($SD = 1.3$), AWS t test $= 1.34$, 99% confidence interval of $[-1.2, 2.0]$. Explain these results, as well as the meaning in terms of the variable names.

SCENARIO 11-I

When infants are born preterm, mothers sometimes leave the hospital before their babies can be discharged. As a result, hospital nurseries use bottles to feed these babies. Babies exposed to both feeding methods sometimes show a preference for bottle-feeding and may reject breast-feeding, the method that many believe is more beneficial for child development. Further, mothers of preterm babies sometimes have trouble initiating breast-feeding. Abouelfettoh, Dowling, Dabash, Elguindy, and Seoud (2008) were interested in ways that preterm babies could be encouraged to breast-feed. They conducted a study in Egypt, where bottle-feeding is common in hospitals, to investigate the possibility of using a different alternative to breast-feeding. Cup-feeding involves providing breast milk to the baby in a cup that does not have a lip. The cup is thought to be different enough from breast-feeding that it would not become the baby's preferred feeding method (although it carries some risks and should be used only under supervision of a health-care professional). According to the Centers for Disease Control and Prevention, a baby is considered full-term when born at 37–42 weeks gestation. The study involved babies who were born at 34–37 weeks gestation, so these babies were considered late preterm. The researchers wanted to compare babies who were cup-fed in the hospital with babies who were bottle-fed in the hospital. The study began with the bottle-fed babies, which was the standard protocol for feeding preterm infants in the hospital where the study was conducted. Thirty babies were enrolled in this group. The next 30 preterm babies whose mothers agreed to join the study received cup-feeding. The intervention was implemented while the babies were in the hospital—but we will focus on an outcome measure after the babies were discharged from the hospital. The Premature Infant Breastfeeding Behavior Scale (PIBBS) is

(Continued)

used to measure the maturity of the baby's breast-feeding behavior. Babies whose behavior is considered more mature will latch on and suckle for longer periods, among other things. Higher numbers on the PIBBS indicate more mature breast-feeding behavior. During a follow-up visit 1 week after discharge, one of the researchers interviewed the mothers and observed the babies while they were breast-fed. One week after discharge, did babies who were cup-fed in the hospital have a higher mean PIBBS score than babies who were bottle-fed in the hospital? 11-36. What kind of research is this, and how do you know? 11-37. What kind of variable is the feeding method while the babies were in the hospital? 11-38. What kind of variable is the PIBBS score? 11-39. What kind of variable is the age of the babies on discharge from the hospital? 11-40. Are the researchers interested in change across time? 11-41. Mothers and babies exist in pairs in this study, but are mothers and babies being measured on the same variable? That is, are there two scores for each pair? 11-42. Following logically from the last question, is there any indication of difference scores being computed for the mother-baby pairs? 11-43. Are the babies who were cup-fed in the hospital paired in any way with the babies who were bottle-fed in the hospital? 11-44. What implication do your answers to the previous questions have on the choice of test statistic for comparing the two groups on their mean PIBBS scores? 11-45. Which two statistics should be considered for testing a hypothesis about the mean PIBBS scores for cup-fed versus bottle-fed babies? 11-46. What detail in the scenario will help you to decide which statistic should be computed? Thus, which statistic will you choose? 11-47. Looking at other measured variables, we see that the cup-fed babies were discharged at an average gestational age of 37.2 weeks $(SD = 0.9)$, compared with the bottle-fed babies being discharged at an average gestational age of 38.1 weeks $(SD = 1.2)$. The researchers reported $t = 3.16$, $p < .01$. What kind of t test was computed, and what would be the numeric value of its numerator? 11-48. Explain these results. 11-49. The study also reported the mean birth weights for the two groups: $M = 2267$ g $(SD = 319)$ for the cup-fed babies and $M = 2033$ g $(SD = 329)$ for the bottle-fed babies, $t = 2.78$, $p < .01$. Which t test was computed, and how would you interpret these results? 11-50. How might you expect age at discharge to affect PIBBS scores for the two groups? 11-51. How might you expect birth weight to affect PIBBS scores for the two groups? 11-52. Can we say whether the feeding method is causally responsible for the difference in mean PIBBS scores for the two groups? In other words, what can you say about the internal validity of this study?

SCENARIO 11-J

When medical students are unaccustomed to being around older adults, they may feel uncomfortable treating people with dementia. Dementia is a general term for a number of conditions that involve a decline in mental abilities to the point of impairing an older person's ability to perform daily tasks. Research has found that when medical students see older

(Continued)

adults in a clinical setting only, they can develop negative attitudes about the elderly. George, Stuckey, and Whitehead (2013) wanted to find out whether nonclinical interactions between medical students and people with dementia would improve the students' levels of comfort and knowledge about dementia. After completing their regular course work, students in a medical school in the eastern United States had to participate in a 4-week humanities elective. The students were enrolled in a course called Creativity and Narrative in Aging. They visited a residential home for people with Alzheimer's disease and related dementias (ADRD), and a number of residents joined a group of students in a storytelling session. The sessions used a program called TimeSlips®, which does not rely on people's memories. Instead, people are shown photos of people and animals that are intended to stimulate the creation of a story. Group facilitators ask questions such as, "What do you think is going on in the picture?" and another person writes down anything said by members of the group. Together, the group puts together a unique narrative about the image. The researchers wrote, "The activity is intended to help persons with ADRD exercise their imaginations—even in the face of memory loss and disorientation. In doing so, TimeSlips sessions underscore the inherent dignity of persons with ADRD by creating a valued social role for them and engendering playful yet substantive interaction" (p. 838). All 22 students enrolled in the humanities course agreed to participate in the study. The main outcome measure was the Dementia Attitudes Scale. Respondents rate 20 statements using a scale from 1 (*strongly disagree*) to 7 (*strongly agree*), then their ratings were summed to create scores. The 20 statements included, "People with ADRD can be creative," and "Difficult behaviors may be a form of communication for people with ADRD." The scale provides two scores: knowledge and comfort. Those with higher scores on the knowledge subscale know more about how dementia affects people, and those with higher scores on comfort are more at ease about interacting with people affected by dementia. The medical students were measured before they were introduced to the residents, then again after they had completed the TimeSlips sessions. Would the attitudes of medical students toward people with dementia improve on average as a result of participating in the class? 11-53. How many groups of participants are described in this scenario? 11-54. In general, how can a researcher measure improvement in anything? 11-55. Based on your answer to the last two questions, why is the paired *t* test the correct inferential statistic for this scenario? 11-56. Write the alternative hypothesis for the paired *t* test on comfort about people with dementias. 11-57. Now write your alternative hypothesis in another way. 11-58. Write your alternative hypothesis in a third way. 11-59. Although we will not use it below, let's practice finding a critical value. Compute *df* for the paired *t* test and, using $\alpha = .05$, find the critical value in Table B.1. 11-60. The researchers reported that the students' comfort scores had a mean of 44.2 (*SD* = 7.4) before the class and a mean of

(*Continued*)

54.9 (*SD* = 5.8) after the class. The journal article reported $p < .001$ for the paired *t* test. Test the null hypothesis, explaining how you reached your decision and the meaning of the results. 11-61. Explain what we say about the TimeSlips program and the students' attitudes.

SCENARIO 11-K

In Chapter 9, we talked about research involving glucosamine (Wilkens, Scheel, Grundnes, Hellum, & Storheim, 2010). The researchers randomly assigned 250 participants with low-back pain to a treatment group or a control group. Both groups took identical-appearing pills. The treatment group's pills contained glucosamine, and the control group's pills were a placebo. The researchers measured both groups of participants at baseline on a large number of variables—age, smoking status, duration of low-back pain, severity of pain, and health-related quality of life. They wanted to make sure the groups were comparable before the intervention began. They chose a significance level of .05 for two-tailed tests. 11-62. How many groups were in this study? Is there any mention of pairs of participants? 11-63. How does the information about sample size help us as we are figuring out which statistic should be used? 11-64. Choose a test statistic and write the alternative hypothesis for the patients' baseline ratings of severity of pain. 11-65. Write the alternative hypothesis for the patients' baseline health-related quality of life. 11-66. The researchers reported that for baseline pain ratings, where a higher number meant more pain, the treatment group had $M = 9.2$ (*SD* = 3.9), and the control group had $M = 9.7$ (*SD* = 4.5). They also said the inferential statistic comparing the means had $p = .37$. Test the null hypothesis using $\alpha = .05$ and explain all of these statistics. 11-67. For the health-related quality of life, the researchers reported $M = 5.8$ (*SD* = 2.2) for the treatment group and $M = 6.4$ (*SD* = 2.0) for the control group, and for the inferential statistic, $p = .02$. Test the null hypothesis using $\alpha = .05$ and explain all of these statistics. 11-68. How does this study mirror the research on life collages in nursing homes, which was described in the section "Limitations on Using the *t* Statistics in This Chapter"? Therefore, why are the statistics in this chapter insufficient for the job of analyzing the effect of glucosamine?

SCENARIO 11-C, Revisited

This was the study by Schell et al. (2010), who wanted to know whether the average blood pressure reading would be the same on the upper arm and the forearm. The results of the study could help nurses to know whether blood pressure readings taken on the forearm are interchangeable with readings taken on the upper arm. The forearm sometimes must be used because of intravenous catheters, injuries, or other health reasons. Readings were taken while the patients were lying flat (supine) and while they were lying with the HOB inclined to 30°. For $N = 70$ patients, systolic and diastolic blood pressure was recorded along with mean arterial

(Continued)

pressure (MAP), which is an average blood pressure over several heart beats. If a patient has a low MAP, the vital organs might not be receiving enough blood flow. If a patient has a high MAP, there is a risk of too much pressure being placed on the brain. The journal article about this research gave the following information: "Paired t test results revealed statistically significant differences between upper-arm and forearm systolic ($t = -5.55$, $p < .0001$ supine; $t = -10.16$, $p < .0001$ HOB 30°), diastolic ($t = -3.48$, $p = .0009$ supine; $t = -7.6$, $p < .0001$ HOB 30°), and MAP ($t = -5.33$, $p < .0001$ supine; $t = -10.6$, $p < .0001$ HOB 30°)." This quotation may look insanely complicated to you. That is how journal articles often are written. But we are going to help you to decipher it. For every paired t test, we will use a null hypothesis that says our sample comes from a population in which the mean for a forearm measure equals the mean for an upper-arm measure. Like these researchers, we will use $\alpha = .05$. 11–69. The quotation says one of the paired t tests $= -5.55$. Notice that this result immediately follows the phrase "significant differences between upper-arm and forearm systolic." So the systolic blood pressure, which is the top number in a blood pressure reading, was the outcome measure that went into the computation of the paired t test $= -5.55$. This $t = -5.55$ is given before the p value, which is followed by the word "supine." Now we can be more specific: with this paired $t = -5.55$, the researchers are comparing the mean systolic blood pressure on the upper arm versus the mean systolic blood pressure on the forearm, with both measures taken while the patients were lying flat. Test the null hypothesis and explain what your decision means. 11-70. What additional information do you need to fully explain the meaning of the paired $t = -5.55$? 11-71. Let's take the next piece of the quotation: $t = -10.16$, $p < .0001$ HOB 30°. Look at where this information appears in the quotation. Which two means are being compared? 11–72. Test the null hypothesis and explain the meaning of your decision. 11-73. The next part of the quotation says "diastolic ($t = -3.48$, $p = .0009$ supine …)" Which two means are being compared? 11-74. A table in the journal article reported that for the readings taken while the patients were supine, the upper-arm diastolic $M = 53.46$ ($SD = 12.07$), and for the forearm diastolic, $M = 56.71$ ($SD = 13.41$). Using information in the previous question, test the null hypothesis, then explain the meaning of the decision. 11-75. The table of results in the article says that when the patients were lying with the HOB inclined at 30°, the upper-arm diastolic blood pressure had a mean $= 55.5$ ($SD = 12.84$), and the forearm had a mean $= 61.71$ ($SD = 13.29$). The paired t test was equal to -7.6. How was the subtraction performed for the numerator of this paired t test? 11-76. Let's use the critical value decision rule to test the null hypothesis for the diastolic blood pressure readings with the patient inclined. Use Table B.1, $\alpha = .05$, and $N = 70$ to look up the information you will need. 11-77. Test the null hypothesis using the critical value decision rule, then explain the meaning of your decision. 11-78. Now explain the rest of the quotation, referring to the paired t tests for MAP.

(Continued)

(Your instructor may wish to assign questions similar to these exercises using the scenario of the tai chi-fibromyalgia study by Wang et al., 2010, which was described in Chapters 1 and 4. The two groups could be compared at baseline on the following variables: Fibromyalgia Impact Questionnaire, Pittsburgh Sleep Quality Index, the mental component of the SF-36, and the physical component of the SF-36. One group's change between two occasions in time also could be analyzed using a paired *t* test. The data set and information about the variables are available via http://desheastats.com.)

References

Abouelfettoh, A. M., Dowling, D. A., Dabash, S. A., Elguindy, S. R., & Seoud, I. A. (2008). Cup versus bottle feeding for hospitalized late preterm infants in Egypt: A quasi-experimental study. *International Breastfeeding Journal, 3,* 1–11. doi:10.1186/1746-4358-3-27

Bauman, W. A., Spungen, A. M., Wang, J., Pierson, R. N., Jr., & Schwartz, E. (1999). Continuous loss of bone during chronic immobilization: A monozygotic twin study. *Osteoporosis International, 10,* 123–127. doi:10.1007/s001980050206

Bloom, M. S., Schisterman, E. F., & Hediger, M. L. (2007). The use and misuse of matching in case-control studies: The example of polycystic ovary syndrome. *Fertility and Sterility, 88,* 707–710. doi:10.1016/j.fertnstert.2006.11.125

Buron, B. (2010). Life history collages: Effects on nursing home staff caring for residents with dementia. *Journal of Gerontological Nursing, 36,* 38–48. doi:10.3928/00989134-20100602-01

Efe, E., & Özer, Z. C. (2007). The use of breast-feeding for pain relief during neonatal immunization injections. *Applied Nursing Research, 20,* 10–16. doi:10.1016/j.apnr.2005.10.005

Fayers, T., Morris, D. S., & Dolman, P. J. (2010). Vibration-assisted anesthesia in eyelid surgery. *Ophthalmology, 117,* 1453–1457. doi:10.1016/j.ophtha.2009.11.025

George, D. R., Stuckey, H. L., & Whitehead, M. M. (2013). An arts-based intervention at a nursing home to improve medical students' attitudes toward persons with dementia. *Academic Medicine, 88,* 837–842. doi:10.1097/ACM.0b013e31828fa773

Grant, A. M., & Hofmann, D. A. (2011). It's not all about me: Motivating hand hygiene among health care professionals by focusing on patients. *Psychological Science, 22,* 1494–1499. doi:10.1177/0956797611419172

Hoffman, P. K., Meier, B. P., & Council, J. R. (2002). A comparison of chronic pain between an urban and rural population. *Journal of Community Health Nursing, 19,* 213–224.

Schell, K., Morse, K., & Waterhouse, J. K. (2010). Forearm and upper-arm oscillometric blood pressure comparison in acutely ill adults. *Western Journal of Nursing, 32*, 322–340. doi:10.1177/0193945909351887

Sowell, E. R., Thompson, P. M., Leonard, C. M., Welcome, S. E., Kan, E., & Toga, A. W. (2004). Longitudinal mapping of cortical thickness and brain growth in normal children. *Journal of Neuroscience, 24*, 8223–8231. doi:10.1038/13154

Stephens, R., Atkins, J., & Kingston, A. (2009). Swearing as a response to pain. *NeuroReport, 20*, 1056–1060. doi:10.1097/WNR.0b013e32832e64b1

Wang, C., Schmid, C. H., Rones, R., Kalish, R., Yinh, J., Goldenberg, D. L., ... McAlindon, T. (2010). A randomized trial of tai chi for fibromyalgia. *New England Journal of Medicine, 363*, 743–754. doi:10.1056/NEJMoa0912611

Wilkens, P., Scheel, I. B., Grundnes, O., Hellum, C., & Storheim, K. (2010). Effect of glucosamine on pain-related disability in patients with chronic low back pain and degenerative lumbar osteoarthritis. *Journal of the American Medical Association, 304*, 45–52. doi:10.1001/jama.2010.893

12

Tests and Estimates for Two or More Samples

Introduction

At various points in a statistics course, students sometimes lose sight of the "big picture"—how did we get to where we are now? Let's take a look at the test statistics and interval estimates we have covered so far and how they have led us to the current topic:

- The *z* test statistic allowed us to compare one sample mean to one population mean using a hypothesized numeric value for μ from the null hypothesis. A critical value from the standard normal distribution was used to test the null hypothesis and compute an interval estimate of the population mean. The z test statistic and its corresponding confidence interval required a hypothesized value for one parameter, μ, as well as knowledge of another parameter, the population standard deviation (or variance).

- Similar to the z test statistic, the one-sample t test was computed using the sample mean and a hypothesized value of a population mean from the null hypothesis. But unlike the z test statistic, the one-sample t test did not require knowledge of the population standard deviation (or variance); instead, it used the sample's standard deviation (SD). Our hypothesis test and our interval estimate of the population mean used a t critical value.

- We learned that we can focus on the *difference* in two means, which frees us from having to know numeric values of population means. If the null hypothesis says two population means are equal, then their difference is zero, regardless of where the population means are located on the number line. Three test statistics and their associated confidence intervals involved a difference in two means. The paired t test was computed on difference scores and had a strong computational link to the one-sample t test, and a confidence interval was computed for the difference in paired means. For two independent groups, we had two test statistics to choose from: the independent-samples t test and the AWS t test. Each of these statistics had an associated confidence interval for the mean difference. The AWS t test was needed because of a potential weakness in the independent-samples t test, which led to our discussion of robustness. All of these two-sample t statistics and confidence intervals used t critical values.

Now we have reached tests and estimates for two or more samples. Although the main statistic in this chapter could be used to compare two independent means, we are focusing on situations where we have more than two independent samples that are measured on a single occasion. If we drew a schematic showing the kind of study we are talking about, it might look like Figure 12.1, in which the boxes represent three independent groups measured on one occasion. The schematic has three boxes representing three groups or conditions: Treatment A, Treatment B, and Placebo.

Researchers in the health sciences do not use the study design shown in Figure 12.1 very often because they usually are interested in change across time for different groups. To measure change across time, we have to measure participants on at least two occasions. If we had a second occasion of measurement, we would need a second box for each group (a total of six boxes). Some studies ask

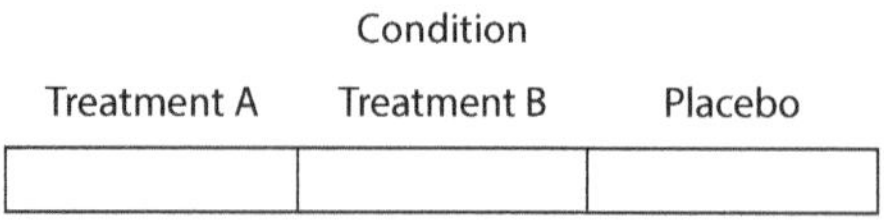

Condition

Treatment A	Treatment B	Placebo

Figure 12.1

This schematic shows one way of representing a study with three independent groups.

questions about the point at which a treatment loses effectiveness, which would require even more occasions of measurement. As you can see, studies can get quite complicated, depending on the research questions being asked.

Even though the main statistic in this chapter appears less frequently in the health sciences literature, it is foundational for understanding a family of statistics known as *analysis of variance*. This term does not refer to only one statistic because there are many tests and estimates within the family called the analysis of variance. We cover the most basic form of analysis of variance: the one-way analysis of variance *F* test.

Going Beyond the Independent-Samples *t* Test

For most of this book, we have tried to use actual research to illustrate the concepts being presented. We think real examples show the relevance of the statistics. Finding good examples from health sciences research can be difficult, especially because neither of your authors is a health-care professional. We have seen scientific journal articles that we could not use in this book because we did not understand the applied content. For example, one of the statistics in this chapter was cited in an article entitled, "Angiotensin receptor regulates cardiac hypertrophy and transforming growth factor-beta 1 expression" (Everett, Tufro-McReddie, Fisher, and Gomez, 1994). If we cannot understand the title, we doubt our ability to understand the content well enough to explain it accurately and succinctly. So let's return to an example used in Chapter 11, then imagine a way to extend the research as an illustration of this chapter's statistics.

Efe and Özer (2007) wanted to know whether the average crying time would be shorter for babies who were breast-feeding when they were given a routine injection, compared with the average crying time for babies receiving the usual care. The researchers randomly assigned babies to groups. The control group received the usual care, which involved being wrapped in a blanket, being placed on a padded examination table, receiving the injection, then being cradled and verbally comforted by their mothers. The babies in the treatment group were breast-feeding when the injection was given, and the mothers were instructed to encourage their babies to resume breast-feeding after the shot. The mean crying time for the breast-feeding babies was half as long as the mean crying time for the babies in the control group. According to the results of an independent-samples *t* test, this difference in mean crying time was statistically significant.

Suppose after reading this article, we think: Would any form of feeding make a difference in mean crying time? Many babies are not exclusively breast-fed. In some cultures, it is common for babies to alternate between being breast-fed and bottle-fed using breast milk. Suppose we want to run a study like the one by Efe and Özer (2007), except we would have a third group: bottle-feeding of breast milk to infants receiving the shot. We could randomly assign babies to one of three groups, then manipulate an independent variable called soothing method. This independent variable would have three levels. (*Level* is a term we defined in Chapter 1 as different values of an independent variable. Soothing method is a variable because it can take on different values, which are categorical.) One level would be breast-feeding during the shot. A second level would be bottle-feeding during the shot, and a third level would be usual care (control). Will the soothing method lead to any difference in the mean crying time?

Obviously, we would have three means, and the independent-samples t test only can tell us whether there is a difference in two means. We might be tempted to use the independent-samples t test repeatedly to compare the mean crying times of two groups at a time. Specifically, we could compare the means for

- The breast-feeding babies versus the bottle-feeding babies
- The breast-feeding babies versus the babies in the control group
- The bottle-feeding babies versus the babies in the control group

The problem with using the independent-samples t test three times is that every time a hypothesis is tested, the researcher is risking a Type I error—that is, rejecting the null hypothesis when the null hypothesis is true in the population. The probability of a Type I error in a given hypothesis test is controlled by making α small, such as .05. If we repeatedly conducted the study and the null hypothesis was true in the population, our $\alpha = .05$ would mean that five times out of 100 studies, we still could get a significant result with an independent-samples t test. By setting α at .05, we are accepting this frequency of Type I errors. But if α is set at .05 for each of our three independent-samples t tests, then the total probability of making at least one Type I error is as much as three times .05 (i.e., as high as .15). To answer our research question (*Will the soothing method lead to any difference in the mean crying time?*), would we want to increase this probability by as much as threefold Probably not. Nor would we want to use three test statistics to answer our research question. How many of those independent-samples t tests would have to be significant? One? Two? All of them? We need a new test statistic.

Before we go into the details of the new statistic, let's talk about the different ways that the soothing method could lead to a difference in mean crying times:

- The breast-feeding babies on average could cry for a shorter time than the bottle-feeding or control-group babies. That is, we could have one mean different from the other means.
- The breast-feeding babies and the bottle-feeding babies on average could cry for a shorter time than the control-group babies. In other words, we again could have one mean different from the other means, but in a different pattern.
- The breast-feeding babies could have a shorter mean crying time than the bottle-feeding babies, who on average also could cry for a shorter time than the control-group babies. That is, all three means could differ from each other.

12. Tests and Estimates for Two or More Samples

The above list does not exhaust the possible patterns of results. We also could observe a pattern of means that contradicts our expectations. We could observe the following differences:

- The control-group babies on average could cry for a shorter time than the breast-feeding babies or the bottle-feeding babies.
- The control-group babies and bottle-feeding babies on average could cry for a shorter time, relative to the mean crying time for the breast-feeding babies.
- The babies in the control group could have the shortest mean crying time, followed by the bottle-feeding babies' mean, with the breast-feeding babies crying the longest on average.

In other words, there are many ways that three means could differ from each other, and we have described only a few of them. Our research question (*Will the soothing method lead to any difference in the mean crying time?*) did not specify one of these possible outcomes. Our research question is asking whether the three means will *vary*. In Chapter 11, we talked about detecting a difference in two independent means. Now that we have more than two means, we will talk about the means *varying*, and we will use one of the statistics that belong in the family called *analysis of variance*. This family of statistics shares the characteristic of examining how much variation exists in a group of means. In this chapter, we will cover only one example of analysis of variance: the one-way analysis of variance *F* test, which is used when we have one categorical independent variable and one quantitative outcome variable measured on one occasion.

Check Your Understanding

12-1. Suppose we added a fourth group to the study: babies who are bottle-fed, but the bottles contain manufactured formula instead of breast milk. List some of the ways that the means for the four conditions could vary.

Suggested Answers

12-1. One possibility is that all four means could differ from each other, but there are several ways that could happen. That is, the magnitude of the four group means could be in different orders. If we labeled the four sample means as A, B, C, and D, they could be in the order $A < B < C < D$, or $A < B < D < C$, or $A < C < B < D$, or $A < C < D < B$, and many other variations. Or one mean could differ from three equal means: $A < B = C = D$, or $A > B = C = D$, or $B < A = C = D$, or $B > A = C = D$ and so forth. In sum, any of these variations would be evidence that the soothing method led to a difference in mean crying time.

Variance Between Groups and Within Groups

The analysis of variance often is abbreviated as ANOVA (ANalysis Of VAriance, usually pronounced in America as "uh-NOH-vuh," although sometimes we hear "AN-oh-vuh"). Because statistics in this family share the same last name (ANOVA), we need to put a first name on any ANOVA or its associated statistics, called F tests. The main statistic that we will use in this chapter is called the one-way ANOVA F test. The term *one-way* means we have one independent variable, the soothing method. The one-way ANOVA F test also sometimes is called an independent-groups ANOVA F or a fixed-effects ANOVA F. *Fixed effects* is a term that has different definitions in different fields. You can think of the term as meaning that the levels of the independent variable can be replicated in another study, based on a specific definition of each level. In our present example, the independent variable is the soothing method. This independent variable could be responsible for differences in mean crying time in our three-group experiment, and the one-way ANOVA F test can tell us whether there is some difference in the means. This statistic is not limited to a study with three groups; we could have four, five, or more kinds of soothing methods in the study. We are using three levels for simplicity's sake. To understand the one-way ANOVA F test, we must know about two kinds of variation that can be observed in the data. Pay attention to the words *between* and *within*, which are crucial in the following discussion of variability. We will italicize the words *between* and *within* to draw your attention to the two kinds of variation that go into the one-way ANOVA F test.

Let's consider a different research example involving babies and mothers. Suppose we are interested in the amount of sleep that mothers of newborns get. Specifically, we want to run a study like the one by Montgomery-Downs, Clawges, and Santy (2010), who compared mothers who exclusively breast-fed their babies, mothers who exclusively bottle-fed formula to their babies, and mothers who used both feeding methods. The outcome variable is the amount of sleep that the mothers get in a 24-hour period. Is there a difference in the average amount of sleep that these mothers get? In other words, is the feeding method related to the mean amount of sleep? These researchers measured mothers once a week for several weeks, but in our imagined study, let's say that we measure the mothers during one 24-hour period when the babies are 2 weeks old. We use an unobtrusive device that provides an objective measure of the total number of minutes of sleep during the 24-hour period. Obviously, not everyone sleeps for the same amount every night, regardless of whether they have a newborn. Some people naturally need more sleep or less sleep than other people. When it comes to mothers of newborns, sometimes the mother is solely responsible for taking care of the baby, while other households have one or more additional caregivers, allowing the mothers to get more sleep. When we are observing mothers who bottle-feed, we can expect to see variability *within* this group in terms of the mothers' amounts of sleep. There also will be variability *within* each of the two other groups' sleep times. But will there be variability in the number of minutes of sleep *between* the groups, depending on the feeding method?

To illustrate this idea of the variation *between* the means and variability *within* the groups, we will describe three different scenarios. Each scenario represents a different possible outcome for the study of three groups of mothers: those who are breast-feeding exclusively, those who are bottle-feeding with formula exclusively, and those who are using both feeding methods. None of these results should be cited elsewhere because we used *fake data*. The only time that it is acceptable to fabricate data is to illustrate concepts in a book like this one, and even then it is important to point out that the data are not real. For real results on such a study, please see the research by Montgomery-Downs et al. (2010). Figure 12.2 shows the first possible outcome, which we labeled Scenario 12-A.

Figure 12.2 shows that all three groups of mothers have the same mean amount of sleep in 24 hours: $M = 300$ minutes, or 5 hours. There is no variability *between* the groups; they might as well have come from three populations with the same mean number of minutes of sleep. There is variability *within* the groups, meaning that not all mothers slept for the same amount of time. That is no surprise. In the breast-feeding group, some mothers got less sleep than their group average and other mothers slept more than their group average. Looking at the bottle-feeding group, we again see that some mothers slept less than their group average and others slept more than their sample mean. The same can be said

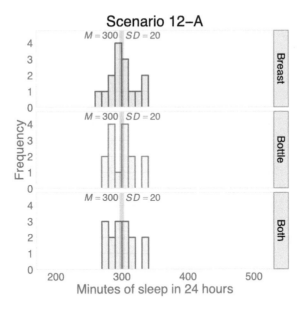

Figure 12.2

Three groups with equal means and tightly clustered scores within groups. We created some numbers to illustrate three possible scenarios for a study that compares the number of minutes of sleep in 24 hours by mothers who are breast-feeding exclusively, mothers who are bottle-feeding with formula exclusively, and mothers who use both feeding methods. We are focusing on how the groups compare on average, as well as the amount of variability within each group.

about the mothers using both feeding methods. In other words, these histograms show that there is variability *within* the groups. The three groups in Figure 12.2 have the same amount of spread because $SD = 20$ for every group. So Scenario 12-A depicts a situation in which there is no variability *between* the groups, but there is variability *within* the groups.

Now let's look at another set of fabricated data. Look at Figure 12.3 and ask yourself: How does Scenario 12-B differ from Scenario 12-A? In Scenario 12-B, we have one histogram that is shifted to the right on the number line. The mothers who used both feeding methods slept an average of $M = 400$ minutes (more than 6½ hours) during the 24-hour observation period. The two other groups appear the same as they were in Scenario 12-A, with $M = 300$. All three groups still have the same amount of spread in the number of minutes of sleep because $SD = 20$ for each sample. Scenario 12-B shows more variability *between* the groups than *within* the groups. As a result, we might suspect that there is a difference in the populations from which these samples were drawn. Perhaps the feeding method affects the average amount of sleep that the mothers receive.

Now let's look at Scenario 12-C, shown in Figure 12.4. How does Scenario 12-C differ from the previous two scenarios? Scenario 12-C is much different. Now each of the histograms is quite spread out. There is a great deal of variability *within* the groups. In fact, each group now has $SD = 130$. What about the variability *between* the groups? Notice that the means are the same in Scenario 12-C (Figure 12.4) as the means in Scenario 12-B (Figure 12.3). The variation *between*

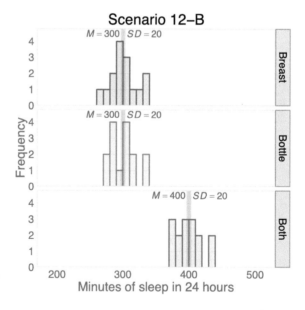

Figure 12.3

Three groups with unequal means and tightly clustered scores within groups. This graph looks similar to Figure 12.2, except now one of the groups is shifted to the right on the number line. There is variability between the means (fabricated data).

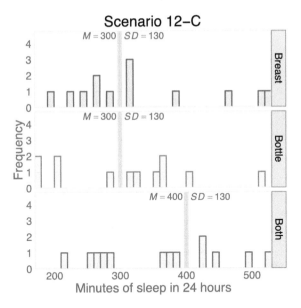

Figure 12.4

Three groups with unequal means and a great deal of variability within groups. This graph shows the same means that we saw in Scenario 12-B. But in Scenario 12-C, the variability within the groups obscures whether the means actually differ in a noteworthy way (fabricated data).

the groups seems less pronounced in Figure 12.4 because of the greater amount of variability *within* the groups. The three means appear to vary in a way that seems to be similar to the variation *within* the groups; it is almost as if the three means could be mistaken for individual scores. Is there a difference in the mean amount of sleep obtained by mothers who are exclusively breast-feeding, exclusively bottle-feeding with formula, or using both feeding methods? It is hard to say; the variation *within* the groups has clouded the situation.

The comparison of the variability *between* the groups and the variability *within* the groups reminds us of the discussion of signal and noise in Chapter 9. We described trying to tune in a radio station on an old-fashioned radio that has a knob. Between stations, there is noise or static. As we dial toward a station, a signal is heard through the noise. We mentioned a signal-to-noise ratio, in which a number is assigned for the signal and a number is assigned for the noise. If there is less signal than noise, then the signal-to-noise ratio is less than 1. If there is more signal than noise, then the signal-to-noise ratio is greater than 1. Let's apply this idea to the three scenarios, which now appear together in Figure 12.5.

In Scenario 12-A (the left column of histograms in Figure 12.5), the three means are identical and have no variation; there is no "signal" telling us that there is any difference in the means for the minutes of sleep in 24 hours for the

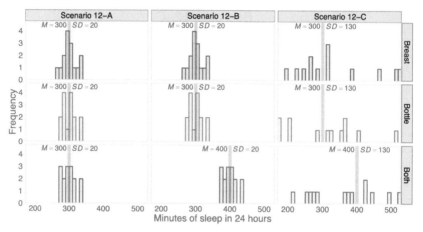

Figure 12.5

Variability between groups, variability within groups. The comparison of the variability between the groups with the variability within the groups is fundamental to this chapter's concepts. Be sure you understand why Scenario 12-C is quite different from Scenario 12-A and Scenario 12-B (fabricated data).

mothers in the three groups. The signal-to-noise ratio would be zero. In Scenario 12-B (the middle column of histograms in Figure 12.5), there appears to be more variability *between* the groups (signal) than *within* the groups (noise). That is, we seem to be receiving more signal than noise, so the signal-to-noise ratio would be greater than 1. In Scenario 12-C (the right column of histograms in Figure 12.5), there is a great deal of noise because the mothers *within* the groups vary so much in the amount of sleep they received. In fact, the signal (the variation *between* the groups) seems to be lost in all the noise (the variability *within* the groups). The signal-to-noise ratio might be somewhere close to 1.

These three scenarios demonstrate the fundamental concept underlying the specific analysis of variance that we are covering in this chapter: a comparison of the variation *between* the groups and the variation *within* the groups. In Chapter 11, you actually did this kind of comparison, but we did not explain it that way. Think back to the independent-samples t test. Its numerator contains a difference in means, which might be written as $M_{treatment} - M_{control}$. The difference between two sample means is like a measure of variation *between* the groups, except when we compute measures of variability, differences usually are squared. The denominator of the independent-samples t test is an estimated standard deviation, which is a measure of variability of the scores *within* the groups. If we squared this estimated standard deviation, we would get a variance. If the two means differed substantially from each other, then we could fit a lot of estimated standard deviations in the gap between them, leading to a statistically significant result. In other words, if the variation *between* the groups substantially exceeded

12. Tests and Estimates for Two or More Samples

the variation *within* the groups, then we would find a significant difference in the means.

When we switched to examples with more than two groups, we had to talk about *variability* of the three means instead of a *difference* between two means. If the feeding method has an effect on the mean amount of sleep, then there will be markedly more variability *between* the groups than *within* the groups. Next we extend the explanation of *between* versus *within* variability to explain the computation of the one-way ANOVA *F* test.

Check Your Understanding

SCENARIO 12-D

People in Japan generally believe gargling can help to prevent upper respiratory tract infections. Satomura et al. (2005) conducted a study to examine whether there would be any clinical evidence to support this idea. Suppose we are running a study similar to theirs, randomly assigning healthy adults to three groups. Then we give them different instructions. Participants in the first group are told to gargle with water at least three times a day. Those in the second group are given an antiseptic solution and are told to gargle with the solution at least three times a day. The members of the control group are told to maintain their usual habits with respect to gargling or not gargling. The study will last for 1 year, and we tell all participants to contact us if they experience symptoms of any kind of upper respiratory infection (cold, sore throat, etc.). When participants contact us for the first time to say they have contracted such an infection, we ask them to rate the severity of various symptoms using a scale from 0 (*not experiencing this symptom*) to 5 (*severe symptom that is interfering with daily life*). We think the gargling routine (water, antiseptic solution, or usual behavior) will lead to a difference in the mean severity of nasal symptoms. 12-2. Use this scenario to explain the concepts of *variability between groups* and *variability within groups*.

Suggested Answers

12-2. We naturally would expect people to differ in their experiences because some people will be sicker than others, even if they are in the same group. So there will be variability within each group in terms of their severity ratings. If the kind of gargling routine affects the severity of nasal symptoms during an upper respiratory infection, we would expect the mean severity ratings to vary for the three groups. The variation between the sample (group) means for the severity ratings might be greater than the variation in the severity ratings within the groups.

One-Way ANOVA F Test: Logic and Hypotheses

Let's return to our imagined example of extending the research on soothing babies. Efe and Özer (2007) compared babies receiving the usual comfort from their mothers with babies who were breast-feeding when the shot was given, with the main outcome measure being the number of seconds that the babies cried. Our imagined example would expand the study to include a third group: babies who are being bottle-fed breast milk. We want to know whether the soothing method has an effect on mean crying times. How do the variability between the groups and the variability within the groups help us to answer our research question? We will explain in the following summary of the logic of the one-way ANOVA F test:

1. We compute two estimates of variability. One estimate is based on the variability *between* the sample means (group means) and the other estimates the variability of scores *within* the groups.
2. Place the two estimates of variability into a ratio (or fraction) called the one-way ANOVA F test, with the between-groups estimate of variability as the numerator and the within-groups estimate of variability as the denominator.
3. If the variation between the sample (group) means is similar to the variation of scores within the groups, then the scores for the different samples may overlap considerably. The variability *between* the groups will be similar to the variability of scores *within* the groups. In this case, the one-way ANOVA F test will be something close to 1. (As a quick numeric example, if the estimate of variability between the groups is equal to 10, and the estimate of variability within the groups is equal to 10.5, then the one-way ANOVA F test would be $10/10.5 \approx 0.95$.)
4. As the variability between the sample means (group means) gets bigger than the variance within the groups, the one-way ANOVA F test will get bigger than 1. (Try demonstrating this idea to yourself by using a numerator of 18 for the variability between the sample means and a denominator of 10 for the within-groups variability.) As the one-way ANOVA F test gets bigger, we begin to suspect that there is some difference in the population means.
5. The more variability in the sample means (i.e., group means), the bigger the one-way ANOVA F test will get. At some point, the one-way ANOVA F test will exceed a critical value, and we will be able to say that there is a significant difference in the means. Another way to say the same thing is that the p value for the one-way ANOVA F test will be smaller than α, and we will reject a null hypothesis.

The logic implies certain hypotheses for the one-way ANOVA F test. Luckily for you, the way we write these hypotheses is much simpler than the hypotheses used with the t tests in Chapter 11. We were careful in the way we wrote our research question: will the soothing method lead to any difference in the mean crying time? This research question implies a certain alternative hypothesis:

$$H_1: \text{some difference in the } \mu\text{'s}$$

In words, we might say

Our samples come from populations in which
there is some difference in the mean crying times,
depending on whether the babies were soothed as usual,
soothed by breast-feeding, or soothed by bottle-feeding.

The one-way ANOVA F test will tell us whether there is *some difference*, not whether a particular group is different from the others. The wording of the alternative hypothesis ("some difference in the μ's") can be used with all one-way ANOVA F tests. Even if only one group differs from the others, there will be more between-groups variation than within-groups variation. Studies in which the between-groups variation greatly exceeds the within-groups variation will have large values of the one-way ANOVA F. What is nice about this way of writing the alternative hypothesis is that it will work no matter how many groups you have. If we added a fourth group with babies who were given a pacifier as the soothing method, we would write the alternative hypothesis the same way.

A statement opposite to the alternative hypothesis is the null hypothesis for the one-way ANOVA F test:

$$H_0: \text{no difference in the } \mu\text{'s}$$

This way of writing the null hypothesis will work no matter how many groups we have. For our current example, we might say

Our samples come from populations in which
there is no difference in the mean crying time,
regardless of which soothing method was used.

The null hypothesis is saying that the population means do not vary at all, which is linked to the idea of the variability between the groups, compared with the variability within the groups. This null hypothesis could be true and we still might get group (sample) means that are not exactly equal, but they would be close to each other. The null hypothesis is saying the population means are all equal, so we could write this null hypothesis as three equal population means, with a different subscript for each group:

$$H_0: \mu_{\text{breast-feeding}} = \mu_{\text{bottle-feeding}} = \mu_{\text{control}}$$

Equivalently, we could label the groups as A, B, and C, and specify that A = the breast-feeding condition, B = the bottle-feeding condition, and C = the control condition. Then we could use A, B, and C as the subscripts. We prefer to write the null hypothesis as "no difference in the μ's" so that there will be no temptation to write the alternative hypothesis with three μ's separated by "not equal to" symbols. That would be a problem because of the many ways that a difference may exist among the means.

Let's go back to the alternative hypothesis for a moment. Notice that H_1 is nondirectional. That is always the case with the one-way ANOVA F test because the statistic compares variation between the groups with the variation within the groups. In other words, any variation in the means will interest us. We cannot make directional predictions with the one-way ANOVA F test. Either there is no variation in the means, or there is some variation in the means. The greater the variation in the means, the larger the test statistic becomes. With previous test statistics, using a nondirectional alternative hypothesis meant that we would perform a two-tailed test. Interestingly, the one-way ANOVA F test is always one-tailed. Why? It goes back to the reliance on estimates of variability. Let's think about a sample statistic like SD, which measures the variability of scores in a sample. As you know, if the sample's scores do not vary (i.e., they are all the same), then the statistic SD would equal zero, indicating no variability. If one score differs from the others, then SD will be bigger than zero—a positive number. The more variability of scores around the mean, the bigger that SD gets. Now consider the one-way ANOVA F test, a ratio of two estimates of variability. If the sample means were equal for all the groups, then there would be no variability *between* the means, and the numerator of the one-way ANOVA F test would be zero, meaning the whole statistic would equal zero. But if there is any variability *between* the sample means for the different groups, the numerator of the one-way ANOVA F test will be a positive number. Any variation in the scores *within* the groups will be represented by a positive denominator. So as the differences between the groups increase, the variability between the means increases, and the one-way ANOVA F test goes in only one direction: it gets bigger. The greater the variation between the groups, the bigger the one-way ANOVA F test, so extreme results for the one-way ANOVA F tests can occur only in the positive tail. Next we will describe the computation of the one-way ANOVA F test, including its degrees of freedom and a table of critical values.

Check Your Understanding

SCENARIO 12-D, Continued

This scenario concerned the three gargling routines (gargling with water, gargling with the antiseptic solution, or usual behavior/control group). We thought the gargling routine might lead to differences in mean ratings of symptom severity when the participants contracted an upper respiratory infection. 12-3. Write the alternative hypothesis for this scenario, using words and symbols. 12-4. Write the null hypothesis for this scenario, using words and symbols.

(Continued)

12. Tests and Estimates for Two or More Samples

12-3. The alternative hypothesis is that our samples come from populations in which there is some difference in the means for the gargling conditions. This alternative hypothesis could be written as H_1: some difference in the µ's. In fact, it must be written only this way. (An example of a way that the alternative hypothesis cannot be written would be the following: the µ for the water-gargling condition is not equal to the µ for the antiseptic-gargling condition, which is not equal to the µ for the usual-habits condition.) We cannot be any more specific than "some difference in the µ's" because the one-way ANOVA F test is not specific. If the test statistic is significant, it will indicate there is some difference in the means. 12-4. The null hypothesis is that our samples come from populations in which the gargling conditions have equal means for the ratings of symptom severity. This statement can be written as H_0: no difference in the µ's.

Computing the One-Way ANOVA *F* Test

We already said that the one-way ANOVA *F* test has a numerator and a denominator, both of which are variances. The numerator itself is a fraction, and the denominator is another fraction, so the formula for the one-way ANOVA *F* test can start to look quite complicated. We will let statistical software perform the calculations; our explanation of the statistic should help you to understand the output from whatever software you may use.

Before we get into the details of the one-way ANOVA *F* test's numerator and denominator, let's review some ideas from Chapter 2. We went through the computation of two variances in that chapter, the sample variance and the unbiased variance. To measure the spread of scores around the mean, we first computed the distance of each score from the mean. We got rid of negative signs by squaring the distances. Then we added up the squared distances. The process of squaring numbers and adding them up was called the *sum of squares*. Both the sample variance and the unbiased variance have the same numerator: the sum of squared distances from the mean. The two variance statistics have different denominators: for the sample variance, the denominator is N, and for the unbiased variance, the denominator is $N - 1$. When we reached Chapter 10, we learned another way to define the denominator of the unbiased variance: it is the degrees of freedom, *df*, for the unbiased variance. So the unbiased variance is a ratio of the sum of squared differences from the mean, divided by its *df*.

Now we are talking about the one-way ANOVA *F* test, which is a ratio of two variances. As in previous chapters, *df* is needed to look up a critical value. Interestingly, *F* statistics contain two different *df* computations, and both are needed to look up one critical value. The idea of a sum of squares divided by degrees of freedom will be found in both the numerator and denominator of the one-way ANOVA F test. So we will need to use subscripts to distinguish the numerator's components from the denominator's components. Let's start with the numerator, which is a variance called the *mean square between*, or MS_B. Similar to the unbiased variance, MS_B is a ratio of sum of squares divided by degrees of freedom. The mean square between equals the sum of squares between (SS_B) divided by the degrees of freedom between (df_B):

$$MS_B = \frac{SS_B}{df_B}$$

The numerator sum of squares, SS_B, is an ugly formula, so we are not showing it to you. Essentially it is based on a measure of the variability of the sample (group) means around the *grand mean*, which is the mean of all the scores in the study. The formula for df_B is

$$df_B = \text{number of groups} - 1$$

In our imagined study of three soothing methods for babies receiving routine immunization shots, the numerator df would be equal to 2 (i.e., three groups − 1). The SS_B and df_B have other names: the *numerator* sum of squares and the *numerator df* because these two numbers are used to compute the numerator of the one-way ANOVA F test.

Now let's turn our attention to the denominator of the one-way ANOVA F test, which is called the *mean square within*, or MS_W. Sometimes this denominator is called the *mean square error*, or the *error term*. Similar to the numerator of the one-way ANOVA F test, the mean square within is a ratio of a sum of squares divided by degrees of freedom—but now we are talking about the sum of squares within (SS_W) and the degrees of freedom *within* (df_W):

$$MS_W = \frac{SS_W}{df_W}$$

The sum of squares formula is big and ugly, so we will not cover it. The formula for df_W is

$$df_w = \text{total sample size minus the number of groups}$$

$$df_w = N - \text{the number of groups}$$

Suppose we have a study with three groups, each containing $n = 32$ participants. That means $N = 96$ (i.e., 3 times 32), and df_W would be 93 (i.e., N − the number of groups = 96 − 3 = 93). The SS_W and df_W sometimes go by other

names: the *denominator* sum of squares and the *denominator* degrees of freedom because SS_W and df_W are used to compute the denominator of the one-way
ANOVA F test.

Putting the numerator and denominator together, the one-way ANOVA F
test is

$$\text{one-way ANOVA } F = \frac{MS_B}{MS_W}$$

The parts of the one-way ANOVA F test (SS_B, df_B, SS_W, df_W, MS_B, MS_W) appear
in the output for statistical software, which is why we wanted to explain these
details. After we explain the table of F critical values and the decision rules for
testing the null hypothesis, we will do a numeric example using the imagined
example of the three methods of soothing babies receiving an injection.

Check Your Understanding

SCENARIO 12-D, Continued

Continuing the scenario of the three groups being given different
instructions with respect to gargling, suppose we determine that 80 participants per group are needed to detect a small yet clinically noteworthy
effect of the gargling routine (water gargling, antiseptic gargling,
or usual habits). 12-5. Compute df_B and df_W for this scenario.

Suggested Answers

12-5. *The df_B equals the number of groups − 1, so $df_B = 2$. The df_W equals the
total sample size minus the number of groups. With three groups of 80 participants each, the total sample size is 240. So $df_W = 240 − 3 = 237$.*

Critical Values and Decision Rules

When we used the z test statistic, we used a critical value from the standard normal distribution. When we computed t test statistics, we used critical values from
t distributions that were defined by their degrees of freedom; each t distribution is slightly different, depending on the *df*. Now that we are computing one-
way ANOVA F tests, we need F critical values from F distributions. You already
have seen that the one-way ANOVA F statistic has two numbers for *df*. Those
two numbers define the shape of the particular F distribution needed for a given
scenario. To give you an idea of the different shapes that an F distribution can
have, Figure 12.6 shows two F distributions. One distribution has $df_B = 2$ and
$df_W = 30$ and the other distribution is $df_B = 5$ and $df_W = 96$; these *df* are shown in
parentheses above the F distributions.

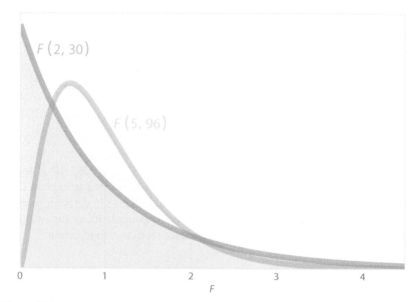

$F(2, 30)$

$F(5, 96)$

0   1   2   3   4

F

Figure 12.6

Two F distributions with different df. The one-way ANOVA F test has different distributions, depending on the two values of the degrees of freedom: df_B and df_W. The blue distribution has $df_B = 5$ and $df_W = 96$. The orange distribution has $df_B = 2$ and $df_W = 30$.

We can work backward from the df numbers to learn more about a study that would use each of these distributions. If a study has $df_B = 2$, then we have three groups because df_B is the number of groups minus 1 (i.e., $3 - 1 = 2$). If the same study has $df_W = 30$, then we have a total $N = 33$ because df_W is the total sample size minus the number of groups (i.e., $33 - 3 = 30$). So the study that would use the F distribution with $df_B = 2$ and $df_W = 30$ would have three groups with 11 subjects in each group, if the sample sizes were equal. The other distribution in Figure 12.6 has $df_B = 5$ and $df_W = 96$. The $df_B = 5$ tells us that this study would have six groups. The $df_W = 96$ is equal to $N - 6$, so the study has a total sample size of $N = 102$. If the groups are equal sizes, then there would be 17 subjects per group.

Similar to a variance statistic, the one-way ANOVA F test has values as small as zero, and the rest of the values are positive numbers. As a result, F distributions have a minimum possible value of zero, and all other possible values are positive. F distributions also are positively skewed, as shown in Figure 12.6. Remember all the squared differences that went into the computation of a variance? When numbers are squared, we get even bigger numbers:

$$2^2 = 4$$
$$3^2 = 9$$
$$4^2 = 16$$

As a ratio of variances, a one-way ANOVA F test can get big quickly as the variation between the group (or sample) means increases.

Table C.1 in the back of the book contains critical values from a large number of combinations of df_B and df_W. Figure 12.7 shows what Table C.1 looks like. Table C.1 is arranged differently from the table of critical values that we used in Chapter 11. To find an F critical value, we need to use both df_B and df_W. Look for "Numerator df" (which is df_B) at the top of the columns and "Denominator df" (which is df_W) as the row labels. There are different critical values for different combinations of the numerator df and denominator df—that is, the intersection of each column (numerator df) and each row (denominator df) gives a different critical value. Further, the table gives us two critical values for each combination of df_B and df_W: one critical value for $\alpha = .05$ and one critical value for $\alpha = .01$.

Table C.1 has a column for $df_B = 1$, which would mean there are two independent groups. Are you surprised? The one-way ANOVA F test can be used to compare the means of two groups, and it will lead to the same decision on the null hypothesis as the independent-samples t test will. Either test statistic can tell us whether two independent means differ significantly. When we covered the independent-samples t test, we talked about only one number for df (i.e., $df = n_1 + n_2 - 2$) because its numerator df always is 1 (i.e., the number of groups minus one is $2 - 1 = 1$). The study by Efe and Özer (2007) concerning the babies receiving injections had two groups (breast-feeding and usual care/control group) and reported the results of an independent-samples t test on the mean crying times of the two groups. These researchers could have used the same data to run a one-way ANOVA F test to test whether the means were equal. The independent-samples t test would be a different number from the one-way ANOVA F test, but the two test statistics have a relationship. When there are two groups, we can square the independent-samples t test (i.e., multiply it by itself), and we will get the one-way ANOVA F test. Efe and Özer reported an independent-samples t test $= 3.64$. If they had analyzed their data with a one-way ANOVA F test to compare the two means, the F test would have been approximately 13.25 (i.e., $t^2 = 3.64^2 \approx 13.25$).

Let's look at an example of finding a critical value in Table C.1. Suppose we have $df_B = 3$ and $df_W = 80$. (Can you use these numbers to figure out how many groups we would have and how many participants would be in each group?) To find a critical value, we find the column for the numerator $df = 3$, then we go down that column until we find the row for the denominator $df = 80$. If we are using $\alpha = .05$, the F critical value will be 2.72. If we are using $\alpha = .01$, the F critical value will be 4.04. Figure 12.8 shows the F distribution with $df_B = 3$ and $df_W = 80$. The critical value of $F = 4.04$ is shown with an orange vertical line that cuts off an orange tail area of $\alpha = .01$. The critical value of $F = 2.72$ has a red vertical line, cutting off a tail area of $\alpha = .05$, which is the entire tail area, including the red area and the orange area.

A one-way ANOVA F test can go in only one direction: getting bigger as one or more means increase in variation from the other means. As a result, the one-way ANOVA F test is a one-tailed, upper-tailed test, which makes the critical value decision rule straightforward:

Table C.1 Critical Values for *F* Distributions

Denominator df	α	1	2	3	4	5	6	7	8
11	.05	4.84	3.98	3.59	3.36	3.20	3.09	3.01	2.95
	.01	9.65	7.21	6.22	5.67	5.32	5.07	4.89	4.74
12	.05	4.75	3.89	3.49	3.26	3.11	3.00	2.91	2.85
	.01	9.33	6.93	5.95	5.41	5.06	4.82	4.64	4.50
13	.05	4.67	3.81	3.41	3.18	3.03	2.92	2.83	2.77
	.01	9.07	6.70	5.74	5.21	4.86	4.62	4.44	4.30
14	.05	4.60	3.74	3.34	3.11	2.96	2.85	2.76	2.70
	.01	8.86	6.51	5.56	5.04	4.69	4.46	4.28	4.14
15	.05	4.54	3.68	3.29	3.06	2.90	2.79	2.71	2.64
	.01	8.68	6.36	5.42	4.89	4.56	4.32	4.14	4.00
16	.05	4.49	3.63	3.24	3.01	2.85	2.74	2.66	2.59
	.01	8.53	6.23	5.29	4.77	4.44	4.20	4.03	3.89
17	.05	4.45	3.59	3.20	2.96	2.81	2.70	2.61	2.55
	.01	8.40	6.11	5.18	4.67	4.34	4.10	3.93	3.79
18	.05	4.41	3.55	3.16	2.93	2.77	2.66	2.58	2.51
	.01	8.29	6.01	5.09	4.58	4.25	4.01	3.84	3.71
19	.05	4.38	3.52	3.13	2.90	2.74	2.63	2.54	2.48
	.01	8.18	5.93	5.01	4.50	4.17	3.94	3.77	3.63
20	.05	4.35	3.49	3.10	2.87	2.71	2.60	2.51	2.45
	.01	8.10	5.85	4.94	4.43	4.10	3.87	3.70	3.56
21	.05	4.32	3.47	3.07	2.84	2.68	2.57	2.49	2.42
	.01	8.02	5.78	4.87	4.37	4.04	3.81	3.64	3.51
22	.05	4.30	3.44	3.05	2.82	2.66	2.55	2.46	2.40
	.01	7.95	5.72	4.82	4.31	3.99	3.76	3.59	3.45
23	.05	4.28	3.42	3.03	2.80	2.64	2.53	2.44	2.37
	.01	7.88	5.66	4.76	4.26	3.94	3.71	3.54	3.41
24	.05	4.26	3.40	3.01	2.78	2.62	2.51	2.42	2.36
	.01	7.82	5.61	4.72	4.22	3.90	3.67	3.50	3.36
25	.05	4.24	3.39	2.99	2.76	2.60	2.49	2.40	2.34
	.01	7.77	5.57	4.68	4.18	3.85	3.63	3.46	3.32
26	.05	4.23	3.37	2.98	2.74	2.59	2.47	2.39	2.32
	.01	7.72	5.53	4.64	4.14	3.82	3.59	3.42	3.29
27	.05	4.21	3.35	2.96	2.73	2.57	2.46	2.37	2.31
	.01	7.68	5.49	4.60	4.11	3.78	3.56	3.39	3.26
28	.05	4.20	3.34	2.95	2.71	2.56	2.45	2.36	2.29
	.01	7.64	5.45	4.57	4.07	3.75	3.53	3.36	3.23
29	.05	4.18	3.33	2.93	2.70	2.55	2.43	2.35	2.28
	.01	7.60	5.42	4.54	4.04	3.73	3.50	3.33	3.20
30	.05	4.17	3.32	2.92	2.69	2.53	2.42	2.33	2.27
	.01	7.56	5.39	4.51	4.02	3.70	3.47	3.30	3.17

Figure 12.7

Excerpt from Table C.1: Critical values for *F* distributions. Table C.1 in the back of the book requires us to know three facts: df_B, df_W, and α.

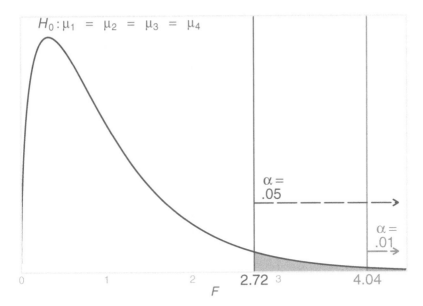

Figure 12.8

F distribution showing two values of α. The one-way ANOVA F test is a one-tailed test. If we had $df_B = 3$ and $df_W = 80$, we would have a critical value of $F = 2.72$ if α = .05. If we chose a significance level of .01, the F critical value for these df would be 4.04.

If the observed one-way ANOVA F test is equal to or

more extreme than the critical value,

then we reject the null hypothesis.

Otherwise, we retain the null hypothesis.

If we reject the null hypothesis, we can conclude that there is some significant difference in the means. If we retain the null hypothesis, we draw the conclusion that the means do not differ significantly. The fact that the one-way ANOVA F test is a one-tailed test also simplifies the p value decision rule:

If the observed p value is less than or equal to α,

then reject the null hypothesis.

Otherwise, retain the null hypothesis.

The observed one-way ANOVA F test cuts off the p value, and the F critical value cuts off α. If the observed test statistic is more extreme than the critical value, then the observed F will be cutting off an area for the p value that is smaller than the significance level, α. So we will reach the same decision using either the critical value or p value decision rule. If we reject the null hypothesis with the one-way ANOVA *F* test, we can conclude only what the alternative hypothesis says: there is some difference in the population means. That means this test statistic is not used for making directional predictions. If we had only two groups and a directional prediction, we probably would be better served by one of the *t* tests (independent-samples or AWS, depending on the sample sizes), which can be used with directional alternative hypotheses. Next we will work through a numerical example.

Check Your Understanding

SCENARIO 12-D, Continued

For our analysis of the ratings of severity of nasal symptoms in our sample of 240 people who were randomly assigned to different gargling conditions, we had $df_B = 2$ and $df_W = 237$. 12-6. Use Table C.1 and $\alpha = .05$ to look up the critical value for the one-way ANOVA *F* test.

Suggested Answer

12-6. In Table C.1, we use df_B to find the column. Here, we use the column labeled 2 because $df_B = 2$. When we try to find the row, we discover that we do not have a listing for $df_W = 237$, so we use the next smaller $df = 200$, giving us a critical value of 3.04.

Numeric Example of a One-Way ANOVA *F* Test

Let's continue with our imaginary study extending the work by Efe and Özer (2007). These researchers compared the mean crying times of babies who were breast-feeding during an injection versus the usual care of babies being cuddled and verbally soothed by their mothers. We suggested adding a third group of babies, who would be given breast milk in a bottle during the shot. The babies would be randomly assigned to one of three groups, and we would manipulate the independent variable, soothing method. We asked the research question, "Will the soothing method lead to any difference in the mean crying time?" Because we have no real data for a study like this one, we have to rely on fabricated numbers for this teaching example. (The fabricated data are available via http://desheastats.com.)

Let's begin by graphing our fabricated data to look for outliers that could influence the means. Figure 12.9 shows boxplots of the crying times for the three groups.

What can we learn from Figure 12.9 about the crying times for the three groups? The circles represent the individual scores, and none of the circles is beyond the reach of a whisker, so it appears there are no outliers. The diamond shape in each boxplot represents the group's mean. It *appears* that the babies in the usual care (control) group cried longer on average because their scores and the diamond representing their mean are farther to the right on the number

12. Tests and Estimates for Two or More Samples

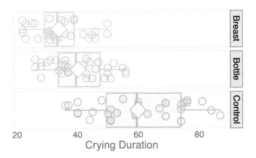

Figure 12.9

Boxplots for fabricated data on crying duration. This graph shows our fabricated data for the imaginary study investigating the crying times of babies whose mothers provided one of three kinds of soothing after an injection. Do you see any outliers? What do you notice about the three sample distributions?

Table 12.1 One-Way ANOVA F Test Results with a Questionable Last Column

Source	Sum of Squares	df	Mean Square	F	Sig.
Between Groups	11,687.7708	2	5,843.8854	47.461	.000
Within Groups	11,451.1875	93	123.1310		
Total	23,138.9583	95			

line, compared with the other groups' data. Notice how much variability there is *within* the groups, particularly the control group. Is the variation between the sample means substantially greater than the variation within the groups? That is a question for the one-way ANOVA F test.

Table 12.1 shows the results of the one-way ANOVA F test for our fabricated data. Different statistical packages produce ANOVA tables that may not look similar to this table; this time we used IBM® SPSS®. We also intentionally used one column label (Sig.) that we do not like so that we can explain it.

Table 12.1 contains a lot of big numbers, and this kind of table can be intimidating, but we are going to lead you through it so that everything is understandable. Let's start with explaining the rows and columns. The Source column tells us about the numerator and denominator of the one-way ANOVA F test; these are the sources of the variability being analyzed, Between Groups and Within Groups. The numerator is based on the variability between the sample (group) means and the denominator measures the variability within the groups. So we have one row for the numerator and one row for the denominator of the one-way ANOVA F test. The last row represents the total variability, which we will explain below. We also have columns for the following: Sum of Squares (SS), df, Mean Square (MS), F, and something called "Sig."

The number 11,687.7708 is the numerator sum of squares, also known as the between sum of squares, SS_B. We know that the numerator of the one-way ANOVA F test is SS_B divided by df_B. The numerator df (or df_B) is the number of groups minus 1. That is where the 2 came from in the row labeled Between Groups because we have three groups in this example. (Some statistical software will label this row Model.) Staying with the Between Groups row, we reach Mean Square Between, or MS_B. We know that any mean square is computed by taking a sum of squares and dividing by df. If you do the math, you will find the numerator of the one-way ANOVA F test:

$$MS_B = \frac{SS_B}{df_B}$$

$$= \frac{11,687.7708}{2}$$

$$= 5,843.8854$$

Now let's look at the row labeled Within Groups and figure out the denominator of the one-way ANOVA F test. (Some statistical packages will label this row Error.) The denominator sum of squares, or SS_W, is 11,451.1875. The denominator df, or df_W, is 93 because we had a total $N = 96$ (i.e., three groups of 32 babies each), and the df_W is the total sample size minus the number of groups, or $96 - 3 = 93$. To get the denominator of the one-way ANOVA F test, which is MS_W, we divide the sum of squares within by the df_W:

$$MS_W = \frac{SS_W}{df_W}$$

$$= \frac{11,451.1875}{93}$$

$$= 123.1310$$

Now that we have found the numerator and denominator, let's compute the one-way ANOVA F test:

$$\text{one-way ANOVA } F = \frac{MSB}{MSW}$$

$$= \frac{5843.8854}{123.1310}$$

$$\approx 47.461$$

You should see this value of the one-way ANOVA F test in Table 12.1. What does it mean? Let's think back to the logic of the one-way ANOVA F test: if the variation between the group means is similar to the variability within the groups, then the one-way ANOVA F test will be something in the vicinity of 1. As the variation between the sample means for the different groups increases, so will the one-way ANOVA F test. It appears that our observed one-way ANOVA F test is far away from 1. If the null hypothesis is true and the population mean crying times are equal for the different soothing methods, what is the probability of getting a one-way ANOVA F test as extreme as or more extreme than 47.461? The answer is our p value. But Table 12.1 does not have a column labeled p. Some statistical packages use a different label. As shown in Table 12.1, the p value is sometimes labeled "Sig." We *hate* this label for two reasons: the results are not always significant, and the observed p value is not the same thing as a significance level, which is α. We recommend always using the letter p, not "Sig." (Some statistical packages such as SAS® will label the p value as "$Pr > F$," meaning the probability of obtaining a result beyond the observed one-way ANOVA F test if the null hypothesis is true.) Notice that the p value in Table 12.1 is listed as .000, but a p value is *never zero*. It is extremely small but not zero. When we see a p value listed as ".000" in statistical output, we know it is something less than .001 that was rounded down to .000. Your statistical software (such as SAS) may report "<.0001" for this p value; widely used publication guidelines suggest reporting three decimal places: $p < .001$.

Let's complete our discussion of Table 12.1 by looking at the last line, labeled Total. (Your statistical software may label it "Corrected Total.") This line gives the total sum of squares and the total *df*. If we add up the SS_B and the SS_W, we get the total sum of squares, 23,138.958. If we add up the df_B and the df_W, we get the total *df*, which equals the total sample size minus 1. Here, $N = 96$, so the total *df* should be $N - 1 = 95$. Let's check by adding up the two *df* numbers:

$$df_B + df_W = 2 + 93$$
$$= 95$$

The total line is interesting mainly for understanding the meaning of the analysis of variance, so let's extend this explanation. If we take the total sum of squares and divide it by the total *df*, we get the unbiased variance for all the scores in the study. For our fabricated data, we have the following:

$$\text{unbiased variance} = \frac{\text{total sum of squares}}{\text{total } df}$$
$$= \frac{23,138.958}{95}$$
$$\approx 243.568$$

If you download our fabricated data for this example and you compute descriptive statistics on the crying times, 243.568 is approximately what you will get for the unbiased variance. The unbiased variance is a measure of variability of the scores around the grand mean—that is, the total variation of scores around the mean of all the scores in the one-way ANOVA design.

By itself, the unbiased variance is not that interesting. But let's remember that the total sums of squares and the total *df* came from adding numbers from the

previous two lines of the ANOVA table. In other words, some of the total variation in the scores is attributable to the fact that the scores are in a particular
group, and the rest of the variation is within the groups. This total variation is a
measure of spread accumulated across participants, whose scores differ from the
grand mean. Let's think about this total variation in terms of explaining one person's score. What might contribute to making the person's score different from
the grand mean for everyone in the study?

Figure 12.10 shows a histogram of all 96 babies' crying times in our made-
up data set. The baby who cried longest (87 seconds) is represented by the
yellow bar and the yellow dotted line. The solid blue line is drawn at 44.73,
which is the grand mean for all 96 crying times. The red dashed line is drawn
at 59.91, which is the mean of the 32 babies in the control group. How can we
explain the crying time for the baby who cried most? Another way of asking
the question: what are the sources of variability for crying time? Figure 12.10
has a yellow arrow to show the *total variation* of this baby's crying duration
versus the grand mean.

This total variation (the distance represented by the yellow arrow in
Figure 12.10) can be broken into two pieces. We know from looking at our data
set that the baby belongs to the control group, which has a mean represented by

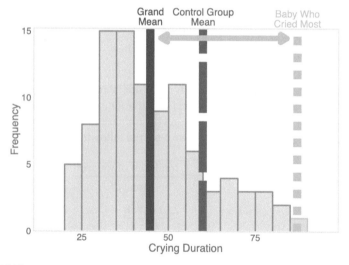

Figure 12.10

Location of control group's mean, relative to the grand mean. Using our 96 fabricated crying times from the example of three soothing methods, we created this
histogram. The crying duration for the baby who cried the longest (87 seconds)
is represented by the yellow box and dotted line. The mean for all 96 scores is the
grand mean (44.73 seconds), shown by the solid blue line. The red dashed line
shows the mean of the control group (59.91 seconds), to which the longest-crying
baby belonged. The arrow represents the total variation of the top crying time from
the grand mean.

12. Tests and Estimates for Two or More Samples

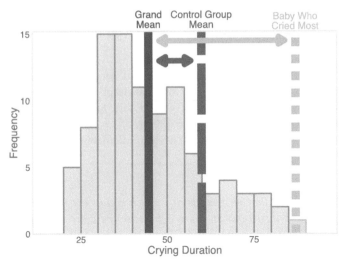

Figure 12.11

Difference between control group's mean and the grand mean. Part of the explanation for the longest-crying baby's crying duration may be the fact that the baby belonged to the control group. The red arrow represents the variation attributable to membership in the control group; that is, there is variability *between* the control group's mean and the grand mean. Notice that the distance between the grand mean and the control group's mean is part of the total variation, shown by the top arrow. The next figure completes this example, which uses fabricated data.

the red dashed line. One possible reason for the long crying time is that the baby was in the control group. In other words, there is an effect of being in the control group, and the effect of this condition is represented by the distance between the blue line and the red dashed line. Figure 12.11 adds a red arrow to represent the difference between the control group's mean and the grand mean for all of the scores. There is variability *between* the sample means around the grand mean, and we can see a gap between the grand mean (blue line) and the control group's mean (red dashed line).

How do we explain the rest of this score's variation from the grand mean? This baby happened to cry the longest; perhaps the distance from this baby's crying time to the control group's mean can be explained by something *within* the baby. Maybe this baby needed a nap, was hungry, or was agitated about being in unfamiliar surroundings when the shot was given. Figure 12.12 adds a bluish-purple arrow to represent the variation of this baby's crying time from the mean for the control group.

In sum, the total variation in the scores can be broken down into two components: the variability *between* the sample means for the groups (around the grand mean) and the variability *within* the groups. Next we will complete this example of the one-way ANOVA *F* test by testing the null hypothesis.

After collecting ratings of the severity of nasal symptoms from all 240 participants in our study of gargling (water, antiseptic, and usual habits), we are prepared to run a one-way ANOVA F test. Suppose we find the following results: $SS_B = 4.0373$, $SS_W = 180.91$, $df_W = 237$. 12-7. Compute the one-way ANOVA F test. (Hint: We need to calculate one more number to complete the computation!)

Suggested Answers

12-7. The number that we need is df_B, which equals the number of groups minus one. For this scenario, $df_B = 2$ because there are three groups. The numerator of the one-way ANOVA F test is the mean square between, which equals $SS_B/df_B = 4.0373/2 = 2.01865$ (do not round yet—wait until all the calculations are done). The denominator is the mean square within, which equals $SS_W/df_W = 180.91/237 = 0.7633333$. The one-way ANOVA F test is the mean square between divided by the mean square within $= 2.01865/0.7633333 = 2.6445197 \approx 2.64$.

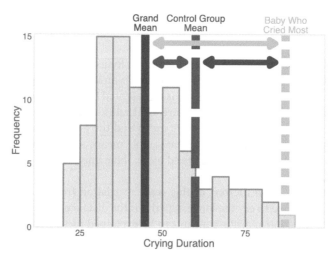

Figure 12.12

Difference between one baby's crying duration and the baby's group mean (fabricated data). Now we have added a bluish-purple arrow to represent the distance between the longest-crying baby's crying duration and the control group's mean. This baby differs from the average of the control group for reasons that may be internal to the child; maybe this baby was tired or hungry before the shot. In other words, the bluish-purple arrow represents the baby's variability *within* the group.

Testing the Null Hypothesis

Let's test the null hypothesis for the example of soothing methods for babies receiving injections. The null hypothesis said the population mean crying times are the same regardless of whether babies were breast-feeding, bottle-feeding, or receiving the usual comfort from their mothers after a shot. We will use $\alpha = .05$ and demonstrate both the critical value decision rule and the p value decision rule, even though it is redundant to do both. Looking in Table C.1, we discover there is not a critical value from an F distribution with $df_B = 2$ and $df_W = 93$. As explained in previous chapters, we cannot give ourselves extra degrees of freedom, so we will look for the critical value for the next smaller df_W. The next smaller denominator df in Table C.1 would be $df_W = 80$. Using the column labeled $df_B = 2$, we look on the line for $\alpha = .05$, and we find a critical value of $F = 3.11$. Figure 12.13 shows the F distribution with $df_B = 2$ and $df_W = 80$, and we have drawn a vertical line through the critical value, 3.11.

Where would our observed one-way ANOVA $F = 47.461$ go in Figure 12.13? Look at the horizontal number line. The left side of the distribution begins at zero, and the critical value is shown at 3.11. So the observed one-way ANOVA F statistic would be so far to the right of the critical value that it would be off the page! Because the observed one-way ANOVA $F = 47.461$ is more extreme than the critical value of $F = 3.11$, we reject the null hypothesis. We will reach the same decision using the p value decision rule. Table 12.1 showed that the p value was something less than .001. Because p is less than the chosen $\alpha = .05$, we reject the null hypothesis. A journal article might summarize this decision as follows: "Soothing method led to a significant difference in mean crying times, one-way ANOVA $F (2, 93) = 47.46, p < .001$." The numbers in parentheses after the F are the degrees of freedom, always represented in the order of "df_B, df_W." (As we have mentioned earlier, some journals report only the p value.) We conclude that the experimentally manipulated soothing method caused a difference in the mean crying times.

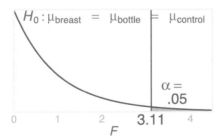

Figure 12.13

F distribution for the example of soothing methods and crying duration. Table C.1 did not list critical values for $df_B = 2$ and $df_W = 93$. Looking at the table's next-smaller values, we found a critical value of $F = 3.11$ for $df_B = 2$ and $df_W = 80$.

But wait—which soothing method is *best* for babies who are getting a shot? The one-way ANOVA *F* test cannot answer that question. It can say, "These means are not significantly different," or it can say, "There is some difference somewhere among these means—but I cannot tell you where it is or what kind of difference it is!" There is variation between the groups, and it may not be a simple difference of one mean versus the other means. The one-way ANOVA *F* test is like a legislative aide who is asked to listen to a committee discussion and report back on the committee's vote. If the committee members voted unanimously, the legislative aide says, "They agreed." (A nonsignificant one-way ANOVA *F* test would say, "The means are statistically indistinguishable.") But if at least one committee member voted differently from the others, the legislative aide would say, "There was a disagreement!" This legislative aide does not say who voted for or against the issue or even which side won. Similarly, a significant one-way ANOVA *F* test does not tell us where the difference is or what kind of difference it is. For all we know, the significant result could mean that one sample mean differs from the *average* of two other group means in a three-group study. We do not know what kind of difference has been detected, only that there is some difference involving the means.

Why would we tell you all about a statistic that is so limited in what it can say? To answer questions such as, "Which soothing method leads to the shortest duration of crying?" we need a set of statistics buried within the statistical software that gives us the one-way ANOVA *F* test results. We will explain a few of those statistics later in this chapter. But first we will complete our discussion of the one-way ANOVA *F* test by talking about its assumptions and robustness.

Check Your Understanding

SCENARIO 12-D, Continued

After analyzing the data on the ratings of severity of nasal symptoms for our sample of 240 people who were randomly assigned to different gargling conditions, we found a one-way ANOVA *F* test = 2.64. The *p* value for our test statistic was .0731. 12-8. Using $\alpha = .05$, test the null hypothesis. 12-9. Just for practice, use the critical value that we found earlier (3.04) and test the null hypothesis using the critical value decision rule.

Suggested Answers

12-8. *Because .0731 is greater than .05, we will retain the null hypothesis.* 12-9. *The observed one-way ANOVA F test = 2.64, which is not more extreme than 3.04, so we will retain the null hypothesis, the same decision we reached with the p value decision rule.*

Assumptions and Robustness

We talked about assumptions and robustness at length in Chapter 11, using the independent-samples t test as an example. As you will recall, inferential statistics come with "owner's manuals"—conditions that are supposed to be met in order for us to be able to rely on theoretical distributions to compute trustworthy p values. Those conditions are the test statistic's assumptions, and the one-way ANOVA F test has the same assumptions as the independent-samples t test:

- Normality: the scores are normally distributed in the populations that provided the samples.
- Independence: the scores are independent of each other.
- Equal population variances: the populations of scores are equally spread out.

As we saw in Chapter 11, the independence assumption definitely must be met. Typically, some kind of random sampling is expected, but researchers usually depend on convenience samples and try to make sure that the scores from each participant are independent of (or uninfluenced by) any other participant's scores. If the independence assumption is violated, it can be disastrous for the study. For example, suppose some babies in our study of soothing methods could hear other babies crying in a neighboring examination room. Hearing other babies crying could influence the babies in our study. Infants sometimes cry when they hear other babies crying, so we would have to make sure that each baby in our study received the immunization shot in a room where they would not hear any other baby crying. Not only could babies overhearing each other violate the independence assumption, but overhearing could be an extraneous variable that would keep us from being able to say whether the soothing methods were responsible for any observed differences in the means.

Suppose the researchers suspect that they sampled from populations that are not normally distributed. Specifically, what if they think the population of crying times for babies in the control group is positively skewed, with some babies crying much longer than others? Sampling from one or more nonnormal populations would be a violation of the normality assumption of the one-way ANOVA F test. Luckily, the statistic is fairly robust to violations of the normality assumption, which was the case with the independent-samples t test's robustness to nonnormality.

So far, the robustness of the one-way ANOVA F test is similar to the robustness of the independent-samples t test. With regard to the assumption of equal population variances, these two statistics differ in their robustness. We said in Chapter 11 that if we have equal and sufficiently large sample sizes, the independent-samples t test is robust to most violations of the assumption of equal population variances. That is *not* true for the one-way ANOVA F test. Even with large and equal sample sizes, the one-way ANOVA F test might not produce a trustworthy p value if the population variances are not equal. How do we know? Statisticians can do experiments using computer simulations. They can mathematically define the populations of scores, take thousands of random samples from those populations, and compute thousands of one-way ANOVA F tests. The computerized experiment for testing the behavior of statistics across many replications, called a *Monte Carlo simulation study*, can manipulate the conditions of the populations being sampled. If the null hypothesis is true, if all the assumptions are met, and if $\alpha = .05$, then we would expect to find a significant result 5 times out of 100 repeated studies. That is, the actual Type I error rate would be exactly what we hoped it would be: $\alpha = .05$. These computer simulations can check whether that is the case under different conditions. The experimenter violates an assumption, makes the assumption not true, and runs the simulation again to see how often a significant result is found when the null hypothesis is true. Suppose we are doing a Monte Carlo simulation study for the one-way ANOVA F test, and we randomly sample 50 scores from each of four normal populations. We set up the experiment so that three of the populations have a standard deviation $= 1$ and one of the populations has a standard deviation $= 4$. This experiment would test the effect of violating the assumption of equal population variances. If the null hypothesis is true, if α is .05, and if the two other assumptions (independence and normality) are met, will we find a significant result 5 times out of 100? That is, will our actual probability of a Type I error be .05? According to Wilcox (1987), the answer is no. He did this experiment and found that the probability of a Type I error was .088. Even with large and equal sample sizes, violating the assumption of equal population variances led to a higher chance of incorrectly finding significance when the null was true. That tells us that the one-way ANOVA F test is not robust to unequal population variances. That is not good, and Wilcox says the situation is worse when there are more groups in a study.

You have read all of these pages about the one-way ANOVA F test, only to learn that the statistic is not robust to unequal variances. Have we wasted your time? Oh, we would not do that. Despite its limitations, analysis of variance remains a widely used family of statistical procedures, and this chapter gives you a foundation for learning more about ANOVA statistics. In addition to being a general statistic in terms of what it can tell us (the means are the same or there is some difference in the means), the one-way ANOVA F test also has the weakness of giving us p values that may not be trustworthy if the population variances are not equal. But we are not too concerned about these weaknesses because we really are looking for more specific answers to our questions. Statistics that can tell us which soothing method is *best* are embedded in the one-way ANOVA procedure in statistical software packages. Next we will introduce you to some fundamental concepts underlying those statistics.

In our study of different gargling methods and ratings of severity of nasal symptoms, we retained the null hypothesis of equal population means for the three conditions (gargling with water, gargling with antiseptic solution, and usual habits of gargling). 12-10. Explain the meaning of the decision to retain the null hypothesis, using the variable names.

Suggested Answers

12-10. There is no significant difference in the mean severity of nasal symptoms for people in the three gargling conditions. In other words, on average the three gargling conditions did not affect the severity of nasal symptoms. This particular hypothesis test would not tell us whether gargling might affect the frequency of upper respiratory infections, another outcome that interested Satomura et al. (2005).

How to Tell Which Group Is Best

Early in this chapter, we speculated that there were many ways that three means could differ from each other. In our imaginary study of three groups of babies being soothed in different ways, there are many possible outcomes. How many ways could one sample mean differ from two other sample means? If we were arranging the means in different orders and they all differed from each other, how many orders could we get? (Find out: label the sample means A, B, and C, then see how many ways you can arrange those three letters.) The one-way ANOVA *F* test is quite limited in what it can say about the means. It can say, "The means are the same" or it can say, "There is some difference among the means, but I cannot tell you where it is or what kind of difference it is." Further, if the samples come from populations that are not equally spread out, we might get an untrustworthy *p* value for the one-way ANOVA *F* test.

All is not lost. In the various statistical software packages, there are options within the one-way ANOVA *F* test procedures. These options will compute additional statistics that can tell us specifically which means differ from each other. For any one-way ANOVA design, we could compare the means two at a time. But we would have to do it multiple times. If we want to know whether two means differ in any way (disregarding the order of the means on the number line), three mean comparisons are possible in the study of soothing method:

- One comparison would be the mean crying time of breast-feeding babies versus the mean crying time of bottle-feeding babies.
- Another comparison would be the mean crying time of breast-feeding babies versus the mean crying time of babies in the control group.

- The third comparison would be the mean crying time of bottle-feeding babies versus the mean crying time of babies in the control group.

This process of comparing all possible combinations of two means at a time and determining whether they differ significantly is called *multiple comparisons*. The statistics that perform these multiple comparisons are called *multiple comparison procedures*. We decide which means to compare based on predetermined hypotheses. Many researchers want to know where all the differences are, so they examine all possible pairs of means (called *pairwise comparisons*) within the one-way ANOVA design. We will take that approach in our examples.

Wait, didn't we discourage you from using the independent-samples t test for these kinds of comparisons? Yes, we did. Every time we would compare two means with an independent-samples t test, we would risk a Type I error with a probability of $\alpha = .05$ (or whatever our significance level is). Luckily for us, there are statisticians who have identified other statistics for comparing two means at a time in a one-way ANOVA design. These ways of doing multiple comparisons are intended to control the probability of a Type I error so that it equals α for the entire set of pairwise comparisons of means. All three pairwise comparisons of means listed above could be done, and the total probability of making a Type I error in that set of all pairwise comparisons would not exceed α.

Not all multiple comparison procedures are equally good. Some do a poor job of controlling the probability of a Type I error for the set of all pairwise comparisons. We will give you some working knowledge of multiple comparison procedures and the hypothesis being tested with each comparison. We will use one of the most commonly used multiple comparison procedures to illustrate how these statistics generally are used. Then you will meet this procedure's better-looking brother—that is, a multiple comparison procedure that we think is better than the most commonly used one. Finally, we will mention a multiple comparison procedure that we think should be used when sample sizes are unequal. All of the multiple comparison procedures we give in this book will do a good job of controlling the probability of a Type I error for the set of all pairwise comparisons.

Check Your Understanding

12-11. If we have four independent groups and we want to compare the means two at a time to see if the means are equal, how many pairwise comparisons are possible?

Suggested Answers

12-11. Six comparisons of means are possible: Group 1 versus Group 2; Group 1 versus Group 3; Group 1 versus Group 4; Group 2 versus Group 3; Group 2 versus Group 4; and Group 3 versus Group 4. (Multiple comparisons generally are nondirectional tests.)

Multiple Comparison Procedures and Hypotheses

Just as ANOVA is a term referring a family of statistics, the term *multiple comparison procedures* also refers to many statistics. Sometimes researchers use other terms to refer to the comparison of pairs of means in an ANOVA: individual comparisons, *post hoc* comparisons, *a priori* comparisons, planned contrasts, orthogonal comparisons, and so forth. These terms differ in meaning, but we will not torture you by explaining the differences. The most general term that encompasses all of them is multiple comparisons. The main idea behind these statistics is that we want to understand the mean differences that may exist in an ANOVA design, yet we do not want to increase the chance of finding significance incorrectly (i.e., we do not want to increase the chance of false positive results). Our examples show all possible pairs of means being compared, but the number of comparisons that you perform in a given research situation should correspond to a study's research questions.

The most common alternative hypothesis used with multiple comparisons is nondirectional and says the means differ. In other words:

Our samples come from populations in which

the mean of one group is not equal to the mean of another group.

The symbolic expression of this alternative hypothesis is written as

$$H_1: \mu_1 \neq \mu_2$$

Perhaps "1" represents the group of breast-feeding babies and "2" the bottle-feeding babies; the numeric subscripts on the population means could be replaced with words. The alternative hypothesis indicates that the mean crying time of the breast-feeding babies would be compared with the mean crying time of the bottle-feeding babies. The corresponding null hypothesis would be

$$H_0: \mu_1 = \mu_2$$

The null hypothesis is translated as follows:

Our samples come from populations in which

the mean crying time for breast-feeding babies

equals the mean crying time for the bottle-feeding babies.

So far, we have covered one comparison. But we have two more comparisons in our study of soothing method. If Group 3 is the control group, then what does the following alternative hypothesis mean?

$$H_1: \mu_1 \neq \mu_3$$

Now we are saying that the samples come from populations in which breast-feeding babies' mean crying time would differ from the control-group babies' mean crying time. The corresponding null hypothesis to be tested would be

$$H_0: \mu_1 = \mu_3$$

If we are going to compare all possible pairs of means in this fictitious study, we need one more alternative hypothesis and its corresponding null hypothesis:

$$H_1: \mu_2 \neq \mu_3$$

This alternative hypothesis says our samples come from populations in which the mean crying time is different for babies who bottle-feed during an injection, compared with babies receiving the usual care. Here is the corresponding null hypothesis for this comparison:

$$H_0: \mu_2 = \mu_3$$

If we reject this null hypothesis, it could mean one of two things: either the bottle-feeding babies on average cried longer than the control group did or the bottle-feeding babies on average cried for a shorter time than the control group did. Typically, researchers have nondirectional alternative hypotheses when they do multiple comparisons so that no matter which group has the higher mean, it is possible to find significance.

All three of these null hypotheses typically can be tested in the same analysis in a statistical software package. But we want to control the total probability of making a Type I error. Specifically, for this set of three pairwise comparisons of means, we want the total probability of a Type I error to equal α. If α is .05, the simplest approach to maintaining control over the Type I error rate would be to divide .05 by the number of comparisons, then use that number as the significance level for each null hypothesis tested. (A statistic such as an independent-samples t test could be used for each hypothesis test, but there are better statistics available for multiple comparisons.) For our first null hypothesis, which said $H_0: \mu_1 = \mu_2$, the significance level could be .05/3 $\approx$ .0166667. The same probability could be used for the significance level for each of the two other null hypotheses. Then the total probability of making a Type I error within our three comparisons would not exceed .05. This approach sometimes is called a *Bonferroni correction*.

We will demonstrate statistics that have been built on the idea of a Bonferroni correction, except they are better statistics. What makes them better? As you may recall from Chapter 9, when we use a smaller α, we tend to lose power—that is, we have a smaller probability of finding a significant result. The cost of dividing α by the number of comparisons is a loss of power. As a result, statisticians have looked for ways to improve the sensitivity of multiple comparison procedures while still controlling the probability of Type I errors for the set of multiple comparisons. Dozens of statistics have been proposed for multiple comparisons in ANOVA designs. We discuss three such statistics, omitting the computational details. One of these statistics may be the most commonly used multiple comparison procedure for comparing independent means, the second one is better than the most commonly used one, and the third one is recommended when sample sizes are unequal.

12-12. The answer to the previous question was that there are six pairwise comparisons of means possible when we have four independent groups. If we were taking a Bonferroni approach and using $\alpha = .05$, how large would α be for *each comparison*?

Suggested Answer

12-12. *Each comparison would be performed using $\alpha/6$, or $.05/6 = .0083333$.*

Many Statistics Possible for Multiple Comparisons

If we have a study in which the one-way ANOVA F test is being computed, we will need only one multiple comparison procedure. More complicated ANOVA designs might require two kinds of multiple comparison statistics, one for independent groups and one for repeated measures. We will restrict our discussion to multiple comparisons in a one-way ANOVA study.

In our example of the three soothing methods, we are interested in testing null hypotheses for all three pairwise comparisons of mean crying times. One of the most commonly used multiple comparison procedures in this kind of situation is called *Tukey's Honestly Significant Difference (HSD)*. Because we are leaving out the details of the computations, let's jump directly into results for our imaginary example of the three soothing methods for babies receiving an injection. (The fabricated data set is available via http://desheastats.com).

Table 12.2 shows a typical way of displaying descriptive statistics in a journal article, with the mean followed by the standard deviation in parentheses for each group. It is tempting to look at these means and make judgments about any apparent differences. But we cannot perform significance tests with our eyeballs. Yes, the mean for the breast-feeding group is a smaller number than the two other means, but perhaps the mean for the breast-feeding group is *statistically indistinguishable* from the mean for the bottle-feeding group. We will not know until we perform a multiple comparison procedure.

Table 12.2 **Means for the Fabricated Data on Crying Times**

Soothing Method	Mean Crying Time (*SD*)
Breast-feeding	34.0 (8.2)
Bottle-feeding	40.3 (9.2)
Usual comfort (control)	59.9 (14.7)

Statistical software programs display the results of multiple comparison procedures in different ways, so our explanation of Tukey's HSD will not rely on a specific statistical package. We used software called SAS® to run this analysis; if you run the Tukey procedure in another software package, you should find the same results, although they may be displayed differently. We found that the mean crying time for the breast-feeding babies ($M = 34.0$ seconds) was not statistically different from the mean crying time for the bottle-feeding babies ($M = 40.3$ seconds). But according to Tukey's HSD, both of those means differed significantly from the mean crying time for the babies receiving the usual care ($M = 59.9$ seconds). Can you match these results with the null hypotheses? We wrote the null hypotheses as follows, using 1 = breast-feeding, 2 = bottle-feeding, and 3 = usual care/control:

$$H_0: \mu_1 = \mu_2$$
$$H_0: \mu_1 = \mu_3$$
$$H_0: \mu_2 = \mu_3$$

The first null hypothesis says our samples come from populations with equal mean crying times for breast-feeding and bottle-feeding babies; this null hypothesis was retained. The second null hypothesis says our samples come from populations with equal mean crying times for breast-feeding and usual-care babies; this null hypothesis was rejected. The third null hypothesis says our samples come from populations with equal mean crying times for bottle-feeding and usual-care babies; this null hypothesis was rejected. So the breast-feeding and bottle-feeding group means did not differ significantly from each other, but both of them differed significantly from the mean for the control group. When we examine the means, we can see shorter average crying times for the two groups of babies who were being fed when they received the shots.

Tukey's HSD can be a good choice when sample sizes are equal, but we know a better multiple comparison procedure, called the *Ryan–Einot–Gabriel–Welsch Q statistic*, which often is abbreviated REGWQ in statistical software packages. We call it the Ryan procedure. Similar to Tukey's HSD, the Ryan procedure is used with equal sample sizes. What makes the Ryan procedure better than Tukey's HSD is that the Ryan procedure provides a tiny bit more power. That is, it is slightly more sensitive when real differences exist in the population means. If you download and analyze the fabricated data on the crying times using the Ryan procedure, you will find that the Ryan procedure will say all three means differ from each other.

We would like to mention one more multiple comparison procedure. Our examples have used equal sample sizes, yet researchers sometimes end up with unequal sample sizes—participants do not show up on the day of the study, they exercise their right to quit the study before it is over, or whatever. If you have a one-way ANOVA kind of study and the groups have unequal numbers of participants, we would recommend a multiple comparison procedure called the *Games–Howell procedure*. As you will recall from Chapter 11, an alternative to the independent-samples *t* test, called the AWS *t* test, was able to accommodate unequal sample sizes. The Games–Howell multiple comparison procedure is similar to the AWS test. Like the other multiple comparison procedures, the

Games–Howell procedure usually is performed as an option in computer software for the ANOVA. More details are available in Toothaker (1991).

You may have noticed that we have not talked about confidence intervals yet in this chapter. We intentionally waited until we had completed this section on multiple comparisons. Next we will talk briefly about confidence intervals for a one-way ANOVA study.

Check Your Understanding

SCENARIO 12-D

In the study comparing the mean severity of nasal symptoms for people in the three gargling conditions, suppose the water-gargling group had a mean of 2.4, the antiseptic-gargling group had a mean of 2.8, and the usual-habits group had a mean of 3.1. 12-13. Write the null hypotheses that Tukey's HSD would test.

Suggested Answers

12-13. Tukey's HSD would test three null hypotheses. If Group 1 is the water-gargling group, Group 2 is the antiseptic-gargling group, and Group 3 is the usual-habits group, then here are the three null hypotheses that would be tested:

$$H_0: \mu_1 = \mu_2$$

$$H_0: \mu_1 = \mu_3$$

$$H_0: \mu_2 = \mu_3$$

Confidence Intervals in a One-Way ANOVA Design

We have tried to be consistent in this book, describing each inferential statistic, its assumptions, and the confidence interval that corresponded to the inferential statistic. That is what we did for the z test statistic, the one-sample t test, and the t statistics in the chapter pertaining to two samples. Now we have been talking about the one-way ANOVA F test, which is used to compare two or more independent means. But there is not a single, widely recommended confidence interval for this situation.

Researchers often report an interval estimate of the population mean for each group in a one-way ANOVA study. But there are different ways of computing that interval estimate. Some software packages report confidence intervals that depend on a one-sample t test's critical value for each group. Other software packages compute the interval estimate for each group by incorporating the denominator of the one-way ANOVA F test. The interval estimate that you obtain for a given study may depend on the software being used to analyze the data.

One other option is available, and it involves multiple comparisons. If we were working on a study involving a one-way ANOVA F test and we planned to submit a paper to a journal that preferred for confidence intervals to be reported, we probably would compute a confidence interval for each pairwise comparison of means. This confidence interval would be similar to the interval estimates presented in Chapter 11, except now the computation of the interval would involve whatever multiple comparison procedure was performed. If you read about a one-way ANOVA study in a health sciences journal, the description of results may make it hard to tell what kind of confidence interval was computed. Most of the time, an interval estimate of the population mean will be reported for each group. Just do not use those intervals in place of multiple comparison procedures.

What's Next

This chapter has introduced you to the analysis of variance, specifically the one-way ANOVA F test, which is used to determine whether two or more independent means vary significantly, relative to the amount of variation with the groups. As described at the beginning of the chapter, there are many ANOVA-type statistics, and our goal was to give you a foundation built on the one-way ANOVA F test. This test statistic can say only two things: either "the means are the same" or "some difference exists in the means, but I cannot explain anything further." That is why we need multiple comparison procedures: to determine which means differ.

This book will not go into more complicated ANOVA designs. In Chapter 13, we return to the idea of measuring linear relationships, this time with hypothesis testing in mind.

Exercises

SCENARIO 12-E

Many people like to listen to music while they exercise. Does the tempo of the music affect how hard people exercise? Suppose we want to modify a study by Waterhouse, Hudson, and Edwards (2010) and randomly assign 42 adult male bicyclists to one of three groups. (Those researchers studied each bicyclist on three occasions, a decision with valid reasons behind it, but we need to change the scenario to fit with the content in this chapter.) We provide each bicyclist with a music player containing popular music that we have chosen, and we ask them to listen to the music while they ride a stationary bike at a steady pace that they can maintain for 30 minutes. The bicyclists think we are measuring the consistency of their self-chosen riding pace during the riding session. What they do not know is that we are manipulating the speed of the music. All three groups receive recordings of the same popular music. Those in Group 1 receive a recording of music that has been sped up by 10%, compared with the original recording speed; the speed is changed in a way that does not affect the pitch of the music, only the speed

(Continued)

of the playback. Those in Group 2 receive the same music, but the music has been slowed down by 10%, compared with the original recording speed. Participants in Group 3 receive the original music without any changes. The stationary bike has an odometer, which tells how far the bicyclists would have traveled if they had been on real bicycles. We ignore the first 5 minutes on the stationary bike as a warm-up period and the last 5 minutes as a cool-down period. Our main outcome measure is the distance traveled in the middle 20 minutes. If people work out harder with faster music, perhaps the faster-paced playback will be associated with greater mean distances and the slowed-down music will lead to shorter mean distances. Will the speed of the music affect the mean distances for the three groups? 12-14. What kind of research is this, and how do you know? 12-15. What kind of variable is the speed of the music? 12-16. What kind of variable is the riding distance during the 20 minutes after the warm-up period? 12-17. What kind of variable is the bicyclists' usual biking distance per week? 12-18. How did the researchers control the extraneous variable of gender? 12-19. Why might we want to analyze the data using a one-way ANOVA *F* test? 12-20. Write the alternative hypothesis for the one-way ANOVA *F* test, using both symbols and sentences. 12-21. Write the null hypothesis for the one-way ANOVA *F* test, using both symbols and sentences. 12-22. Explain the logic of the one-way ANOVA *F* test, using this scenario as a basis for the explanation. 12-23. The scenario stated, "If people work out harder with faster music, perhaps the faster-paced playback will be associated with greater mean distances and the slowed-down music will lead to shorter mean distances." What kind of analysis will determine whether there is evidence for this statement?

SCENARIO 12-F

(Inspired by the study of Wilson, McGrath, Vine, Brewer, Defriend, and Masters, 2010. Details of this scenario may differ from the actual research.) Suppose we are researchers interested in the eye–hand coordination of surgeons, particularly those who perform laparoscopic techniques. Laparoscopy sometimes is called minimally invasive surgery and often is used with abdominal surgery. Small incisions are made, the abdomen is inflated with carbon dioxide gas, and a tiny video camera is inserted, providing the surgeon with a view of the area to be operated on. Thin instruments may be inserted through other small incisions. Suppose we have reviewed the research literature on ways to train surgeons to perform laparoscopic procedures, and we are familiar with virtual reality training simulators, which allow the surgeons-in-training to gain experience without placing patients at risk. What interests us most is gaze control. Suppose we think surgeons who are experienced with laparoscopic procedures will have better eye–hand coordination. Specifically, we think the experts will be efficient in controlling their gaze, focusing on the targets in the area to be operated on and not being distracted by the surgical tools. We also think experts will complete tasks more quickly than novices. To become a surgeon in the United States, a person must

(Continued)

complete at least 5 years of residency after graduating from medical school. We obtain the virtual reality training simulator, and we recruit surgeons at four levels of expertise. The first group will consist of 14 doctors who are in the first year of their surgical residency. The second group will include 14 doctors in the last year of their surgical residency. The third group will be 14 surgeons who have led up to 40 laparoscopic procedures. The fourth group will be 14 surgeons who have led 80 or more laparoscopic procedures. Each surgeon will complete a training task with the virtual reality training simulator. Before each surgeon does the simulation, we fit him/her with unobtrusive head gear that records the dominant eye's gaze. The apparatus is able to determine what the eye is looking at and for how long. In this way, we are able to record the total amount of time that the dominant eye was fixed on a target. The virtual reality simulator shows a three-dimensional field of balls, and the training task requires the doctor to move virtual surgical instruments to touch different balls. The doctor is signaled to touch a ball using a particular virtual instrument (the one operated by the left hand or the right hand) when the simulator makes the ball change color and flash. One color requires the use of the left-hand instrument and a second color means the right-hand instrument should be used. The simulator records how long it takes to complete the entire simulation. Thus, we have two outcome variables: the total number of seconds with the gaze fixed on targets and the completion time for the entire simulation. 12-24. For a one-way ANOVA F test analyzing the total gaze fixation time, write the alternative hypothesis in words. Then write the null hypothesis in words. 12-25. Write the alternative hypothesis in words for the simulation completion time. Then write the null hypothesis in words. 12-26. Suppose we have 14 doctors in each of the four groups. Compute df_B and df_W. 12-27. Using $\alpha = .05$, look up the critical value for the one-way ANOVA F test in Table C.1. 12-28. Why would the same critical value be used for both of the one-way ANOVA F tests (one F test per criterion variable)? 12-29. Why would it be important to graph the data for each group? 12-30. If we find statistical significance with both one-way ANOVA F tests, what conclusion can we draw about expertise? 12-31. Name an analysis that we should perform if we want to know which level(s) of expertise had significantly longer gaze times. 12-32. Suppose we find the following results for the mean simulation completion times: for doctors at the beginning of their residency, $M = 74.5$ seconds ($SD = 13.4$); for doctors at the end of their residency, $M = 69.8$ seconds ($SD = 12.1$); for surgeons who have led up to 40 laparoscopic procedures, $M = 58.7$ seconds ($SD = 14.9$); and for surgeons who have led 80 or more laparoscopic procedures, $M = 51.5$ seconds ($SD = 11.0$). What can we say about these means? 12-33. If we want to compare all pairs of means for a given criterion variable, how many comparisons will we perform? 12-34. Suppose we use these labels for each group: Group A = the doctors at the beginning of their residency, Group B = doctors at the end of their residency, Group C = surgeons who have led as many as 40 laparoscopic procedures, and Group D = surgeons who have led 80 or more

(*Continued*)

procedures. Write out all the pairwise comparisons of means that are possible for these groups. 12-35. Suppose $\alpha = .05$ was set for each comparison and we did all of the comparisons that you enumerated in Question 12-34, except we made the mistake of using independent-samples t tests for each comparison. What would be the maximum of our total probability of making at least one Type I error in this group of pairwise comparisons? 12-36. Suppose three surgeons in Group D had agreed to participate in the study, but then they had to drop out of our study because of a scheduling conflict. Now that one group has lost three surgeons, what procedure should be used to perform all pairwise comparisons of means?

SCENARIO 12-F, Continued

Suppose we have run a one-way ANOVA F test on the completion times for the virtual reality simulator, with data coming from 56 doctors (14 in each of the four groups). We printed the ANOVA results on an inkjet printer in a computer lab, brought home the output, and left the paper on a sofa overnight. In the morning, we discover our bloodhound, Sir Drools-a-Lot, has slobbered all over the printout, smearing some of the ink. 12-37. Using your knowledge of the formula for the one-way ANOVA F test, fill in the blanks in Table 12.3. 12-38. Test the null hypothesis for the total simulation completion time, using the critical value decision rule and $\alpha = .05$. 12-39. Just for practice, test the same null hypothesis, using the p value decision rule and $\alpha = .05$. 12-40. Explain the meaning of the decision about the null hypothesis, using the variable names.

Table 12.3 Incomplete ANOVA Table (for Exercises 12-37 through 12-40)

Source	Sum of Squares	df	Mean Square	F	p
Between groups	4587.345	_____	_____	_____	.0001
Within groups	8696.740	_____	_____		
Total	_____	_____			

References

Efe, E., & Özer, Z. C. (2007). The use of breast-feeding for pain relief during neonatal immunization injections. *Applied Nursing Research, 20,* 10–16. doi:10.1016/j.apnr.2005.10.005

Everett, A. D., Tufro-McReddie, A., Fisher, A., & Gomez, R. A. (1994). Angiotensin receptor regulates cardiac hypertrophy and transforming growth factor-beta 1 expression. *Hypertension, 23,* 587–592. doi:10.1161/01.HYP.23.5.587

Montgomery-Downs, H. E., Clawges, H. M., & Santy, E. E. (2010). Infant feeding methods and maternal sleep and daytime functioning. *Pediatrics, 126,* e1562–e1568. doi:10.1542/peds.2010-1269

Satomura, K., Kitamura, T., Kawamura, T., Shimbo, T., Watanabe, M., Kamei, M., …Great Cold Investigators-I. (2005). Prevention of upper respiratory tract infections by gargling: A randomized trial. *American Journal of Preventive Medicine, 29,* 302–307. doi:10.1016/j.amepre.2005.06.013

Toothaker, L. E. (1991). *Multiple comparisons for researchers.* Newbury Park, CA: Sage Publications.

Waterhouse, J., Hudson, P., & Edwards, B. (2010). Effects of music tempo upon submaximal cycling performance. *Scandinavian Journal of Medicine & Science in Sports, 20,* 662–669. doi:10:1111/j.1600-0838.2009.00948.x

Wilcox, R. R. (1987). *New statistical procedures for the social sciences.* Hillsdale, NJ: Lawrence Erlbaum Associates.

Wilson, M., McGrath, J., Vine, S., Brewer, J., Defriend, D., & Masters, R. (2010). Psychomotor control in a virtual laparoscopic surgery training environment: Gaze control parameters differentiate novices from experts. *Surgical Endoscopy, 24,* 2458–2464. doi:10:1007/s00464-010-0986-1

Tests and Estimates for Bivariate Linear Relationships

Introduction

The relationships between variables are at the center of most research. In an experiment, one or more independent variables may be investigated as causally affecting on one or more dependent variables. If the research is nonexperimental, the study would examine the influence of one or more predictor variables on one or more criterion variables. These concepts should be quite familiar by now. Let's look at another way of describing the relationships between variables.

The last couple of chapters have described studies in which the investigators wanted to know whether a categorical independent (or predictor) variable had an effect on a quantitative dependent (or criterion) variable. For instance, an example in Chapter 12 involved babies receiving a shot and the effect of soothing method on their duration of crying. The categorical variable was soothing method, and it had three levels: breastfeeding, bottle-feeding, or usual comfort. The quantitative outcome variable was the number of seconds of crying after the shot. Could a *quantitative* independent (or predictor) variable influence a *categorical* dependent (or criterion) variable? Yes, such a study is possible. For example, researchers in Australia wanted to know whether the amount of fruit and vegetable intake might explain whether people will get certain kinds of cancer (Annema, Heyworth, McNaughton, Iacopetta, & Fritschi, 2011). This book will not cover the kinds of statistics that would be used in a study with a quantitative predictor and a categorical outcome variable. What about combining two categorical variables (one as the predictor and one as the criterion)? Yes, that is also possible. For instance, Vaccarino et al. (2013) studied whether having a history of post-traumatic stress disorder (yes/no) was related to the development of coronary heart disease (yes/no). We will describe some statistics for this kind of study in Chapter 14.

What about the combination of two quantitative variables? Yes, we can have a quantitative predictor variable and a quantitative criterion variable. These variables can be independent and dependent variables, but usually the relationship between two quantitative variables is studied in nonexperimental research. The analysis of two quantitative variables was discussed at length in Chapter 5 on correlation. You may recall Chapter 5's example of 51 locations in the United States and the linear relationship between the locations' rates of food hardship and obesity. States with higher percentages of adults saying they could not afford food at least once in the last year tended to have higher rates of obesity, and states with lower rates of food hardship generally had lower percentages of adults who could be categorized as obese. Now that you have learned about hypothesis testing, we will return to correlation and explain how researchers can test a null hypothesis about the strength of a linear relationship, and we will use Pearson's correlation coefficient, r, as a test statistic. We will go beyond Pearson's r in this chapter. We no longer will have to imagine a line going through a scatterplot to describe the linear relationship between two variables—we actually will draw such a line, based on a statistical analysis. We will explain some estimates and tests involving this straight line, which will be used to make predictions from one quantitative variable to another. Pearson's correlation coefficient told us about the degree to which two quantitative variables shared a linear relationship, but with these additional tests and estimates, we will specify that one of those variables is a predictor and the other variable is an outcome that can be predicted.

This chapter is called "Tests and Estimates for *Bivariate Linear* Relationships" because we have *two* quantitative variables involved in the relationship, and we are limiting our discussion to straight-line (linear) relationships. We will begin by refreshing your memory about the example of food hardship and obesity. We will use this example to explain how researchers can test a null hypothesis about r, a measure of the strength and direction of a linear relationship (positive or negative) between two variables. After that, we will compute the formula for a line that is used to make predictions about an outcome variable, based on values of a predictor variable. The analysis will include tests and estimates about the linear relationship between the predictor and criterion variables.

Hypothesizing About a Correlation

We return now to the 2011 food hardship and obesity rates for the 50 states and the District of Columbia. We obtained the data on food hardship from the Food Research and Access Center, which defined food hardship as the percentage of adults in representative samples who said they lacked money to feed their families on at least 1 day in the previous 12 months. Obesity rates were obtained from the Centers for Disease Control and Prevention (CDC). The CDC's Behavioral Risk Factor Surveillance System (BRFSS) operates at the state level and involves surveys of representative samples. The obesity rate for each location was defined as the percentage of respondents whose self-reported weight and height yielded a body mass index (BMI) of 30 or greater. The unit of analysis in this example is the location (the 50 states and the District of Columbia).

Let's pretend we have not analyzed this data set. Suppose we have been reading about food hardship. People who have food insecurity often buy cheap food that is high in calories and low in nutrition. They may live in neighborhoods with stores that do not sell fresh, unprocessed food. They also may lack transportation to get to a good grocery store or to take their children to participate in organized sports. Unsafe neighborhoods may keep them from getting outside to exercise. (You can read more about people with food insecurity at http://frac.org.) From what we have read, suppose we suspect that locations in the United States with higher rates of food hardship also would tend to have higher rates of obesity, while states with lower rates of food hardship generally would have lower obesity rates. We are describing a positive linear relationship, and we can translate this speculation into an alternative hypothesis.

To refresh your memory about Pearson's correlation coefficient, the r statistic is a measure of the degree of linear relationship between two variables. Its strongest values are -1 (indicating a perfect negative linear relationship) and $+1$ (meaning a perfect positive linear relationship), and $r = 0$ means there is no linear relationship between the two variables. How far from zero must r get before it is statistically noteworthy? It depends on the sample size and other details related to hypothesis testing. The hypothesis test will involve statements about the population correlation, ρ (this symbol is the lowercase Greek letter rho). Pearson's r is an estimate of ρ, and researchers can use this exact same r statistic as an inferential statistic, testing a null hypothesis about a population correlation.

Let's write some hypotheses about correlation. If we suspect there is a positive linear relationship between food hardship and obesity rates, then we could state this speculation as follows:

> Our sample comes from a population in which food hardship and obesity rates share a positive linear relationship.

To translate this suspicion into a directional alternative hypothesis written in symbols, we must think about values of the population correlation that would correspond to a positive linear relationship: these values of ρ would be positive. If the population correlation is greater than zero, then there is a positive linear relationship between the two variables in the population. This idea can be written as follows:

$$H_1: \rho > 0$$

This alternative hypothesis would have the following corresponding null hypothesis:

Our sample comes from a population in which food hardship and obesity rates have no linear relationship or a negative linear relationship.

The null hypothesis can be written as follows:

$$H_0: \rho \leq 0$$

Just to be complete, let's look at other ways of writing the hypotheses about a correlation. If we had reason to believe that a negative linear relationship existed between food hardship and obesity, then we would write the alternative hypothesis as

$$H_1: \rho < 0$$

This directional alternative hypothesis would correspond to the following null hypothesis:

$$H_0: \rho \geq 0$$

(Your instructor may prefer to write both of the above null hypotheses as $\rho = 0$.)

It is quite common for researchers to predict a direction for a correlation, based on their understanding of prior research on those variables. But it is possible to have a nondirectional alternative hypothesis:

$$H_1: \rho \neq 0$$

This alternative hypothesis says

Our sample comes from a population in which there is some linear relationship between food hardship and obesity, but we do not know whether the relationship will be positive or negative.

With this alternative hypothesis, we could detect a statistically noteworthy degree of linear relationship that turns out to be positive *or* that happens to be negative. The corresponding null hypothesis would say that our sample comes from a population in which there is no linear relationship between the variables:

$$H_0: \rho = 0$$

Next we will explain how to test a null hypothesis about a correlation. The steps will be familiar to you: checking whether the results are in a predicted direction, comparing a p value with our chosen significance level ($\alpha = .05$), and so forth. (A note about the use of the word *sample* in relation to the food hardship example: you may be uncomfortable with the idea of the 50 states and DC being a sample. Isn't this a population? We would say no because each state's rates of food hardship and obesity relied on samples of respondents, and we are looking at only one year's results. If we obtained different samples providing the responses used to compute the locations' rates of food hardship and obesity, we would have different results. There also is variation to be expected between samples and across different years of samples.)

Check Your Understanding

SCENARIO 13-A

Obesity researchers have investigated whether visual cues may influence how much a person eats. Wansink, Painter, and North (2005) recruited volunteers for a study involving soup consumption. The participants were led to believe that the researchers were studying how people's perceptions of taste were affected by the color of the heavy bowls in which they were served soup. In fact, the researchers were manipulating something instead of bowl color. Some participants were served soup in large, ordinary soup bowls, while other participants received soup in similar bowls that were rigged to refill themselves to a certain level as the people ate—that is, they had "bottomless bowls." Those who used ordinary bowls were able to see that the soup was disappearing, while those using the rigged bowls were getting no such visual cues. The main research question was about the difference in the amount of soup consumed for people in the two groups, but this chapter is not about mean differences, so let's look at another idea that the study explored. For those who used the rigged bowls, the researchers wanted to know whether there was a linear relationship between the amount of soup that the participants *estimated* that they had consumed and the amount of soup that they *actually* consumed. 13-1. Write the alternative hypothesis, using words and symbols. 13-2. Write the null hypothesis, using words and symbols.

Suggested Answers

13-1. *The scenario does not predict a direction for r. The alternative hypothesis is H_1: $\rho \neq 0$, meaning there is some kind of linear relationship (positive or negative) between the participants' estimates of their soup consumption and their actual amounts of soup consumed. 13-2. The null hypothesis is H_0: $\rho = 0$, meaning there is no linear relationship between the participants' estimated soup consumption and their actual soup consumption.*

Testing a Null Hypothesis About a Correlation

Based on what we have read about food hardship and its relationship with obesity, we would be justified in predicting that locations with lower rates of food hardship would have generally lower obesity rates, while locations with higher rates of food hardship would tend to have higher obesity rates. To show how to test a null hypothesis about a correlation between food hardship and obesity rates, we will use the directional alternative hypothesis that said H_1: $\rho > 0$. But first let's look at a scatterplot of the data and see if any obvious outliers might be present. As you know from Chapter 5, outliers can dampen or strengthen Pearson's r. Figure 13.1 reproduces Figure 5.1, showing the relationship between food hardship and obesity.

This data set is available via http://desheastats.com, if you would like to try to create a scatterplot like ours. As we said in Chapter 5, the horizontal line at the bottom of the graph in Figure 13.1 is called the X axis. It corresponds to a predictor variable and its numeric values. The vertical line forming the left border of the graph is called the Y axis, and it corresponds to the criterion variable and its numeric values. These two number lines are called the *axes*, pronounced AX-eez in the United States. Where the axes meet will become important in interpreting

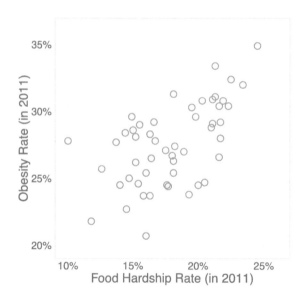

Figure 13.1

Scatterplot of food hardship and obesity rates. Food hardship is the predictor variable, appearing on the X axis, and obesity rate is the criterion variable, appearing on the vertical (Y) axis. (Food hardship data from "Food hardship in America 2011: Data for the nation, states, 100 MSAs, and every congressional district," by the Food Research and Action Center, 2012, February, retrieved from http://frac.org. Obesity data from "Adult obesity facts," by the Centers for Disease Control and Prevention, 2012, August 13, retrieved from http://www.cdc.gov/obesity/data/adult.html.)

one of the statistics in this chapter. As we scan the scatterplot from left to right, we see that the point cloud appears to go uphill. The location with the highest food hardship rate also has the highest prevalence of obesity, and the state with the lowest food hardship rate may be a bit outside the rest of the point cloud. But no obviously extreme points appear to be influencing the linear relationship. (Boxplots can show whether an extreme score is an outlier for one variable at a time, but not whether scores in a scatterplot are extreme. There are ways of assessing the effect of extreme points in a scatterplot, but this book does not cover them.)

Let's proceed to hypothesis testing. Where could we get a probability associated with our observed r statistic? We could imagine creating a sampling distribution for Pearson's r. Such a distribution could be computed by taking all possible samples of the same size from the same population, computing Pearson's r on the data from every sample, and arranging all those values of r in a distribution. As we have seen before, mathematical statisticians can tell us what our r distribution will look like, if certain assumptions are met. Then we can use a theoretical r distribution in place of a sampling distribution. We will come back to those assumptions shortly; for now, let's talk about these theoretical r distributions. Think back to the other test statistics we have learned:

- For the various t tests that we computed, we used t distributions to find critical values.
- When we computed one-way analysis of variance (ANOVA) F tests, we needed critical values from F distributions.
- Now that we have r test statistics, we will need r distributions for hypothesis testing. (We said *distributions* because the degrees of freedom will determine exactly which r distribution to use in a particular situation.)

When the null hypothesis is true and there is no linear relationship between the variables in the population, we can know what a distribution of r looks like; it may remind you of a normal distribution. There are different r distributions, depending partly on the numeric value of the degrees of freedom. The formula for df for Pearson's r is

$$df = N - 2$$

In this formula, N is the number of pairs of scores or the number of units of analysis. In our study of food hardship and obesity rates, we have data from 50 states plus the District of Columbia, so $N = 51$, meaning that $df = 51 - 2 = 49$. After we explain the hypotheses, we will show a graph of a theoretical r distribution with $df = 49$.

Like the t test statistics, the r test statistic can be a one-tailed test (if we have a directional alternative hypothesis) or a two-tailed test (if we have a nondirectional alternative hypothesis). If we were using a critical value decision rule, we would look at the alternative hypothesis to determine whether we have a one-tailed or a two-tailed test. Alpha would go into one tail if we had a directional alternative hypothesis, and it would be divided between the tails if we had a nondirectional alternative hypothesis. Then we could look in a table of r critical values using $df = 49$ for a one-tailed or two-tailed test, then determine whether our observed r statistic had equaled or exceeded a critical value. We are not presenting the critical value decision rules for the test of correlation because we want you to rely in this chapter on p value decision rules. Researchers rarely use critical value decision rules. They analyze their data with statistical software, which provides p values. For many of the inferential statistics in this book, we have given enough information that you could use a calculator to compute the test statistic and then compare the result with a critical value from a table that we provided. But we did not give you the big, ugly formula for Pearson's r. We are relying on statistical software to perform the computations. Therefore, we are going to follow the example of researchers and present only p value decision rules in this chapter.

For our example of food hardship and obesity, suppose prior research leads us to believe that there will be a positive linear relationship between the rates of food hardship and obesity. Our alternative hypothesis is

$$H_1: \rho > 0$$

When we compute the r statistic for food hardship and obesity, we find $r = .581$, indicating a possible linear relationship between the variables. We say "possible" because at this point, we do not know whether .581 is statistically noteworthy. If it is not significant, then the observed r statistic would be statistically indistinguishable from zero, meaning no linear relationship. The p value decision rule for a one-tailed test using Pearson's r is

If the observed test statistic is in the predicted direction and if the one-tailed

p value is less than or equal to alpha, reject the null hypothesis.

Otherwise, retain the null hypothesis.

If we had a nondirectional alternative hypothesis, we would use the following p value decision rule for a two-tailed test of correlation:

If the two-tailed p value is less than or equal to alpha,

reject the null hypothesis.

Otherwise, retain the null hypothesis.

We will use the p value decision rule for a one-tailed test for the example of food hardship and obesity because our alternative hypothesis is $H_1: \rho > 0$. This hypothesis is saying that the population correlation is greater than zero, meaning a positive linear relationship between food hardship and obesity rates. We computed $df = 49$, which defines the exact shape of the theoretical r distribution. Figure 13.2 shows this r distribution. Figure 13.2 is drawn as if the null hypothesis is true, so it is centered on zero. Alpha $= .05$ is shown in the upper tail because the alternative hypothesis predicts an outcome in that direction. This distribution

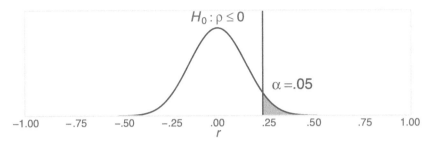

$H_0 : \rho \le 0$

$\alpha = .05$

| -1.00 | -.75 | -.50 | -.25 | .00 | .25 | .50 | .75 | 1.00 |

r

Figure 13.2

Distribution of r with $df = 49$ and a directional prediction. The r distribution may look like it is sitting on the horizontal axis, but in fact the curve extends to the limits of the r statistic: –1 and +1. We predicted a positive linear relationship, so α goes in the upper tail.

looks a lot like a standard normal distribution, and the curve appears to sit on the horizontal axis. In fact, the blue curve does not touch the horizontal axis except at two points: $r = -1$ on the left and $r = +1$ on the right, the smallest and largest possible values of r.

Because we computed a positive Pearson's $r = .581$ for our data, the results are in the predicted direction, so we have met the first part of the p value decision rule. Now we go to the second part, comparing a p value with α. Our statistical software (SAS®) reported a one-tailed p value $= .000008$, which is less than $\alpha = .05$. Therefore, we reject the null hypothesis and conclude that there is a statistically significant positive linear relationship between food hardship and obesity rates.

Correlation goes hand-in-hand with another kind of analysis, called *regression*. Before we extend our example of food hardship and obesity into the topic of regression, we will explain the assumptions of Pearson's r as a test statistic.

Check Your Understanding

SCENARIO 13-A, Continued

This scenario concerned a correlation between participants' estimates of how much soup they had consumed and the actual amount of soup they had eaten. Obesity researchers (Wansink et al., 2005) served soup in large, ordinary bowls to some participants, while other participants were given bowls that were rigged to be "bottomless" and refill themselves as the people ate. All participants were told not to move the bowls as they ate. For those who used the bottomless bowls, the researchers wanted to know whether there was a linear relationship between participants' estimates for how much soup they ate and the actual amount of soup consumed, as measured by the researchers. The alternative hypothesis was H_1: $\rho \neq 0$, meaning that the sample came from a population in which there is some linear

(*Continued*)

relationship (positive or negative) between the participants' estimated soup consumption and their actual soup consumption. Suppose the researchers computed $r = .12$ and two-tailed $p = .551$ for those using bottomless bowls. 13-3. Using $\alpha = .05$, test the null hypothesis, then explain the meaning of your decision in terms of the variable names. 13-4. Suppose the correlation for the usual-bowl group was $r = .67$, two-tailed $p = .0001$. Test the null hypothesis for this group and explain the results.

Suggested Answers

13-3. We are doing a two-tailed test because the alternative hypothesis was nondirectional. Therefore, we do not have to check whether the results were in a predicted direction. Because $p > .05$, we retain the null hypothesis. For those who had no visual cues about how much they had consumed, there was no linear relationship between the estimated consumption and the actual consumption. In other words, for those deprived of the visual cues about their consumption while they ate from "bottomless" bowls, there was no significant linear relationship between how much they thought they had eaten and how much they actually consumed. 13-4. Because $p < .05$, we reject the null hypothesis for the usual-bowl group and conclude that there was a significant linear relationship between the estimated and actual amounts of soup. The correlation was positive, so we can explain the results in these terms: having the visual cues meant that those who thought they had eaten very little actually did tend to eat less soup, and those who gave greater estimates generally had higher actual soup consumption.

Assumptions of Pearson's *r*

We can know what the distribution of Pearson's *r* looks like if its assumptions are met. The test statistic has two assumptions:

- The pairs of scores are independent of each other.
- The scores have a bivariate normal distribution in the population.

The independence assumption pertains to the participants or units of analysis, not the variables. Obviously we believe the two *variables* are related; that is why we are interested in computing the correlation between the variables. This assumption says the two scores for one participant must be independent of the other participants' pairs of scores. How might participants have scores that are related to other participants' scores? It could happen if participants influenced each other's scores. In the study of soup consumption, described in Scenario 13-A, the researchers had to take steps to make sure that participants who were eating together were not close friends because prior research said people eat more food in the presence of familiar versus unfamiliar companions. If several friends were participating in the study and they sat together while eating the soup, their results for actual soup consumption could be influenced by each other's presence, which would violate the independence assumption of Pearson's r. As we have seen with other test statistics, a violation of the independence assumption is Very Bad News and would make us distrust the results of the study. Random sampling usually is expected to assure independence, but most researchers make a judgment about whether any participant has influenced other participants' scores. Researchers typically will decide whether their use of careful research methods has ensured that the assumption is met.

The assumption of bivariate normality can be harder for some students to grasp. It is *not* merely that the population of scores is normal for the predictor variable, and it is *not* merely that the population of scores for the criterion variable is normal. Bivariate normality means that the scores from both variables (X and Y) together form a three-dimensional normal distribution in the population. The distribution looks like a mountain, and if we walked around the mountain, it would look normal from any direction. The bivariate normality assumption often is violated, but the consequences usually are not severe. Unless there is extreme skewness in one or both variables, the results of the r test statistic typically will be trustworthy.

Next we will describe how researchers can take an identified linear relationship between two variables and make predictions for a new person or unit.

Using a Straight Line for Prediction

As a descriptive statistic and as an inferential statistic, Pearson's correlation coefficient can tell us only about the strength and direction of a *linear* relationship between two variables. We said in Chapter 5, "We can imagine drawing a line through the point cloud to summarize the relationship between food hardship and obesity." Researchers actually use such lines to make predictions. The purpose of a *regression analysis* is to compute a line through the point cloud and use the line to predict a value for the outcome variable. The line that we compute is called the *regression line*. Before we can explain what it means to make predictions, we have several concepts to cover first.

The statistical term *regression* seems at odds with the purpose of this analysis: prediction. When we are predicting, it seems as if we should be looking forward; *regression* sounds as if we are looking backward. In a way, we are looking backward. For example, suppose we want to identify variables that would explain why patients with the same condition have longer or shorter hospital stays. We could identify patients within a certain age range—for example, 65 and older—who were hospitalized after a hip fracture. Many lab results, demographic characteristics, and comorbid conditions might explain why some patients had to stay in the hospital for more days than other patients. We are looking backward at explanatory variables that would predict the criterion variable, which is the length of stay. Brown, Olson, and Zura (2013) conducted such a study, and the patient's general health is one variable that predicted length of stay. Scores for general health were assigned, with lower scores indicating healthier people with fewer health problems. Not surprisingly, generally healthier people were able to leave the hospital sooner than less healthy patients. The researchers looked backward at general health and its ability to predict length of stay. Doctors could use that relationship to predict the length of stay for a patient who has just arrived at the hospital with a hip fracture, as well as plan for any additional care for less healthy patients.

Certain terminology sometimes comes up in relation to regression, and other times the term *regression* is used in a way that is unrelated to linear prediction. When researchers run a regression analysis, they sometimes say that the criterion variable was *regressed on* the predictor variable. That is just another way of saying that a regression analysis was run using that predictor and criterion variable. Another phrase that you may encounter is *regression to the mean*. This phrase is not related directly to the analysis that we are preparing to explain, but it has a historic connection with this topic. It is best explained with an example. If you have really tall parents, chances are that you are shorter than they are. Or if you have really short parents, most likely you are taller than they are. If your parents' heights are far away from average, your height probably will be closer to the mean for people of the same gender and heritage—and that's regression to the mean. Let's see how regression to the mean can play a role in research. In the health sciences, interventions sometimes are aimed at people who are at risk of bad health outcomes. For example, suppose a study was interested in identifying people for an intervention intended to bring high blood pressure under control. If people are recruited into a study at a time that their blood pressure is particularly high, there is a real possibility that the next time their blood pressure is measured, it will be closer to average, even without an intervention. That is one good reason to have a control group to compare with the group receiving an intervention: to make sure that the treatment is effective and not the result of regression to the mean.

Now that we have covered that terminology, let's talk about regression analysis. Both of your authors have taught this material many times. What we have found is that many students have some trouble understanding the topic of regression because it can be a painful reminder of bad experiences in math classes. Even the term *regression* may sound like a reaction to painful memories! Instead of launching directly into this oddly named statistical analysis, let's ease into the topic by looking at some graphs. Figure 13.3 shows a regression line going through some points, with no scatter of points around the line. The horizontal axis is the X axis, corresponding to the predictor variable, and the vertical axis is the Y axis, corresponding to the criterion variable.

Figure 13.3 has been drawn using 10 pairs of numbers. For every point, the value of X is the same as the value of Y. Suppose you are going out of town, and you subscribe to a newspaper, but instead of stopping the delivery, you ask a neighbor to pick up the newspaper every day and keep it for you. So X would be the number of days that you are out of town and Y would be the number of newspapers that the neighbor picked up. If you are gone only one day, there will be only one newspaper waiting for you at the neighbor's house; this combination of $X = 1$ and $Y = 1$ is shown as the first point near the bottom left corner of the

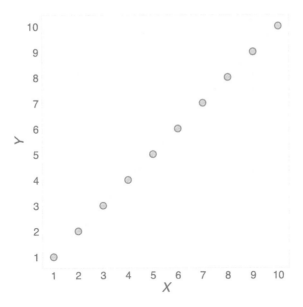

Figure 13.3

Ten pairs of numbers forming a line. Unlike the scatterplot of food hardship and obesity rates, the points in this graph form a perfect line.

graph in Figure 13.3. If you are gone for two days, there will be two newspapers waiting for you; this outcome is the second point from the left on the regression line. And so on. (This explanation of Figure 13.3 may seem quite elementary to you, but we do not want to lose anyone at this point, when some math anxiety could be triggered.)

Now let's imagine building a staircase on this line. Figure 13.4 shows such a staircase. To climb this staircase requires movement both vertically and horizontally. Another way to say it: to get from one point on the regression line to another point on the line, we have to travel vertically, and we have to travel horizontally. That is, there is some vertical *rise* (change on the Y variable) and some horizontal *run* (change on the X variable). In this example, the amount of rise is equal to the amount of run. To get from one point to the next closest point, we step up one unit (rise = 1) and we move forward one unit (run = 1). For the newspaper example, every additional newspaper being picked up by the neighbor (rise = 1) means one more day that you have been out of town (run = 1). We can think about the relationship between the two variables in terms of the vertical change on the Y variable *relative to* the amount of horizontal change on the X variable. We can express the relationship as a ratio or fraction as follows:

$$\frac{\text{Change in } Y}{\text{Change in } X}$$

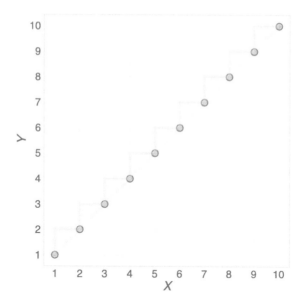

Figure 13.4

Traveling along the line by moving vertically and horizontally. How tilted is the line? We can describe the tilt according to the relationship between vertical movement and horizontal movement as we travel from left to right on the line, using this staircase.

Another way to express this relationship is as follows:

$$\frac{\text{Rise}}{\text{Run}}$$

The ratio of "rise over run" is the *slope* of the regression line, or a numeric value for how tilted the line is. In Figure 13.4, the rise is the same as the run. One more newspaper picked up means that you have been gone one more day. The slope of the regression line in Figure 13.4 can be computed as

$$\frac{\text{Rise}}{\text{Run}} = \frac{1}{1} = 1$$

The slope is the same regardless of which two points we compare. Think of it this way: the staircase in Figure 13.4 does not become steeper or less steep as we climb; the angle remains the same for our entire climb. Suppose we took two other points that are not side by side and we computed the slope. Figure 13.5 shows a comparison of the two most distant points on the regression line. Now the vertical rise is 9, meaning a change from $Y = 1$ to the level of $Y = 10$. The horizontal run is 9, meaning a change from $X = 1$ to $X = 10$. So we could express the slope in terms of these two points as follows:

$$\frac{\text{Rise}}{\text{Run}} = \frac{9}{9} = 1$$

13. Tests and Estimates for Bivariate Linear Relationships

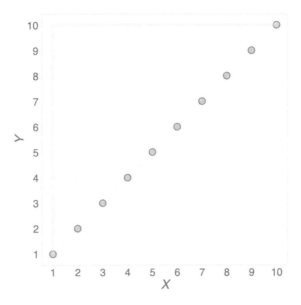

Figure 13.5

Computing slope by using any two points on the line. It does not matter which two points are used to illustrate the slope. If we are traveling from left to right on the line, the slope is the ratio of the distance we move vertically to the distance we move horizontally. The slope computation is the same, whether we compare two adjacent points or two distant points.

As you can see, the slope is still 1 because the number of newspapers picked up by your neighbor is directly related to the number of days you have been out of town.

What if the regression line is going downhill? Let's change the newspaper example a bit (we are going to use American dollars in this example). Suppose your neighbor has an 8-year-old son who wants to earn some money. You decide to pay him $1 for every day that he picks up your newspaper and saves it for you. Because you might be gone for two weeks and you want to keep him motivated, you give the parent a fund of $20. Your adult neighbor agrees to give the boy $1 every time he brings in your newspaper. As the days go by, the total amount of money left in the fund decreases by $1. Figure 13.6 shows the relationship between the amount of money left in the fund and the number of newspapers picked up.

Notice the thin black lines in Figure 13.6. The vertical black line represents the point where the number of newspapers picked up is zero, and the horizontal black line represents the point where $0 is left in the fund. The left border and bottom border of the graph are close to but not exactly the same as the X and Y axes. The difference will become important shortly. As we scan this graph from left to right, we can see that the points are on a line that is going downhill. How does this fact affect the "rise over run?" Let's do a numeric example. In Figure 13.7, we have chosen two points on the regression line; we know we do not have to choose two points that are next to each other because the slope is the same for the entire line.

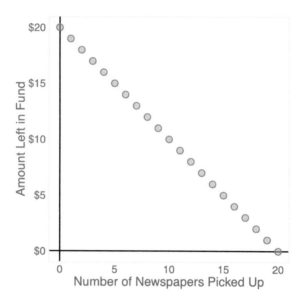

Figure 13.6

Slope when the line goes downhill. We "read" the line from left to right. As we scan the line from left to right, our eyes follow the line downhill in this graph. The slope of this line is negative.

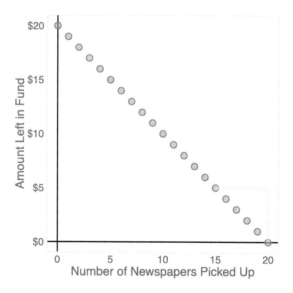

Figure 13.7

Comparing two points to compute the slope of a downhill line. The idea behind the slope is the same for a line that goes downhill from left to right. The "run" is the distance forward horizontally. The "rise" in this example is negative because the line goes downhill. So the slope is negative.

13. Tests and Estimates for Bivariate Linear Relationships

We are comparing two points:

- the point where there is $5 left in the fund and 15 newspapers have been picked up, and
- the point where there is no money left in the fund and 20 newspapers have been picked up.

How much rise and run occur between these two points? The run is 5; this is the horizontal change required to move forward from the point on the left to the point on the right in our large step in Figure 13.7. The rise is—well, you would think it is not a rise at all, because we are going down! Let's consider the change in Y to be a *negative* rise; we have to go down by 5. So rise $= -5$. Let's compute the slope:

$$\text{Slope} = \frac{\text{Rise}}{\text{Run}}$$

$$= \frac{-5}{5}$$

$$= -1$$

The regression line in Figure 13.7 has a slope of -1. The line is going downhill at the same rate as the line went uphill in Figure 13.5. The slope is one of two numbers needed to graph a line. What is the other number? Look at the arrow in Figure 13.8. Now the thin black lines are important. The vertical black line is the exact location of the Y axis. The arrow shows a point on the Y axis where no newspapers have been picked up yet, so the entire $20 remains in the fund. That is, if $X = 0$, there is $20 in the fund. This point is called the *Y-intercept*. It is the point where the regression line runs into the Y axis, which is the same thing as the predicted value of Y when $X = 0$. To draw a regression line through any point cloud, we need to know where the line runs into the Y axis and how much to tilt the line up or down. That is why it is so important to know *exactly* where the Y axis is.

We showed the thin black lines because the Y-intercept may not make sense unless the borders of the scatterplot approximate these exact locations of the X and Y axes. In Figure 13.8, you can see that the left edge and bottom edge of the graph are close to the axes' exact locations. The left border of the graph may appear to be the Y axis, and the bottom border of the graph may appear to be the X axis, but these number lines really serve only as references to the numeric values for X and Y. The thin black lines form a 90° angle and cross each other at a point called the *origin* (the point where $X = 0$ and $Y = 0$). In Figure 13.8, the bottom edge and left edge of the graph form the bottom left corner, which is close to the origin. Other scatterplots, like Figure 13.1, zoom in on the point cloud. As a result, the left and bottom edges of the graph no longer are close to the exact location of the X and Y axes, and they do not form the bottom left corner close to the origin. If you glance back at Figure 13.1, you will see that the bottom left corner is not close to the point where the food hardship rate $= 0\%$ and the obesity rate $= 0\%$. As you

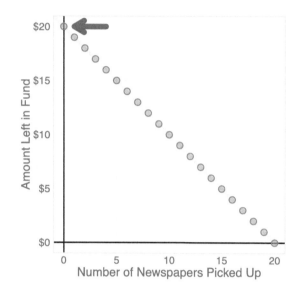

Figure 13.8

Exact locations of the axes and the point where the line crosses the Y axis. The slope is one number needed to define the line. The other number is the point where the line crosses the Y axis, shown here with the dark blue arrow. The black vertical line is the Y axis. It goes through the point on the X axis where $X = 0$. That is, no newspapers have been picked up yet, so the full $20 remains in the fund.

will see shortly, the numeric value of the Y-intercept for this data set will not make sense on that graph.

Notice that we are talking about different concepts involving one line crossing another. We have talked about a regression line crossing or running into the Y axis at a point called the Y-intercept. We interpret the Y-intercept as the predicted value of Y when $X = 0$. We also have talked about the exact location of the X and Y axes, with these two axes crossing each other at the origin. And we described the left and bottom borders of a graph running into each other to form the bottom left corner of the graph. If those borders meet at a point that is not close to the origin, it can be hard to see how the numeric value of the Y-intercept fits into the graph.

Let's show one more graph to bring together the concepts of Y-intercept and slope. Going back to the first newspaper example, suppose your adult neighbor has agreed to pick up the newspaper every day that you are out of town—but then forgets to do so until the third day. The first two days of newspapers have gotten wet because it rained, so the neighbor throws away those newspapers. Figure 13.9 shows a graph of this situation. On day 2 of your trip, no newspapers have been picked up; the first newspaper gets picked up on day 3. The neighbor consistently picks up the newspaper on every day after that. So the slope of the regression line is still 1: every additional newspaper being picked up corresponds to one more day that you have been out of town. But where is the Y-intercept? Figure 13.9 shows the regression line extended downward from the data, and we discover

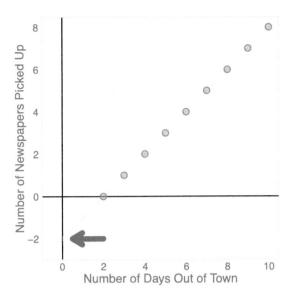

Figure 13.9

A negative *Y*-intercept. The *Y*-intercept can be a negative number, as we show in this example. But it may not be an interpretable number.

that the *Y*-intercept is a negative number: –2. It does not seem to make sense—if you have been out of town 0 days, how could a negative number of newspapers be picked up at that point? (And how can a negative number of newspapers be picked up at all?!) The *Y*-intercept mathematically is just telling us where to start the regression line and often is not interpretable. Once the line is anchored at the *Y*-intercept, then we can tilt it to the degree that we want.

The regression line has a mathematical formula. Before we give you the formula, we need a way of saying, "This formula equals the line." So what are we going to call the line? Generally the regression line is used to predict a numeric value for the outcome variable, *Y*. So the line can be called the *Predicted Y*. Continuing with our desire to free you from as many symbols as possible, for now we will write the formula using words:

$$\text{Predicted } Y = Y\text{-intercept} + (\text{slope} * \text{some value of the predictor variable})$$

The asterisk is another symbol meaning "multiply." Let's use our last example, where the neighbor did not start picking up the newspaper until the third day that you were out of town. The *Y*-intercept is –2, and the slope is 1. If we replace "some value of the predictor variable" with the symbol for the predictor variable, *X*, then the formula for the regression line in Figure 13.9 is

$$\text{Predicted } Y = -2 + (1 * X)$$

$$= -2 + X$$

Oh dear, does the symbol X give you bad memories of algebra? Let's think through the meaning of this formula. If we want to know how many newspapers will be waiting for you after the trip, we plug in the number of days that you will be gone. If you will be gone for six days, then $X = 6$. That means there will be $-2 + 6 = 4$ newspapers waiting for you. If you will be gone 12 days, then there will be $-2 + 12 = 10$ newspapers waiting for you. For any number we plugged into the formula, we can come up with a point on the regression line. Depending on what value of the X variable is plugged into the formula, we will find different points on the line. So the entire line can be explained by this formula, and a point on the line can be located for any value of X.

We will give one more example in the following Check Your Understanding question. Now that we have provided an informal explanation of slope and Y-intercept, next we will explain how a regression line can be used for prediction with a real data set, where the points do not form a perfect line.

Check Your Understanding

SCENARIO 13-B

Suppose you want a friend to come over and feed your cat while you are out of town. The cat is affectionate with you but hates everyone else, so he does not need anything besides water and his dry food while you are out of town. Your friend, Barb Dwyer, can be a little prickly herself and does not like cats, but she is willing to stop by every other day and make sure there is plenty of food and water in the cat's bowls. You tell Barb that you will pay her $20 simply for agreeing to check on the cat, and then for every day that she stops by your home, you will pay her another $10. So on the day you leave town, you will owe her $20. Two days later, Barb fills the cat's food and water bowls, so you now owe her a total of $30. Another two days later, she fills the bowls again, so now you owe her $40. On day 6, she checks on the bowls again ($50 owed), and on day 8, she checks one more time ($60 owed), then you come back to town. 13-5. On the day that you leave town ($X = 0$), how much do you owe Barb? 13-6. Sketch the X and Y axes. For this example, $X =$ the number of days that you are out of town, and $Y =$ the amount owed to Barb. 13-7. Draw a point in your sketch to represent the amount owed to Barb at the moment you leave town (0 days out of town). Then draw points to show the amount owed if you are out of town 2 days, 4 days, 6 days, or 8 days. Then draw a line through the points. 13-8. Sketch a stair step from the first point to the third point on the line. 13-9. What is the rise for your stair step? 13-10. What is the run for your stair step? 13-11. What is the slope of your regression line? 13-12. What is the Y-intercept of your regression line? 13-13. Write the formula for your regression line.

(Continued)

Suggested Answers

13-5. You owe $20 on day zero. 13-6, 13-7, and 13-8. See Figure 13.10. 13-9. *The rise is 20 because we are going up vertically from $20 owed to $40 owed. 13-10. The run is 4 because we are going horizontally from 0 to 4 days. Notice that the rise of 20 takes up about the same amount of space on this graph as the run of 4. By comparison, many of our graphs have had equal distances for rise and run; in those graphs, a run of 4 would appear to be the same distance on the horizontal number line as a rise of the same value, 4, on the vertical number line. It is common for scatterplots to have different scales for the X and Y variables. 13-11. The slope is rise over run, or 20/4 = 5. 13-12. The Y-intercept is 20 because on day zero, you owe Barb $20 for agreeing to feed the cat. Can you see how these values of slope and Y-intercept are depicted in Figure 13.10? 13.13. The formula for your line is 20 + 5X. If you are gone 10 days, then X = 10, and you would owe Barb 20 + (5 × 10) = 20 + 50 = $70.*

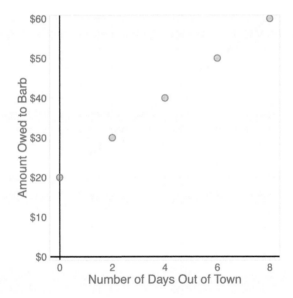

Figure 13.10

Graph of the line for Scenario 13-B. This regression line does not have a slope = 1, like our earlier examples. Notice that the distance between numbers on the X axis is different from the distance between numbers on the Y axis. That is, the graph shows a range of numbers from 0 to 8 on the X axis, and for about the same amount of distance on the page, the Y axis shows a range of numbers from 0 to 60.

Linear Regression Analysis

The previous section may have seemed quite elementary to you. Or perhaps you are grateful that we did not make assumptions about your memory of prior mathematics training. If you had any trouble with the previous section, we recommend that you take time to reread that section *now*. You need to make sure that the concepts of slope and intercept are clear before continuing with this chapter.

In regression, the slope and Y-intercept are both called *regression coefficients*; these two numbers define the regression line and specify where it is drawn. Let's think about the lines through the scatterplots in the previous section. The lines went through every point, and there were no points scattered around the lines. Those examples showed perfect linear relationships. But that is not the kind of scatterplot we saw for food hardship and obesity rates. Figure 13.1 actually had scattered points. Suppose we have the food hardship rate for a new location in the United States. Could we use the point cloud to estimate what the obesity rate would be for that location? It would be hard to decide where in the cloud we might find an estimated obesity rate for that location. But based on the linear relationship between the two variables, we could draw a regression line through the point cloud, and the line would give us a prediction for the new location's obesity rate. This predicted obesity rate would be a point on the regression line.

Linear regression analysis involves predicting an outcome based on (1) a known value for a predictor variable as well as (2) a linear relationship between the predictor variable and criterion variable. If we know a new location's food hardship rate, we can rely on the equation for a regression line through a scatterplot of data to estimate or predict the new location's obesity rate. The line's equation is called a *regression equation*, *regression formula*, or *prediction equation*. You might hear the analysis in this chapter referred to as *simple regression*. This term is not a criticism; "simple regression" means there is only one predictor variable and one criterion variable.

Suppose we wanted to draw a regression line through the scatterplot for food hardship and obesity. If we were relying on our best guess, we could print out the scatterplot and draw a line using a pen and a straight edge. We would try to make the line go through the middle of the point cloud and somehow be representative of the linear relationship between food hardship and obesity rates. The problem with this approach is that every person with a pen and a straight edge could draw a slightly different line. Many researchers use a particular mathematical approach called *ordinary least squares linear regression*, and that is what we will use in this book. We will explain later what *ordinary least squares* means. For now, let's take a look at a regression line that has been mathematically determined by our statistical software.

Figure 13.11 shows the scatterplot for the food hardship and obesity rates. This graph is similar to the first graph in this chapter, but now there is a regression line drawn through the scatterplot. Statistical analysis defines where the green line is located in Figure 13.11. We can think of this regression line as being made up of many points, a continuous series of predicted obesity rates, so we will call this line the predicted obesity rate. For any rate of food hardship (X) on the horizontal

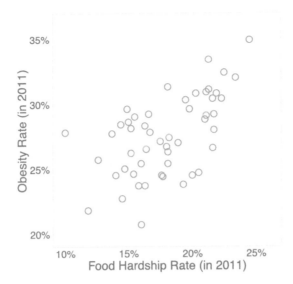

Figure 13.11

Regression line for food hardship and obesity rates. We have computed a regression line for the food hardship/obesity data set. The tilted green line is plotted based on the results of our analysis. (Food hardship data from "Food hardship in America 2011: Data for the nation, states, 100 MSAs, and every congressional district," by the Food Research and Action Center, 2012, February, retrieved from http://frac.org. Obesity data from "Adult obesity facts," by the Centers for Disease Control and Prevention, 2012, August 13, retrieved from http://www.cdc.gov/obesity/data/adult.html.)

number line, we can go straight up to the regression line; that point on the regression line will be a predicted obesity rate. Let's do an example. The state with the highest rate of food hardship in 2011 was Mississippi, represented by the point closest to the top right corner of Figure 13.11. Mississippi's food hardship rate was 24.5%, meaning about one in four adults surveyed in 2011 said that at some point in the previous year, they lacked money to buy food for their family. We know what Mississippi's reported obesity rate was for 2011: 34.9%, the highest in the country. What would this regression line have suggested as a *predicted* obesity rate for Mississippi? Let's examine Figure 13.12.

From the horizontal axis where food hardship is 24.5%, we could draw a vertical line like the one shown in Figure 13.12. This blue vertical line runs into the regression line. If the blue vertical line were to continue past the regression line, it would run into the point representing Mississippi. When the vertical line reaches the regression line, it is indicating the predicted obesity rate. We have to look at the numbers on the vertical axis to find the numeric value for Mississippi's predicted obesity rate. The blue horizontal line running from the regression line to the Y axis shows that the predicted obesity rate for Mississippi would be about 31%.

We can be more exact than this visual estimate of 31%, but we showed you Figure 13.12 because we wanted you to understand what the regression line is for.

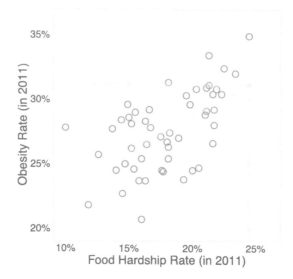

Figure 13.12

Predicting Mississippi's obesity rate. To find a predicted obesity rate for Mississippi, we are looking for a point on the regression line. The vertical blue line goes straight up from Mississippi's food hardship rate (24.5%). Where the vertical blue line runs into the green regression line, we have found Mississippi's predicted obesity rate. To find out that predicted obesity rate, we have to look at the Y axis, so we follow the horizontal blue line to the left. When we use the equation for the regression line, we will discover Mississippi's predicted obesity rate is 31.1%. (Food hardship data from "Food hardship in America 2011: Data for the nation, states, 100 MSAs, and every congressional district," by the Food Research and Action Center, 2012, February, retrieved from http://frac.org. Obesity data from "Adult obesity facts," by the Centers for Disease Control and Prevention, 2012, August 13, retrieved from http://www.cdc.gov/obesity/data/adult.html.)

When we analyzed the data on food hardship and obesity rates using statistical software, we found a Y-intercept = 17.793 and a slope = 0.545, which means the formula for the line is

$$\text{Predicted Obesity Rate} = 17.793 + (0.545 * \text{Food Hardship})$$

Before we go any further, look again at Figure 13.12. Do you see where the green regression line seems to meet the vertical axis? It is between 20% and 25%. But the Y-intercept is 17.793. This is an example of the numeric value of the Y-intercept not making sense in a graph that does not show the exact locations of the X and Y axes, with their intersection at the origin. Imagine extending the regression line toward the bottom left corner, pushing the left border until it showed where food hardship = 0%, and pushing the bottom border until it showed where obesity = 0%. There would be a lot of white space on the page, but if we extended the regression line, it would cross the Y axis at the Y-intercept = 17.793.

Let's simplify the way this regression equation is written. Instead of using an asterisk, we can leave out the multiplication sign; any number shown right next to the X means that the number and some value of X will be multiplied. Instead of "Food Hardship," we can use the symbol X to stand for any food hardship rate that might interest us. Now we can rewrite the regression equation as

$$\text{Predicted Obesity Rate} = 17.793 + 0.545X$$

For any value of food hardship that we insert in place of X, we can do the math and get a predicted obesity rate—that is, a predicted score on the criterion variable.

Let's interpret the meaning of the slope, 0.545. Remembering that this statistic is "rise over run," what is "rise" in this example? It would be the vertical change in the Y variable, obesity rate. What would be the "run?" It would be the horizontal change in the X variable, food hardship. So we could write the slope as

$$0.545 = \frac{\text{Change in obesity rate}}{\text{Change in food hardship rate}}$$

What if we divide 0.545 by 1? It will not change the numeric value of the slope, but maybe it would help us to interpret the results:

$$\frac{0.545}{1} = \frac{\text{Change in obesity rate}}{\text{Change in food hardship rate}}$$

We may interpret the slope as saying that a 1% change in food hardship is associated with a 0.545% change in the obesity rate. Because the slope is positive, we can state these changes as numeric increases: for every 1% increase in food hardship, we can expect obesity rates to increase more than one-half of 1%.

We have chosen not to torture you with the equations for the Y-intercept and slope. The formula for the Y-intercept is small and cute, but the formula for the slope is big and ugly, so we will let the statistical software do the work. But we can use the regression equation to predict an obesity rate. Let's do two examples, starting with Mississippi, where the food hardship rate was 24.5%. We will insert 24.5 in place of the X in the following equation:

$$\text{Predicted Obesity Rate} = 17.793 + 0.545X$$

$$= 17.793 + 0.545(24.5)$$

$$= 17.793 + 13.3525$$

$$= 31.1455$$

$$\approx 31.1\%$$

Can you see how the food hardship rate of 24.5% corresponds to the blue vertical line and this predicted obesity rate of 31.1% corresponds to the blue horizontal line in Figure 13.12? There was no need to compute a prediction for Mississippi, though, because we already know Mississippi's actual obesity rate was 33.4% in 2011. When *would* we want to compute a predicted obesity rate? Consider this example: suppose the U.S. Congress granted statehood to Puerto Rico, currently a U.S. territory, and we wanted to predict the prevalence of obesity in Puerto Rico, based on its food hardship rate. We could not find the actual rate of food hardship for Puerto Rico, so let's pretend it is 18%, which would mean that 18% of adults lacked money for food at some point in the previous year. We could predict Puerto Rico's obesity rate, using the regression equation that was based on the linear relationship for 51 other locations in the United States. Here is the computed prediction:

$$\text{Predicted Obesity Rate} = 17.793 + 0.545(18)$$

$$= 17.793 + 9.81$$

$$= 27.603$$

Based on the linear relationship between food hardship and obesity, we might predict that a location with 18% food hardship would have an obesity rate of about 27.6%. How trustworthy is this kind of prediction? Good question, and we will provide some general answers when we cover statistical significance and confidence intervals for regression. But maybe the best answer is that the results of any study should be considered only one small piece of the picture that researchers are trying to draw about a phenomenon. If we used the rates of food hardship and obesity for a different year, most likely we would come up with slightly different correlation and regression results. The rates depend on who answers the surveys in the different locations, and different people are surveyed every year. There also may be societal changes across time, affecting the rates. If the positive linear relationship between food hardship and obesity rates actually exists in the population, then we should see correlation and regression results similar to ours in subsequent years, which would lend credibility to a prediction equation like ours.

One of the most valuable aspects of regression analysis is that we can test hypotheses about whether a linear relationship is statistically significant. If Pearson's *r* is significant, then the linear relationship will be significant when the simple regression analysis is performed. The difference is that Pearson's *r* did not distinguish between the predictor and criterion variables; that is, it did not matter to Pearson's *r* whether food hardship predicted obesity rates, or obesity rates predicted food hardship. In simple regression, it matters—the slope and *Y*-intercept will be different when $X =$ food hardship versus $X =$ obesity. Researchers should rely on theory and prior research in specifying which is the predictor variable and which is the criterion variable. Next, we will talk about the mathematical determination of the best-fitting line for the data, then hypothesis testing with simple regression analysis. We will conclude the chapter by talking about confidence intervals in simple regression.

SCENARIO 13-C

Falvo and Earhart (2009) conducted a study of the characteristics of patients with Parkinson's disease. Participants were measured on many variables that could affect the mobility of people with this disease. Among these variables were tests called the Six-Minute Walk Distance (6MWD) and the Timed Up and Go (TUG). The 6MWD measures the number of feet that a person can walk in six minutes at a normal pace, using any assistive device (like a cane) that the person ordinarily may use. The TUG measures how many seconds it takes a person to get up from a chair, walk 3 m, return to the chair, and sit down. These researchers graciously shared some of their data with us; you may download the data on these two variables via http://desheastats.com. Suppose we want to use the TUG (a measure of mobility) as a predictor of 6MWD (a measure of walking capacity). Shorter TUG times indicate greater ease of mobility, and longer 6MWDs are interpreted as indicative of greater walking capacity. Suppose we have run a simple regression analysis using the data from Falvo and Earhart, and we have found a slope $= -60.96$ and a Y-intercept $= 1945.319$. 13-14. Write the regression formula for predicting 6MWD, where $X =$ TUG score in seconds. 13-15. Suppose we are physical therapists and we have a new patient whose TUG score is 20 seconds. We think this patient is similar to those in the study by Falvo and Earhart. Predict how far this patient could walk in 6 minutes.

Suggested Answers

13-14. Predicted 6MWD $= 1945.319 + (-60.96)X$. This formula can be rewritten as Predicted 6MWD $= 1945.319 - 60.96X$. 13-15. This patient's predicted 6MWD $= 1945.319 - 60.96(20) = 1945.319 - 1219.2 = 726.119$ ft. Based on the results from Falvo and Earhart, we would predict that our patient could walk a little over 726 feet in six minutes.

Determining the Best-Fitting Line

As we suggested earlier, different people could look at the scatterplot for food hardship and obesity rates, then draw different lines that they thought represented the linear relationship between these two variables. The line that we drew through the scatterplot for food hardship and obesity was not arbitrary. The determination of the best-fitting line is made mathematically, and there are different criteria that could be used to say, "Here is the best line." The criterion that we are using is called the *ordinary least squares criterion*, and it gives us a kind of average line. If you read a journal article that mentions "OLS regression," then this criterion has been used to decide the best prediction equation for the data.

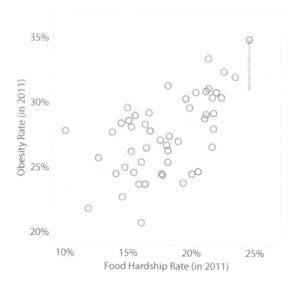

Figure 13.13

Distance between Mississippi's actual obesity rate and its predicted obesity rate. The tan line connects the dot for Mississippi to the regression line. If we extended the tan line down to the X axis, we would reach Mississippi's food hardship rate, 24.5%. As we saw in Figure 13.12, Mississippi's predicted obesity rate is the point on the regression line directly above Mississippi's food hardship rate. The tan line represents the gap between the actual obesity rate and the predicted obesity rate. (Food hardship data from "Food hardship in America 2011: Data for the nation, states, 100 MSAs, and every congressional district," by the Food Research and Action Center, 2012, February, retrieved from http://frac.org. Obesity data from "Adult obesity facts," by the Centers for Disease Control and Prevention, 2012, August 13, retrieved from http://www.cdc.gov/obesity/data/adult.html.)

Before we explain the ordinary least squares criterion, we need to introduce another concept first. Figure 13.13 shows the scatterplot for the food hardship and obesity data, except now a tan vertical line has been added. The tan line connects the regression line and the point representing Mississippi. The tan vertical line represents the distance between Mississippi's actual obesity rate in 2011 (34.9%) and the state's predicted obesity rate (31.1%, based on our computations above). This distance is called an *error* or *residual* in regression; we will use these terms interchangeably. An error in regression is computed by taking the actual Y score minus the predicted Y score. For Mississippi, the residual is 34.9 – 31.1 = 3.8. So the actual obesity rate for this state is 3.8% higher than the predicted obesity rate, based on our regression formula. If a state has a point below the regression line, the error will be a negative number because the actual obesity rate would be less than the predicted obesity rate.

We could write a chapter about the residuals in regression, but this book is not intended to be a sleep aid. We need this concept to explain the ordinary least squares criterion. Many statistics in this book have involved squaring numbers.

Let's build on what you already know: we squared numbers in Chapter 2 when we computed variances. From each score, we subtracted the mean to find how far the score was from the mean. We could not use the sum of the distances as a measure of spread because they always summed to zero. So we squared those distances and computed an average squared distance (for the sample variance) or almost the average squared distance (for the unbiased variance).

From our discussion of squaring numbers, you may surmise correctly that the ordinary least squares criterion has something to do with squaring numbers. Specifically, it has to do with squaring the errors. If the error (residual) for Mississippi is squared, it almost would be like computing the area of a square rug. That is, we could envision creating a square of area associated with Mississippi's residual, as shown in Figure 13.14. We could do the same thing for every point in Figure 13.14: we could find each state's residual, represented by a vertical line from each point to the regression line, then we could imagine drawing a square for each state, with the residual forming one of the four equal sides. We also could add up the area contained in those squares.

Why would we want to draw a bunch of squares on the scatterplot, and why would we want to add up the areas in those squares? We do not actually do these

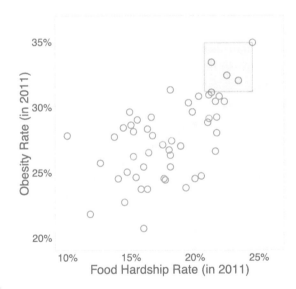

Figure 13.14

Illustrating Mississippi's squared residual. If we squared Mississippi's residual, it would be like finding the area of a rug, with each side being equal to this state's residual. The ordinary least squares criterion is related to squared residuals. (Food hardship data from "Food hardship in America 2011: Data for the nation, states, 100 MSAs, and every congressional district," by the Food Research and Action Center, 2012, February, retrieved from http://frac.org. Obesity data from "Adult obesity facts," by the Centers for Disease Control and Prevention, 2012, August 13, retrieved from http://www.cdc.gov/obesity/data/adult.html.)

things—we are using the concept to explain the ordinary least squares criterion. Here it is:

The ordinary least squares criterion says the sum of the squared errors is a number that is smaller than any similar number could be computed for any other line through the same data set.

In other words, the sum of the areas inside those squares drawn around the ordinary least squares regression line will be a smaller sum, compared with the squares that could be drawn around any other line through that data set. No other line results in less area being contained in those squares. The statistical software that computes the formula for our regression line for food hardship and obesity is finding the line that will result in the smallest sum of squared errors. (Remember the term *sum of squares*? Here is another example of numbers being squared and added up.) We mentioned earlier that the ordinary least squares regression line was a kind of average line. As it turns out, the sum of the errors (without squaring) is zero—just as the distances of scores from the mean summed to zero.

Next we will explain how to test a hypothesis about a linear relationship between two variables using regression analysis.

Check Your Understanding

13-16. You now know that the sum of the residuals is zero for the regression line determined by the ordinary least squares criterion. What would it mean if the sum of *squared* residuals was 0?

Suggested Answer

13-16. If the sum of squared errors = 0, then all of the data points form a perfect line, like our examples illustrated by Figures 13.3 through 13.10. If the points are on the line, there is no error in prediction; the actual Y score equals the predicted Y score.

Hypothesis Testing in Bivariate Regression

We started this chapter by talking about Pearson's correlation coefficient r as a test statistic. The significance of a linear relationship between two variables also can be determined with a bivariate regression analysis. We can check whether the slope of the regression line is significantly different from zero—that is, the slope differs significantly from a flat horizontal line with zero tilt. The hypotheses in regression analysis typically are nondirectional, and the confidence intervals that we will present later also are almost always two sided.

To make our explanation of this hypothesis test similar to the way we presented other hypothesis tests in this book, we must think about the slope as a statistic that estimates a parameter. The slope in the sample may be symbolized by the letter b, and it estimates the slope in the population, which may be symbolized as β, the lowercase Greek letter beta. You have seen the symbol β before; it was used as the symbol for the probability of a Type II error. On behalf of statisticians everywhere, we apologize for the fact that the same symbol is being used for radically different concepts. But we cannot change the symbol to something that you would be unlikely to find in any other statistics book. If our alternative hypothesis says there is some linear relationship between food hardship and obesity in the population, then we could state this hypothesis as follows:

$$H_1 : \beta \neq 0$$

The corresponding null hypothesis would say that our sample came from a population with no linear relationship between the rates of food hardship and obesity. This null hypothesis can be written as follows:

$$H_0 : \beta = 0$$

We can compare the sample slope with this hypothesized value for the slope in the population. Because we are going to be computing another kind of t test, let's think back to the one-sample t test. Its numerator is the difference between the sample mean and the hypothesized population mean, and this difference was divided by an estimated standard deviation. We are going to compute a similar test statistic in regression, except now the numerator will have a sample slope minus the hypothesized population slope. But the null hypothesis says the population slope is zero, which will simplify the test statistic. The denominator will be a kind of standard deviation. Like other statistics, the slope for the sample has a sampling distribution. If we took all possible samples of the same size from the same population and computed a slope on every sample, we could arrange those slopes (b statistics) into a distribution. That distribution would have a standard deviation, called the standard error of the slope, which will be the denominator of our test statistic.

This test statistic can be called the t test for the slope. As we said earlier, both the Y-intercept and the slope are called regression coefficients, so does that mean we could perform a t test for the Y-intercept? Yes, but it rarely will make sense to do that test. In the food hardship example, the Y-intercept was 17.793, meaning that if the relationship that we observed in the data were extended to the situation in which a location had zero food hardship, then the obesity rate would be 17.793%. The t test for the slope would test whether this percentage was different from zero. That makes no sense to us. Besides, it is a bad idea to generalize the results of a regression analysis beyond the range of the data that were used to create the regression line. Suppose there were an impoverished island that came under the United States' protection, and the island had a food hardship rate of 60%. Would it be a good idea to use our regression equation to estimate the obesity rate on that island? We don't think so, because our data set had a maximum food hardship rate of 24.5%. Thus, we have no information in our data set about the relationship between food hardship and obesity in locations where more than half of the adults last year experienced one or more occasions of lacking money to buy food. We suspect that extreme rates of food hardship would not be related to ever-increasing rates of obesity.

Back to our *t* test for the slope. We already have described the formula for this statistic: the numerator is the sample slope, *b*, minus the hypothesized value for the population slope, β, and the denominator is the standard error for the slope statistic. This standard error can be abbreviated as SE_b. So the formula for the *t* test for the slope is

$$t = \frac{b - \beta}{SE_b}$$

Because the null hypothesis says the population slope is zero, this formula simplifies to the following:

$$t = \frac{b}{SE_b}$$

To test the null hypothesis, we use the following *p* value decision rule:

If the observed *p* value for the *t* test for the slope is less than or equal to alpha, then reject the null hypothesis. Otherwise, retain the null hypothesis.

Let's look at part of the results from one statistical software package (SAS®) for the simple regression analysis of the food hardship/obesity data. Table 13.1 shows these results.

The first column lists Intercept and Food Hardship. The first row of results is associated with the Y-intercept, and the second row of results is for the other regression coefficient, the slope of the regression line representing the effect of food hardship on our outcome variable. The column labeled Parameter Estimate lists the two statistics for the regression equation: the Y-intercept and the slope. The rounded results for the Y-intercept and the slope were mentioned earlier in this chapter. We are going to ignore the rest of the numbers on the Intercept line, focusing instead on the Food Hardship line. If we take the unrounded slope of 0.54537 and divided it by its standard error, 0.10926, we will get the (rounded) *t* test for the slope, *t* = 4.99, shown in the column labeled "*t* value." The last column of the same row shows a two-tailed *p* value. This *p* value is labeled "Pr > |*t*|," an abbreviation that we could explain, but we do not want to torture you. Because the *p* value is extremely small and less than any typical significance level, such as our usual .05, we can reject the null hypothesis and conclude that there is a significant linear relationship between food hardship and obesity rates. Another

Table 13.1 Regression Analysis of Food Hardship and Obesity Rates

Variable	Parameter Estimate	Standard Error	*t* Value	Pr > |*t*|
Intercept	17.79331	1.99268	8.93	<.0001
Food Hardship	0.54537	0.10926	4.99	<.0001

13. Tests and Estimates for Bivariate Linear Relationships

way to say the same thing is that the slope of the regression line for food hardship and obesity is significantly different from zero. Because the slope is a positive number, we can conclude that food hardship had a significant positive linear relationship with obesity rates for the 51 locations in the United States in 2011. This is the same conclusion that we drew when we used Pearson's r to test a null hypothesis about the linear relationship between these variables.

Researchers often report a confidence interval for the slope. Next, we will conclude our discussion of simple regression by giving a brief description of this confidence interval and what it can tell us about the linear relationship between two variables.

Check Your Understanding

SCENARIO 13-C, Continued

This scenario concerned the data from Falvo and Earhart (2009), who measured people with Parkinson's disease. The researchers measured how far the patients could walk in six minutes (6MWD) and how long it took them to get up from a chair, walk 3 m, and return to sit in the chair (TUG). Suppose we think that TUG time will be a significant predictor of walking capacity. We have analyzed these researchers' data and found the following results (Table 13.2).

13-17. Write the null hypothesis for the t test for the slope. 13-18. Test the null hypothesis, using $\alpha = .05$. 13-19. Explain the meaning of your decision on the null hypothesis.

Suggested Answers

13-17. The scenario does not imply a directional prediction. The null hypothesis can be stated as H_0: $\beta = 0$. This statement says that our sample comes from a population where there is no linear relationship between the TUG times for patients with Parkinson's disease and the distance they can walk in six minutes. 13-18. Because $p < .05$, we reject the null hypothesis. 13-19. The rejection of H_0 means that there is a significant linear relationship between TUG times and walking distance in six minutes. The slope is negative (−60.96), which means the relationship is inverse. Participants who take more time to get up from a chair, walk 3 m, and return to a sitting position in the chair tend to walk shorter distances in six minutes, while those who can do the TUG task more quickly generally could walk greater distances in six minutes.

Table 13.2 Regression Results for Timed Up and Go as a Predictor of Walking Distance

Variable	Parameter Estimate	Standard Error	t value	Pr > $|t|$
Intercept	1945.319	96.219	20.218	<.0001
TUG	−60.960	8.543	−7.136	<.0001

Confidence Intervals in Simple Regression

Statistical software packages typically have an option in simple regression, allowing the data analyst to obtain confidence intervals—that is, interval estimates of both the population Y-intercept and the population slope. We just explained the t test for the slope, so it will not be a surprise for you to learn that a t critical value is used in the calculation of the margin of error for these confidence intervals. Our focus here will be on the interpretation of the confidence interval, not the computational details.

The sample slope is a point estimate of the population slope. But across repeated samples, we could get different values for the sample slope. The confidence interval is an interval estimate of the population slope, and different samples would produce different intervals. Like other confidence intervals presented in this book, our 95% confidence interval for the slope may or may not contain the slope for the population. But for 95% of the samples that we could draw from the same population, we would get confidence intervals that bracket the true value of the population slope.

When we analyzed the food hardship and obesity data, we calculated a 95% confidence interval of [0.33, 0.76] for the slope. Like other confidence intervals, this interval estimate has a range of values, with the point estimate in the middle; the slope for our data was 0.545. Consider the meaning of the interval [0.33, 0.76]. We know there is variability from sample to sample, and this interval attempts to quantify the sampling variability. We do not expect the sample slope to equal the population slope. The confidence interval gives a reasonable range of values where the true population slope might be.

Let's test a null hypothesis using this confidence interval. We will test the same null hypothesis that we stated earlier, $H_0: \beta = 0$. This null hypothesis says our sample comes from a population with no linear relationship between food hardship and obesity. Does the interval [0.33, 0.76] contain zero? No, it does not. Therefore, we reject the null hypothesis and conclude there is a significant linear relationship between food hardship and obesity. Because the slope is positive, we can say that locations with lower rates of food hardship tend to have lower prevalence of obesity, while locations with higher rates of food hardship generally have higher obesity prevalence.

As we near the end of our discussion of regression, we will provide a few cautionary statements about regression analysis.

Check Your Understanding

SCENARIO 13-C, Continued

We return to the research by Falvo and Earhart (2009), who measured people with Parkinson's disease. The 6MWD was predicted by the TUG, the amount of time required to stand up from a chair, walk 3 m, and return to a seated position in the chair. Suppose we have computed the following

(Continued)

95% confidence interval for the slope: [−77.986, −43.934]. 13-20. Explain the meaning of this interval. 13-21. What null hypothesis could be tested with this interval? 13-22. Test the null hypothesis. 13-23. Explain the meaning of your decision on H_0.

Suggested Answers

13-20. This confidence interval is an interval estimate of the population slope for the linear relationship that TUG has with 6MWD. This interval may or may not contain the true population slope, but 95% of the time that we compute a confidence interval like ours across repeated samples from the same population, we get intervals that do contain the true population slope. 13-21. We could test a null hypothesis that says the population slope is zero, indicating no linear relationship between the two variables. That is, H_0: $\beta = 0$. 13-22. Because the interval [−77.986, −43.934] does not contain zero, we reject the null hypothesis. 13-23. This decision means TUG has a significant linear relationship with 6MWD. Because the slope was negative (which we can tell because both the upper and lower limits of the confidence interval are negative), we conclude that TUG shares a significant negative linear relationship with walking distance.

Limitations on Using Regression

Chapter 5 on correlation listed many factors to consider when using Pearson's *r*. The same considerations apply to simple regression:

- Only linear relationships can be detected.
- A linear relationship being detected does not imply that a causal relationship exists between the variables.
- Outliers can influence the regression coefficients.
- The regression coefficients can be affected if a limited range of scores is used.
- Combining groups of scores can affect the regression coefficients.
- If a participant or unit of analysis is missing a score on one or both variables, that person or unit of analysis typically is omitted from the computation of the regression coefficients.

We already mentioned that it matters in regression which variable is designated as the predictor variable and which variable is the criterion variable. The numeric values of the *Y*-intercept and slope will change if the variables are reversed. We also mentioned that we must avoid generalizing about a possible linear relationship beyond the range of the data. The simple regression equation's two coefficients (*Y*-intercept and slope) are computed based on the linear relationship that is evidenced by the data. The food hardship rates in our data set ranged from 10% to 24.5%. It would be unwise to try to predict an obesity rate for a place with a food hardship rate less than 10% or greater than 24.5%. The data are like flood lights in a park at night; we can "see" a linear relationship only in the area being brought to light by the data. Food hardship rates less than 10% or greater than 24.5% are like the dark areas of the park; we do not know what might be going on between food hardship and obesity rates where we have no data.

Regression is a huge area within statistics, and entire courses are taught on this topic alone. We hope this introduction will serve as a solid basis for understanding descriptions of regression in research articles and any future statistics courses you may take. (Oh, you know you want to take more stats classes!)

What's Next

There are many kinds of regression beyond what we have presented in this chapter. We focused on bivariate linear regression—*bivariate* meaning we had only one predictor variable and one criterion variable, and *linear* because that is the kind of relationship being analyzed. This analysis also has been called simple regression or ordinary least squares regression. It is possible to have *multiple regression*, meaning multiple predictors of one criterion variable. Suppose we thought poverty rates and food hardship rates together would predict obesity rate in the United States. If so, we could conduct a multiple regression analysis and try to determine which predictor variable had the strongest influence on the outcome variable. We even could have *multivariate* multiple regression, with many predictors of many outcomes, but that analysis gets pretty messy. The aforementioned analyses are variations on *bivariate* regression; what about variations on the term *linear*? Yes, we can have regression involving nonlinear relationships, including regression that involves an outcome variable with only two outcomes, such as "the person has the disease" or "the person does not have the disease."

In Chapter 14, we will talk about relationships between variables, but instead of quantitative variables, categorical variables will be the focus of the analysis.

Exercises

SCENARIO 13-D

(Inspired by Murdock, 2013. Details of this scenario may differ from the actual research.) Does the widespread use of text messaging with mobile phones affect people's sleep? Suppose we want to know whether the number of text messages that college freshmen usually send in a week will predict how much trouble they have with sleeping. We recruit a sample of 104 first-year college students, who give us permission to count the number of text messages that they send in a week. We create a cell-phone app that gives us

(Continued)

the total number of texts without revealing the content of any messages. The students fill out a survey that asks them about many variables, such as the number of credit hours they are carrying. Our main interest is on their score for a measure of sleep trouble, where a lower score means less sleep trouble, and a higher number means more trouble falling asleep and staying asleep. Suppose we have analyzed the data from the study of text messaging and sleep trouble. Here are our results: $r = .46$, two-tailed p value $= .0006$, mean number of texts $= 285$, slope $= 0.23$, Y-intercept $= 18$, and the standard error of the slope, $SE_b = 0.0632$. 13-24. What kind of research is this? 13-25. What kind of variable is *sleep trouble score*? 13-26. What kind of variable is *number of text messages sent in a week*? 13-27. Using words and symbols, write the alternative hypothesis for a test of correlation for this scenario. 13-28. Test the null hypothesis for the correlation using $\alpha = .05$, then explain the meaning of your decision, using the variable names. 13-29. Using words and symbols, write the alternative hypothesis for the population slope. 13-30. Suppose we are examining the output from statistical software, and we see that the t test for the slope equals 3.64, with a two-tailed $p = .0006$. Test the null hypothesis about the population slope, then explain the meaning of your decision, using the variable names. 13-31. Write the regression equation, using the results given previously. 13-32. Suppose there are two teenagers in your extended family who will start college next year. You know that one of these teenagers, Mona Tone, seems to balance her responsibilities and social life, while getting sufficient rest and exercise. You hope that she will take care of her physical health after she starts college. You tell her about this research, and you ask her to install the cell-phone app to count the number of text messages she sends in a week. The number turns out to be 240. What would you predict for her sleep trouble score? 13-33. Suppose you are thinking about the other teenager in your extended family, Corey Lation, who also will start college next year. Corey installs the cell-phone app for you, and he later reports that he sent 1,255 text messages in a week. Based on the results of the previous study, why would we discourage you from computing Corey's predicted sleep trouble score? 13-34. What can we say about the internal validity of the study of text messages and sleep trouble?

SCENARIO 13-E

(Inspired by Stamps, Bartoshuk, & Heilman, 2013. Details of this scenario may differ from the actual research.) Dementia researchers were looking for a quick, inexpensive way to screen people for possible impairment of the first cranial nerve. Such impairment can be associated with Alzheimer's disease (AD). If the first cranial nerve is impaired, then patients' sense of smell also can be reduced. Peanut butter is considered a "pure odorant" that stimulates the first cranial nerve. If people have trouble smelling peanut butter, then they may have impairment in that nerve, possibly an early indication of cognitive decline. Suppose we are conducting a study in which we are using a peanut butter smell test, and we have recruited 94 older adults to participate.

(*Continued*)

Each person is tested individually with eyes and mouth closed. The participant is instructed to breathe normally. We open a small jar containing 14 g of peanut butter and hold it at one end of a 30-cm ruler. The ruler is held horizontally in front of the participant, with one end touching the jar of peanut butter and the other end touching the left nostril; the participant holds the right nostril closed. The subject is instructed to say whenever s/he smells something. If the participant does not smell something when the jar is 30 cm from the nostril, then during the next exhale, we move the container 1 cm closer to the nostril. If the subject does not smell anything, the jar is moved another centimeter during the next exhale. This process continues until the subject reports smelling something, at which point the distance from the nostril is recorded. After the participant completes the smell test, we administer the Mini-Mental State Exam (MMSE), a widely used questionnaire used to screen for cognitive impairment. Lower scores indicate lower abilities related to mathematical reasoning, orientation, and memory, while higher scores indicate better abilities. The neurologists on our research team predict that shorter left-nostril smell distances generally will correspond to lower MMSE scores (indicating worse cognitive abilities), while longer left-nostril smell distances generally will be associated with higher MMSE scores (indicating better cognitive abilities). 13-35. Write the alternative hypothesis for the test using Pearson's r, using words and symbols. 13-36. Suppose we analyze the data from this study and we find $r = .338$, one-tailed $p = .0004$. Test the null hypothesis using $\alpha = .05$, then explain the results, using the variable names. 13-37. Suppose we have found a slope $= 0.441$ and the standard error of the slope $= 0.1277$. Compute the t test for the slope. 13-38. Why must this t test be statistically significant, given the answers to the other questions about this scenario?

SCENARIO 13-F

(Inspired by Noble, Fifer, Rauh, Nomura, & Andrews, 2012. Details of this scenario may differ from the actual research.) Babies who are born between 37 and 41 weeks of gestation typically are considered full term and have been studied as a homogeneous group, meaning these babies are all similar. A neuroscientist was interested in the later academic achievement of these children. For those who were categorized as having been born full term, would achievement vary in accordance with variation in gestational age? She obtained access to data from thousands of children, whose gestational ages and standardized third-grade reading scores were available. Suppose we are analyzing her data and we want to determine whether there is a linear relationship between gestational age and reading scores. Let's say that we obtain the following results: Y-intercept $= 46.5$, slope $= 0.8$. 13-39. What does the Y-intercept mean? 13-40. Interpret the slope in terms of "rise over run," using the variable names. (Hint: A positive change in gestational age can be thought of as "an additional week in the womb.") 13-41. What kind of relationship appears to exist between gestational age of full-term babies and third-grade reading achievement scores, and how do you know?

13. Tests and Estimates for Bivariate Linear Relationships

References

Annema, N., Heyworth, J. S., McNaughton, S. A., Iacopetta, B., & Fritschi, L. (2011). Fruit and vegetable consumption and the risk of proximal colon, distal colon, and rectal cancers in a case-control study in western Australia. *Journal of the American Dietetic Association, 111,* 1479–1490. doi:10.1016/j.jada.2011.07.008

Brown, C. A., Olson, S., & Zura, R. (2013). Predictors of length of hospital stay in elderly hip fracture patients. *Journal of Surgical Orthopaedic Advances, 22,* 160–163. doi:10.3113/JSOA.2013.0160

Falvo, M. J., & Earhart, G. M. (2009). Six-minute walk distance in persons with Parkinson disease: A hierarchical regression model. *Archives of Physical Medicine and Rehabilitation, 90,* 1004–1008. doi:10.1016/j.apmr.2008.12.018

Murdock, K. K. (2013). Texting while stressed: Implications for students' burnout, sleep, and well-being. *Psychology of Popular Media Culture, 2,* 207–221. doi:10.1037/ppm0000012

Noble, K. G., Fifer, W. P., Rauh, V. A., Nomura, Y., & Andrews, H. F. (2012). Academic achievement varies with gestational age among children born at term. *Pediatrics, 130,* e257–e264. doi:10.1542/peds.2011-2157

Stamps, J. J., Bartoshuk, L. M., & Heilman, K. M. (2013). A brief olfactory test for Alzheimer's disease. *Journal of the Neurological Sciences, 333,* 19–24. doi:10.1016/j.jns.2013.06.033

Vaccarino, V., Goldberg, J., Rooks, C., Shah, A. J., Veledar, E., Faber, T. L. ... Bremner, J. D. (2013). Post-traumatic stress disorder and incidence of coronary heart disease: A twin study. *Journal of the American College of Cardiology, 62,* 970–978. doi:10.1016/j.jacc.2013.04.085

Wansink, B., Painter, J. E., & North, J. (2005). Bottomless bowls: Why visual cues of portion size may influence intake. *Obesity Research, 13,* 93–100. doi:10.1038/oby.2005.12

14

Analysis of Frequencies and Ranks

Introduction

If you were hungry between meals, what kind of snack would you prefer: a piece of fruit or a candy bar? Perhaps you would give one answer if the question related to future eating and a different answer if you were offered the snack immediately. Suppose we run an experiment in which we randomly assign people to two groups, both of which will be offered a snack. The independent variable will be the difference in how the snacks will be offered. One group will be asked to choose between two snacks (an apple or a slice of cake) that they would like to receive tomorrow afternoon. The other group will be shown these two options in person and will choose the snack immediately. Our question is whether the snack choices will differ for people who are asked to state a preference for the future versus people who choose the snack immediately. Maybe people are more likely to choose a healthful snack for the future, but an unhealthful snack for immediate consumption.

This scenario, which is a variation on a study by Weijzen, de Graaf, and Dijksterhuis (2008), differs from the scenarios in previous chapters because it does not involve means or linear relationships. The kind of snack is a categorical variable with two levels (apple and cake). The independent variable, timing of the offered snack, also is a categorical variable with two levels (being offered an immediate snack or a future snack). Our data are frequencies: within each group, we count how many people chose the apple and how many people chose the slice of cake. Then we could ask whether the distributions of people across snack categories are the same for the future-snack group and the immediate-snack group.

The data from our scenario could be analyzed statistically, even though the results are frequencies in categories. *Categorical data analysis* is a family of statistics that analyze the frequencies for nonnumeric variables like snack choice. There are dozens of statistics in this family, and we will present only a few of them. Another family of test statistics does not rely on categorical frequencies or on numeric scores like food hardship and obesity rates. *Rank tests* involve computations performed on ranks. We will provide a brief explanation of rank tests in this chapter. Statistics that use frequencies or ranks follow similar steps as other inferential statistics: stating the hypotheses, choosing a significance level, computing a test statistic, and using critical values and p values in decision making. The main difference is that the data will be frequencies, proportions, or ranks.

One-Sample Proportion

Researchers sometimes are interested in percentages or proportions for one sample. For example, suppose we are interested in the eating behaviors of adults with diabetes living in Oklahoma. We might ask: in terms of fruit and vegetable consumption, are these people typical of Americans in general? The Centers for Disease Control and Prevention (CDC) says that in 2009, 23.4% of Americans consumed five or more servings of fruit/vegetables daily (CDC, 2014). If we want to know whether Oklahoma adults with diabetes were similar to Americans in general in 2009, what is a point estimate that we could compute for their fruit and vegetable consumption? If a point estimate is possible, then an interval estimate also could be computed. This section will explain the most basic kind of categorical data analysis, estimating a population proportion.

Percentages and proportions are numbers, but they are based on frequencies or counts in categories. For example, we are looking at the percentage of American adults who consume 5+ servings of fruit/vegetables daily. To find the percentage, state-based representative surveys asked people about their fruit/vegetable intake and counted everyone who responded positively to questions about eating 5+ fruit/vegetable servings daily. These people share an attribute or characteristic about their eating habits, so they are categorized together and counted. The number of people in that category is divided by the total number of people surveyed.

Let's be clear about how percentages and proportions are related. We stated a national rate of 23.4%. If we drop the percentage sign and divide by 100, we will get a proportion:

$$\text{Proportion} = \frac{\text{percentage}}{100}$$
$$= \frac{23.4}{100}$$
$$= .234$$

For this example on fruit/vegetable consumption, we will treat .234 as a *population proportion*. The population proportion is the number of people who share some attribute divided by the total number of people in the population. The fraction 23.4/100 is the same as 234/1000, which means that out of every 1,000 survey respondents, there are 234 people who said they consume five or more servings of fruit/vegetables daily. The original rate of 23.4% from the CDC was computed based on survey answers from hundreds of thousands of respondents, not 234 and 1,000. We are more accustomed to thinking in terms of percentages, but proportions and fractions are related to percentages; 234/1000 is the same thing as .234, and we can multiply .234 by 100 to get a percentage, 23.4%. Be sure it is clear to you that proportions and percentages are not the same thing, but we can easily transform proportions into percentages, and vice versa.

In this example, the proportion .234 is being used as a population parameter. The population may be defined as adults in the United States, and for every 1,000 people in the population, 234 of them meet the criteria of consuming 5+ servings of fruit/vegetables daily. Now let's think about adults in Oklahoma who have diabetes. These Oklahomans are a subset of the population of American adults; we will treat them as a sample. In fact, we have obtained data for a sample of Oklahomans with diabetes. Out of this sample, how many people eat 5+ fruit/vegetable servings daily? Now we are talking about a *sample proportion*, which is the number of people in a sample who share an attribute divided by the total number of people in the sample. We consulted the 2009 Behavioral Risk Factor Surveillance System (BRFSS) representative survey of Oklahomans (Oklahoma State Department of Health, 2014). The survey asked people if they had diabetes, and it asked about their fruit and vegetable consumption. Figure 14.1 shows data that we obtained from the Oklahoma State Department of Health's OK2SHARE website.

The category or attribute of interest is the consumption of 5+ fruit/vegetable servings daily; responses from 196 Oklahomans with diabetes placed them in this category. The total sample size in Figure 14.1 is $N = 1,102$ Oklahomans with diabetes. So the sample proportion is

$$\frac{196}{1102} = .1778584$$

Daily Fruit/Vegetable Consumption	Number of Respondents Whose Doctor Has Said They Have Diabetes
5+ Servings	196
<5 Servings	906
Column Total	1,102

Figure 14.1

Fruit/vegetable consumption by Oklahoma adults with diabetes, 2009. (Data from "Adults with diabetes and fruit/vegetable consumption, Behavioral Risk Factor Surveillance System," by the Oklahoma State Department of Health, 2014, March 13, retrieved from http://www.health.state.ok.us/stats/index.shtml.)

This sample proportion is a point estimate. As a subset of the population, are the people in our Oklahoma sample similar to the population of Americans? If so, then maybe the sample proportion of Oklahomans with diabetes consuming 5+ fruit/vegetable servings per day is within sampling variability of the population proportion, .234. We can compute an interval estimate of the population proportion and see if a confidence interval would contain the national rate, .234. If an interval estimate brackets .234, then the sample proportion of Oklahomans with diabetes consuming 5+ servings per day would be statistically the same as the population proportion for Americans in general.

Let's say we are wondering how the Oklahoma sample compares with the population. We happen to know that Oklahoma is low on the list of states in terms of the rate of fruit/vegetable consumption. Perhaps for our sample of Oklahomans, their knowledge of having diabetes has led to improved eating habits. Or maybe our sample is like other Oklahomans and eats fewer fruit/vegetables. In other words, we have some reason to believe that adult Oklahomans with diabetes may differ from the population of American adults in terms of their fruit/vegetable consumption. We could state this conjecture as an alternative hypothesis that says

Our Oklahoma sample comes from a population in which
the proportion of people eating 5+ servings per day
differs from .234 (or 23.4%).

The alternative hypothesis may be written as

H_1: population proportion ≠ .234

The corresponding null hypothesis would say

Our Oklahoma sample comes from a population in which
the proportion of adults consuming 5+ servings daily is .234 (or the
percentage is 23.4%).

14. Analysis of Frequencies and Ranks

The null hypothesis can be written as follows:

$$H_0: \text{population proportion} = .234$$

It is possible to have a directional alternative hypothesis for a proportion, but because researchers usually rely on confidence intervals for a proportion, we are presenting only the nondirectional alternative hypothesis. We could compute a test statistic comparing the sample proportion to the population proportion, but health sciences journals tend to report a confidence interval for a proportion instead of a test statistic for a proportion. Next we will demonstrate a confidence interval for a proportion.

Check Your Understanding

SCENARIO 14-A

When we accessed the data on fruit/vegetable consumption and diabetes for people who responded to the 2009 BRFSS survey, we also found information about Oklahoma adults who did not have diabetes and their reported daily fruit/vegetable intake. We are interested in the proportion of these respondents who say they consume fewer than five servings of fruit/vegetables daily. 14-1. Suppose we know that 76.6% of Americans consume fewer than five fruit/vegetable servings per day. If we think the proportion of Oklahomans without diabetes who eat fewer than five fruit/vegetable servings daily may be different from the proportion for the American population, how would we write our alternative hypothesis (in words)? 14-2. We found the following information on the OK2SHARE website: out of the 6,562 Oklahoma adults surveyed who said they did not have diabetes, 5,553 of them said they ate fewer than five fruit/vegetable servings a day. Compute the sample proportion.

Suggested Answers

14-1. Our alternative hypothesis may be stated as follows: our sample comes from a population in which the proportion of people eating fewer than five servings of fruit/vegetables daily differs from .766. 14-2. The sample proportion is the number of people in the sample who share the attribute of eating fewer than five servings a day, divided by the number of people in the sample, or 5553/6562 = .8462359 ≈ .846.

Confidence Interval for a Proportion

An interval estimate can quantify the sampling variability inherent in our sample proportion, and we can check whether the interval contains a hypothesized population proportion. In our present example, we have been talking about the proportion of Americans who consume five or more servings of fruit/vegetables daily. If the population proportion of .234 is contained within our confidence interval, then our sample proportion would be statistically the same as the population proportion. If the confidence interval does not bracket the parameter, then our sample proportion would be significantly different from the population proportion.

There are many ways of computing a confidence interval for a proportion; we are presenting a way that is appropriate for large samples. (Researchers who want to estimate a population proportion that is hypothesized to be very small, like .03, will need larger samples than researchers who hypothesize about a proportion like .50. See Samuels & Lu, 1992, for guidelines.) To compute this confidence interval, we need to find the margin of error, which will follow the pattern of other margins of error that we have seen: it will be a critical value times the standard deviation of a sampling distribution. Like any statistic, the sample proportion has a sampling distribution. If we could draw all possible samples of the same size from the same population and compute each sample's proportion of people sharing an attribute, we would have a pile of sample proportions. These proportions could be arranged to form a sampling distribution. Like any sampling distribution, this distribution is beyond our ability to actually create, so we would like to use a theoretical distribution instead. So what does the sampling distribution of the sample proportion look like? For large samples, researchers often use a standard normal distribution to approximate the shape of the sampling distribution of a proportion. We will use a critical value from the standard normal distribution in our calculation of the margin of error for our confidence interval.

Now we need the standard deviation of the sampling distribution of the sample proportion. Mathematical statisticians have determined the formula for the standard deviation of this sampling distribution. This formula is small and cute, if you use symbols. Unfortunately, the letter p or P often is used in statistics books in connection with proportions. By now, you probably are accustomed to this letter being used to represent a p value, and we refuse to use that letter to represent anything else in this book. The cost of our decision is that we will have to describe the standard deviation of the proportion's sampling distribution and not use a small, cute formula made up of symbols. Please be aware, however, that the letter p can represent other statistics besides a p value.

Based on the proportion of people sharing an attribution, we can compute the proportion of people who do *not* have that attribute. For example, if 23.4% of Americans consume 5+ fruit/vegetable servings a day, then 76.6% of Americans do not consume 5+ servings of fruit/vegetables daily. To convert these numbers into proportions, the population proportion in our null hypothesis is .234; the rest of the population is 1 minus the population proportion $= 1 - .234 = .766$. These two proportions are contained within the following formula for the standard deviation of the sample proportion's sampling distribution:

$$\sqrt{\frac{\text{population proportion} \times (1 - \text{population proportion})}{N}}$$

Our sample size of $N = 1,102$ Oklahomans with diabetes and the population proportion of .234 result in the following standard deviation of the sample proportion's sampling distribution:

$$\sqrt{\frac{.234 \times .766}{1102}}$$

$$= \sqrt{\frac{.179244}{1102}}$$

$$= \sqrt{.0001626533}$$

$$= .01275356$$

Now that we have the standard deviation of the sample proportion's sampling distribution, we can use it to compute the interval estimate of the population proportion. The margin of error will be a critical value times this standard deviation. For large samples, the critical value will come from a standard normal distribution. Let's compute a 95% confidence interval. As we saw in Chapter 8, a standard normal distribution gives us two critical values when $\alpha = .05$ for a two-tailed test. Figure 14.2 is similar to a figure from Chapter 8 and shows that $z = -1.96$ cuts off $\alpha/2 = .025$ in the lower tail, and $z = 1.96$ cuts off $\alpha/2 = .025$ in the upper tail.

To find the margin of error, we multiply 1.96 by the standard deviation that we computed for the sample proportion's sampling distribution:

$$\text{Margin of error} = 1.96 \times .01275356$$

$$= .02499698$$

To remind you, the sample proportion is the number of Oklahoma adults with diabetes who said they consume 5+ servings of fruit/vegetables daily, divided by the total number of Oklahoma adults with diabetes surveyed:

$$\frac{196}{1102} = .1778584$$

To find the lower limit of the 95% confidence interval, we subtract the margin of error from the sample proportion:

$$\text{Lower limit} = \text{sample proportion} - \text{margin of error}$$

$$= .1778584 - .02499698$$

$$= .15286142$$

$$\approx .153$$

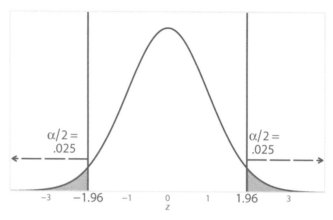

Figure 14.2

Standard normal distribution showing alpha for a two-tailed test. A standard normal distribution can provide critical values for confidence intervals that estimate a population proportion if a large sample has been used. This figure is a reminder of the fact that two critical values (−1.96 and +1.96) contain 95% of a standard normal distribution.

To find the upper limit of the confidence interval, we add the margin of error to the point estimate:

$$\text{Upper limit} = \text{sample proportion} + \text{margin of error}$$
$$= .1778584 + .02499698$$
$$= .20285538$$
$$\approx .203$$

The 95% confidence interval is [.153, .203]. This interval may or may not bracket the true population proportion, but across repeated samples that could be used to compute an interval like this one, 95% of the time the intervals would capture the true population proportion. Does this interval contain the hypothesized population proportion, .234? No, it does not. We may conclude that the proportion of Oklahoma adults with diabetes who consume 5+ servings of fruit/vegetables daily differs significantly from the population proportion of .234. Now that we have rejected the null hypothesis, we can consider the direction in which our results turned out. The point estimate was the sample proportion, which was approximately .178, or 17.8%. The percentage of Oklahoma adults with diabetes consuming 5+ servings of fruit/vegetables per day is significantly lower than the national rate for all adults, 23.4%.

A sample proportion or percentage allows us to consider how many people are in one category; our example described the number of people who shared the attribute of consuming 5+ servings of fruit/vegetables daily. There were two possibilities: either people had the attribute or they did not. A categorical variable can have more than two possible outcomes. Next we will look at an example in which people are spread out across several categories. Later, we will explain how we could compare two proportions, such as the proportions of Oklahomans with and without diabetes who consume 5+ servings of fruit/vegetables daily.

Check Your Understanding

SCENARIO 14-A, Continued

The previous question in this scenario involved the computation of a sample proportion of Oklahoma adults without diabetes who said they ate fewer than five servings of fruit/vegetables daily. This proportion was .8462359, and the sample size was $N = 6,562$. Our alternative hypothesis said the sample came from a population with a proportion different from .766. 14-3. Compute the margin of error for a 95% confidence interval for the population proportion. 14-4. Compute the 95% confidence interval. 14-5. What can we conclude about Oklahoma adults without diabetes?

Suggested Answers

14-3. We need to compute the standard deviation of the sample proportion's sampling distribution. First, we multiply the hypothesized .766 by $1 - .766$ $= .766 \times .234 = .179244$. Second, we divide this number by N, and we get .179244/6562 $\approx$ .00002732. Third, we take the square root to get the standard deviation: .00522642. Fourth, we multiply this number by 1.96, a critical value from the standard normal distribution when $\alpha = .05$: margin of error $= 1.96 \times .00522642 = .01024378$. 14-4. The lower limit is the sample proportion minus the margin of error: .8462359 $-.01024378 = .8359921 \approx .836$. The upper limit is the sample proportion plus the margin of error: .8462359 $+$ .01024378 $= .85647968 \approx .856$. The 95% confidence interval is [.836, .856]. 14-5. We may conclude that our sample of Oklahoma adults without diabetes has a significantly higher proportion of people who consume fewer than five servings of fruit/vegetables daily, compared with the proportion of .766 for all Americans. Notice how small the margin of error became with the huge sample size. Researchers must think about the size of an observed difference and what might be clinically noteworthy. Statistical significance does not correspond necessarily with clinical or practical significance. The statistically significant difference here probably would be judged to have practical significance and could be evidence supporting current public health education efforts in Oklahoma.

Goodness of Fit Hypotheses

Sometimes researchers want to know whether a categorical variable has a distribution that looks like a previously identified or theoretical distribution. When we say *distribution* here, we are talking about the frequencies for the different levels of the categorical variable. The categorical variable could be the days of the week, and our research question could involve the number of people who died of a heart attack each day. Are sudden cardiac deaths evenly distributed across the days of the week, or are people more likely to die suddenly of a cardiovascular event on certain days, like on Mondays? Witte, Grobbee, Bots, and Hoes (2005) conducted a study of this question about the days of the week and cardiac deaths, using data from a larger epidemiological study conducted in the Netherlands. A registry was created for recording sudden deaths in one city during a two-year period, and the researchers identified all sudden cardiac deaths. When such deaths were counted for each day of the week, would the researchers find a greater proportion of these deaths occurring on Mondays? If the day of the week did not have any relationship with sudden cardiac deaths, then the proportions should be the same for all seven days of the week; that is, the cardiac deaths should be evenly distributed across the seven days. We will use this example to illustrate a statistic that can answer a question about the way frequencies are spread out across the levels of a categorical variable.

The data from this study could be analyzed using a statistic called the *chi-square test for goodness of fit*. This statistic also is sometimes called a *one-way chi-square test* because there is one categorical variable; here, it is the day of the week. There are many chi-square statistics, and we know that "goodness of fit" is an odd term; this chi-square is looking at how well the data fit or match a specified theoretical distribution for one categorical variable. Chi is a Greek letter that is pronounced like the first two letters in the word *kite*. The chi-square symbol is χ^2, which often appears in journal articles. In our current example, the theoretical distribution reflects the notion that cardiac deaths are evenly distributed across the days of the week.

Table 14.1 shows this proposed distribution, with one-seventh of the sudden cardiac deaths occurring every day. If we looked at 700 sudden cardiac deaths, we would expect 100 of them to occur each day. Suppose we think sudden cardiac deaths are not equally likely on any day of the week. This belief would be reflected in an alternative hypothesis. We might state the alternative hypothesis as follows:

> Our sample comes from a population in which the frequency
> of sudden cardiac deaths depends on the day of the week.

Table 14.1 Theoretical Distribution of Sudden Cardiac Deaths across Days of the Week

Day of the Week						
Sunday	Monday	Tuesday	Wednesday	Thursday	Friday	Saturday
1/7	1/7	1/7	1/7	1/7	1/7	1/7

14. Analysis of Frequencies and Ranks

The population that we are describing is thought to be different from a theoretical distribution of evenly distributed deaths across the days of the week. We could summarize this difference as follows:

$$H_1: \text{distribution}_{\text{population}} \neq \text{distribution}_{\text{theory}}$$

Let's look at another way to translate the alternative hypothesis, a restatement that follows the statistical H_1 more closely:

> Our sample comes from a population
> with sudden cardiac deaths distributed across
> the days of the week (distribution$_{\text{population}}$)
> in a way that differs from the theoretical
> distribution (distribution$_{\text{theory}}$)
> of evenly distributed sudden cardiac deaths
> across the days of the week.

If our alternative hypothesis is true, the population distribution of cardiac deaths across days will not match the theoretical distribution in which such deaths are equally likely every day. Another way of saying it is that there will be an effect of the day of the week on the frequency of sudden cardiac deaths. We do not have directional hypotheses with the chi-square test for goodness of fit.

Our alternative hypothesis would correspond to an opposite statement, a null hypothesis that could be stated as follows:

> Our sample comes from a population in which
> the frequency of sudden cardiac deaths
> does not depend on the day of the week.

The population that we are describing in the null hypothesis would be the same as the theoretical distribution of evenly distributed deaths across the days of the week. We could summarize this null hypothesis as follows:

$$H_0: \text{distribution}_{\text{population}} = \text{distribution}_{\text{theory}}$$

Another way of stating this null hypothesis is as follows:

> Our sample comes from a population in which
> the distribution of sudden cardiac deaths across days of the week
> matches or fits the theoretical distribution
> in which such deaths are equally likely every day of the week.

Another way of saying it is that there will not be an effect of the day of the week on the frequency of sudden cardiac deaths. Next we will introduce the computational details of the statistic that will be used to test our null hypothesis.

Check Your Understanding

SCENARIO 14-B

Human blood is categorized in many ways. You probably have heard of people having a blood type like "O positive," or O+, which reflects two classification systems. For the purpose of blood transfusions, two major classification systems are the ABO system and the Rhesus (Rh) system. The blood types A, B, AB, and O represent different molecules on the red blood cells. Rh is a protein that may be present or absent; if you are Rh positive (+), then the protein is present in your blood, but if you are Rh negative (−), the protein is absent. Suppose we read about the frequencies of different blood types in the United States on the website http://unitedbloodservices .org/learnMore.aspx. Further, suppose we have been helping with a series of blood drives, and we are talking to the organizer of the blood drives, Angie O'Plasty. Angie says she doubts whether the blood drives have resulted in a sample that is representative of the population distribution of blood types described on the above website. We tell Angie that we can use a chi-square test for goodness of fit to address her concern. 14-6. Explain why this test statistic is appropriate for this situation. 14-7. Write the alternative hypothesis for this scenario. 14-8. Write the null hypothesis for this scenario.

Suggested Answers

14-6. The website can provide a theoretical distribution for the frequencies that we could expect for different blood types. The actual number of units of each blood type collected in the blood drives can be compared to the theoretical distribution from the website to see whether the frequencies of different blood types came from a population that looks like the theoretical distribution. 14-7. The alternative hypothesis may be stated as follows: Our sample comes from a population in which the frequency of occurrence of different blood types differs from the theoretical distribution from the website. 14-8. The null hypothesis may be stated as follows: Our sample comes from a population in which the frequency of occurrence of different blood types matches or fits the theoretical distribution from the website.

Goodness of Fit Statistic

We already acknowledged that "goodness of fit" is an odd term. Let's think through this term in connection with the null and alternative hypotheses. In the alternative hypothesis for the heart attack mortality example, we think there will be an effect of the day of the week. In other words, our sample comes from a population in which the distribution of sudden cardiac deaths does not fit a theoretical distribution, which specifies equal numbers of sudden deaths from cardiac events across the days of the week. If we find data in support of this alternative hypothesis, then it would appear that the sample came from a population with a distribution that *does not fit* the theoretical distribution of evenly distributed deaths. So if the chi-square test for goodness of fit is significant, then it would mean there is a *poor fit* between the data and the theoretical distribution. If the null hypothesis is retained, then it would mean there is a *good fit* between the data and the theoretical distribution. It may seem odd that the statistic's name contains "goodness of fit" when a significant result would indicate a poor fit between the data and a theoretical distribution. Just remember that the null hypothesis reflects the idea of a *good* fit, and we always *test* the null hypothesis, which corresponds to the statistic's name: chi-square *test* of *goodness* of fit.

The chi-square test for goodness of fit is computed using the observed frequencies from the sample, compared with frequencies that would reflect the theoretical distribution or a distribution of frequencies that has been found in previous research. In our example of the sudden cardiac deaths, the theoretical distribution will reflect the null hypothesis, which indicates no relationship between the day of the week and frequency of deaths. One out of every seven sudden cardiac deaths would be expected on any day of the week, according to the null hypothesis. The article by Witte et al. (2005) said 1,828 people died suddenly of cardiac events during the two-year observation period in the city in the Netherlands. Based on the theoretical distribution in the null hypothesis, how many of these cardiac deaths would we expect to have occurred on any given day of the week? This number will be the *expected frequency*, the number of occurrences that we predict in each category based on our theory or previous research. We can find the expected frequency per day by taking the proportion (1/7) times the sample size, N. The letter E is used to represent the expected frequency:

$$E = \frac{1}{7} \times 1828$$
$$= .1428571 \times 1828$$
$$= 261.14286$$

Obviously, cardiac deaths are counted in whole numbers, not fractional numbers like 261.14286, but we are averaging the number of deaths across the days of the week. For our next calculations, we need to keep those decimal places. According to our null hypothesis, if there is no relationship between the day of the week and the number of sudden cardiac deaths, we could expect the 1,828 deaths to be distributed as shown in Table 14.2, which contains the expected frequencies.

It is possible to have different expected frequencies for each day of the week, which we will show in a Check Your Understanding section. The article by Witte

Table 14.2 Expected Frequencies of Sudden Cardiac Deaths across the Days of the Week

	Day of the Week					
Sunday	Monday	Tuesday	Wednesday	Thursday	Friday	Saturday
261.14286	261.14286	261.14286	261.14286	261.14286	261.14286	261.14286

Table 14.3 Fabricated Data for Sudden Cardiac Deaths across the Days of the Week

	Day of the Week					
Sunday	Monday	Tuesday	Wednesday	Thursday	Friday	Saturday
262	313	251	270	240	243	249

et al. (2005) did not state the actual numbers of deaths for each day of the week, because their research questions were more refined than ours. Therefore, we *fabricated* the frequencies shown in Table 14.3; these numbers are *not real* and are being used only for teaching purposes.

The data are called the *observed frequencies*, the number of occurrences of the event being counted for each category. Each number in Table 14.3 is an observed frequency, or O, and the categories are the days of the week. It is obvious that Mondays during the two-year observation period had the largest number of sudden cardiac deaths (313). But we do not know yet whether the distribution across the days of the week differs significantly from a distribution of equal numbers for the expected frequencies. The chi-square test for goodness of fit will give us an answer.

The basic idea behind this test statistic is to take the observed frequency for a category and subtract the expected frequency to look at the difference between them. Then we will do some other math to accumulate those differences across the categories. Let's take a short detour that we hope will help you to understand the computational details of the chi-square test for goodness of fit.

Remember when we were computing a sample variance? We wanted to use each score's distance from the mean as a measure of spread. But the distances *below* the mean balanced out the distances *above* the mean, so the distances summed to zero. As a result, we had to square the distances before averaging them. We need to do something similar now, but we will not be using a mean because we do not have a quantitative variable. Suppose we have the following observed frequencies in three categories (Table 14.4).

Table 14.4 Numeric Example of Observed Frequencies in Three Categories

	Category	
A	B	C
100	200	300

Table 14.5 Numeric Example of Expected Frequencies, Based on Theory

	Category	
A	B	C
300	200	100

Table 14.6 Numeric Example of Observed Frequency Minus Expected Frequency

Category	O (From Table 14.4)	E (From Table 14.5)	O - E
A	100	300	$100 - 300 = -200$
B	200	200	$200 - 200 = 0$
C	300	100	$300 - 100 = 200$
Totals	600	600	0

Further, suppose we had the expected frequencies shown in Table 14.5, based on theory. What would happen if we took each category's observed frequency (O) and subtracted the expected frequency (E), then added up the differences? Table 14.6 shows the results. The $O - E$ differences between the observed and expected frequencies sum to zero. But clearly, there is a different pattern of observed frequencies in Table 14.4, compared with the pattern of expected frequencies in Table 14.5. Therefore, we need to do something similar to what we did when we computed a sample variance: we will square the $O - E$ differences. The result will be squared differences between the observed and expected frequencies. Each squared difference then will be divided by its corresponding expected frequency; it is as if we are making the squared difference relative to what we expected the frequency to be. Then those numbers will be added up, and the result will be the chi-square test for goodness of fit. This is the end of our short detour.

The formula for the chi-square test for goodness of fit is a bit ugly, but it is manageable. We just described it above; now let's list the steps for computing the statistic:

- For each category, take the observed frequency (O) and subtract the expected frequency (E).
- Square the $O - E$ differences.
- Divide each squared difference by its corresponding expected frequency (E).
- Add up those results.

That doesn't sound too bad, does it? If the observed frequencies were perfectly equal to the expected frequencies, then every $O - E$ would equal zero, and the squared differences would be zero. As a result, the smallest possible number for the chi-square test for goodness of fit is zero. Any difference between the observed and expected frequencies will lead to positive squared differences, so the test statistic can only get bigger.

Like other statistics, the chi-square test for goodness of fit has a sampling distribution. We could compute chi-square tests for multiple samples and arrange them in a distribution. Each of these numbers could equal zero or any positive number, so the chi-square test for goodness of fit is a one-tailed test. If certain assumptions are met, then we can use a theoretical χ^2 distribution instead of having to create a sampling distribution. We will come back to the assumptions and robustness of this inferential statistic shortly. Like the many theoretical t distributions and the many theoretical F distributions, there are many theoretical χ^2 distributions, each defined by a different value of degrees of freedom. For the chi-square test for goodness of fit, the degrees of freedom are

$$df = \text{number of categories} - 1$$

If we had three categories, df would equal 2. If we had 9 categories, df would equal 8. Figure 14.3 shows five chi-square distributions, with df ranging from 2 to 8. As you can see, theoretical chi-square distributions differ in appearance for different values of df. Let's see what a chi-square distribution would look like for our example of sudden cardiac deaths. We have $df = 6$, because 7 days of the week minus $1 = 6$. The chi-square distribution with $df = 6$ is shown in Figure 14.4.

Figure 14.4 shows a positively skewed χ^2 distribution. When we test our null hypothesis about the cardiac deaths, we might choose a significance level of $\alpha = .05$. Because the chi-square is a one-tailed test, we will put all of alpha in the upper tail. The significance level of .05 is the area cut off by a critical value of $\chi^2 = 12.59$. Where did this critical value come from? Table D.1 in the back of the

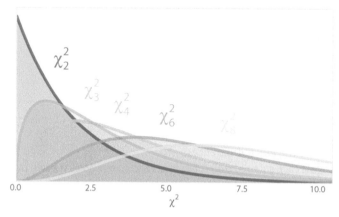

Figure 14.3

Chi-square distributions with different values of df. Theoretical chi-square distributions are defined by their degrees of freedom. This figure shows five distributions, each with a different value of df shown as a subscript on the chi-square symbol.

14. Analysis of Frequencies and Ranks

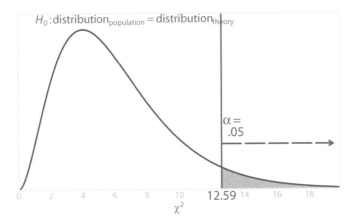

H_0:distribution$_{population}$ = distribution$_{theory}$

$\alpha =$
.05

0   2   4   6   8   10   12.59 14   16   18

χ^2

Figure 14.4

Chi-square distribution with $df = 6$. For the example of sudden cardiac deaths, we will use a chi-square for goodness of fit with six degrees of freedom. The figure shows a χ^2 critical value = 12.59 and $\alpha = .05$.

book contains a table of critical values for chi-square tests. Figure 14.5 shows an excerpt from Table D.1.

Table D.1 is easy to use. The chi-square test is a one-tailed test, so we compute df, then find the row for our df. Then we look in the column for our chosen significance level. When we look at the row for $df = 6$ and the column for $\alpha = .05$, we find the critical value of 12.59, shown in Figure 14.4.

The decision rules for this test statistic are straightforward and will seem identical to the decision rules for the one-way ANOVA F test. The difference is that a chi-square test has been computed, so a chi-square distribution is consulted. Like the one-way ANOVA F test, the alternative hypothesis for the χ^2 for goodness of fit is nondirectional, so no direction has been predicted even though the test is one-tailed. The critical value decision rule is

> If the observed test statistic is equal to or more extreme than the critical value, reject the null hypothesis.
> Otherwise, retain the null hypothesis.

The p value decision rule is

> If the observed p value is less than or equal to alpha,
> then reject the null hypothesis.
> Otherwise, retain the null hypothesis.

Next we will calculate the observed χ^2 for goodness of fit for the example of sudden cardiac deaths and test the null hypothesis.

Table D.1 Critical Values for χ2 Distributions

	α for One-Tailed Test			
df	.10	.05	.01	.001
1	2.71	3.84	6.63	10.83
2	4.61	5.99	9.21	13.82
3	6.25	7.81	11.34	16.27
4	7.78	9.49	13.28	18.47
5	9.24	11.07	15.09	20.52
6	10.64	12.59	16.81	22.46
7	12.02	14.07	18.48	24.32
8	13.36	15.51	20.09	26.12
9	14.68	16.92	21.67	27.88
10	15.99	18.31	23.21	29.59
11	17.28	19.68	24.72	31.26

Figure 14.5

Excerpt from Table D.1, which appears in the back of the book. This table provides critical values for the chi-square test for goodness of fit and other χ^2 statistics.

Check Your Understanding

SCENARIO 14-B, Continued

This scenario concerned the relative frequencies of blood types expected to exist in the population, in which the blood types are categorized according to eight blood types. The research question was whether a series of blood drives had resulted in a distribution of blood types that mirrored the expected frequencies in the eight categories of blood type. 14-9. Compute df for this scenario. 14-10. Find the critical value in Table D.1, using α = .05.

Suggested Answers

14-9. We have eight categories of blood types in this scenario, so df = 8 – 1 = 7.
14-10. Using the row for df = 7 and the column for α = .05, we find a chi-square critical value = 14.07.

Computing the Chi-Square Test for Goodness of Fit

Rather than show you an ugly formula, we described how this test statistic is computed:

- For each category, take the observed frequency (O) and subtract the expected frequency (E).
- Square the difference computed for each category.
- Divide each squared difference by its corresponding expected frequency (E).
- Add up those results.

Table 14.3 contained the (fabricated) observed frequencies for the example of cardiac deaths, and Table 14.2 showed the expected frequencies if sudden cardiac deaths were equally likely for every day of the week. Let's break down the computation of the chi-square for goodness of fit and show *some* of the steps in Table 14.7. We will show the remaining steps in another table. The first column shows the day of the week. The second column shows the (fabricated) observed frequencies. Every day of the week has the same expected frequency (261.14286), so for the sake of space, we are not showing a separate column for E. The third column shows the difference between the observed and expected frequency $(O - E)$ for each day of the week. This table will take us partway through the computation of the chi-square test for goodness of fit.

Now that we have computed the differences, we need to square each difference and divide the squared difference by the expected frequency, $E = 261.14286$. Table 14.8 shows these calculations.

The last step is to add up the results in the third column of Table 14.8 to get the chi-square test for goodness of fit:

χ^2 for goodness of fit

$= 0.00281336 + 10.2976699 + 0.393951452 + 0.30040618$

$\quad + 1.711785377 + 1.260472406 + 0.564629831$

$= 14.53173$

≈ 14.53

Table 14.7 Computing χ^2 for Goodness of Fit Using Fabricated Data

Day of the Week	Observed Frequency (O)	$O - E$
Sunday	262	$262 - 261.14286 = 0.85714$
Monday	313	$313 - 261.14286 = 51.85714$
Tuesday	251	$251 - 261.14286 = -1.14286$
Wednesday	270	$270 - 261.14286 = 8.85714$
Thursday	240	$240 - 261.14286 = -21.14286$
Friday	243	$243 - 261.14286 = -18.14286$
Saturday	249	$249 - 261.14286 = -12.14286$

Table 14.8 Computing the Squared Differences on Fabricated Data

Day of the Week	Squared Differences, $(O - E)^2$	Squared Differences Divided by E
Sunday	$0.85714^2 = 0.73468898$	$0.73468898/261.14286 = 0.00281336$
Monday	$51.85714^2 = 2689.162969$	$2689.162969/261.14286 = 10.2976699$
Tuesday	$-1.14286^2 = 102.877609$	$102.877609/261.14286 = 0.393951452$
Wednesday	$8.85714^2 = 78.44892898$	$78.44892898/261.14286 = 0.30040618$
Thursday	$-21.14286^2 = 447.020529$	$447.020529/261.14286 = 1.711785377$
Friday	$-18.14286^2 = 329.163369$	$329.163369/261.14286 = 1.260472406$
Saturday	$-12.14286^2 = 147.449049$	$147.449049/261.14286 = 0.564629831$

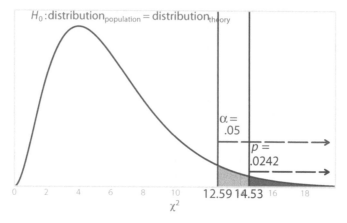

H_0: distribution$_{population}$ = distribution$_{theory}$

$\alpha = .05$

$p = .0242$

0   2   4   6   8   10   12.59 14.53 16   18

χ^2

Figure 14.6

Observed chi-square test statistic for the example of sudden cardiac deaths. We computed the observed $\chi^2 = 14.53$ for the example of sudden cardiac deaths. This observed test statistic had $p = .0242$, according to statistical software.

As usual, we do not round until we have the final answer. Now that we have the observed test statistic, we can test the null hypothesis, which said our sample comes from a population in which the frequency of sudden cardiac deaths does not depend on the day of the week. That is, the frequency of sudden cardiac deaths will be the same every day of the week. We will illustrate both the critical value and p value decision rules. Figure 14.6 shows our theoretical chi-square distribution with $df = 6$ and $\alpha = .05$. Now we have included the observed test statistic and its corresponding p value in the graph.

Applying the critical value decision rule, we ask whether the observed test statistic is equal to or more extreme than the critical value. Because 14.53 is more extreme than the critical value of 12.59, we reject the null hypothesis and conclude that the day of the week had a significant effect on the frequency of sudden cardiac deaths. We will reach the same conclusion using the p value decision rule. As shown in Figure 14.6, the observed p is .0242, which is less than alpha = .05, so we reject the null hypothesis and draw the same conclusion.

Our results indicate that sudden cardiac deaths are not equally likely on every day of the week. Does this decision mean that there is one day of the week on which people are more likely to die of a sudden cardiac event? We cannot say based on this hypothesis test. Like the one-way ANOVA F test, the chi-square test for goodness of fit sometimes is too general. There is a difference in the frequencies, but the chi-square test for goodness of fit cannot tell us whether any particular day's frequency differs from any other day's frequency. To find those differences, we would need a multiple comparison procedure. We wouldn't use the same multiple comparison procedures mentioned in Chapter 12 because those statistics involved means, and the present example has frequencies. As an introductory book, this text obviously is limited in the statistics that can be presented, so we must leave out the multiple comparisons needed to determine which days of the week differed.

Next we will conclude our discussion of the chi-square test for goodness of fit by explaining its assumptions and robustness. Then we will introduce you to another chi-square that appears frequently in health sciences research.

Check Your Understanding

SCENARIO 14-B, Continued

This scenario pertained to the distribution of eight blood types. Suppose we are helping with a series of blood drives. The organizer of the drives, Angie O'Plasty, has been looking at a website that lists the relative frequency of various blood types. She doubts whether the sample came from a population that matches the distribution of blood types as described in Table 14.9. 14-11. If 902 units of blood were collected in the blood drives, what would be the expected frequencies? 14-12. This chapter's exercises will include the observed frequencies for this fabricated example. The observed chi-square test for goodness of fit is $\chi^2 = 5.16$. In a previous Check Your Understanding, you found the critical value, which was 14.07. Test the null hypothesis, using the critical value decision rule. 14-13. Suppose we have chosen alpha =.05, and the p value for the observed chi-square equals .64. Test the null hypothesis, using the p value decision rule. 14-14. Explain the meaning of the decision about the null hypothesis, using the variable names.

Suggested Answers

14-11. The expected frequencies can be found by converting the percentages in Table 14.9 into proportions (i.e., the proportion = the percentage divided by 100), then multiplying each proportion by the number of units in the sample, N = 902, to find that blood type's expected frequency. In the same order as shown from left to right in Table 14.9, the expected frequencies are 306.68, 90.2, 36.08, 333.74, 54.12, 18.04, 9.02, and 54.12. 14-12. Because the observed test statistic (5.16) is not more extreme than the critical value (14.07), we retain the null hypothesis. 14-13. Because .64 is not less than or equal to .05, we retain the null hypothesis. 14-14. We may conclude that our sample came from a population that has a distribution of the frequency of blood types that matches (or is not significantly different from) the distribution of frequencies based on the proportions from the website.

Table 14.9 Theoretical frequencies of blood types, based on a website's report

			Blood Type				
A+	B+	AB+	O+	A–	B–	AB–	O–
34%	10%	4%	37%	6%	2%	1%	6%

Goodness of Fit: Assumptions and Robustness

The chi-square test for goodness of fit has three assumptions:

- Observations are independent.
- Categories are mutually exclusive.
- Categories are exhaustive.

First, the assumption of independence of observations means we are assuming that every person in our example who died suddenly of a cardiac event was affected independently. That is, there was not some major event that caused extreme stress to many people on the same day, resulting in all of them having heart attacks. Each person also must be counted only once. If a patient died exactly at midnight, someone had to make a decision about the death date; the death could not be recorded as happening on two days.

Counting people in one and only one category is the idea behind the second assumption: the categories are mutually exclusive. Being counted among the deaths on a Monday eliminates the possibility of being counted on any other day. You probably can see how the assumption of mutually exclusive categories is linked with the independence assumption. If someone died of a sudden cardiac event exactly at midnight and we counted the person as a Monday death *and* as a Tuesday death, then the independence assumption would be violated because the same person was counted twice. The assumption of mutually exclusive categories also would be violated, because being counted as a Monday death should have eliminated the possibility of being counted on any other day.

The third assumption is that the categories are exhaustive, which means there must be a category for everyone in the study. For the example of cardiac deaths, obviously we would include every day of the week. But if we were interested only in weekday cardiac deaths, then anyone who died on a weekend would be excluded from the study. Having read about other assumptions of independence, you will not be surprised to learn that this test statistic is not robust to violations of the independence assumption.

One other issue comes up with chi-square statistics. As you will recall, each squared difference between the observed and expected frequencies is divided by an expected frequency. If an expected frequency is less than 5, then the theoretical χ^2 distribution may not reflect what the sampling distribution of the statistic actually looks like, meaning the p value would be untrustworthy. Statistical software often will warn the user if there are expected frequencies less than 5. There are statistical options when that occurs; we will not go into those details.

We mentioned earlier that the chi-square test for goodness of fit also is known as a one-way chi-square. Does that mean there is a two-way chi-square? Yes, there is, and it also goes by other names as well. Next we will introduce you to that statistic.

SCENARIO 14-C

Suppose a nursing school encourages its students to donate blood every other month. Some people may not be eligible for blood donation because of illness or other reasons. We think 70% of the 210 currently enrolled nursing students should be able to give blood every other month. 14-15. Would it be a good idea to use the chi-square test for goodness of fit to determine whether 70% of the 210 current nursing students donated blood during four months (September, November, January, March)? That is, can we use this test statistic to compare the nursing school's actual numbers of blood donors during those months to an expected frequency of 147 (or 70% of the 210 enrolled students) for each of those months?

Suggested Answer

14-15. It would be a bad idea to use the chi-square test for goodness of fit in this way. The independence assumption and the assumption about mutually exclusive categories would be violated because the same students would be observed on four occasions.

Chi-Square for Independence

Sometimes researchers want to analyze the frequencies in categories that are defined by two variables. For example, Schauer et al. (2012) wanted to compare three groups of patients with uncontrolled type 2 diabetes who underwent treatment for obesity. All patients received intensive medical therapy. One group received only the medical therapy. In addition to the medical therapy, a second group received gastric bypass surgery, which reduces the size of the stomach and connects it directly to the small intestine. A third group received medical therapy and sleeve gastrectomy, another kind of weight-loss surgery that reduces the size of the stomach to form a tube-like structure. The researchers were interested in whether the patients would achieve control over their blood sugar levels. A blood test called a hemoglobin A1c (or HbA1c) is a measure of how well controlled the blood sugar has been in the previous three months. A reading of 5.6% or less is typically considered normal. These researchers decided to categorize patients as successfully controlling their blood sugar if they had an HbA1c reading less than or equal to 6% one year after treatment began. Figure 14.7 illustrates the design of the study.

As you can see in Figure 14.7, there are two categorical variables being studied: kind of treatment and whether the blood sugar was under control 12 months later. The research question for this study could be stated as follows: Is there a relationship between the treatment for obesity and control of blood sugar levels a year later? We can restate this question another way: Are the distributions of people across the categories of blood sugar control different for the various treatment

		Treatment		
		Medical Therapy Only	Gastric Bypass	Sleeve Gastrectomy
Blood Sugar	Controlled			
	Uncontrolled			

Figure 14.7

Design of the study by Schauer et al. (2012)

groups? Another way to think about this question is to consider the proportion of people in each group with controlled blood sugar at the one-year mark. A proportion is related to the way people are distributed between two categories; here, the categories would be "controlled" and "uncontrolled." Within each treatment group, the proportion would be computed as the number in the controlled category divided by the total number of people who received that treatment. Are the proportions different, depending on the kind of treatment? (The idea of whether proportions are equal or different is useful only when one of the categorical variables has two levels.)

The data will be the observed frequencies or counts of the people in each combination of treatment and blood sugar control. These data can be analyzed using a *chi-square test for independence*, also known as a *chi-square test for contingency tables* or a *two-way chi-square test*. A contingency table is a schematic that combines the levels of categorical variables; Figure 14.7 shows a contingency table. All of our examples will be limited to two variables, but it is possible to have a chi-square with more than two variables. The term *chi-square for independence* does not refer to an independence assumption; it refers to a null hypothesis that would say the two categorical variables are independent of each other. That is, there is no relationship between the two variables. Next we will explain the hypotheses for the chi-square for independence.

Check Your Understanding

SCENARIO 14-D

Marbella, Harris, Diehr, Ignace, and Ignace (1998) wanted to learn about Native Americans' consultations with Native American healers. They recruited a convenience sample of 150 patients of an urban Indian Health Service clinic. The patients, who represented 30 tribes, were asked whether they tended to seek medical care only from a physician or tended to seek medical care from both a physician and a Native American healer. The researchers categorized patients into two age groups: 18–39 years and 40–83 years. 14-16. What kind of research question could these researchers answer, using the chi-square for independence?

(Continued)

14. Analysis of Frequencies and Ranks

14-16. The researchers may want to know whether there is a relationship between age group and the kind of care being sought by the patients. Specifically, they may ask: are the proportions of patients seeking care from both a physician and a Native American healer the same for the younger age group and the older age group?

Hypotheses for Chi-Square for Independence

The example of obese patients with type 2 diabetes implied the kind of hypotheses that could be tested using the chi-square test for independence. The hypotheses can be stated in multiple ways. The researchers suspected there might be a relationship between the kind of treatment (medical therapy only, gastric bypass, or sleeve gastrectomy) and blood sugar status 12 months after treatment (controlled or uncontrolled). The alternative hypothesis can be stated in a similar way:

> Our sample comes from a population in which
> blood sugar status one year after treatment is related to the kind of
> treatment that obese patients with type 2 diabetes received.

Another way to state H_1 is say that the categorical variables are not independent of each other; that is, the likelihood of achieving controlled blood sugar (or not) depends on what treatment was received. The alternative hypothesis could be stated in terms of distributions. Within each group, think about the proportion of people achieving control of their blood sugar. If there is a relationship between the kind of treatment received and blood sugar status 12 months after treatment, then the proportions will not be the same. That is, there will be some difference in the distributions across blood sugar categories for the patients in different treatment groups. The alternative hypothesis can be summarized in many ways:

> H_1: some difference in the distributions
> (med therapy, gastric bypass, sleeve gastrectomy)
> H_1: kind of treatment is related to blood sugar control 12 months later
> H_1: blood sugar control is not independent of kind of treatment

The variables can be reversed in the last two statements of H_1; it does not matter which variable is mentioned first. It sometimes can make sense which variable may be stated first. In this study, patients were randomly assigned to groups, and the researchers manipulated the independent variable of treatment, so it would make sense to say that the treatment was responsible for the subsequent blood sugar status. Even if the study were nonexperimental, the time between the treatment and the outcome may influence the order in which the variables are listed.

Like the alternative hypothesis, the null hypothesis can be stated in different ways:

Our sample comes from a population in which
the kind of treatment that obese patients with type 2 diabetes received
is unrelated to blood sugar status 12 months after treatment.

Or we could say that the variables are independent of each other; the likelihood of reaching a status of controlled blood sugar does not depend on the kind of treatment that was received 12 months earlier. The null hypothesis also can be written in terms of distributions. Thinking about each group's proportion of people achieving control of their blood sugar, we might say that if the treatment is unrelated to blood sugar, then the proportions would be equal. That is, if there is no relationship between blood sugar status and treatment, then there will be no difference in the distributions across blood sugar categories for the patients in different treatment groups. The null hypothesis can be summarized as follows:

$$H_0: \text{distribution}_{\text{med therapy}} = \text{distribution}_{\text{gastric bypass}} = \text{distribution}_{\text{sleeve gastrectomy}}$$
$$H_0: \text{kind of treatment is unrelated to blood sugar control 12 months later}$$
$$H_0: \text{blood sugar control is independent of kind of treatment}$$

Notice that we did not show an alternative hypothesis with "not equal to" symbols in it. If we reject the null hypothesis above with the equals signs in it, there are many ways that the distributions could have some difference. Perhaps only one distribution differs from the others; we saw a similar issue with the hypotheses for the one-way ANOVA F test. Next we will describe the calculation of the chi-square test for independence.

Check Your Understanding

SCENARIO 14-D, Continued

This study by Marbella et al. (1998) involved Native Americans and their consultations with Native American healers. The patients were asked about the source of their medical care: a physician only, or both a physician and a Native American healer. Patients were categorized according to age: 18–39 years and 40–83 years. 14-17. The researchers conducted a chi-square for independence using the source of medical care and age group as the two variables. Write the alternative hypothesis in two ways. 14-18. For each way of writing H_1, write the corresponding null hypothesis.

(Continued)

Check Your Understanding (*Continued*)

Suggested Answers

14-17. One way to write the alternative hypothesis is: The sample comes from a population in which there is a relationship between source of medical care and age group. Another way to write H_1: The source of medical care is not independent of age group. 14-18. One way to write the null hypothesis is: The sample comes from a population in which there is no relationship between source of medical care and age group. Another way to write H_0: The source of medical care is independent of age group.

Computing Chi-Square for Independence

The steps for computing the chi-square test for independence are the same as the steps for the previous chi-square test that you learned. The difference is in the way that the expected frequencies are reached. It is possible to have a theoretical distribution of expected frequencies based on prior research, but researchers almost always compute the expected frequencies based on the observed frequencies in a two-way chi-square. Each cell in the contingency table has an expected frequency, *E*, which is computed as follows:

$$E = \frac{\text{row total} \times \text{column total}}{N}$$

Even though the formula is the same for every expected frequency, each cell in the table has a different *combination* of row and column totals in the table of observed frequencies. Figure 14.8 shows the actual observed frequencies from the study by Schauer et al. (2012). Figure 14.8 includes the column totals for the number of people in each treatment group, as well as the row totals for the number of people who did or did not achieve control of their blood sugar. These row and column totals are the ones that go into the computation of the expected frequencies. For example, the expected frequency for the top left cell in Figure 14.8 would be

	Medical Therapy Only	Gastric Bypass	Sleeve Gastrectomy	*Row Totals*
Controlled	5	21	18	44
Uncontrolled	36	29	31	96
Column Totals	41	50	49	$N = 140$

Figure 14.8

Observed frequencies in study of obesity treatments and blood sugar control. (Data from "Bariatric surgery versus intensive medical therapy in obese patients with diabetes," by P. R. Schauer et al., 2012, *The New England Journal of Medicine, 366*, 1567–1576.)

$$E = \frac{\text{row total} \times \text{column total}}{N}$$

$$= \frac{44 \times 41}{140}$$

$$= \frac{1804}{140}$$

$$= 12.885714$$

We keep as many decimal places as possible because these expected frequencies will be used in the computation of the chi-square test for independence. We are showing the number of decimal places that you might see in a basic calculator, but if you are using statistical software, greater precision would be gained. No other cell in Figure 14.8 has the same combination of row and column totals, so a different E must be computed for each cell. It is possible for two cells to have the same expected frequencies, but that is not the case with this data set. Figure 14.9 shows all six expected frequencies for this scenario. (For practice, you might see if you can replicate these E's.) The computed expected frequencies will be proportionally equal across the categories. Using the expected frequencies in Figure 14.9, you might try to compute the proportion of people in each treatment who achieved blood sugar control and see for yourself.

Now that we have the observed and expected frequencies, the steps for computing this chi-square are very similar to the steps for computing the previous chi-square test in this chapter.

- For each cell in the contingency table, take the observed frequency (O) and subtract the expected frequency (E).
- Square the difference.
- Divide each squared difference by its corresponding expected frequency (E).
- Add up those results to get the chi-square test for independence.

Aside from the computation of the expected frequencies, the steps for computing this test statistic are so similar to the chi-square test for goodness of fit that we will not show them here. An exercise at the end of the chapter will ask for those calculations.

	Medical Therapy Only	Gastric Bypass	Sleeve Gastrectomy	Row Totals
Controlled	12.885714	15.714286	15.4	44
Uncontrolled	28.114286	34.285714	33.6	96
Column Totals	41	50	49	N = 140

Figure 14.9

Expected frequencies for obesity treatment and blood sugar control. (Data from "Bariatric surgery versus intensive medical therapy in obese patients with diabetes," by P. R. Schauer et al., 2012, *The New England Journal of Medicine, 366*, 1567–1576.)

This test statistic has a sampling distribution, which could be created by taking all possible samples from the same population, computing the chi-square test for independence on every sample, and arranging the statistics in a distribution. The smallest number possible for this χ^2 test is zero, which would occur if all the observed and expected frequencies were equal. If there is any difference between an observed and expected frequency, the chi-square for independence will be greater than zero. This statistic has the same assumptions and general robustness as the chi-square for goodness of fit. The assumptions are

- Observations are independent.
- Categories are mutually exclusive.
- Categories are exhaustive.

As a reminder, the independence assumption is about the individuals being counted, not the two variables. We are expecting the variables to be related and not independent of each other, according to our alternative hypothesis. If the assumptions are met, we can use a theoretical χ^2 distribution in place of the statistic's sampling distribution. For the chi-square test for independence, the degrees of freedom are

$$df = (\text{number of rows} - 1) \times (\text{number of columns} - 1)$$

For our scenario with the obese patients with type 2 diabetes, there are two rows (controlled and uncontrolled blood sugar at 12 months) and three columns (the treatments). So the degrees of freedom would be

$$df = (2-1) \times (3-1)$$
$$= 1 \times 2$$
$$= 2$$

We use the same table in the back of the book to look up critical values for this chi-square test as the previous chi-square test. In Table D.1, on the row for $df = 2$ and the column for $\alpha = .05$, we find a critical value of 5.99. Like the previous chi-square, the χ^2 test for independence is a one-tailed test even though the alternative hypothesis always is nondirectional. The critical value 5.99 cuts off .05 in the upper tail, as shown in Figure 14.10.

The decision rules for the chi-square test for independence are exactly the same as the decision rules for the earlier chi-square test. For our current example, the null hypothesis said our sample comes from a population in which blood sugar status one year after treatment is unrelated to the kind of treatment that obese patients with type 2 diabetes received. Suppose we are using $\alpha = .05$, and we compute the observed chi-square test for independence to be approximately 10.27, which our statistical software says has $p = .006$. Using the critical value decision rule, what do we decide about H_0? Because 10.27 is more extreme than the critical

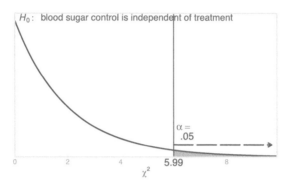

Figure 14.10

Chi-square distribution with $df = 2$. The example of obesity treatments and blood sugar control had two degrees of freedom and $\alpha = .05$. Table D.1 in the back of the book shows a critical value = 5.99.

value of 5.99, we reject the null hypothesis. The same decision is reached with the p value decision rule: because $p = .006$, which is less than $\alpha = .05$, we reject the null hypothesis. We may conclude that blood sugar control is significantly related to the kind of treatment received 12 months earlier.

You may be wondering which treatment was *best* for reaching a status of controlled blood sugar. Obviously, many factors go into a decision about the best treatment for a given patient, but can we make a general statement about the best treatment for reaching HbA1c control? Like the previous chi-square and the one-way ANOVA F test, the chi-square test for independence can be quite general in terms of the conclusions we can draw. The statistic can tell us that there is some difference in the distributions of controlled versus uncontrolled blood sugar across the three treatment groups, but our results cannot tell us specifically which treatment differs from any other treatment. We can answer the question by doing a kind of multiple comparison procedure in which two proportions are compared. There are many ways of doing multiple comparisons in a two-way chi-square. One way involves this same test statistic, the chi-square test for independence. Let's see how this test statistic can be used to compare two independent proportions.

We mentioned the possibility of comparing two independent proportions when we were talking about the confidence interval for a single proportion. The example involved adults in Oklahoma with diabetes, and we were interested in the proportion of these people reporting consumption of five or more servings of fruit/vegetables. What if we computed a similar proportion for Oklahoma adults who do not have diabetes? A chi-square test for independence could be computed to compare these two proportions. The table would have two rows and two columns. The row variable could be the consumption of 5+ servings a day (yes/no). The column variable could be diabetes (yes/no). The chi-square test would tell us whether there was a relationship between having diabetes and consuming 5+ servings a day. If there is such a relationship, then the proportions are statistically different.

This idea of comparing two proportions also underlies a common way of performing multiple comparison procedures in a chi-square design that is larger than 2 × 2. Imagine limiting our focus to two groups at a time in the example of obesity treatments and blood sugar control. For each kind of obesity treatment, we could compute the proportion of people who achieved blood sugar control. Then we could compare the proportions for these groups:

- Gastric bypass versus medical therapy
- Sleeve gastrectomy versus medical therapy
- Gastric bypass versus sleeve gastrectomy

We could apply a Bonferroni-type correction on alpha, then compute three 2 × 2 chi-square tests. There are more sophisticated analyses possible, but you may see this approach in journal articles.

The two-way chi-square test also is used in research to compare people who completed a study versus people who did not complete a study. Researchers often hope a chi-square for independence will be nonsignificant when comparing completers with noncompleters on some important categorical variable. If there is no relationship between study completion (yes/no) and the important categorical variable, then researchers feel more confident about making generalizations based on the results from the completers. In contrast, if the chi-square for independence is significant, then the people who stayed in the study may differ in important ways from the people who quit the study. Consequently, the results may generalize only to part of the population of interest. Next we will link contingency tables with Chapter 6 on probability, then provide some examples of relative risk and odds ratios.

Check Your Understanding

SCENARIO 14-D, Continued

This study by Marbella et al. (1998) concerned Native Americans' consultations with Native American healers. Among other research questions, the researchers asked whether there was a relationship between age group (18–39 years or 40–83 years) and source of medical care (consultation with a physician only versus consultation with both a physician and a Native American healer). Our analysis of the data results in $\chi^2 \approx 4.44$, $p = .035$. 14-19. Compute df for this study. 14-20. Use $\alpha = .05$ and Table D.1 to find a critical value. 14-21. Use the critical value decision rule to test a null hypothesis that says the patients' source of medical care is independent of age group. 14-22. For more practice, use the p value decision rule to test the null hypothesis. 14-23. Using the variable names, explain the meaning of the decision about the null hypothesis.

(Continued)

Suggested Answers

14-19. We have two rows and two columns, so df = (number of rows − 1) × (number of columns − 1) = (2 − 1) × (2 − 1) = 1 × 1 = 1. 14-20. On the row for df = 1 in Table D.1, we find a critical value of 3.84 in the column for α = .05. 14-21. Because the observed chi-square for independence = 4.44, which is more extreme than the critical value of 3.84, we reject the null hypothesis. 14-22. Because p = .035, which is less than α = .05, we reject the null hypothesis. 14-23. We can conclude there is a significant relationship between age group and source of medical care in this sample. The preference for consulting a physician only versus both a physician and Native American healer is not independent of the age group.

Relative Risk

The contingency tables in this chapter may remind you of Chapter 6 on probability and risk, where similar tables were shown. We talked about risk as a probability of an undesired outcome, and relative risk as a statistic that quantifies how people with a risk factor differ from people without a risk factor. We will continue our coverage of the analysis of frequencies by reviewing relative risk, describing a confidence interval that sometimes is reported for estimating the relative risk in the population, and then explaining another statistic that often appears in health sciences studies about risk factors for diseases or conditions.

When epidemiologists study disease risk, they want their estimates of risk to be good estimates of the actual risk in the population. After all, if we are trying to understand the risk of contracting a disease, we do not want to base this understanding on a small sample of people. That is why risk and relative risk usually are seen in studies of extremely large samples, which often are obtained using complex sampling procedures. These samples also may be observed over a number of years to see whether certain health conditions emerge more frequently for people exposed to a risk factor, compared with people who have not been exposed to that risk factor. In other words, relative risk tends to be used in cohort studies, which we defined in Chapter 1 as research that identifies people exposed or not exposed to a potential risk factor and compares these people by observing them across longer periods.

As we said above, relative risk is a statistic that quantifies how people with a risk factor differ from people without the risk factor—more specifically, how the groups differ in their likelihood of an event, such as contracting a disease. When interpreting a relative risk, it may help to focus on the name of the statistic: relative risk. Two risks are being *compared*, with one of those risks being *relative to* another risk. The numerator of the relative risk tells us about the risk of disease given exposure to a risk factor, and the denominator tells us about the risk of the same disease in the absence of exposure to the same risk factor.

Let's look at an example. A study by He et al. (2001) reported on risk factors in the United States for congestive heart failure, a condition in which the heart is unable to provide adequate oxygenated blood to the rest of the body. These researchers relied on data from an ongoing study called the National Health and Nutrition Examination Survey (NHANES), which uses complex sampling procedures to obtain representative samples of Americans across all ages. The researchers divided the subjects into groups: those who had developed congestive heart failure in the years since the study began and those who had not developed this heart condition. At that point, they could ask about everyone's risk of developing congestive heart failure, which would be a probability. But what interested them most were risk factors, including current cigarette smoking. The people with and without congestive heart failure could be divided further into those who are current cigarette smokers and those who are not current cigarette smokers. The study by He et al. (2001) did not list the frequencies for the four combinations of current cigarette smoking (yes/no) and diagnosis of congestive heart failure (yes/no). Based on their statistics and the number of people in different groups (e.g., current smokers vs. nonsmokers), we have *fabricated* some numbers that will provide us with statistics that are similar to theirs. Figure 14.11 shows our fabricated numbers. In this example, we are ignoring the effect of age, whereas He et al. took into account such variables that could explain differences in heart failure rates.

Based on our fabricated data in Figure 14.11, what would be the risk of a randomly chosen patient developing congestive heart failure, given that the patient currently smokes? This question is asking for a conditional probability. We find the denominator first—the "given" part. Here, we are being given 4,775 people who currently smoke. Next we find the numerator: out of those 4,775 people, we are looking for the ones who have congestive heart failure. There are 615 such patients. So the risk is 615/4775 = .1287958, or about .13. About 13 out of every 100 patients who smoke could be expected to develop congestive heart failure, according to our fabricated data set.

But how does that risk compare to the risk of getting congestive heart failure for people who do not smoke? Maybe smoking is unrelated to congestive heart failure, and we would find the same risk for nonsmokers. Let's compute the risk of congestive heart failure, given that the patient is a nonsmoker. Again, we are asking about a conditional probability, and we first find the denominator: the 8,868 nonsmokers

		Patient Has Congestive Heart Failure		
		Yes	No	Row Totals
Patient Currently Smokes	Yes	615	4,160	4,775
	No	767	8,101	8,868
Column Totals		1,382	12,261	N = 13,643

Figure 14.11

Fabricated data on congestive heart failure and current smoking

in our fabricated data set. Then we look for the subset of those 8,868 nonsmokers with congestive heart failure; there are 767 such patients. So the risk of congestive heart failure, given the person is not a current smoker, would be 767/8868 = .0864908 or about .09. About 9 out of every 100 patients who do not smoke could be expected to develop congestive heart failure. Now that we have the two risks of congestive heart failure—the risk for current smokers and the risk for nonsmokers—we can compute the relative risk, which often is abbreviated in journal articles as RR. One way that relative risk can be defined is as follows:

$$RR = \frac{\text{probability of disease given exposure to the risk factor}}{\text{probability of disease given no exposure to the risk factor}}$$

For our example, the risk factor is current smoking. The numerator of our relative risk will be the probability of congestive heart failure, given the patient is a current smoker. The denominator will be the probability of congestive heart failure, given the patient is a nonsmoker. Plugging in the probabilities that we computed above, we find

$$RR = \frac{.1287958}{.0864908}$$

$$RR = 1.489128$$

$$\approx 1.49$$

If the two probabilities were equal, we would have gotten a relative risk = 1, meaning the probability of getting the disease given the exposure to the risk factor is equal to the probability of getting the disease given no exposure to the risk factor. Because the relative risk that we computed above is greater than 1, the probability of developing this condition is greater for smokers than nonsmokers. Our $RR \approx 1.49$ can be interpreted as follows: the risk of the condition is 49% higher for current smokers than nonsmokers. It is possible to have a relative risk less than 1. Sometimes the exposure being studied is beneficial to people's health, such as exercise, and the people who exercise have a lower probability of getting a disease, compared with the likelihood of getting the disease for people do not exercise. A relative risk = 0.5 would mean that people who were exposed to the (positive) factor were half as likely to get the disease, compared with people who were not exposed. It is theoretically possible for an exposure to a positive effect to eliminate the risk of the health condition entirely, which would lead to a relative risk = 0, but this is unlikely to occur in applied research.

It is important to recognize what a relative risk does *not* say. A relative risk of 1.39 means those who are exposed to the risk factor are 39% more likely to get the disease, compared with the likelihood for people who were not exposed to the risk factor. It does *not* mean that 39% more *people* will get the disease. What happens to relative risk if the *risk* (likelihood) is small? Let's consider the Women's Health Initiative, which studied the effect of hormone therapy on women who have gone through menopause. Hormone therapy sometimes is used to relieve intense symptoms of menopause, such as hot flashes. Health-care professionals used to assume that hormone therapy provided some health benefits, such as a reduced risk of heart disease, but they did not have evidence. Researchers wanted to know whether hormone therapy carries health risks that might outweigh the benefits. This question motivated the Women's Health Initiative, a randomized controlled trial of different hormone therapies versus a placebo.

Let's look at one part of this study, involving more than 10,000 women who had undergone a hysterectomy and were taking estrogen only. The National Heart, Lung, and Blood Institute, a division of the National Institutes of Health and sponsor of the study, said the women who took estrogen had an increased risk of stroke, compared with women who took the placebo. The Institute said that for every 10,000 women per year who take estrogen only, 44 of them could be expected to have a stroke, compared with 32 out of 10,000 women taking a placebo. In other words, taking estrogen meant 12 more cases of stroke per year out of every 10,000 women. This effect was statistically significant. But how would a woman facing the decision on hormone therapy view this difference in risks? Any given woman may or may not have these same risks. If stroke runs in her family, she may be at a higher risk of stroke already and may want to avoid anything that could increase her chances of suffering a stroke. If she is otherwise healthy and has no known risk factors for stroke or other serious conditions, she may decide that in her case, the benefits of hormone therapy outweigh the risk.

"Omada-GDP Per Capita" (6' × 8'), by Gary Simpson, used with permission. One inch of brass rod equals 10 million people in this close-up image. This grouping totals 202 million people who live in countries with a per capita gross domestic product (GDP) from $33,000 to $36,000.

Before we continue with relative risk, we want to point out again the need to be cautious about drawing conclusions based on only one study. The Women's Health Initiative was an ambitious study with several limitations. The Endocrine Society put together a task force of experts to review all recent research on hormone therapy and published a 60-page analysis of the benefits and risks (Santen et al., 2010). The first line of the executive summary accompanying the analysis is noteworthy: "A sound understanding of the actual benefits and risks of menopausal hormone therapy (MHT) requires interpretation of a complex body of existing data" (p. S1). Among other concerns, the Endocrine Society pointed out that only a small fraction of the women in the Women's Health Initiative were in their early 50s, the approximate ages when most women may be making decisions about starting hormone therapy. The results may generalize only to older menopausal women, unless an adjusted relative risk is computed, taking into account the ages of the women in the study. Whenever we are weighing scientific evidence, we must consider whether we are looking at a body of research or only one study. Sound science requires replication across multiple samples to fill in the gaps in our understanding of complex phenomena.

Let's return now to relative risk. Relative risk statistics often are compared to a value of $RR = 1$, which means there is no difference in the chance of getting the disease for people who were exposed versus not exposed to the risk factor. As we have seen with other statistics, a relative risk statistic will vary across repeated samples from the same population; in other words, there is sampling variability. Is a relative risk statistic of 1.49 roughly the same as 1 after we take into account the sampling variability? Or is our RR significantly different from 1? To quantify the sampling variability, we can compute a confidence interval for a relative risk. But the formula for this confidence interval is complex, so we will skip the computations and go directly to the interpretation of a confidence interval for a relative risk. He et al. (2001) reported the following 95% confidence interval for the relative risk of congestive heart failure for current cigarette smokers, compared with nonsmokers: [1.3, 1.7], $p < .001$. We can say that 95% of confidence intervals computed like this one will bracket the true population relative risk. This interval does not bracket 1, so we would doubt that the true population relative risk would be 1. The small p value would lead us to conclude that current smoking is a significant risk factor for the development of congestive heart failure. (These authors' results involved more complex analyses than we have presented here. Among other things, their analyses adjusted for factors like age and race.)

Relative risk is not appropriate for case-control studies, where the researchers already know who does and does not have a condition and may have chosen patients without the condition (i.e., the controls) based on their demographic similarities to the people with the condition (i.e., the cases). Case-control studies have a different purpose than cohort studies, which are better suited for estimating disease risk in the population. How do researchers talk about probabilities for different groups when they have a case-control study and not a cohort study? We will answer that question next.

SCENARIO 14-E

A 2013 report from the American Heart Association described numerous statistics related to heart disease and stroke. A stroke is when blood stops flowing to part of the brain, leading to the death of brain cells. An ischemic stroke occurs when a blood clot stops the blood flow. A hemorrhagic stroke occurs when a blood vessel breaks open in the brain and the person bleeds in or around the brain. The report described results of a systematic review of studies that considered risk factors for these different kinds of major stroke. After taking into account age, lifestyle, and other risk factors, the systematic review looked at the relative risk of ischemic stroke for people who were obese relative to people who were normal weight. The report said the relative risk was 1.64, with a 95% confidence interval of [1.36, 1.99]. 14-24. Explain the meaning of this relative risk and associated confidence interval.

Suggested Answers

14-24. The relative risk of 1.64 means that people who are obese have a risk of ischemic stroke that is 64% higher than the risk for people who are normal weight. This confidence interval is an estimate of the population relative risk and quantifies the variability that we could expect across repeated samples. Out of 100 confidence intervals, we would expect 95% of them to contain the true population relative risk. This confidence interval does not bracket 1, a value that would have indicated equal risk of ischemic stroke for obese individuals, relative to normal-weight individuals. We can say that people who are obese have a significantly higher risk of ischemic stroke, relative to the risk for people who are normal weight.

Odds Ratios

Sometimes researchers want to estimate the probability of a condition for different groups in a case-control study, which can involve the analysis of frequencies. The focus is not on disease surveillance or estimation of population risk. Instead, the risk factors are studied within the limited time frame of a case-control study, which involves the comparison of similar people, except one group has a disease or condition, and the other group does not. For example, Hernandez et al. (2011) described a study of 1,103 patients with shingles and 523 controls treated at a clinic in Texas. Already we can see the difference between this study and the kind of study used to estimate disease risk in a population. This study was limited to one location, and the researchers already knew who had shingles and who did not, meaning they were not watching groups of people across time to see who developed the disease after having been exposed to a potential risk factor. The *cases* in a case-control study have a disease or condition. The *controls* in this kind of study do not have the disease or condition, and they were chosen to be similar to the cases on some important extraneous variables. The controls were chosen from the records of patients treated for skin conditions in 1992–2005. They were selected because they resembled the patients with shingles in terms of age, sex, and race/ethnicity. (The two groups were similar in their proportions on these extraneous variables; participants were not matched in a pairwise manner.) In 2006, a shingles vaccine became available for older Americans, but in this study, no one had received the vaccine. The researchers' focus was whether people with shingles, also known as herpes zoster, were more likely than controls to have a family history of shingles.

Suppose we are running a study similar to the one by Hernandez et al. (2011) and we obtain the data shown in Figure 14.2. (We used the frequencies from the study by Hernandez et al., but we are categorizing participants in a different way than the researchers did.)

Let's use Figure 14.12 to review some concepts from Chapter 6. As you may recall, the odds of something happening is *not* the same thing as a probability. We defined odds as follows:

$$\text{Odds} = \frac{\text{probability of something happening}}{1 - \text{probability of something happening}}$$

Looking at Figure 14.15, let's find the probability of a randomly chosen participant having relatives with a history of shingles. This probability would have 535 in the numerator and 1,578 in the denominator, so the probability is approximately .339. (We are using rounded numbers for simplicity, but statistical software would use unrounded figures.) What is the probability of a randomly chosen participant having no relatives who had shingles? This probability has a numerator of 1,043 and a denominator of 1,578, so the probability is approximately equal

		Disease Group		
		Cases	Controls	Row Totals
Suspected Risk Factor	Has Relatives with History of Shingles	480	55	535
	No Relatives with History of Shingles	575	468	1,043
Column Totals		1,055	523	N = 1,578

Figure 14.12

Frequencies for shingles diagnosis and family history of shingles. We used numbers that appeared in a research study, but we combined some of the categories used by the researchers. (Data from "Family history and herpes zoster risk in the era of shingles vaccination," by P. O. Hernandez et al., 2011, *Journal of Clinical Virology*, 52, 344–348.)

to .661 (or 1 − .339). What are the odds of a random participant in the study having relatives with a shingles history? The answer is

$$\text{Odds of relatives with shingles history} = \frac{.339}{.661}$$
$$= 0.5128593$$
$$\approx 0.51$$

The odds of 0.51 means a randomly chosen participant is about half as likely to have relatives with a shingles history as to have no relatives with a shingles history. When the odds computation places the two probabilities in the fraction, this is the step where many people begin to lose touch with the meaning of the odds. Then the odds ratio is one step farther from probability: two odds computations are placed in a fraction or ratio to form the odds ratio. Notice how many fractions are embedded within other fractions to get the odds ratio. Fortunately for you, some of those fractions-within-fractions have denominators that cancel each other out, as we will demonstrate, and we will explain an easier way to compute an odds ratio. But the odds ratio is quite different from a probability and even gets misinterpreted in some journal articles.

Focus on the name of the statistic: odds ratio. There are two *odds* computed, and they are placed in a *ratio*. But *each* of those odds is a comparison of two probabilities. To demonstrate the simpler way of computing the odds ratio, let's go back to Figure 14.12, setting aside the numbers momentarily. In place of the frequencies, we will use letters, as shown in Figure 14.13.

Let's bring in the effect of family history on the likelihood of getting shingles and do a little math that will simplify the eventual computation of an odds ratio. We need to use conditional probabilities to compute the odds of getting shingles (or not), given that the person has one or more relatives with a history of shingles:

$$\text{Odds} = \frac{\text{probability of shingles, given relatives with history}}{\text{probability of no shingles, given relatives with history}}$$

		Disease Group		
		Disease Cases	Group Controls	Row Totals
Suspected Risk Factor	Has Relatives with History of Shingles	a	b	a + b
	No Relatives with History of Shingles	c	d	c + d

Figure 14.13

Simplifying the odds ratio calculation using the shingles example

Both the numerator and denominator of the odds are conditional probabilities. Let's start with the numerator:

$$\text{Probability of shingles, given relatives with history} = \frac{a}{a+b}$$

Now the denominator:

$$\text{Probability of no shingles, given relatives with history} = \frac{b}{a+b}$$

The odds will be the first conditional probability divided by the second conditional probability—but don't worry, this fraction is going to simplify:

$$\text{Odds of shingles, given family history} = \frac{a/a+b}{b/a+b}$$

When one fraction is divided by a second fraction, we can flip over the second fraction and multiply instead:

$$\text{Odds of shingles, given family history} = \frac{a}{a+b} \times \frac{a+b}{b}$$

Now the "$a + b$" in the numerator cancels out the "$a + b$" in the denominator, because something divided by itself equals 1. We are left with the following:

$$\text{Odds of shingles, given family history} = \frac{a}{b}$$

So the odds of shingles, given a family history of shingles, can be defined as a fraction that takes the number of people with a family history who did get shingles and divides it by the number of people with a family history who did not get shingles. In other words, how are the people with a family history spread out between the number of cases (a) and the number of controls (b)? Look back at Figure 14.3. The odds of getting shingles, given a family history of at least one person with shingles, has simplified to being a fraction: the number of cases relative to the number of controls in the first row of the table. If people were equally likely to get shingles or not, given at least one relative who has had shingles, then a would equal b, and the odds would equal 1. Is that the situation in our example? Let's use the numbers from Figure 14.12 and find out:

$$\text{Odds of shingles, given family history} = \frac{480}{55}$$

$$\approx 8.727$$

The odds of getting shingles given that family members have had the disease equaled about 8.727. This number means a person is more than eight times more likely to be a case (having shingles) in this study than a control (not having shingles) if there is a family history of shingles.

Now let's compute the odds of shingles, given no family history. Based on what we just learned using the first row of Figure 14.13, we can use this formula:

$$\text{Odds of shingles, given no family history} = \frac{c}{d}$$

Now we are looking at the people who do not have a family history of shingles, and we are comparing the number who did versus did not get shingles. If there is no family history, what are the odds of getting shingles? Are people equally likely to get shingles or not, if there are no relatives who previously had the disease? Let's plug in the numbers from Figure 14.12 and find out:

$$\text{Odds of shingles, given no family history} = \frac{575}{468}$$
$$\approx 1.229$$

The odds of getting shingles, given there is no family history, equaled a number much closer to 1 than the previous odds that we computed. It appears that people in this study were almost equally likely to be a case or a control if there was no family history of shingles.

So far, these computations probably make sense to you. Here is the step that can make odds ratios harder to understand: we are going to divide the first odds by the second odds to estimate the relationship between getting shingles (or not) and having a family history of shingles (or not). The odds ratio compares the odds of getting shingles given a family history with the odds of getting shingles given no such family history. Even we have trouble wrapping our brains around that concept! Consider our present example. The odds were nearly even (about 1.229) of getting shingles or not, given no family history. What about the odds of getting shingles or not, given there *is* a family history? That number was about 8.727.

The odds ratio will take the first odds we computed (about 8.727) and divide it by the second odds that we computed (about 1.229). Each of those computations has rounding error, which could get worse with yet another division, so let's look at a way of computing the odds ratio that will have only one division. We can use the letters in Figure 14.3 and show the odds ratio as a fraction, with the numerator being the odds of shingles given a family history, and the denominator being the odds of shingles, given no family history. The odds ratio often is abbreviated as *OR*:

$$OR = \frac{a/b}{c/d}$$

Once again, we have one fraction divided by another fraction, which means we can flip over the second fraction and multiply:

$$OR = \frac{a}{b} \times \frac{d}{c}$$

$$= \frac{ad}{bc}$$

Now we have only one division. Let's do the computation using the frequencies from Figure 14.2:

$$OR = \frac{480 \times 468}{55 \times 575}$$

$$= \frac{224,640}{31,625}$$

$$\approx 7.103$$

This odds ratio means that the *odds* of getting shingles for those with a family history versus no family history is more than seven times the *odds* of not getting shingles when comparing those with and within the family history. Can you see how different the odds ratio is from probability? We did *not* say that someone is seven times more likely to get shingles if they have a family history. We can think of the odds ratio this way: if there is no relationship between the risk factor and the disease, then the odds ratio will be close to 1. In other words, the way people with the risk factor are spread out between cases and controls is similar to the way people without the risk factor are spread out between cases and controls. In this case-control study, similar people with and without shingles were compared. For people with a family history of shingles, the odds of getting shingles is more than seven times the odds for people without a family history. It appears that people with a family history of shingles may be vulnerable to the disease.

How can we tell if the relationship between family history and shingles is significant? A confidence interval can be computed for the odds ratio, but we will omit the hairy computational details. Like other confidence intervals we have covered, this confidence interval is interpreted in terms of repeated sampling from the same population. For a 95% confidence interval, we would say that 95% of intervals computed on repeated samples from the same population would bracket the true population odds ratio. If the interval does not bracket 1, then we would say that the risk factor is significantly related to the odds of being case versus being a control. If the interval is less than 1, then we would say the risk factor is positive or protective against the disease. An interval greater than 1 would mean the risk factor led to greater odds of being a case (getting the disease), compared with the odds of being a case for those who do not have the risk factor. For our example of shingles, we will pretend that we obtained a 95%

confidence interval of [5.13, 8.98]. The interval does not bracket 1 and the interval is greater than 1. The odds of getting shingles for people with a family history are significantly greater than the odds of getting shingles for people without a family history.

It is important to remember the context of the odds ratios that you may see in research studies. The cases (people with a disease or condition) and the controls (people without the disease or condition) did not get into those groups as a result of random assignment, nor was an independent variable manipulated. Another case-control study could show different results. Just because a complex computation is involved does not mean that the research has revealed a relationship between a risk factor and a disease that will be confirmed by future research. Something else to consider when you read about odds ratios is that they often are computed as part of a larger analysis involving multiple predictors of the disease. This larger analysis may mean that the numeric results are dependent on the presence of other risk factors being analyzed simultaneously. For the shingles study, the numeric value of the odds ratio probably would be different if the researchers controlled for other factors. The complexity of interpretation increases when odds ratios come out of that kind of analysis. When reading about such studies, try to notice whether the study describes an analysis that controls for some demographic variables or other risk factors. If so, the study might describe adjusted odds ratios or adjusted likelihood statistics. If a relationship between a risk factor and the odds of disease appears to be significant, you can think to yourself: this effect is persisting even after controlling for these demographic variables or other risk factors.

Next we will shift from the analysis of frequencies to a brief explanation of statistics that can be used to analyze ranked data.

Check Your Understanding

SCENARIO 14-F

Suppose we want to compare two odds: the odds of being a shingles patient for those who have multiple family members with a history of the disease versus the odds of shingles for those with only one family member with a history of shingles. 14-25. For odds of being a shingles patient, given multiple relatives with a history, suppose we find the odds = 23.0. Explain these odds. 14-26. For the odds of being a shingles patient, given only one relative with a history, suppose we compute the odds = 6.98. Explain these odds. 14-27. Using these two odds, compute the odds ratio and explain it.

(Continued)

14-25. The odds of being a shingles patient, given multiple relatives with shingles, equaled 23.0, meaning people are 23 times more likely to get shingles than not to get shingles if they have multiple relatives with a history. 14-26. The odds of being a shingles patient, given a single relative with shingles, equaled 6.98, meaning people are almost 7 times more likely to get shingles than not to get shingles if they have only one relative with a history. So the odds of getting shingles for people with only one relative who had the disease is bigger than 1, but smaller than the odds for those with multiple relatives who had shingles. 14-27. The odds ratio can be computed by dividing the first odds by the second odds: 23/6.98 ≈ 3.295. The odds of getting shingles are more than three times greater if patients have multiple relatives with a history of the disease versus a single relative who has had shingles. In other words, there appears to be a relationship between getting shingles and having multiple relatives (vs. one family member) with a shingles history. We would need to compute a confidence interval for the odds ratio to determine whether the relationship is statistically significant.

Analysis of Ranks

You may have noticed that the hypotheses for the two chi-square tests differed markedly from the hypotheses you saw earlier in this book. For example, the scenario about the number of sudden cardiac deaths on different days of the week used the chi-square for goodness of fit. This scenario had a null hypothesis that said our sample came from a population in which the distribution of sudden cardiac deaths across days matched a theoretical distribution in which such deaths were equally likely every day. The null hypotheses for the chi-square for goodness of fit and the chi-square for independence did not contain any parameters, like population means. That is why these chi-square tests belong to a category of test statistics called *nonparametric statistics*; their hypotheses do not contain parameters. A statistic that tests a null hypothesis that does contain a parameter is called a *parametric statistic*.

Nonparametric statistics generally free researchers from an assumption of normality. You may have noticed that the assumptions for the two chi-square tests in this chapter did not include an assumption about the shape of the population distribution. Not all nonparametric statistics rely on frequencies in categories. Many nonparametric statistics involve the analysis of ranked data. What do we mean by ranked data? Let's think about the example of food hardship and obesity rates for the 50 states plus the District of Columbia. A correlation was computed, as well as a regression analysis, using these rates. It is possible to convert the food hardship and obesity rates into ranks. You may recall that the state with the highest obesity rate in our data set was Mississippi (34.9%); Colorado had the lowest obesity rate (20.7%). We could assign the smallest rank to the state with the lowest obesity rate and the highest rank to the state with the highest obesity rate, so then the numbers of the ranks and the numbers of the obesity rates would be in the same order. Colorado would be ranked 1, meaning the lowest rate of adult obesity has the smallest rank number. The largest rank, 51, would be assigned to Mississippi, and all other locations also would be placed in order and ranked. Similarly, we could assign ranks for food hardship: North Dakota had the lowest food hardship rate (10%), and Mississippi had the highest food hardship rate (24.5%). Every state could be ranked in terms of obesity rates, from North Dakota's lowest number, 1, to Mississippi's highest number, 51. Then we could set aside the percentages and use the ranks to compute a correlation between food hardship and obesity. The correlation computed on ranks is called *Spearman's rho*.

Why would someone want to analyze ranks instead of using the actual data? Sometimes researchers expect to sample from populations that violate a normality assumption, and nonparametric statistics do not have assumptions of normality. For instance, Pearson's r has an assumption of bivariate normality, but Spearman's rho does not. If we have reason to expect that the normality assumption has been violated severely, we may need to use a nonparametric test that relies on ranks. Let's consider another statistic with a normality assumption and its nonparametric alternative. In Chapter 11, we talked at length about the robustness of the independent-samples t test, which assumed the data came from two populations that are normally distributed. We said the independent-samples t test usually is robust when it encounters data from nonnormal populations, except for oddly shaped populations. We wrote: "If we have reason to believe that there would be a clump of outliers in one tail of a population distribution, then the independent-samples t test may have a sampling distribution that doesn't look like a theoretical t distribution." If the sampling distribution does not match the theoretical t distribution, then the p values for our test statistic will not be trustworthy, and we may make errors in hypothesis testing. A distribution with a clump of outliers in one tail would be a situation in which a nonparametric statistic might be a better option.

A skewed distribution also would violate normality, but would be no problem for ranked data. Let's recall the study by Efe and Özer (2007) on different soothing methods for infants receiving injections. The researchers randomly assigned 66 babies to groups. Babies in the treatment group were breast-fed during the shot, and mothers were told to encourage the babies to continue breast-feeding after the shot. Babies in the control group received the usual care, with the mothers being told to cuddle and talk soothingly to the babies after the shot. The dependent variable was the number of seconds of crying after the shot. We want to show a limited numeric example of ranks, so let's imagine a very small study like this one. Instead of having 33 babies per group, we have only 5 babies per group. Table 14.10 shows some made-up numbers for the two groups. Crying times (in seconds) are given, along with the corresponding ranks for those crying times.

Table 14.10 Fabricated Data on Crying Times to Illustrate the Creation of Ranks

	Group		
Treatment		Control	
Crying Time	Rank	Crying Time	Rank
40	4.5	60	9
50	7	180	10
25	1	40	4.5
30	2	55	8
35	3	45	6

The crying time of 25 seconds is the shortest duration, so this score has a rank = 1. The next longer crying time of 30 seconds has a rank = 2. The third-longest crying time of 35 seconds has a rank = 3. The next longest crying time is 40 seconds, which appears twice. The next two ranks, 4 and 5, are averaged, and both appearances of 40 seconds are ranked 4.5. Look at the crying time of 180 seconds, which has a rank = 10. This crying time might be a couple of standard deviations above the mean of the control group and probably has a fairly large effect on that mean. But compare the rank of 10 to the rank of 9; now we see that the distance between the two highest *ranks* is 1, regardless of how extreme the top-ranked *score* is. This small numeric example helps to illustrate why the analysis of ranks can be valuable.

If we could avoid a problem with nonnormal data by using nonparametric statistics, why not use them all the time? There are a few reasons:

- The hypotheses tested by nonparametric statistics do not match exactly with the hypotheses tested by parametric statistics, and researchers may not want to modify their hypotheses. For example, the independent-samples *t* test tests a null hypothesis about two population means. Its nonparametric counterpart, the Mann–Whitney *U* test, has a null hypothesis about equal distributions, which usually comes down to equal medians, according to the way that most researchers use this test.
- Even though a nonparametric statistic may be used to test a certain null hypothesis, the statistic may be significant if the groups differ in another way. For example, the Mann–Whitney *U* test can be significant because of a difference in variances. If you wanted to know whether two groups had different medians, would you want your test statistic to be significant when the two groups have equal medians but unequal variances? We doubt it.
- Parametric statistics sometimes provide more power than their nonparametric alternatives, all else being equal. In specific cases, such as an analysis of numeric data that are not continuous, nonparametric statistics can have more power (Nanna & Sawilowsky, 1998). A nonparametric statistic also may be more powerful when sampling from populations with a lump of outliers in one tail, a situation we mentioned previously.

Table 14.11 List of Some Parametric Statistics in This Book and Their Nonparametric Counterparts

Kind of Test Statistic	
Parametric	Nonparametric
Pearson's *r*	Spearman's rho
Independent-samples *t* test	Mann–Whitney *U* test
Paired *t* test	Wilcoxon signed-ranks test
One-way ANOVA *F* test	Kruskal–Wallis test

Table 14.11 lists the names of a few rank tests that are nonparametric alternatives to some of the parametric statistics we have covered. You may encounter these nonparametric statistics in journal articles.

The nonparametric statistics in Table 14.11 should *not* be considered exactly equivalent to their parametric alternatives. The parametric and nonparametric statistics will test different null hypotheses. Entire books have been written on nonparametric statistics. Our goal here simply has been to introduce you to the names of some statistics you might encounter and the parametric statistics that are used in similar situations. This knowledge might help you to understand research that uses these nonparametric statistics.

What's Next

We have come a long way in this book, starting with an overview of the context in which statistics are used. We talked about computing descriptive statistics, graphing data, measuring relative location of a score in a distribution, and working with normal distributions. Bivariate correlations were described, then later we explained how predictions can be made based on linear relationships. We relied on your intuitive understanding of probability to introduce this crucial concept. The pivotal role of sampling distributions was described as a lead-in to hypothesis testing, and we covered test statistics and estimates for one population mean and for two population means. For comparing two or more means, we explained one kind of analysis of variance and a few multiple comparison procedures. Finally, this chapter covered some statistics used for analyzing categorical data and briefly described statistics that are computed on ranked data.

One more chapter awaits, and the purpose of Chapter 15 is to summarize what you have learned and give you some practice in deciding which statistic should be used in a given situation. Choosing the correct test statistic or interval estimate is a skill that can require a lot of practice. We hope Chapter 15 will provide you with some hints on how to choose from among the statistics presented in this book, along with practice applying what you have learned to a number of research scenarios.

SCENARIO 14-G

Expanding on the idea of the proportion of Oklahomans who consume five or more servings of fruit/vegetables per day, we returned to the Oklahoma State Department of Health's OK2SHARE website. We wanted to know about different age groups of respondents in the 2009 BRFSS survey. If we are looking at people who are 75 years or older, what proportion of respondents say yes to the question about eating 5+ servings per day? Suppose we want to test a null hypothesis that says our sample comes from a population in which 14% of respondents say they consume 5+ servings of fruit/vegetables daily. The website said 1,234 Oklahomans who were 75 years and older answered the question, and 252 of them said yes. 14-28. Compute the proportion of respondents in this sample who answered yes. 14-29. Compute the standard deviation of the sampling distribution of the sample proportion. 14-30. Compute the margin of error for a 95% confidence interval for the population proportion. 14-31. Compute the confidence interval. Does the interval contain the hypothesized proportion of .14? Interpret the confidence interval.

SCENARIO 14-B, Continued

This scenario concerned the distribution of the 902 units across eight blood types. A Check Your Understanding question concerned the expected frequencies. The population percentages from a website were applied to create a distribution of expected frequencies. Suppose we are ready to compare the expected and observed frequencies, shown in Table 14.12.

14-32. Compute the squared differences between the observed and expected frequencies. 14-33. Divide each of the squared differences by its corresponding expected frequency. 14-34. Add up the numeric answers to the last question to obtain the chi-square test for goodness of fit. 14-35. We previously established that the null hypothesis said the sample came from a population with a distribution of blood types that is the same as the website's distribution, and we found a critical value of 14.07 for this scenario ($\alpha = .05$). Test the null hypothesis, then explain the meaning of your decision, using the variable names.

Table 14.12 Fabricated Data on Blood Types

	Blood Type							
	A+	B+	AB+	O+	A−	B−	AB−	O−
Expected	306.68	90.2	36.08	333.74	54.12	18.04	9.02	54.12
Observed	303	95	37	323	55	27	9	53

(Continued)

Respondent's Role	Non-punitive reporting environment for nurses witnessing disruptive behavior?		
	Yes	No	*Row Totals*
Physician	136	17	153
Nurse	438	123	561
Executive	22	2	24
Column Totals	596	142	*N* = 738

Figure 14.14

Survey of hospital nurses, physicians, and executives. (Data from "Nurse-physician relationships: Impact on nurse satisfaction and retention," by A. H. Rosenstein, 2002, *American Journal of Nursing, 102*, 26–34.)

SCENARIO 14-H

(Previously shown as Scenario 6-D.) Rosenstein (2002) conducted a survey of U.S. hospital professionals about their perceptions of disruptive behavior by physicians. The study defined disruptive physician behavior as "any inappropriate behavior, confrontation, or conflict, ranging from verbal abuse to physical and sexual harassment." Data came from nurses, physicians, and hospital executives at 84 hospitals around the country, ranging from small, rural not-for-profit hospitals to large, urban academic centers. One question asked whether they felt their hospital provided a "non-punitive reporting environment for nurses who witness disruptive behavior." Figure 14.14 contains some of the survey's results.

14-36. What kind of research is this? 14-37. What kind of variable is *respondent's role*? 14-38. How would you describe the external validity of this study? 14-39. How would you describe the internal validity of this study? 14-40. Suppose the researchers had planned to compute a chi-square statistic for this scenario. Which chi-square would they choose, and why? 14-41. Write the alternative hypothesis for this chi-square test. 14-42. Write the null hypothesis for this chi-square test. 14-43. What would be the *df* for this chi-square test? 14-44. Using α = .05, look up a critical value in Table D.1. 14-45. Compute the expected frequencies. 14-46. Compute the difference between the observed and expected frequencies. 14-47. Square the differences computed in the previous question. 14-48. Divide each squared difference by its corresponding expected frequency. 14-49. Add up the answers from the previous question to get the chi-square test for independence. 14-50. Test the null hypothesis using the critical value decision rule. 14-51. Explain the meaning of your decision on the null hypothesis, using the variable names. 14-52. Do these results tell us which kind of respondent (physician, nurse, or executive) is more likely to report working in a nonpunitive reporting environment? If not, what would be required to answer that question?

SCENARIO 14-G, Continued

Returning to the proportion of Oklahomans who consume five or more servings of fruit/vegetables per day, we obtained the following data from the Oklahoma State Department of Health's OK2SHARE website. Now we have

(*Continued*)

		Age Group	
		18–24 Years	75+ Years
Consume 5+ Fruit/	Yes	27	252
Vegetable Servings Daily	No	215	982

Figure 14.15

Younger and older Oklahomans with diabetes and fruit/vegetable consumption. (Data from "Adults with diabetes and fruit/vegetable consumption, Behavioral Risk Factor Surveillance System," by the Oklahoma State Department of Health, 2014, March 13, retrieved from http://www.health.state.ok.us/stats/index.shtml.)

the possibility of comparing two proportions by computing a chi-square for independence: the proportion for people in the youngest age group and the proportion for those in the oldest age group. Figure 14.15 shows the frequencies.

14-53. An earlier question involved computing the proportion of people in the oldest age group who consume 5+ servings of fruit/vegetables daily. Now compute the proportion of the youngest adults who said yes to this question about fruit/vegetable intake. 14-54. This chapter said a chi-square for independence could be used to compare two proportions. Write the alternative hypothesis for this scenario. 14-55. Write the null hypothesis for this scenario. 14-56. Compute the expected frequencies. 14-57. Compute the differences between the observed and expected frequencies, then square the differences. 14-58. Divide the squared differences by their corresponding expected frequencies. 14-59. Add up your answers to the last question to get the chi-square test for independence. 14-60. Compute *df*. 14-61. Using $\alpha = .05$ and *df*, look up the critical value in Table D.1. 14-62. Test the null hypothesis, then explain its meaning, using the variable names. 14-63. What does this test tell us about the two proportions that we computed (the one for the oldest age group and the one for the youngest age group)?

SCENARIO 14-I

Suppose researchers wanted to know whether there was a relationship between the lunar phase and the experience of seizures by patients with epilepsy. (This topic has been studied by researchers, but this scenario is not based on any single study.) The study would be conducted by a large university-based program for monitoring the seizure activity experienced by hundreds of patients. The researchers say they will collect data every month for two years, recording whether patients experience seizures (yes/no) during each of four moon phases: new moon, first quarter, full moon, and last quarter. 14-64. What might be the alternative hypothesis for this study? 14-65. Which statistic might be used to test the null hypothesis? 14-66. Why should the researchers be concerned about the independence assumption?

(*Continued*)

14. Analysis of Frequencies and Ranks

	Medical Therapy Only	Gastric Bypass	Sleeve Gastrectomy	*Row Totals*
Controlled	5	21	18	*44*
Uncontrolled	36	29	31	*96*
Column Totals	41	50	49	*N = 140*

Figure 14.16

Data on obesity treatments and blood sugar control. (Data from "Bariatric surgery versus intensive medical therapy in obese patients with diabetes," by P. R. Schauer et al., 2012, *The New England Journal of Medicine, 366,* 1567–1576.)

SCENARIO 14-J

The observed frequencies from an earlier table are reproduced here as Figure 14.16. This scenario concerned patients with diabetes who underwent one of three treatments for obesity: medical therapy only, gastic bypass surgery, or sleeve gastrectomy. Schauer et al. (2012) categorized the patients 12 months later as having achieved control of their blood sugar or not. 14-67. Compute the expected frequencies. (The rest of the calculations were shown in this chapter. For more practice, you might try to reproduce those results.)

SCENARIO 14-D, Continued

This study by Marbella et al. (1998) concerned Native Americans' consultations with Native American healers. Among other research questions, the researchers asked whether there was a relationship between age group (18–39 years or 40–83 years) and source of medical care (consultation with a physician only vs. consultation with both a physician and a Native American healer). A Check Your Understanding question gave you the numeric result for the chi-square test for independence. The data published by these researchers are shown in Figure 14.17. 14-68. Compute the expected frequencies. 14-69. Compute the difference between the observed and expected frequencies. 14-70. Square the differences computed in the previous question. 14-71. Divide each squared difference by its corresponding expected frequency. 14-72. Add up the answers from the previous question to get the chi-square test. 14-73. Test the null hypothesis using the critical value decision rule and $\alpha = .05$. 14-74. Explain the meaning of your decision on the null hypothesis, using the variable names.

SCENARIO 14-K

Liu et al. (2000) examined the risk of cardiovascular disease (CVD) for women and the potential effect of fruit and vegetable intake. Their cohort study included information from nearly 40,000 health-care professionals.

(*Continued*)

| | | Source of Health Care | | |
		Physician Only	Physician and Native American Healer	*Row Totals*
Age Group	18–39 Years	54	23	*77*
	40–83 Years	39	34	*73*
Column Totals		93	57	*N* = 150

Figure 14.17

Source of health care for two age groups of Native Americans. (Data from "Use of Native American healers among Native American patients in an urban Native American health center," by A. M. Marbella, 1998, *Archives of Family Medicine, 7,* 182–185.)

The women had no history of CVD or cancer at the beginning of the study. Over the years of tracking these women, the researchers observed which women suffered heart attacks, strokes, and other cardiovascular-related events. These observations were part of a larger study involving low-dose aspirin and vitamin E, with women having been randomly assigned to conditions, so the analyses about fruit and vegetable intake had to account for group membership. Relative risk statistics were computed to look at the relationship between fruit/vegetable intake and later CVD. Liu et al. reported, "The *RR* of CVD adjusted for age and randomized treatment was 0.68 (95% CI: 0.51, 0.92), when the highest and lowest quintiles were compared." If we arranged the women according to the number of fruits and vegetables they ate on a typical day, then we divided them into five equally sized groups, we would have five quintiles. This relative risk compares the women with the highest and lowest intake of fruits and vegetables, with fruit/vegetable intake being considered as a positive risk. 14-75. Explain the meaning of the relative risk = 0.68. 14-76. The researchers thought the analysis might have been clouded by the inclusion of women with other conditions, such as high blood pressure or diabetes, that might have made them susceptible to CVD. They ran the analysis again with only those women who did not have such comorbid conditions. They reported, "… the age- and treatment-adjusted RR was 0.33 (95% CI: 0.17, 0.64) …." Explain the meaning of this relative risk.

SCENARIO 14-L

Rauh et al. (2013) conducted a study drawing on a medical database for members of the U.S. armed forces. The researchers wanted to identify service members who had undergone combat-related amputation and to compare those with and without traumatic brain injury (TBI). They identified

(*Continued*)

		TBI Status	
		TBI	No TBI
Cellulitis Infection	Yes	12	19
	No	44	147

Figure 14.18

Traumatic brain injury status and incidence of cellulitis infection among service members with amputation. (Data from "Effect of traumatic brain injury among U.S. service members with amputation," by M. J. Rauh et al., 2013, *Journal of Rehabilitation Research and Development, 50*, 161–172.)

546 service members who served in combat between 2001 and 2006 and suffered combat wounds requiring amputations. This group included 127 people with a diagnosis of TBI. The researchers found that those with a TBI diagnosis were at greater odds of having complications, indicating a possible need for additional support from health-care professionals. One complication of amputation is cellulitis, which is a bacterial skin infection that can be extremely painful and life-threatening. The researchers compared the odds of getting cellulitis for the service members with a TBI versus the odds for those without a TBI. Figure 14.18 shows their results.

14-77. Compute the odds ratio for the relationship between TBI status and cellulitis. 14-78. Explain the meaning of the odds ratio. 14-79. The researchers performed another analysis that took into account the severity of injuries and location of the amputation. They found the following 95% confidence interval for the odds ratio after adjusting for differences in injury severity and amputation location: [1.1, 6.3]. Explain this confidence interval.

References

Centers for Disease Control and Prevention (2014, April 13). Adults who have consumed fruits and vegetables five or more times per day. *Behavioral Risk Factor Surveillance System Prevalence and Trends Data*, 2009. Retrieved from http://tinyurl.com/phgupsf

Efe, E., & Özer, Z. C. (2007). The use of breast-feeding for pain relief during neonatal immunization injections. *Applied Nursing Research, 20*, 10–16. doi:10.1016/j.apnr.2005.10.005

He, J., Ogden, L. G., Bazzano, L. A., Vupputuri, S., Loria, C., & Whelton, P. K. (2001). Risk factors for congestive heart failure in U.S. men and women: NHANES I epidemiologic follow-up study. *Archives of Internal Medicine, 161*, 996–1002. doi:10.1001/archinte.161.7.996

Hernandez, P. O., Javed, S., Mendoza, N., Lapolla, W., Hicks, L. D., & Tyring, S. K. (2011). Family history and herpes zoster risk in the era of shingles vaccination. *Journal of Clinical Virology, 52*, 344–348. doi:10.1016/j.jcv.2011.08.014

Liu, S., Manson, J. E., Lee, I.-M., Cole, S. R., Hennekens, C. H., Willett, W. C., & Buring, J. E. (2000). Fruit and vegetable intake and risk of cardiovascular disease: The Women's Health Study. *The American Journal of Clinical Nutrition, 72*, 922–928.

Marbella, A. M., Harris, M. C., Diehr, S., Ignace, G., & Ignace, G. (1998). Use of Native American healers among Native American patients in an urban Native American health center. *Archives of Family Medicine, 7*, 182–185.

Nanna, M. J., & Sawilowsky, S. S. (1998). Analysis of Likert scale data in disability and medical rehabilitation research. *Psychological Methods, 3*, 55–67. doi:10.1037/1082-989X.3.1.55

Oklahoma State Department of Health (2014, March 13). Adults with diabetes and fruit/vegetable consumption. *Behavioral Risk Factor Surveillance System, 2009*. Retrieved from http://www.health.state.ok.us/stats/index.shtml

Rauh, M. J., Aralis, H. J., Melcer, T., Macera, C. A., Sessoms, P., Bartlett, J., & Galarneau, M. R. (2013). Effect of traumatic brain injury among U.S. servicemembers with amputation. *Journal of Rehabilitation Research & Development, 50*, 161–172. doi:10.1682/JRRD.2011.11.0212

Rosenstein, A. H. (2002). Nurse-physician relationships: Impact on nurse satisfaction and retention. *American Journal of Nursing, 102*, 26–34.

Samuels, M. L., & Lu, T.-F. C. (1992). Sample size requirements for the back-of-the-envelope binomial confidence interval. *The American Statistician, 46*, 228–231.

Santen, R. J., Allred, D. C., Ardoin, S. P., Archer, D. F., Boyd, N., Braunstein, G. D., …Endocrine Society. (2010). Postmenopausal hormone therapy: An Endocrine Society scientific statement. *Journal of Clinical Endocrinology & Metabolism, 95*, S1–S66. doi:10.1210/jc.2009-2509

Schauer, P. R., Kashyap, S. R., Wolski, K., Brethauer, S. A., Kirwan, J. P., Pothier, C. E., …Bhatt, D. L. (2012). Bariatric surgery versus intensive medical therapy in obese patients with diabetes. *The New England Journal of Medicine, 366*, 1567–1576. doi:10.1056/NEJMoa1200225

Weijzen, P. L. G., de Graaf, C., & Dijksterhuis, G. B. (2008). Discrepancy between snack choice intentions and behavior. *Journal of Nutrition Education and Behavior, 40*, 311–316. doi:10.1016/j.jneb.2007.08.003

Witte, D. R., Grobbee, D. E., Bots, M. L., & Hoes, A. W. (2005). Excess cardiac mortality on Monday: The importance of gender, age and hospitalisation. *European Journal of Epidemiology, 20*, 395–399. doi:10.1007/s10654-004-6594-4

15

Choosing an Analysis Plan

Introduction

Students sometimes are overwhelmed by details when they reach the end of a statistics course. The purpose of this chapter is to help you to organize what you have learned. This chapter will review some of the research scenarios used in the book and lead you through a thought process for choosing an analysis plan. A series of questions will form a kind of decision tree. We are not experts in decision analysis, so our series of questions will not be illustrated using the symbols (squares, circles, triangles) that decision analysts use to create a formal decision tree; we are using this term loosely. Our approach will be similar to tax-preparation software, which asks one question at a time to lead the user through a process. Our process will involve deciding which of the tests and estimates in this book are best for a given research scenario.

To get a "big picture" view of the material, you might go back briefly to Chapter 1 and find Figure 1.1, which illustrates the cyclical nature of quantitative research. Notice that Step 4 begins with "design and run the study." As we said in Chapter 1, *research design* refers to the process of making decisions about the plans for the study, such as determining whether a one-time study would answer the research questions or whether a longitudinal study is required. These decisions must precede the decision on how to analyze the data. For example, Chapter 11 described research on the effect of swearing on pain tolerance. We described a study in which people were randomly assigned to groups, then the instructions to participants were manipulated. While holding a hand in near-freezing water for as long as they could, participants were told either to curse or to refrain from cursing. Would people be able to tolerate the pain longer while swearing or not swearing? Having two independent groups implies the possibility of certain statistics (such as the independent-samples t test or the Aspin–Welch–Satterthwaite [AWS] t test) to analyze the results. In fact, the researchers who ran the study of swearing and pain (Stephens, Atkins, & Kingston, 2009) did not randomly assign participants to groups. They measured the same participants twice: in a swearing condition and in a nonswearing condition. When one group is measured on two occasions, it is possible to use the paired t test. But these researchers also were interested in differences between men and women, so they had slightly more complicated research questions and used different statistics. We simplified our explanation of the research to fit our teaching goals.

Our point is that research planning must begin with the research questions. A study design must logically follow from the research questions. In turn, the statistics that will be used to analyze the data must follow logically from the study design. The figure below illustrates how quantitative research design is informed by prior research … which leads to research questions … which drive the study design … which determines the statistics that may be computed. Can you see how the figure is expanding on some of the ideas presented in Figure 1.1, which showed the circular nature of quantitative research? The activities in this figure tend to be sequential, beginning with the potential research topic and ending

Encounter a potential research topic. Explore the scientific literature to gain an understanding of the prior studies on the topic.

Identify research questions to be explored. Refine them based on existing theory and research.

Determine the study design needed to answer the research questions. (What do you need - Multiple groups? Repeated measures? Cohorts? Matched pairs?)

Identify how variables will be measured. Do you have categorical variables? Quantitative variables? Both kinds?

Given the study design and kinds of variables, which statistics should be used to answer the reseach questions?

with the choice of statistics, all of which come before quantitative research is conducted. The process of reading about prior research sometimes reveals that the proposed topic already has been studied, so the research questions may have to be reformulated.

Research questions motivate everything that follows. What kind of study design is needed? It depends on the research questions. Which statistics are needed? It depends on the study design, which depends on the research question. These details are logically linked. Researchers should know before they run a study how they will analyze the results. This chapter will focus on the last box in the figure. We will revisit many research scenarios in the book and step through a series of questions to figure out the best statistics for each scenario. Instead of having Check Your Understanding questions in this chapter, we will have a large number of Practice Scenarios, which you will read and assess using our series of questions to choose from among the statistics covered in this book. We hope the process will help to bring some organization to the details that may be swimming in your mind.

Statistics That We Have Covered

There are many ways that we could work through the process of deciding how to analyze the data for a given research situation. The series of questions that we propose is only one way of choosing the best analysis plan, and these questions only will be useful for the following list of statistics covered in this book:

- The z test statistic and its associated confidence interval (CI) for estimating one population mean when a population variance or standard deviation is known
- The one-sample t test and its associated CI for estimating one population mean when a population variance or standard deviation is unknown
- The paired t test and its associated CI for estimating the difference in paired population means
- The independent-samples t test and its associated CI for estimating the difference in two independent population means
- The AWS t test and its associated CI, used when we have unequal sample sizes, small sample sizes, or both small and unequal sample sizes
- The one-way analysis of variance (ANOVA) F test, multiple comparison procedures, and the CIs that accompany the multiple comparison statistics
- Pearson's correlation coefficient, r
- Simple linear regression, including a t test for the slope and its associated CI for the slope
- CI for a single proportion
- Chi-square test for goodness of fit
- Chi-square test for independence

- Relative risk and its associated CI
- Odds ratio and its associated CI

Your instructor may not have covered some of these statistics. Notice that our list does not include rank tests. We described rank tests at the end of Chapter 14, but we did not give you enough exposure to them to justify including them on this list. The list also does not include the descriptive statistics that you have learned—the mean, median, unbiased variance, and so on. Descriptive statistics may be used in nearly any quantitative study to describe the samples. The intent of the series of questions in our decision tree will be to choose from among the tests and estimates that may answer a research question. We will spend the rest of this chapter developing the decision tree and practicing the decision process.

Three important disclaimers about the decision tree:

1. Almost all research studies have complex research questions, requiring more complicated analyses than the ones in this book, so the decision tree is limited to the test statistics and interval estimates that we covered. As we will show you at the end of the chapter, research questions that require statistics beyond the scope of this book will not fit into this decision tree. This disclaimer may give you a "why bother" feeling about the decision tree. But there is much to be gained by learning one way of making decisions about analysis plans and organizing your knowledge about statistics. The skills involved in analyzing these research scenarios will transfer to more complex research studies.

2. The scenarios in this chapter will focus on only one statistic at a time, but that is not what real research is like. Studies almost always require many different test statistics, corresponding to a multitude of research questions. The main research question may involve one-way ANOVA F tests with multiple comparisons, with one F test and set of multiple comparisons per outcome variable. But another part of the research question may require categorical data analysis. In addition, other statistics may be needed to assess whether randomization effectively equalized the groups on certain extraneous variables. Researchers conducting a longitudinal or repeated measures study also may need to determine whether those who completed the study are similar to those who dropped out of the study, requiring even more statistics.

3. In real-world studies, decisions about the analysis plan can be much less cut-and-dried than the process presented in our decision tree. Studies can be quite complex, and the same research questions could lead two statisticians to recommend different yet equally valid analysis plans. Remember, our goal here is to organize what you have learned so far, not to give you a decision process that would generalize beyond this book's collection of tests and estimates.

Instead of dumping the decision tree fully formed into this chapter, we will describe one way of organizing the list of statistics that have been covered in the book. Then we will build one part of the decision tree at a time so that you can understand our reasoning.

Organizing Our List: Kind of Outcomes, Number of Samples

Let's begin by recognizing that there are different ways of categorizing statistics. We could talk about statistics that can be used with one sample, two samples, or multiple samples. Another way of categorizing statistics is according to the kind of data being analyzed. Glance back at our list of tests and estimates. The list includes statistics that focus on means, linear relationships, frequencies, and so forth. The first six bullet points in our list showed statistics that are connected with hypotheses about population means. The focus on means implies that the data are from quantitative variables. The next two bullet points described analyses associated with linear relationships. Again, data from quantitative variables would be analyzed. The last five bullet points described statistics involving frequencies for categorical variables. To distinguish among these statistics, the first question we could ask about any research scenario is

Do we have numeric or
categorical outcomes?

Numeric   Categorical

Let's follow the numeric branch first, then we will return to the categorical branch. If we have numeric outcomes in a scenario, then we just eliminated the last five statistics on our list of options because they all involve categorical data.

The next question needs to help us to sort out the analyses that use quantitative data.

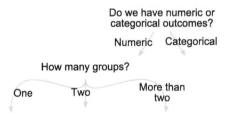

Do we have numeric or
categorical outcomes?

Numeric   Categorical

How many groups?

One   Two   More than
two

We have covered test statistics for use when we have one sample, two samples, or more than two samples. For now, let's consider the options when we have only one sample. We could be interested in asking questions about one population mean. Or we could have one group of participants that is measured twice. Or we could have one sample measured on two quantitative variables that we suspect may share a linear relationship. Let's build the branch for one sample. The next question can decide whether correlation and regression are needed.

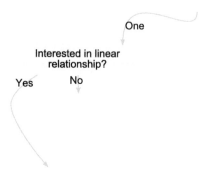

If the answer is yes, then we have arrived at an answer: the appropriate tests and estimates would be Pearson's *r*, simple linear regression, a *t* test for the slope, and a CI for the slope.

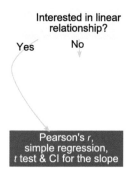

Some of the statistics in this answer are redundant: if a CI for the slope is computed, it can be used to test the same null hypothesis that would be tested by the *t* test for the slope. The simple regression analysis would provide us with the formula for the regression line, if needed, and Pearson's *r* could describe the correlation between the two variables without specifying which one is the predictor and which one is the criterion variable.

If we are *not* interested in a linear relationship, then we eliminate the correlation and regression analyses and our list of potential statistics for analyzing our one sample of data becomes smaller. If you have been keeping track, we have only three sets of statistics left on our list: a *z* test statistic and its associated CI; a one-sample *t* test and its associated CI; and a paired *t* test for repeated measures on one group and its associated CI. Now we will add another question to decide whether to keep the paired *t* test on the list of one-sample statistics (although the paired *t* test will come up again when we have two samples). If we do have repeated measures, then we have arrived at an answer for our one-sample case: the paired t test and its associated CI for estimating the difference in paired population means.

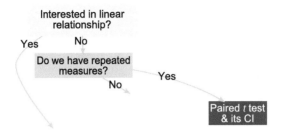

We reversed the order of the "Yes" and "No" arrows after the question about repeated measures because of the location of this statistic in the final version of the decision tree. (As you will see, the paired t test could be the best choice for a study with two groups, such as a study of twins.)

If the answer on repeated measures is no, then we have one sample, one occasion of measurement, no interest in linear relationships, and two sets of statistics left on our list: the z test statistic and its associated CI, or the one-sample t test and its associated CI. The one-sample t test is used instead of the z test statistic when we do not know the population standard deviation or variance. Therefore, our final question for this part of the decision tree will be about those parameters.

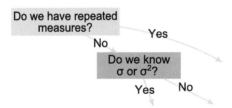

If we know one of those parameters, then we will use the z test statistic and its associated CI. If the answer is no, then we will use the one-sample t test statistic and its associated CI.

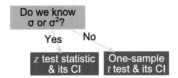

We have covered a lot of questions and options for the answers. The figure below puts together the questions and answers for one sample of participants with quantitative outcomes.

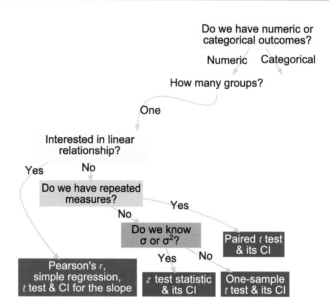

Notice that we have not created the "categorical" side of the decision tree, nor have we explored the questions associated with scenarios that have two or more groups of participants. Before we explore those parts of the decision process, let's practice using this part of the decision tree by revisiting a few scenarios presented earlier in the book.

PRACTICE SCENARIO 1

Schell, Morse, and Waterhouse (2010) were interested in blood pressure readings taken on the upper arm and the forearm. Sometimes the blood pressure cannot be taken on the upper arm because of the patient's physical condition. Before relying on readings from the forearm, the researchers wanted to establish whether such readings differ from upper-arm readings. Do people have a different mean blood pressure reading on their upper arm versus their forearm?

The blood pressure scenario involves quantitative measurement of blood pressure, so this is a numeric outcome. Let's consult the decision tree to analyze this scenario.

- *How many groups of participants do we have?* The scenario appears to imply that one sample will be studied, with measurements being taken on both the forearm and the upper arm.
- *Are we interested in a linear relationship?* The scenario does not say anything about a linear relationship. We would expect that people with high blood pressure on the upper arm also would have high blood pressure on

the forearm, and people with a low blood pressure reading on the upper arm still would have low blood pressure on the forearm. But that is not the kind of question being posed by the scenario. It mentioned the mean for the blood pressure taken on the two locations. So we are not interested in a linear relationship.

* *Do we have repeated measures on the same people?* Yes, we do. The scenario says the blood pressure readings will be taken on the forearm and the upper arm of the participants, so the same variable is being measured twice on everyone. We have arrived at the answer: the paired *t* test and its associated CI.

PRACTICE SCENARIO 2

Falvo and Earhart (2009) ran a study examining the characteristics of patients with Parkinson's disease. Many factors can affect how much these patients are able to walk. Suppose we want to use a measure of mobility as a predictor of walking capacity. The Timed Up and Go, a measure of mobility, is the number of seconds it takes a patient to get up from a chair, walk 3 m, return to the chair, and sit down. The Six-Minute Walk Distance, a measure of walking capacity, is the number of feet that the patient can walk in six minutes at a normal pace.

Both the Timed Up and Go and the Six-Minute Walk Distance are numeric variables. Using our partly built decision tree, we have this series of questions:

* *How many groups of participants do we have?* The scenario implies that we have one group of participants.
* *Are we interested in a linear relationship?* The scenario talks about the Timed Up and Go as a predictor of Six-Minute Walk Distance, so it does appear that to focus on a linear relationship. We will perform an analysis involving correlation and regression.

You may have been expecting to jump down to the question about repeated measures. After all, the same participants are being measured on Timed Up and Go as well as Six-Minute Walk Distance. The question about repeated measures would lead to the paired *t* test and its associated CI. But remember, the paired *t* test compares two means *on the same variable*. We went through a paired *t* test example above, with the blood pressure being measured both on the forearm and the upper arm of the same patients. This scenario about patients with Parkinson's disease uses two different variables. We would not use a paired *t* test, because it does not make sense to compare an average time to an average distance.

We have been reading a report from FantasyLand Studies that says the average 1-year-old boy in the United States weighs 25.5 lb (about 11.6 kg), with a standard deviation of 4.1 lb. We are concerned about some 1-year-old boys in foster care who are suspected of being underweight. Is the average weight of these 31 children significantly less than the average of 1-year-old boys in the United States?

The main variable of interest is weight, which is numeric. Let's work through our questions in the figure:

- *How many groups of participants do we have?* The scenario mentions only one sample.
- *Are we interested in a linear relationship?* Only one quantitative variable has been mentioned, so the answer is no.
- *Do we have repeated measures?* No, the scenario implies that we are talking about how much the boys weigh right now. If the scenario had talked about weight gain or weight loss, then change across time would have required repeated measures of weight.
- *Do we know the population standard deviation or population variance?* Yes, we do. The practice scenario says the weights of 1-year-old American boys have a standard deviation = 4.1 lb. That is a population standard deviation because it is described as applying to all American boys, not a limited number in a sample. We have arrived at the answer: the z test statistic and its associated CI.

You may have been tempted to say that there were two groups: the ones in foster care and 1-year-old American boys. But we do not have a *sample* of 1-year-old American boys in general; we have only one sample of boys in foster care. We do have some information about 1-year-old boys in the United States, and those boys form our population. When we are deciding how many groups of participants there are, we can look for sample sizes to indicate that the people being described actually belong to a sample, not a population.

Suppose we are medical researchers, and we have some reason to suspect that the traditional number for normal human body temperature, 98.6°F, actually is not the right number for the average healthy adult's body temperature. We obtain a data set that includes temperature measurements for 245 healthy military enlistees. Is the average body temperature for this sample significantly different from 98.6?

Temperature is quantitative, so we have numeric outcomes. Let's go through this part of our decision tree once more before we build new branches:

- *How many groups of participants do we have?* Only one sample: the 245 healthy military enlistees.
- *Are we interested in a linear relationship?* Only one quantitative variable has been mentioned, so the answer is no.
- *Do we have repeated measures?* No, the enlistees appear to have had their temperatures measured only once.
- *Do we know the population standard deviation or population variance?* No, we do not. We have a number, 98.6, which appears to be a norm that could be used in a null hypothesis about a population mean, but we have no information about a population standard deviation or variance. Our answer is the one-sample t test and its associated CI.

In the next section, we will build the part of the decision tree pertaining to two samples.

Adding to the Tree: Two Samples

We return now to the question about the number of groups of people when we have quantitative outcome variables. We already filled out the decision tree when there is one group of participants. Let's see which statistics on our list are potential options if we have two groups of participants:

- The paired t test and its associated CI
- The independent-samples t test and its associated CI
- The AWS t test and its associated CI

It is possible to use the one-way ANOVA F test when there are two independent groups and the researcher wants to ask questions about two population means. But as we saw in Chapter 12, the one-way ANOVA F test computed on data from two samples gives us the same thing as a squared value of the independent-samples t test. So we will leave the one-way ANOVA F test with multiple comparisons and their associated CIs for situations in which we have more than two groups.

The three statistics listed above need to be distinguished from each other. Both the independent-samples t test and the AWS t test involve independent samples and their means, whereas the paired t test involves two means computed on data from participants who are in pairs, and such pairing creates two dependent samples. We already have seen the paired t test in our decision tree, when we had one sample measured twice on the same variable. As you know, the paired t test also is used with two samples, such as naturally occurring pairs like twins or researcher-matched pairs. So this statistic needs to appear again in the decision tree for situations when we have two groups of participants.

The first question in this part of the tree will determine whether we need the paired *t* test.

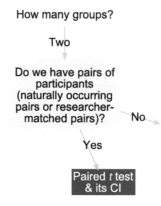

Previously, the paired *t* test was used with pairs of *scores* for one sample, in which each participant was measured twice. We still are talking about pairs of scores, but now each score comes from a separate person, with the two people linked in a pairwise manner. We may have participants who have a connection that predates the research, such as a study of lung cancer that compares twins, with one twin having cancer and the other twin not having cancer. Or we may have pairs that the researcher formed by matching cases and controls; in a study of lung cancer, this pairing could be done by taking each person with cancer (not a twin) and pairing him or her with an unrelated person who is similar on many extraneous variables (age, gender, smoking history, family history of cancer, etc.). The two people's results would be analyzed as a pair.

If we do not have pairs of participants, then we have two options left on our list for two-sample studies: the independent-samples *t* test and the AWS *t* test. The next questions will help us to choose between these statistics. As you will recall from Chapter 11, we use the AWS *t* test when we have small sample sizes (i.e., fewer than 15 people per group), unequal sample sizes, or both small and unequal sample sizes. That is because the independent-samples *t* test needs its inoculation against unequal variances, with the inoculation being equal sample sizes of 15 or larger. We know it is common to sample from populations with unequal variances, so the independent-samples *t* test's assumption of equal variances often is violated. If we have at least 15 people per group and equal sample sizes, then violating the equal variances assumption is no problem for the independent-samples *t* test. But if we do not have at least 15 people per group and equal sample sizes, then the independent-samples *t* test may have a sampling distribution that looks different from a theoretical *t* distribution, and its *p* value may not be trustworthy. So we would switch to the AWS *t* test, which has no assumption of equal variances. Therefore, we add the question about equal and large sample sizes to this part of the decision tree.

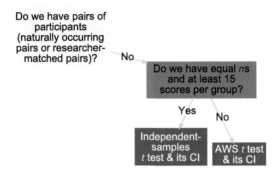

Now that we have completed the part of the decision tree that pertains to quantitative outcomes for two samples, let's practice making decisions about some scenarios presented earlier in the book.

PRACTICE SCENARIO 5

Inspired by Grant and Hofmann (2011), we are investigating whether two wings of a hospital will use different amounts of hand sanitizer if different signs are placed above the dispensers of the sanitizing gel. Each wing has 30 dispensers and a history of using the same amount of gel every month. We decide to run a quasi-experiment in which we manipulate the messages on the signs above the dispensers. In one wing we place signs that read, "Hand hygiene prevents you from catching diseases." In the other wing we place signs that read, "Hand hygiene prevents patients from catching diseases." The outcome variable is the amount of gel used at each dispenser in a month.

You may recall that in this scenario, the unit of analysis is the dispenser; we are not looking at the amount of sanitizer used by different human participants, but instead the amount of sanitizer dispensed by each dispenser. The amount of sanitizer is measured numerically. Let's work through the decision tree:

- *How many groups of participants do we have?* Our "participants" are the dispensers, and we have two groups, which are the two wings of the hospital.

- *Do we have pairs of participants?* No, the scenario said nothing about pairing of dispensers.
- *Do we have equal sample sizes with at least 15 scores per group?* Yes, we have 30 dispensers in each wing, so each sample has $n = 30$. We can compare the mean amount of sanitizer used in each wing by computing an independent-samples t test and its associated CI.

PRACTICE SCENARIO 6

We want to know whether pain ratings will be different for people who receive an injection in a room with soft music playing quietly versus people in an identical room with no music playing. A power analysis has indicated that 74 participants (37 in each group) would give us enough power to detect a clinically meaningful difference in mean pain ratings. One person who initially volunteered learns from the informed consent document that needles are involved and decides not to join the study after all. For the 73 participants who sign the informed consent document, we randomly assign them to two groups, give the injections to the people in the separate conditions, and collect pain ratings on a scale from 0 (no pain) to 10 (worst pain imaginable).

The pain ratings are on a numeric scale. Try to step through the decision tree on your own and find an analysis plan, then return here to continue reading.

Our series of questions can be answered as follows:

- *How many groups of participants do we have?* Two groups: participants in the soft-music condition and participants in the no-music condition.
- *Do we have pairs of participants?* No, the scenario said nothing about pairing.
- *Do we have equal sample sizes with at least 15 scores per group?* We do have more than 15 people per group, so we would have more than 15 scores per group, but the sample sizes will be unequal because one person decided not to complete the study, leaving us with 73 people to randomly assign to the two groups. Therefore, we would use the AWS t test and its associated CI.

The next scenario did not appear earlier in the book. After you read the scenario, consult the decision tree and see if you can come up with a possible analysis plan.

Nurses at an assisted living facility in the United States noticed a possible difference in mood between residents whose individual rooms had a window facing the south and residents across the hall whose windows faced north and were shaded by thick trees. The south-facing rooms let in a lot more sunshine, and the nurses speculated that the residents who had more exposure to sunshine were less depressed than residents whose rooms had very little sunshine coming through the windows. The nurses took their speculation to a nurse scientist who does research on older people. The researcher said, "We can use a widely used quantitative scale for measuring depression, but we may need to control another variable: how close the rooms are to the common area. Those who are close to the common area may have more opportunity for social interaction, which could affect depression. Because everyone has a room to herself or himself, let's match each resident on the sunny side of the hall with the resident directly across the hall on the shady side. So we will be controlling for proximity to the place where they may socialize by pairing the residents who live an equal distance from the common area, but on opposite sides of the hall. We can compare the mean depression scores of the 16 sunny-side residents with the mean depression scores of the 16 shady-side residents."

The scenario says a depression scale that produces quantitative data will be used. Have you tried to work through the decision tree? Here is how we would assess this scenario:

- *How many groups of participants do we have?* Two groups: residents on the sunny side of the hall and residents on the shady side of the hall.
- *Do we have pairs of participants?* Yes, the researcher said we need to control for an extraneous variable of proximity to the common area, so each person on the sunny side of the hall is paired with the resident directly across the hall on the shady side. We would need the paired *t* test and its associated CI.

The next section will build the decision tree for situations in which we have more than two independent groups of participants.

Adding Again to the Tree: More Than Two Samples

There is little to add to this part of the decision tree because this book has covered only one set of statistics for situations with more than two independent groups: the one-way ANOVA *F* test and multiple comparison procedures, with CIs being associated with the mean differences that were examined with the multiple comparisons. So, if the decision tree leads us to saying there are more than two groups of participants, this book gave only one answer.

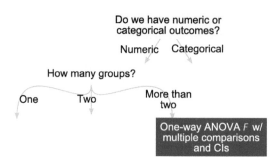

Let's take a look at a scenario that would lead us to the one-way ANOVA *F* test and multiple comparisons.

PRACTICE SCENARIO 8

Inspired by Montgomery-Downs, Clawges, and Santy (2010), we are plan-ning a study in which we will measure the total number of minutes of sleep that new mothers get in a 24-hour period three weeks after giving birth to their first baby. The study will involve three groups of mothers: those who are breast-feeding, those who are bottle-feeding, and those who are using both feeding methods.

Obviously, we have only one set of analyses covered in this book that would fit this situation, but let's step through the questions in the decision tree:

- *Do we have numeric outcomes or categorical outcomes?* Numeric, because the number of minutes of sleep in a 24-hour period is a quantitative outcome.
- *How many groups of participants do we have?* We have three groups of mothers: those who are breast-feeding, those who are bottle-feeding, and those who are using both methods. We can test a null hypothesis about the groups' means for the number of minutes of sleep using the one-way ANOVA *F* test, and we could determine which means differ by performing multiple comparison procedures and computing their associated CIs.

The final major branch of our decision tree will lead us to the statistics that are used with categorical data.

Completing the Tree: Analysis of Categories

Sometimes it can be hard to tell whether the data are quantitative or categorical. When we have categories, the data are frequencies, which are numbers. But the variable itself is based on categories. For example, Chapter 14 described a scenario in which people were offered different kinds of snacks: either an apple or a slice of cake. We can count how many people said they preferred each kind of snack, then we can compute the proportion of people in the sample who preferred an apple and the proportion of people who preferred cake. But these counts and the proportion are based on the chosen category. The *variable* is the kind of snack, with the categories being apple and cake. Participants are counted in categories based on their snack preference.

This book covered the following statistics associated with categorical variables:

- CI for a single proportion
- Chi-square for goodness of fit
- Chi-square for independence
- Relative risk and its associated CI
- Odds ratio and its associated CI

A CI for a single proportion may be used to compare a sample proportion of people sharing some characteristic with a hypothesized population proportion of people sharing that characteristic. A chi-square for goodness of fit can be used to test whether the frequencies are the same for all levels of one categorical variable or whether the distribution of sample participants across the categories fits a theoretical distribution, like the example of blood types in Chapter 14. A chi-square for independence looks at whether there is a relationship between two categorical variables; this statistic also can be used to compare two independent proportions.

Relative risk, odds ratios, and their associated CIs all are associated with risk. Relative risk usually compares a group that has been exposed to a risk factor with another group not exposed to the risk factor. The comparison usually involves the incidence of a disease. So people are categorized as exposed or not, and they are further categorized as having the disease or not. Relative risk is more likely to appear in cohort studies, in which groups of people (exposed and unexposed to the risk factor) are observed longitudinally for the emergence of a disease. By comparison, odds ratios usually are used in case-control studies, in which the researchers already know who has the disease and then the researchers identify a comparison group of people without the disease. The two groups are compared in terms of a potential risk factor—specifically, the odds ratio compares how people are spread out between cases and controls when the risk factor is present, versus how people are spread out between cases and controls when the risk factor is absent. In this way, the relationship between the disease and the risk factor can be assessed within the limited time frame of a case-control study.

To organize these statistics, we begin again at the top of the decision tree.

Do we have numeric or
categorical outcomes?

Numeric   Categorical

We now will build the branches that come off the answer "Categorical." The next question determines whether we will need a statistic that is associated with risk.

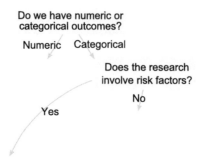

This question would be far too general for use in real-world research planning, because the definition of *risk factors* could apply to many variables. In this book, however, we have covered only a few statistics for categorical data analysis when risk factors are prominent features of a study—specifically, relative risk, odds ratios, and their associated CIs. Let's follow the "Yes" branch and add a question that could help us to decide between these two statistical options.

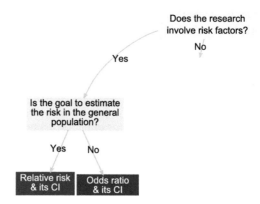

Again, the question about whether the goal is to estimate risk in the general population refers to disease surveillance and identification of risk factors—that is, whether the goal would fit with a cohort study. If we were thinking beyond the purpose of this decision tree, we would not necessarily turn to odds ratios for just any study about risk that was not a cohort study. This question again shows the limitation with this decision tree: its purpose is to organize the statistics that were presented in this book, not to generalize to all possible options in statistics. Having only two sets of statistics dealing specifically with risk in this book, we need only one question to distinguish between them.

Let's try to use this part of the decision tree with a couple of scenarios.

Rauh et al. (2013) identified U.S. military service members with and without traumatic brain injury (TBI) who had undergone combat-related amputation. The 546 service members in the study served in combat between 2001 and 2006. A bacterial skin infection called cellulitis is a complication of amputation, and it can be extremely painful and life-threatening. How do service members with diagnosed TBI compare with service members without a TBI diagnosis in terms of their incidence of cellulitis? Is there a relationship between TBI and cellulitis?

All participants were service members who underwent combat-related amputation. They were categorized according to whether they had been diagnosed with a TBI, and then they were categorized further according to whether they had contracted cellulitis. Let's step through the questions in the decision tree to assess this scenario:

- *Does the research involve risk factors?* Although this scenario does not use the term *risk factor*, it does involve the incidence of a disease, cellulitis. So yes, it does seem to involve risk factors.
- *Is the goal to estimate risk in the general population?* Even if we defined "general population" as the service members who must undergo combat-related amputation, the goal of the study does not seem to be the estimation of a population risk of cellulitis for those with and without a TBI diagnosis. Cohorts have not been observed across time to see whether a diagnosis of cellulitis emerged. Instead, this study involved data from a shorter time frame and the comparison of groups with and without certain diagnoses (TBI and cellulitis). The odds ratio and its CI appear to be the best choice from among the statistics covered in this book.

Let's do another scenario.

Liu et al. (2000) wanted to know about the effect of fruit/vegetable intake and women's risk of cardiovascular disease. Nearly 40,000 female health-care professionals participated in the study. The women had no history of cardiovascular disease or cancer at the beginning of the study. Is there a relationship between fruit/vegetable intake and the eventual experience of cardiovascular-related events, such as stroke and heart attack?

We have categorical outcomes because the main variable of interest is the eventual incidence of cardiovascular events. Try to find an answer by looking at the decision tree on your own. Then read our assessment below, based on the decision tree:

- *Does the research involve risk factors?* Although the fruit/vegetable intake would appear to be a positive influence on health, the undesirable outcome of cardiovascular events like stroke and heart attack would lead us to conclude that risk factors are the main focus of the study. In addition, the researchers' question pertained to risk.
- *Is the goal to estimate risk in the general population?* Only women are being studied, but the extremely large sample of women (nearly 40,000 of them) would lead us to believe that the study is attempting to estimate a risk in the population of women. Thus, we would assert that relative risk and its CI probably are the statistics on our list that best fit this scenario.

We will complete the decision tree so that our series of questions will lead us to one of the remaining statistics on our list: CI for a single proportion, chi-square for goodness of fit, or chi-square for independence.

Completing the Tree: The Remaining Categorical Analyses

The part of the decision tree pertaining to categorical data already has been divided into statistics that involve risk factors and those that do not. The remaining questions to be added to the tree will distinguish between a CI for a one-sample proportion and the two chi-square statistics that we covered. Let's add a question that will determine whether one proportion is the focus.

If the goal is not to compute an interval estimate of one population proportion, then we need a question that would distinguish between the two remaining statistics on our list: the chi-square test for goodness of fit and the chi-square test for independence. The goodness-of-fit chi-square also is called the one-way chi-square test, and the chi-square test for independence also is known as a two-way chi-square test. These names for the chi-square statistics can lead us to the question to decide which chi-square is needed.

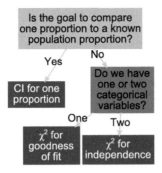

Now we will look at some scenarios to practice using this part of the decision tree. After you read a scenario, try to find the answer yourself before you read our series of questions.

PRACTICE SCENARIO 11

Witte, Grobbee, Bots, and Hoes (2005) investigated whether there was a relationship between the day of the week and sudden cardiac deaths. Using data from a registry for sudden deaths in one city in the Netherlands, the researchers identified sudden cardiac deaths on different days of the week for a two-year period. Are sudden cardiac deaths more likely on some days than others?

We have categorical outcomes because the categories are the days of the week, and deaths are being counted for each day of the week. Here are our questions from the decision tree:

- *Does the research involve risk factors?* A risk factor often is a variable that could be influenced through an intervention. Although some days of the week may be associated with more heart attack deaths, the variable *day of the week* cannot be influenced by an intervention. We would say no, this scenario does not involve a risk factor.
- *Do we have one categorical variable or two categorical variables?* There is one categorical variable: day of the week. The statistic on our list that best fits this scenario is the chi-square for goodness of fit.

PRACTICE SCENARIO 12

The Centers for Disease Control and Prevention (CDC) reported that 23.4% of American adults meet a criterion of consuming 5+ servings of fruit/vegetables daily. We are interested in a representative sample of adults with diabetes living in Oklahoma in 2009 and whether these people are similar to American adults in general in terms of fruit/vegetable consumption.

We have categorical outcomes. Either the people in the sample do consume 5+ servings of fruit/vegetables daily or they do not, and we can compute the percentage or proportion by counting how many in the sample meet the criterion. After you try to find an answer with the decision tree, take a look at our answers to the questions:

- *Does the research involve risk factors?* Fruit/vegetable consumption may be considered a positive influence on disease risk, but the focus is not on the eventual development of disease. The sample of Oklahoma adults already has been identified as having diabetes, and they were surveyed once in 2009. It would appear that risk is not the main focus of the study.
- *Is the goal to compare one proportion to a known population proportion?* Yes, the CDC's reported rate for adult Americans is given as 23.4%, which would correspond to a population proportion of .234. The best analysis on our list would be a CI for one proportion.

Two more practice scenarios follow.

PRACTICE SCENARIO 13

Is there a relationship between treatment for obesity and the control of blood sugar 12 months later? Schauer et al. (2012) compared three groups of patients who received intensive medical therapy for obesity, with two of the groups also receiving surgery. One group received only the medical therapy, a second group also received gastric bypass surgery, and a third group underwent sleeve gastrectomy. One year later, the researchers performed a blood test to determine whether the patients' blood sugar was under control. Is there a relationship between the type of treatment and blood sugar control?

We have categorical outcomes. Participants are counted in categories defined by two categorical variables: (1) the kind of treatment and (2) whether they achieved blood sugar control (yes/no). Here are our answers to the questions in the decision tree:

- *Does the research involve risk factors?* Clearly, obesity is a risk factor for many poor health outcomes, but the study is not looking into disease incidence. The answer is no, the focus was not on risk factors. All of the participants were obese at the beginning of the study, and an intervention was given.
- *Is the goal to compare one proportion to a known population proportion?* No, there are multiple groups and many possible proportions could be computed.

- *Do we have one categorical variable or two categorical variables?* We have two categorical variables: the kind of treatment (with three levels) and blood sugar control (yes/no) one year later. The best choice on our list of statistics is the chi-square test for independence.

PRACTICE SCENARIO 14

Suppose we want to continue our investigation of fruit/vegetable consumption by adults in Oklahoma. We wonder whether young adults (ages 18–24 years) will differ from older adults (ages 75 and up) in terms of the percentages of people consuming 5+ servings of fruit/vegetables per day.

We have two age categories, and everyone can be classified as meeting or not meeting the criterion of 5+ servings of fruit/vegetable daily. Our questions are:

- *Does the research involve risk factors?* There is no indication that risk factors are the main interest in the scenario.
- *Is the goal to compare one proportion to a known population proportion?* We know a population proportion from a previous scenario, but this scenario does not refer to a parameter. It talks about comparing two groups.
- *Do we have one categorical variable or two categorical variables?* We have two categorical variables: age group (18–24 years or 75+ years) and meeting the criterion of 5+ daily servings of fruit/vegetables (yes or no). The proportions for the two age groups can be compared by computing a chi-square test for independence.

The entire decision tree is too big for one figure. The figure given below shows how we will split the tree into two figures (Figures 15.1 and 15.2), which appear before the exercises.

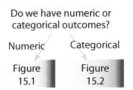

Let's take a look at the limited utility of our decision tree by reviewing a study that was mentioned in Chapters 1 and 11. Buron (2010) conducted a quasi-experiment in which he interviewed nursing home residents, then met with a graphic design artist, who created life history collages for those residents. The researcher wanted to compare two groups of nurses: 18 nurses who interacted with residents who had life history collages in their rooms and 18 nurses who interacted with

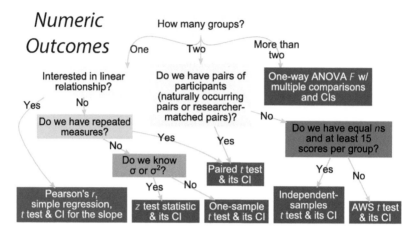

Figure 15.1

The part of the decision tree covering the book's statistics involving numeric outcomes.

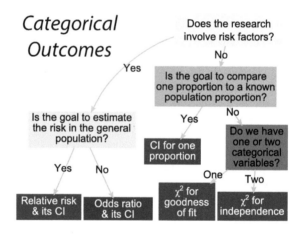

Figure 15.2

The part of the decision tree covering the book's statistics involving categorical outcomes.

residents who did not have collages. The researcher recruited two nursing homes to participate, and he manipulated which nursing home's residents received the collages. Among other variables, the nurses' knowledge of residents' lives was measured twice: before the collages were put in place and one month after the collages were placed in one of the nursing homes. How would the presence of collages affect nurses' knowledge of the residents' lives, compared with the knowledge of nurses in the nursing home without collages?

Nurses' knowledge of residents' lives was measured quantitatively, so let's see what would happen if we tried to analyze this scenario using the decision tree:

- *How many groups of participants do we have?* We have two groups: (1) the nurses working in the nursing home where life history collages were placed and (2) the nurses working in the nursing home with no collages.
- *Do we have pairs of participants (naturally occurring pairs or researcher-matched pairs)?* No, we do not have pairs of participants.
- *Do we have equal sample sizes and at least 15 scores per group?* Yes, the description above said there were 18 nurses in each nursing home who participated. The decision tree seems to lead us to the independent-samples *t* test.

The choice of the independent-samples *t* test *would be wrong*. This scenario clearly described repeated measures, with the nurses being measured on their knowledge of residents' lives on two occasions, before and after the placement of the collages. But this decision tree did not allow for that possibility. The reason is that this book did not cover a statistic that would allow the analysis of change across time for more than one group. We covered statistics that would compare two independent groups' means on one occasion. We also covered statistics that would allow us to compare one group's mean before the collages were placed and the same group's mean one month later. But we did not cover statistics when there are two (or more) groups, each measured two (or more) times. This scenario calls for statistics that are beyond the scope of this book. (One way that the data could be analyzed is a groups-by-trials repeated measures ANOVA.) Please do not try to use our decision tree to determine the correct analysis plans for research scenarios that are outside of this book. The decision tree was intended only to organize your understanding of statistics presented in this text and to help you to develop some skills in choosing from among those statistics.

What's Next

We have concluded every chapter with a section called "What's Next." For some of you, this is the last statistics class you will take, and we hope you will remember enough to be able to make judgments about different kinds of research described in the popular media. For others, this course is just the beginning of your adventure in exploring "the frontier between knowledge and ignorance," as we quoted Neil deGrasse Tyson in the first chapter.

If you do continue to study statistics, what's next? It depends on the course and the instructor. One course could take you beyond what you have learned about simple regression and introduce you to multiple regression, in which multiple predictors of one quantitative outcome variable are assessed. Beyond multiple regression, there is hierarchical linear modeling, nonlinear modeling, survival analysis, and all sorts of other ways of analyzing the relationships between variables. Another instructor may expand on what you have learned about the one-way ANOVA *F* test and multiple comparisons, introducing you to two-way ANOVA, repeated measures ANOVA, and analysis of covariance. Perhaps you will study meta-analysis, a statistical analysis that involves accumulating evidence from many studies of the same topic. Someday you might explore the statistical field of resampling procedures, with which your authors have some experience, or Bayesian statistics, about which your authors know relatively little. We have only scraped the surface of statistics, even though you have accomplished so much by completing this course. We hope you have realized that you *can* learn statistics and that you have learned enough of the language of quantitative research to begin to explore new scientific terrain.

Exercises

Instructions: Some of the following scenarios have appeared in previous exercises. For each scenario: (a) Identify whether the scenario describes an experiment, a quasi-experiment, or observational research. (b) Identify the variables mentioned in the scenario (independent, dependent, predictor, criterion, extraneous). (c) Use the decision tree in Figures 15.1 and 15.2 to determine which tests and estimates covered in this book would be best for analyzing the data.

SCENARIO 15-A

A package insert accompanying a shingles vaccine (Zostavax®) describes a randomized controlled trial with double-blinding. Adults received one dose of the vaccine or a placebo, then they were tracked for up to two years to see whether they contracted shingles. The vaccinated group consisted of 11,211 adults who were 50–59 years old, and the control group consisted of 11,228 adults in the same age range. In the treatment group, 30 people reported contracting shingles, compared with 99 cases of shingles in the control group.

SCENARIO 15-B

(Inspired by Johnson et al., 1999. Details of this scenario may differ from the actual research.) Researchers conducted a study of blood flow in certain parts of the brain and how extraverted (or outgoing) people are. They found that lower scores on extraversion corresponded to greater blood flow to certain parts of the brain. Participants who had higher scores on extraversion tended to have lower readings on blood flow to that part of the brain.

SCENARIO 15-C

(Inspired by Harrison et al., 2013. Details of this scenario may differ from the actual research.) The hormone progesterone is produced by the ovaries. Oral contraceptives that prevent ovulation also reduce blood levels of progesterone. Suppose we have been reading about progesterone's effect on acute mountain sickness, the most common form of altitude sickness,

(Continued)

which can involve flu-like symptoms. We think higher levels of progesterone tend to protect younger women from the effects of acute mountain sickness and that those who are taking oral contraceptives may be more likely to experience this sickness. Suppose we are assisting with a research study in Antarctica involving 248 people on a research team. The research team includes 50 women under 45 years of age. During their deployment to the South Pole, the team will travel by airplane from a station at sea level to another station at an altitude of 2,835 m (about 9,300 ft) in four hours. We will ask the women whether they are taking oral contraceptives, and we will record who gets acute mountain sickness in the first week after arriving at the higher altitude to determine whether there is a relationship between oral contraceptives and altitude sickness.

SCENARIO 15-D

(Inspired by Rose, Koperski, & Golomb, 2010. Details of this scenario may differ from the actual research.) We are researchers who study adults' chocolate consumption, and from our observations, we suspect that people with depression eat more chocolate on average than the general population. We know from our years of research that American adults consume an average of five servings of chocolate per month. We screen a large number of adults and identify 128 people who meet the criteria for a diagnosis of depression but who are not taking antidepressant medications. We ask them about their chocolate consumption. We compute their mean of 8.4 servings consumed per month.

SCENARIO 15-E

(Inspired by Harrington et al., 2012. Details of this scenario may differ from the actual research.) After reading several studies on soothing infants after injections, we want to investigate whether another method of soothing babies, known as the 5 S's, would lead to shorter crying times after the shots. The 5 S's are "swaddling, side/stomach position, shushing, swinging, and sucking" (Harrington et al., 2012, p. 815). Suppose we are running a study at an urban health clinic that mainly serves babies and children from low-income families. We receive appropriate institutional review for our study, obtain parental consent, and randomly assign 1-month-old full-term babies to one of three soothing methods. All babies were scheduled to receive an injection as part of a routine immunization schedule for most babies. Babies in the first group receive the usual comfort measures that their parents normally offer. Babies in the second group receive the 5 S's, with 2 ml of water given for the sucking part of the 5 S's. Babies in the third group receive the 5 S's, with 2 ml of sugared water for the sucking part of the 5 S's. We measure the number of seconds elapsed between the moment of the injection until the babies stop crying.

(*Continued*)

SCENARIO 15-F

Lou Menary is a famous breakfast-cereal researcher. Over the last three decades, he has accumulated a great deal of data about sugar in American breakfast cereals and has tested tens of thousands of samples of cereal. He says that on average, a cup of cereal has 10 g of sugar, and the standard deviation of the sugar content of those thousands of tested cereals is 3.5 g. He suspects that the amount of sugar has increased in the last two years and wants to investigate whether he is right. He decides to randomly sample 82 brands from a list of nearly 400 breakfast cereals available in the United States and buy these cereals at grocery stores. He performs a chemical analysis that determines the amount of sugar per cup of each kind of cereal in his sample. Is his suspicion correct about the sugar content of today's cereals?

SCENARIO 15-G

(Inspired by Wilkens, Scheel, Grundnes, Hellum, & Storheim, 2010. Details of this scenario may differ from the actual research.) Does glucosamine relieve low back pain? Suppose we are conducting a study involving 250 patients with low back pain who have been randomly assigned to two groups, each containing 125 patients. The treatment group will take glucosamine, and the control group will take a placebo. When the participants are signed up for the study, they complete a survey that provides a great deal of data. We want to know whether the randomization process successfully equated the groups on two important variables: history of osteoarthritis and number of months of low back pain. (Hint: You will need to choose a different statistic for each of these two variables.)

SCENARIO 15-H

Suppose we are reading a U.S. governmental report containing several years of health statistics about large representative samples. The report includes information on the resting heart rates of children who were 6–8 years old. Suppose the report says that 20 years ago, the mean for children (ages 6–8 years) was 82 beats per minute. We think about the current problem of childhood obesity, and we wonder whether American children today might have a different average resting heart rate than children 20 years ago.

SCENARIO 15-I

For people who are "night owls" (preferring to stay up late at night and rise later in the day), is it easier to perform cognitive tasks in the evening or during the day? We recruit nursing students who say they are night owls to take a drug calculation test twice. They first take the test on a Tuesday evening and then take a similar test (same format, different numbers) the next day at noon. We compare their average score on the evening test with their average score on the daytime test.

(Continued)

15. Choosing an Analysis Plan

SCENARIO 15-J

(Inspired by Waterhouse, Hudson, & Edwards, 2009. Details of this scenario may differ from the actual research.) We are interested in the physiological effects of music during submaximal exercise. We recruit 60 healthy young male volunteers who typically ride a bicycle for about 30 miles per week. In our study, they will ride stationary bikes while listening to music on earbuds plugged into a music player, which we will provide. They will be instructed to ride at a moderate pace. We secretly randomize them to groups and manipulate the music. The riders in Group 1 will listen to upbeat popular music. The riders in Group 2 will listen to the same music being played 10% faster than the original recording. The riders in Group 3 will listen to the same music played 10% slower than the original recording. Everyone will ride the stationary bikes for 30 minutes. The stationary bikes will record a pedaling speed midway through the 30-minute period. Does the speed of the music make a difference in the pedaling speed?

SCENARIO 15-K

(Inspired by Landon, Reschovsky, & Blumenthal, 2003. Details of this scenario may differ from the actual research.) Researchers conducted a nationally representative telephone survey of physicians in two different years. Let's say that different physicians were surveyed on the two occasions, with 368 respondents in 1997 and 403 respondents in 2001. The survey asked several questions; the one that interests us is a question about the physicians' general satisfaction with their medical career, rated on a scale from 1 (*very dissatisfied*) to 5 (*very satisfied*). In 2001, were physicians on average equally satisfied with their careers, compared with physicians surveyed in 1997?

SCENARIO 15-L

(Inspired by Holas, Chiu, Notario, & Kapral, 2005. Details of this scenario differ from the actual research.) Children with sore throats often do not want to eat or drink, and the throat pain may make them resist taking medicine orally. But unless they take the medicine, their illnesses may be prolonged. Some medicines that are mixed with a liquid (called an oral suspension) may not hurt the throat, but they can taste bad. Pharmaceutical companies have created flavorings to add to oral suspensions to make them more palatable. Suppose we are nurses in a pediatric unit of a hospital, and we want to compare children's acceptance of medications that are mixed with one of two flavorings: Groovy Grape and Banana Bongo. We approach parents of children scheduled for a tonsillectomy and ask them to allow the children to participate in our study. Children are randomly assigned to groups. If children say they are in pain after the tonsillectomy, the nurses are authorized to offer a pain medication in an oral suspension; we know that this form of the pain medicine tastes bad. Children in one group are given the oral suspension mixed with Groovy Grape. Children in the other group receive the

(Continued)

oral suspension mixed with Banana Bongo. After taking the medicine, the children are asked for their opinion of the flavor. They use a scale that shows seven cartoon drawings of children's faces with different expressions. The cartoons range from a frowning, crying face (which is recorded as a 1 on the scale) to a happy face with a big grin (which is recorded as a 7 on the scale). We ask each child to point to the face that shows how they feel about the taste of the medicine. A power analysis indicates that we need 28 children per group to achieve power = .85 to detect a clinically noteworthy difference in the ratings. When we run the study, three children in the Banana Bongo group spit out the medicine and refuse to provide a rating of its flavor. Did children receiving the pain medicine mixed with Groovy Grape give higher ratings on average than the children receiving the medicine mixed with Banana Bongo?

SCENARIO 15-M

(Inspired by Horsted, Rasmussen, Meyhoff, & Nielsen, 2007. Details of this scenario may differ from the actual research.) How do people view their quality of life after suffering a cardiac arrest outside of a hospital setting? Cardiac arrest occurs when an irregular heart rhythm causes the heart to stop; it is different from a heart attack, which usually involves a blockage of blood flow although the heart keeps trying to pump blood. Researchers in Denmark identified people who suffered sudden cardiac arrest and, after being resuscitated by emergency medical personnel, were taken to a hospital. Six months after being released from the hospital, 33 patients were contacted and asked to complete a questionnaire that produced a numeric score for their quality of life. The researchers wanted to know whether these patients' mean quality of life differed from a national norm for people of the same age.

SCENARIO 15-N

(Inspired by Raphael et al., 2012. Details of this scenario may differ from the actual research.) People who clench or grind their teeth in their sleep sometimes suffer from pain related to the muscles in the head, neck, and shoulders associated with jaw movements. Nighttime teeth-grinding is called sleep bruxism, and sometimes the result is myofascial pain associated with the temporomandibular joints (TMJ). But not all TMJ pain comes from sleep bruxism. Researchers wanted to determine whether 24 patients who had TMJ pain (i.e., the cases) experienced more sleep bruxism than 24 people who had not been diagnosed with TMJ pain (i.e., the controls). The researchers measured rhythmic masticatory muscle activity (RMMA) episodes, which are jaw movements; *masticatory* means "related to chewing." These episodes were counted and timed with special equipment. The researchers computed the mean duration of RMMA episodes for each group.

(Continued)

SCENARIO 15-O

(Inspired by Harris et al., 1994. Details of this scenario may differ from the actual research.) Are "maternity blues," a mild form of postpartum depression, related to changes in a hormone called cortisol after delivery? Healthy first-time mothers who carried babies to term were studied two weeks after giving birth. The women responded to a scale that measured maternity blues; higher scores meant more symptoms of the blues. A saliva sample was used to test cortisol levels.

SCENARIO 15-P

(Inspired by Rosenstein, 2002. Details of this scenario may differ from the actual research.) A survey asked hospital professionals about disruptive behavior by physicians. Is there a difference in the rate of agreement for people in different roles? The researchers defined disruptive physician behavior as "any inappropriate behavior, confrontation, or conflict, ranging from verbal abuse to physical and sexual harassment." Each respondent was asked to self-identify as a nurse, physician, or hospital executive. One question asked whether the respondent's hospital provided a "non-punitive reporting environment for nurses who witness disruptive behavior (yes/no)." Are the distributions of responses proportionally the same for all three kinds of professionals?

SCENARIO 15-Q

(Inspired by Price, Amini, & Kappeler, 2012. Details of this scenario may differ from the actual research.) Suppose we have been reading various reports about the birth weight of full-term babies in affluent Western cultures like the United States. We speculate that the population mean birth weight for these infants is 3,400 g (almost 7.5 lb) and that the population standard deviation is 375 g (about 13 oz). We are analyzing data from a study in which the mothers exercised during pregnancy. Will the average birth weight of babies born to the mothers who exercised regularly differ from the population mean?

SCENARIO 15-R

(Inspired by Murdock, 2013. Details of this scenario differ from the actual research.) Does the widespread use of text messaging with mobile phones affect people's sleep? Suppose we want to know whether the number of text messages that students studying physical therapy (PT) or occupational therapy (OT) usually send in a week will predict how much trouble they have with sleeping. We recruit a sample of 104 PT and OT students, who give us permission to count the number of text messages that they send in one week. All participants will be analyzed together; that is, we are not interested in comparing PT students versus OT students. We create a cell-phone app that

(*Continued*)

tells us the total number of texts without revealing the content of any messages. The students fill out a survey that asks them about many variables, such as the number of credit hours they are carrying. Our main interest is their score for a measure of sleep trouble, where a lower score means less sleep trouble, and a higher number means more trouble falling asleep and staying asleep.

SCENARIO 15-S

(Inspired by Samper & Schwartz, 2013. Details of this scenario differ from the actual research.) Do Americans interpret a higher drug price as implying that the drug treats a condition that they are unlikely to get? We recruit a diverse sample of $N = 140$ adults with health insurance to participate in a study. We tell them we want their opinion of a flyer about immunizations—where they are available, how much they cost, and so on. We randomly assign the participants to two same-sized groups. Half of the subjects will read a flyer that says a pneumonia shot costs \$25. The other half will read a flyer that says a pneumonia shot costs \$100. Both flyers say, "If you have health insurance, we guarantee that you will pay nothing out of pocket for the pneumonia shot." The two flyers are identical except for the price. Each participant answers many questions about the flyer—its attractiveness, the clarity of writing, and so on. Our main question asks participants to rate how likely they are to catch pneumonia in the next five years, from 0% (*no chance*) to 100% (*guaranteed to catch it*). Will the difference in the flyers (i.e., the different prices of the pneumonia shot) lead to different means on the ratings of the likelihood of catching pneumonia?

SCENARIO 15-T

(Inspired by Pronk, Katz, Lowry, & Payfer, 2012. Details of this scenario differ from the actual research.) Imagine we have conducted a study in which office workers were given adjustable desks that would allow them to either sit at their computers as usual or to raise the computers, allowing them to work while standing up. The offices are equipped with a device that measures how long the person works while standing up. The device records the time for as long as someone is typing or moving the mouse while the desk at standing height. The timer turns off whenever there is a pause for more than 30 seconds and resumes whenever typing or mouse movement starts again. Over the course of two weeks, each person gets one score for the amount of time spent working while standing. On the last day of the two-week period, we ask each person to rate how much fatigue they have felt during the two weeks, where the ratings range from 0 (*no fatigue*) to 100 (*complete fatigue*). Is there a relationship between the amount of time spent working while standing and the amount of reported fatigue?

(Continued)

SCENARIO 15-U

(Inspired by Dhaliwal, Welborn, & Howat, 2013. Details of this scenario differ from the actual research.) Suppose we are conducting a longitudinal study of cardiovascular health. For 20 years, we have tracked a representative sample of nearly 19,000 middle-aged adults who began the study with no history of heart disease. Among many other variables studied, we collected data on recreational physical activity. We have identified which participants have suffered some sort of cardiac-related event (heart attack, cardiac arrest, stroke, etc.) since the study began. We want to know whether people who exercised moderately had a lower risk of suffering a cardiac event, compared to people who were sedentary.

SCENARIO 15-V

(Inspired by Moulton et al., 2005. Details of this scenario may differ from the actual research.) We have been reading about the prevalence of arthritis among older Americans, as reported by the CDC, which has conducted surveillance studies of chronic conditions like arthritis for many years. The CDC says 23% of American adults say they have received a diagnosis of arthritis. Next we read a study about the prevalence of chronic disease among older Native Americans and Alaska Natives. The report describes a data set in which 9,403 elders in 171 tribal nations were surveyed. The results show that 47% of these elders had arthritis. Does the arthritis rate for Native American/Alaska Native elders differ from the national rate for all Americans reported by the CDC?

SCENARIO 15-W

(Inspired by Morikawa et al., 2014. Details of this scenario differ from the actual research.) Suppose we read a study about the incidence of pregnancy-induced high blood pressure in Japan. The study shows a seasonal pattern to this condition: women who gave birth in the winter and early spring had higher rates of pregnancy-induced high blood pressure, compared with other months of the year. Suppose we want to know whether a similar pattern exists in Oklahoma, which is where your authors live. We arrange to obtain data from all hospitals in the state and the Oklahoma State Department of Health for a one-year period. Specifically, we will identify how many women gave birth each month and, out of that number, how many of these women had pregnancy-induced high blood pressure. Will we find the same rate of the condition every month?

References

Buron, B. (2010). Life history collages: Effects on nursing home staff caring for residents with dementia. *Journal of Gerontological Nursing, 36,* 38–48. doi:10.3928/00989134-20100602-01

Dhaliwal, S. S., Welborn, T. A., & Howat, P. A. (2013). Recreational physical activity as an independent predictor of multivariable cardiovascular disease risk. *PLOS ONE, 8,* 1–6. doi:10.1371/journal.pone.0083435

Falvo, M. J., & Earhart, G. M. (2009). Six-minute walk distance in persons with Parkinson disease: A hierarchical regression model. *Archives of Physical Medicine and Rehabilitation, 90,* 1004–1008. doi:10.1016/j.apmr.2008.12.018

Grant, A. M., & Hofmann, D. A. (2011). It's not all about me: Motivating hand hygiene among health care professionals by focusing on patients. *Psychological Science, 22,* 1494–1499. doi:10.1177/0956797611419172

Harrington, J. W., Logan, S., Harwell, C., Gardner, J., Swingle, J., McGuire, E., & Santos, R. (2012). Effective analgesia using physical interventions for infant immunizations. *Pediatrics, 129,* 815–822. doi:10.1542/peds.2011-1607

Harris, B., Lovett, L., Newcombe, R. G., Read, G. F., Walker, R., & Riad-Fahmy, D. (1994). Maternity blues and major endocrine changes: Cardiff puerperal mood and hormone study II. *British Medical Journal, 308,* 949–953.

Harrison, M. F., Anderson, P., Miller, A., O'Malley, K., Richert, M., Johnson, J., & Johnson, B. D. (2013). Oral contraceptive use and acute mountain sickness in South Pole workers. *Aviation, Space, and Environmental Medicine, 84,* 1166–1171. doi:10.3357/ASEM.3717.2013

Holas, C., Chiu, Y.-L., Notario, G., & Kapral, D. (2005). A pooled analysis of seven randomized crossover studies of the palatability of cefdinir oral suspension versus amoxicillin/clavulanate potassium, cefprozil, azithromycin, and amoxicillin in children aged 4 to 8 years. *Clinical Therapeutics, 27,* 1950–1960. doi:l0.1016/j.clinthera.2005.11.017

Horsted, T. I., Rasmussen, L. S., Meyhoff, C. S., & Nielsen, S. L. (2007). Long-term prognosis after out-of-hospital cardiac arrest. *Resuscitation, 72,* 214–218. doi:10.1016/j.resuscitation.2006.06.029

Johnson, D. L., Wiebe, J. S., Gold, S. M., Andreasen, N. C., Hichwa, R. D., Watkins, G. L., & Boles Ponto, L. L. (1999). Cerebral blood flow and personality: A positron emission tomography study. *American Journal of Psychiatry, 156,* 252–257.

Landon, B. E., Reschovsky, J., & Blumenthal, D. (2003). Changes in career satisfaction among primary care and specialist physicians, 1997-2001. *Journal of the American Medical Association, 289,* 442–449.

Liu, S., Manson, J. E., Lee, I.-M., Cole, S. R., Hennekens, C. H., Willett, W. C., & Buring, J. E. (2000). Fruit and vegetable intake and risk of cardiovascular disease: The Women's Health Study. *The American Journal of Clinical Nutrition, 72,* 922–928.

Montgomery-Downs, H. E., Clawges, H. M., & Santy, E. E. (2010). Infant feeding methods and maternal sleep and daytime functioning. *Pediatrics, 126,* e1562–e1568. doi:10.1542/peds.2010-1269

Morikawa, M., Yamada, T., Yamada, T., Cho, K., Sato, S., & Minakami, H. (2014). Seasonal variation in the prevalence of pregnancy-induced hypertension in Japanese women. *Journal of Obstetrics and Gynaecology Research, 40,* 926–931. doi:10.1111/jog.12304

Moulton, P., McDonald, L., Muus, K., Knudson, A., Wakefield, M., & Ludtke, R. (2005). *Prevalence of chronic disease among American Indian and Alaska Native elders.* Grand Forks, ND: Center for Rural Health. Retrieved from http://ruralhealth.und.edu/projects/nrcnaa/pdf/chronic_disease1005.pdf

Murdock, K. K. (2013). Texting while stressed: Implications for students' burnout, sleep, and well-being. *Psychology of Popular Media Culture, 2,* 207–221. doi:10.1037/ppm0000012

Price, B. B., Amini, S. B., & Kappeler, K. (2012). Exercise in pregnancy: Effect on fitness and obstetric outcomes—A randomized trial. *Medicine & Science in Sports & Exercise, 44,* 2263–2269. doi:10.1249/MSS.0b013e318267ad67

Pronk, N. P., Katz, A. S., Lowry, M., & Payfer, J. R. (2012). Reducing occupational sitting time and improving worker health: The Take-a-Stand Project, 2011. *Preventing Chronic Disease, 9,* 1–9. doi:10.5888.pcd9.110323

Raphael, K. G., Sirois, D. A., Janal, M. N., Wigren, P. E., Dubrovsky, B., Nemelivsky, L. V., ...Lavigne, G. J. (2012). Sleep bruxism and myofascial temporomandibular disorders: A laboratory-based polysomnographic investigation. *Journal of the American Dental Association, 143,* 1223–1231. doi:10.14219/jada.archive.2012.0068

Rauh, M. J., Aralis, H. J., Melcer, T., Macera, C. A., Sessoms, P., Bartlett, J., ... Galarneau, M. R. (2013). Effect of traumatic brain injury among U.S. service members with amputation. *Journal of Rehabilitation Research & Development, 50,* 161–172. doi:10.1682/JRRD.2011.11.0212

Rose, N., Koperski, S., & Golomb, B. A. (2010). Mood food: Chocolate and depressive symptoms in a cross-sectional analysis. *Archives of Internal Medicine, 170,* 699–703. doi:10.1001/archinternmed.2011.2100

Rosenstein, A. (2002). The impact of nurse-physician relationships on nurse satisfaction and retention. *American Journal of Nursing, 102,* 26–34.

Samper, L. A., & Schwartz, J. A. (2013). Price interferences for sacred vs. secular goods: Changing the price of medicine influences perceived health risk. *Journal of Consumer Research, 39,* 1343–1358. doi:10.1086/668639

Schauer, P. R., Kashyap, S. R., Wolski, K., Brethauer, S. A., Kirwan, J. P., Pothier, C. E., ... Bhatt, D. L. (2012). Bariatric surgery versus intensive medical therapy in obese patients with diabetes. *The New England Journal of Medicine, 366,* 1567–1576. doi:0.1056/NEJMoa1200225

Schell, K., Morse, K., & Waterhouse, J. K. (2010). Forearm and upper-arm oscillometric blood pressure comparison in acutely ill adults. *Western Journal of Nursing, 32,* 322–340. doi:10.1177/0193945909351887

Stephens, R., Atkins, J., & Kingston, A. (2009). Swearing as a response to pain. *NeuroReport, 20,* 1056–1060. doi:10.1097/WNR.0b013e32832e64b1

Waterhouse, J., Hudson, P., & Edwards, B. (2009). Effects of music tempo upon submaximal cycling performance. *Scandinavian Journal of Medicine & Science in Sports, 20*, 662–669. doi:10.1111/j.1600-0838.2009.00948.x

Wilkens, P., Scheel, I. B., Grundnes, O., Hellum, C., & Storheim, K. (2010). Effect of glucosamine on pain-related disability in patients with chronic low back pain and degenerative lumbar osteoarthritis. *Journal of the American Medical Association, 304*, 45–52. doi:10.1001/jama.2010.893

Witte, D. R., Grobbee, D. E., Bots, M. L., & Hoes, A. W. (2005). A meta-analysis of excess cardiac mortality on Monday. *European Journal of Epidemiology, 20*, 401–406. doi:10.1007/s10654-004-8783-6

Suggested Answers to Odd-Numbered Exercises

Chapter 1

1-23. Quantitative experimental research because we have random assignment to groups, manipulation of an independent variable (speed of music), and statistical replication.

1-25. Dependent variable.

1-27. Quantitative experimental research—specifically, the study follows a randomized block design. The researcher blocked on gender, and then randomly assigned subjects to groups (sleep interruption). The study also has statistical replication because the effect is studied on multiple participants.

1-29. Frustration is the measured outcome. Because this is an experiment, it is called the dependent variable.

1-31. We should be able to draw causal conclusions about sleep interruption's effect on frustration because of the random assignment to interruption groups.

1-33. It could be mixed-methods research. A few superficial measures such as "bothersomeness" scores are collected on a large number of people, but the participants' diaries might be analyzed qualitatively for themes.

1-35. External validity is limited by the fact that the participants were recruited in one city in the United States, and there was no mention of random sampling or other steps taken to ensure that the sample was drawn without bias. The sample may be biased and not representative of all menopausal women who have hot flashes.

1-37. The quantitative portion is an experiment because the women were randomly assigned to groups and the researchers manipulated whether they participated in meditation or were placed in the wait-list control group.

1-39. An extraneous variable that would be controlled by randomization.

1-41. No, because randomization should control the extraneous variables associated with participants, including surgical history.

1-43. External validity should be strong. The quoted section indicates several measures were taken to ensure a sample that should be representative of the population. The purpose of this study was to use samples to represent trends in the population, so external validity would be crucial for these researchers.

1-45. It is impossible to randomly assign people to career paths, so we could not run an experiment to compare these professionals' career satisfaction. An observational study could answer such a research question, but no causal conclusions could be drawn.

1-47. The discharge experience (pharmacist-assisted or usual care) is the independent variable that we are manipulating. Because this is a quasi-experiment, we might refer to readmission rate as either a criterion variable or a dependent variable. We have little ability to make causal inferences about the discharge experience's effect on readmission rate because we have manipulation but no random assignment to groups, so extraneous variables may interfere with the relationship between the independent variable and the measured outcome.

Chapter 2

2-15. The scenario said nothing about randomly assigning participants to groups, so this study cannot be an experiment. All stimuli were offered to all participants, so there was no manipulation of an independent variable. Therefore, this study amounted to observational research with statistical replication (multiple participants being studied).

2-17. Engagement duration is the criterion variable.

2-19. No. Causal conclusions cannot be drawn from a nonexperimental study.

2-21. External validity (quality of inference about generalization of results from the sample to the population) cannot be assessed without further information. There was no mention of random sampling from a population; in fact, the researchers used a convenience sample of residents in two nursing homes. We probably can generalize the results only to participants like the ones described in the article.

2-23. The scenario said 44 out of the 56 residents were women, so the mode for the variable gender is "female." Because 35 of the 56 residents were widowed, the mode for marital status is "widowed." "High school or above" would be the mode for educational status because 47 of the 56 residents were in this category.

2-25. With the youngest participant being 61 and the oldest being 101, having a mean age of 87 could lead the reader to wonder whether the age of the oldest participant is inflating the mean. If the article had given the median, the reader could judge whether the higher age had affected the mean.

2-27. No, the range would be 21 (i.e., high score minus low score $= 21 - 0 = 21$).

2-29. Treatment (glucosamine or placebo) is the independent variable.

2-31. Usual therapy is an extraneous variable, which could interfere with the researchers' attempts to observe a relationship between the treatment and the dependent variable. But random assignment should control this kind of extraneous variable.

2-33. After a year it appeared that the control group had higher disability because its mean RDMQ $= 5.5$, compared with the treatment group's mean $= 4.8$. The researchers reported that the difference was not statistically remarkable.

2-35. Mean SBP $= 152.5$, median SBP $= 151$, mean and median DBP $= 70$, and mean and median HR $= 71$.

2-37. Subtracting the mean SBP $= 152.5$ from every score gives these results: $-2.5, -0.5, -2.5, 11.5, 3.5, -9.5$. Squaring the distances: $6.25, 0.25, 6.25, 132.25, 12.25, 90.25$. Sum of squares $=$ numerator of the unbiased variance $= 247.5$. Denominator of the unbiased variance $= N - 1 = 6 - 1 = 5$. So the unbiased variance for SBP $= 247.5/5 = 49.5$.

2-39. Subtract the mean HR from each score: $1, -1, -3, 3, -2, 2$. Square the distances: $1, 1, 9, 9, 4, 4$. Sum of squares $=$ numerator of the unbiased variance $= 28$. Denominator of the unbiased variance $= N - 1 = 6 - 1 = 5$. So the unbiased variance for HR $= 28/5 = 5.6$.

2-41. For SBP, $SD \approx 7.04$. For DBP, $SD \approx 9.38$. For HR, $SD \approx 2.37$. SD is reported instead of the unbiased variance because SD is in the original units of measure, whereas the unbiased variance is in squared units.

2-43. The cases had been suffering from the pain for more than 10 years on average, but the middle score for the number of months was 84, or 7 years.

2-45. The quoted material refers to a 5% trimmed mean. If 5% of scores are trimmed from each end of the distribution, the middle 90% of scores would remain to be averaged.

Chapter 3

3-7. Kind of stimulus is a nonnumeric, discrete categorical variable.

3-9. The scenario described the measurement of the amount of time that the residents engaged with each stimulus. A bar graph could be created with a separate bar for each stimulus. The mean amount of engagement time for each stimulus could be displayed.

3-11. A bar graph could be used to show how many residents refused each stimulus. A separate bar would represent each stimulus, and the heights of the bars would show how many people refused the different stimuli.

3-13. An outlier may stand out from the other scores in a histogram, but different people may look at a graph in different ways and disagree on whether a high score or a low score can be called an outlier. A boxplot provides a way to define whether an extreme score is an outlier.

3-15. Power score is a quantitative outcome of this experiment, so it is a dependent variable. Number of previous pregnancies is an extraneous variable, which would have been controlled by randomization to groups.

3-17. The appearance of graphs is dependent upon the software used. Images are being omitted intentionally.

3-19. The appearance of graphs is dependent upon the software used. Images are being omitted intentionally.

3-21. A bar graph can be created.

3-23. The appearance of graphs is dependent upon the software used. Images are being omitted intentionally.

3-25. The appearance of graphs is dependent upon the software used. Images are being omitted intentionally.

Chapter 4

4-11. The purpose of computing a standard score is to measure the relative location of a score within a distribution. A standard score takes away the units of measures and reports how many standard deviations are between a score and its mean. A positive standard score shows that the score is above its mean, and a negative standard score shows that the score is below its mean.

4-13. Experimental research because participants were randomly assigned to groups and the researchers manipulated whether the participants took a tai chi class or an education/stretching class. The researchers also used statistical replication.

4-15. Kind of class is an independent variable because it is manipulated by the researcher and comes first in time.

4-17. The distribution appears negatively skewed, with a minimum score $= 6$ and a maximum score $= 19$. Your graph may show a gap with no scores $= 7$. The most frequently occurring score appears to be 14.

4-19. First, we need to multiply the unbiased variance by $(N - 1)/N = 32/33 = 0.969697$. The biased sample variance will equal this product: $0.969697 \times 9.808712 = 9.5114783$. Second, to get the standard deviation based on the biased sample variance, we take the square root of 9.5114783, which is $3.0840685 \approx 3.08$.

4-21. This person's z score $= (19 - 13.94)/3.08 = 5.06/3.08 = 1.642857 \approx 1.64$. (More accurate results will be found if the unrounded mean and standard deviation are used.)

4-23. Alan's z score $= (30.5 - 25.5)/3.7 = 5/3.7 = 1.35135 \approx 1.35$.

4-25. Alan's z score is above the mean. Half (or .5) of the 1-year-old boys' weights would be below the mean. We need to take .5 and add it to the proportion of boys whose weights are between the mean and Alan's $z = 1.35$. By looking at the table of areas for the standard normal distribution, we find 1.35 in the first column labeled z or $-z$, and then we look in the second column for the area between $z = 0$ (the mean) and our z of interest. This area is .4115. So the proportion of 1-year-old boys who weigh less than Alan is .9115. (There are other ways to find the same answer.)

4-27. Anna's z score $= (21.5 - 24.1)/3.51 = -2.6/3.51 = -0.74074 \approx -0.74$.

4-29. Now we are looking for a middle area between the mean and our z of interest. The answer is .2704. (We also know that half of the distribution is below the mean, and we already accounted for a tail area of .2296, so we could compute the answer: $.5 - .2296 = .2704$.)

4-31. We need to find all of the area below our z of interest. Alan's z score is positive, indicating that his weight is above the mean. Half of the distribution $= .5$, and half of the boys weigh less than the mean. So now we need to find a middle area between $z = 0$ and our z of interest. By looking for $z = 1.05$ in the table, we can find a middle area $= .3531$. Adding this middle area and .5, we get the answer: .8531.

Chapter 5

5-11. Experimental research because participants were randomly assigned to groups and the researchers manipulated the kind of bowl, plus the research had statistical replication (multiple people in each group).

5-13. Dependent variable.

5-15. The 54 participants had $r = .31$ for estimated and actual number of calories consumed. This statistic indicates a positive linear relationship between the variables. Participants using a normal soup bowl had a stronger linear relationship, $r = .67$, which we know is stronger because .67 is closer to 1 than .31 is. Those who used the bottomless bowl had a weaker linear relationship between estimated and actual number of calories consumed, $r = .12$. We can say that those who received accurate feedback (using normal bowls) had more accurate estimates of their soup consumption. The normal-bowl users who ate little soup generally gave lower estimates, and the normal-bowl users who had a lot of soup tended to give higher estimates. But for the bottomless-bowl users, those who ate little soup may have given higher or lower estimates, and those who ate a lot of soup did not consistently give estimates that were higher.

5-17. When the correlation was computed for both groups together, the researchers had no ability to tell whether the relationship between actual and estimate soup consumption differed for those eating from the normal bowls versus those eating from bottomless bowls. When the groups were analyzed separately, the normal-bowl group had a higher correlation and the bottomless-bowl group had a lower correlation, which the researchers interpreted as indicating the importance of visual feedback about food consumption. Visual information influenced how much people ate. Combining groups also can influence the value of a correlation coefficient, as we saw in this example.

5-19. Positive.

5-21. We square $r = .471$ and get .221841, so about 22.2% of the variance in the measure of sleep problems was accounted for by the amount of life disruption and pain from fibromyalgia.

5-23. Without graphing the data, we cannot know what patterns may be present in the data. Perhaps there is a curve in the point cloud or outliers affecting the correlation.

5-25. Observational research, also known as descriptive or nonexperimental research. We know because the scenario does not mention random assignment to groups or manipulation of an independent variable.

5-27. For UPDRS motor scores and six-minute walk distance, $r = -.27$, indicating that participants with fewer effects of Parkinson symptoms tended to walk farther, and those who were more affected by the disease tended to walk shorter distances. With $r^2 = .0729$, about 7.3% of the variance in walking distances was accounted for by the effects of Parkinson disease. For UPDRS motor scores and TUG scores, $r = .19$, a positive correlation, meaning that more symptoms generally corresponded to longer times to complete the TUG, and fewer symptoms were related to shorter times to complete the TUG. With $r^2 = .0361$, about 3.6% of the variance in TUG times was related to the motor effects of Parkinson disease. For TUG scores and six-minute walk distances, $r = -.64$, indicating that shorter TUG times corresponded to generally longer distances walked in six minutes, and longer times to complete the TUG were paired mostly with shorter distances walked in six minutes. With $r^2 = .4096$, we can say that almost 41% of the variance in six-minute walk distances was explained by the speed of completing the TUG. This correlation is the strongest because this pair of variables shares the most variance. TUG times and walking distances have a stronger linear relationship, which makes sense because both scores are measuring walking: speed and distance.

Chapter 6

6-13. Asthma is a categorical variable. In this example, it is impossible to tell with certainty whether it is considered a predictor or criterion variable.

6-15. Nothing. This is not an experiment.

6-17. $7,000/7,769 = .9010169 \approx .901$.

6-19. This is a joint probability. To count someone in the numerator, both facts must be true. So the answer is $306/7,769 = .0393873 \approx .039$.

6-21. This is a conditional probability. We are being given the people who never smoked (3,859) for the denominator. Out of this subset, we need the numerator to be the number of people with asthma (306). The answer is $306/3,859 = .0792952 \approx .079$.

6-23. This is another conditional probability: the probability of a randomly chosen person not having asthma, given that we are looking at the former smokers.

6-25. This appears to be an "or" probability: the probability of a randomly chosen person not having asthma or having never smoked.

6-27. True positives would be the 40 images that App #2 said "melanoma" when the expert had confirmed the lesion was truly melanoma.

6-29. There were 80 images that the app said were melanoma, but the expert had confirmed they were benign.

6-31. Sensitivity $= 40/58 \times 100 \approx 68.97\%$.

6-33. Specificity $= 47/127 \times 100 \approx 37\%$.

6-35. Positive predictive value $= 40/120 \times 100 \approx 33.3\%$.

6-37. Negative predictive value $= 47/65 \times 100 \approx 72.3\%$.

6-39. Observational research because the researchers did not randomly assign patients to groups, and then seal teeth for one group of children but not for the other. That is, they did not manipulate an independent variable.

6-41. The relative risk of 0.998 is a ratio of two risks: the risk of tooth decay for formerly sealed teeth versus the risk of tooth decay for never-sealed teeth. This relative risk is close to 1, indicating that the risks of tooth decay are about the same for teeth that were never sealed versus teeth that used to be sealed, but then the sealant became loose and fell off. The concern was that loose sealant would trap bacteria and increase the chances of decay.

6-43. The probability of losing a sealant would be .3. The probability of not losing a sealant would be $1 - .3 = .7$. The odds would be $.3/.7 = .4285714 \approx .429$. The chance of losing a sealant is less than half of the chance of not losing a sealant.

Chapter 7

7-17. The variation in numeric values that a statistic will have across repeated samples.

7-19. Scores.

7-21. Numerical values of a statistic.

7-23. We would choose a certain sample size, identify a population from which to sample, draw a random sample from that population, and measure some variable (e.g., blood sugar level) on each participant in the sample. Then we would compute the unbiased variance on those scores. We would repeat this process thousands of times until we had thousands of unbiased variance statistics in a pile. We would arrange the statistics in a distribution like a histogram, and the statistics would form a sampling distribution of the unbiased variance.

7-25. No, the Central Limit Theorem pertains only to the sample mean's sampling distribution.

7-27. This study is an experiment because the researchers randomly assigned participants to groups, and then manipulated whether the group engaged in an exercise program or remained sedentary during pregnancy.

7-29. Activity during pregnancy is an independent variable with two levels: active and sedentary.

7-31. The numerator would be $1083.05 - 1100 = -16.95$. The denominator would be the standard error, computed in question 7-30. So, $z = -16.95/25.400025 = -0.6673222 \approx -0.67$. Notice that you do not need the sample SD to answer this question. It was included intentionally to see if you could pick out the information that you did need.

7-33. It seems as if a sample mean that is two-thirds of a standard error below the population mean would be relatively close to the population mean, but so far we have computed only a point estimate. The answer to this question would be better informed by a confidence interval.

7-35. The lower limit would be the sample mean minus the margin of error = $1083.05 - 49.78405 = 1033.266$. The upper limit would be the sample mean plus the margin of error = $1083.05 + 49.78405 = 1132.834$.

7-37. We are interested in a sample mean, and the Central Limit Theorem gives us information about the sample mean's sampling distribution. The sample size is large enough that we could be assured that the mean's sampling distribution would approximate a normal distribution. The z test statistic's sampling distribution will have the same shape as the sample mean's sampling distribution, so we can use a standard normal distribution in this scenario.

7-39. The values +1.645 and −1.645 will contain 90% of the standard normal distribution. We multiply +1.645 by the standard error of the mean to obtain the margin of error. The standard error = $3.7/\sqrt{78} = 3.7/8.8317609 = 0.4189425$. After multiplying the critical value and the standard error, we get 0.6891604. The lower limit is $24.3 - 0.6891604 = 23.61084 \approx 23.61$. The upper limit is $24.3 + 0.6891604 = 24.93916 \approx 24.94$. The 90% CI means that we are 90% confident that this interval contains the true mean of the population that provided our sample of boys who were weighed within a month of their first birthday.

Chapter 8

8-29. This is the null hypothesis: Our sample comes from a population with a mean equal to 60.

8-31. Alpha would be placed in the upper tail of the distribution because we always look at the alternative hypothesis to tell us this answer. Because $H_0: \mu \leq 45$, the alternative hypothesis would be $\mu > 45$. The prediction is that we will sample from a population with a mean greater than 45. If we get a sample mean that is greater than 45, then the z test statistic will be in the upper tail. We need alpha to go there so that we can detect significance in that direction.

8-33. The answers to a and c reflect the fact that one-tailed tests will be performed. The alternative hypothesis must be consulted to determine whether alpha goes in the upper tail or the lower tail of the standard normal distribution. The critical values would be: a. $z = 1.645$. b. $z = -1.96$ and $z = +1.96$. (Half of alpha goes in each tail.) c. $z = -1.645$. d. $z = -1.96$ and $z = +1.96$.

8-35. a. The two-tailed p value $= .617$. b. The two-tailed p value $= .0124$. c. The two-tailed p value $= .617$. d. The two-tailed p value $= .9602$.

8-37. Because the alternative hypothesis is H_1: $\mu \neq 45$, there is no predicted direction for the results. So we can compare the p value directly with alpha.

a. Because the two-tailed p value $= .617$, which is greater than alpha, we retain the null hypothesis. The conclusion is that there is no significant difference between the sample mean and the population mean ($\mu = 45$).

b. Because the two-tailed p value $= .0124$, which is less than alpha, we reject the null hypothesis and conclude that there is a statistically significant difference between the sample mean and the population mean ($\mu = 45$).

c. Same as 8-37, part a.

d. The two-tailed p value $= .9602$, which is greater than alpha, so we retain the null hypothesis and conclude that there is no statistically significant difference between M and the population mean ($\mu = 45$).

8-39. This study is observational/descriptive/nonexperimental research because there is no random assignment of participants to groups and no manipulation of an independent variable.

8-41. No causal conclusion can be drawn because this study is not an experiment. Descriptive studies like this one have lower internal validity, which limits our ability to infer a causal relationship between variables. The lack of randomization to groups and manipulation of an independent variable mean that extraneous variables are uncontrolled and may be responsible for any observed effect.

8-43. Our sample comes from a population where the mean IQ of 7-year-old children who were preterm at birth is greater than 95.

8-45. Our sample comes from a population where the mean IQ of 7-year-old children who were preterm at birth is less than or equal to 95 (or, if your instructor prefers, equal to 95).

8-47. The distribution should have a critical value of 1.645 cutting off $\alpha = .05$ in the upper tail of the standard normal distribution.

8-49. The p value decision rule: The results are in the predicted direction (i.e., the z test statistic is positive because the sample mean is greater than the population mean). Further, because $p < .05$, we reject the null hypothesis. The critical value decision rule: Because the observed z test statistic $= 4.88$, which is more extreme than the z critical value of 1.645, we reject the null hypothesis.

8-51. The 90% confidence interval for μ first requires the computation of the margin of error. In this case, we take the critical value of 1.645 and multiply it by the standard error of the mean, which was 1.2700013, the denominator of the z test statistic. So the margin of error is 1.645 times 1.2700013 = 2.0891521. The sample mean was 101.2. The lower limit is 101.2 − 2.0891521 = 99.110848 ≈ 99.11, and the upper limit is 101.2 + 2.0891521 = 103.28915 ≈ 103.29. So the 90% confidence interval is [99.11, 103.29].

8-53. The 90% confidence interval for μ is [99.11, 103.29], which does not bracket μ = 95. Further, the interval is shifted higher on the number line than 95, as predicted by the alternative hypothesis. So we would reject the null hypothesis and conclude that a significant difference exists between the sample mean and the population mean.

8-55. Our null hypothesis is that our sample comes from a population with a mean BMD that equals 0.88 g/cm². In symbols, we would write: H_0: μ = 0.88.

8-57. We are told that μ = 0.88 and σ = 0.6. Further, the results show that M = 1.03 for N = 36 women. (We can ignore the sample median = 0.99 and SD = 2.1 for now because they are not relevant to the computation of the z test statistic.) The numerator of the z test statistic is $M − μ$ = 1.03 − 0.88 = 0.15. The denominator of the z test statistic is $σ/\sqrt{N}$ = $0.6/\sqrt{36}$ = 0.6/6 = 0.1. So the z test statistic = 0.15/0.1 = 1.5.

8-59. Because .1336 is greater than .05, we retain the null hypothesis.

8-61. A 95% confidence interval would be appropriate because α = .05 and we conducted a two-tailed test, corresponding to the nondirectional alternative hypothesis. As a result, 2.5% of the standard normal distribution was cut off by the lower critical value (z = −1.96), and 2.5% of the standard normal distribution was cut off by the upper critical value (z = 1.96).

8-63. For all of the confidence intervals that we could compute by repeatedly sampling from the same population, 95% of these intervals would capture the true mean BMD for the population being sampled. Although we cannot say for certain whether this confidence interval contains the true value of the population mean, we are 95% confident that it does.

Chapter 9

9-15. Type I error.

9-17. The cost of a Type I error would be the stress on the patient having to undergo additional testing, as well as the cost of the tests required to rule out breast cancer.

9-19. One correct decision would be to reject the null hypothesis ("no breast cancer") when in fact the person has breast cancer. The other correct decision would be to retain the null hypothesis when in fact the person does not have breast cancer. We said the probability of a Type II error was .20, meaning the probability of retaining the null hypothesis when the null hypothesis is false in the population. If the null hypothesis is false, then .80 would represent the probability of correctly rejecting the null hypothesis.

9-21. Alpha = .05. It is the small probability chosen in advance. The quotation implies that the researchers will decide that a statistically significant result has been found if the observed test statistic has a p value of .05 or less. (The article should have used the symbol alpha for the probability = .05.)

9-23. The decision to correctly retain the null hypothesis, when the null hypothesis is true in the population. In this scenario, its probability is $1 - \alpha = .95$.

9-25. Because power = .90, we can restate this fact as $1 - \beta = .90$. Beta is the probability of a Type II error, which is retaining the null hypothesis when H_0 is false in the population. If $1 - \beta = .90$, then $\beta = .10$.

9-27. BMD is an outcome variable, which in nonexperimental research is called a criterion variable.

9-29. The null hypothesis would be that our sample came from a population with a mean BMD = 0.88. The 95% confidence interval of [0.834, 1.226] contains this value, so we would retain the null hypothesis.

9-31. All else being equal, the power would increase, meaning we would have a greater probability of finding statistical significance.

9-33. A larger sample size tends to make confidence intervals more narrow.

9-35. Sample size calculations are performed to achieve a certain amount of power. One important factor to include in the sample size calculations is the smallest effect size that the researchers believe is clinically noteworthy to detect. The context would have been the researchers' description of their decision-making process about the sample size.

Chapter 10

10-11. PSQI is the main outcome variable, which is called a criterion variable in observational research.

10-13. The internal validity (quality of inference about a causal relationship between variables) is weak because this is not an experiment. Causality cannot be inferred.

10-15. The key phrase to answer this question is this sentence from the scenario: "We want to know whether the patients in the tai chi study started out with significantly worse sleep quality than adults in general who are about the same age." This statement reflects the alternative hypothesis. The scenario says Buysse et al.'s mean PSQI for a diverse adult community sample was 6.3. Remembering that higher scores mean worse sleep quality, we would write the alternative hypothesis as follows: our sample comes from a population with a mean PSQI greater than 6.3. In symbols, we would write $H_1: \mu > 6.3$.

10-17. The appearance of the graphs will depend on the software used to create them, so we are omitting the graphs here.

10-19. If you obtained different statistics, check whether you have used the PSQI scores for the first occasion of measurement.

10-21. Because $N = 66$, $df = N - 1 = 65$.

10-23. The numerator is $M - \mu = 13.7 - 6.3 = 7.4$. The denominator is $SD/\sqrt{N} = 3.39/\sqrt{66} = 3.39/8.1240384 = 0.4172802$. (Do not round yet.) So the one-sample $t = 7.4/0.4172802 = 17.733889 \approx 17.73$.

10-25. The evidence supports the notion that our sample of patients with fibromyalgia came from a population with a significantly worse sleep quality than the diverse adult sample studied by Buysse et al.

10-27. Assuming that we are using a two-sided confidence interval, we would look in Table B for a critical value for a two-tailed test. We want a 95% confidence interval, so alpha must be .05. The table does not list critical values for $df = 32$, so we use the next smaller $df = 30$. The critical value is $t = 2.042$.

10-29. For the control group: $SD/\sqrt{N} = 17.9/\sqrt{33} = 17.9/5.7445626 = 3.11599 =$ estimated standard error of the mean. Margin of error $= 3.1159901 \times 2.042 = 6.3628517$.

10-31. Treatment: The lower limit of the confidence interval is $M -$ margin of error $= 34.3 - 7.2870648 = 27.012935$. The upper limit is $M +$ margin of error $= 34.3 + 7.2870648 = 41.587065$. So the 95% confidence interval is approximately $[27.01, 41.59]$.

10-33. This is an interval estimate of the mean of the population from which the treatment group was drawn. Through repeated samplings, we would expect that 95% of confidence intervals like this one to bracket the true population mean.

10-35. Our confidence intervals appear to correspond to the "I"-shaped error bars atop each bar of Figure 10.4.

Chapter 11

11-21. The computation is the same for all ways of obtaining pairs of scores: $df =$ the number of pairs of scores minus 1. For the single group measured twice, each person has one pair of scores (such as pretest and posttest).

11-23. If the order is not randomized, then the results could be influenced by the order of the conditions. We would not know if the swearing was responsible for differences in pain tolerance or if the order of exposure to the conditions influenced the pain tolerance scores.

11-25. We have a directional alternative hypothesis, so we look for alpha for a one-tailed test $= .05$. With $N = 56$, we have $df = N_{pairs} - 1 = 55$. The t critical value shown in Table B is 1.673. The alternative hypothesis specifies a larger mean at posttest. If we plan to compute the difference scores as "pretest minus posttest," then we are saying the mean difference will be negative. So the critical value would need to be negative too.

11-27. The pretest mean, 48, is greater than the posttest mean, 34, which is not what we predicted in the alternative hypothesis. The results are not in the predicted direction, so we retain the null hypothesis.

11-29. This interval may or may not contain the true population mean difference. Across repeated samples, 95% of the time, such a confidence interval would capture the true population mean difference. So we can say we are 95% confident that our interval would bracket the population mean difference. But the observed mean difference was not in the predicted direction, so the null hypothesis is retained, the same decision made above.

11-31. H_1: $\mu_{music} \neq \mu_{no\ music}$. H_1: $\mu_{music} - \mu_{no\ music} \neq 0$. Our samples come from populations where there is some difference in the mean pain ratings for people who receive an injection in a room with soft music versus people who receive a shot in a room without music.

11-33. Having equal and large sample sizes allows us to use the independent-samples t test, which now will be inoculated against the effects of unequal population variances that it may encounter.

11-35. The mean for the music group appears to be slightly lower than the mean for the no-music group. The scores for the music group are slightly less spread out than the scores for the no-music group, judging by the standard deviations. The AWS t test indicates that there are about 1.34 estimated standard errors of the mean difference between the two means. The confidence interval indicates that zero would be a plausible value for the population mean difference. We would retain the null hypothesis and conclude that there is no significant difference in the mean pain ratings for those who received a shot in a room with soft music playing and those who received the shot in a room without music.

11-37. Feeding method is the independent variable.

11-39. Gestational age upon discharge is an extraneous variable.

11-41. No, they are not being measured on the same variable. A paired t test is not the appropriate analysis, which requires two measurements on the same variable, either on the same person or on two people who are linked naturally (e.g., twins) or by the researcher.

11-43. No.

11-45. Independent-samples t test or the AWS t test.

11-47. This statistic appears to be an independent-samples t test, comparing the mean gestational ages for the two groups. The independent-samples t test is positive, so the smaller mean must have been subtracted from the bigger mean: $M_{bottle} - M_{cup} = 38.1 - 37.2 = 0.9$.

11-49. It appears that the independent-samples t test was used, if the equal sample sizes of 30 per group were maintained. The cup-fed babies had a higher mean birth weight than the bottle-fed babies. Assuming no directional prediction and a typical value of alpha (.01 or .05), we could say that the difference in mean birth weight was statistically significant. The cup-fed babies weighed significantly more than the bottle-fed babies.

Perhaps the cup-fed babies' higher birth weight would explain why they were able to go home earlier than the bottle-fed babies, in terms of gestation age (the focus of Question 11-48).

11-51. We would expect babies who are thriving to have higher PIBBS scores, so babies who were born bigger could be expected to have more mature breast-feeding behavior.

11-53. One group of students, who are the participants being measured.

11-55. If we have one group measured on two occasions (pretest/posttest), we can use the paired t test to compare their means on the two times.

11-57. $H_1: \mu_{post} > \mu_{pre}$. (Your answer here may differ because of the many ways of writing this alternative hypothesis.)

11-59. There were 22 students, so $df = 22 - 1 = 21$. We have a directional alternative hypothesis and $\alpha = .05$, so we look for a critical value for a one-tailed test. The row of Table B for $df = 21$ shows that the critical value is $t = 1.721$. We might need to use a negative critical value, depending on the direction of subtraction of the means.

11-61. It appears that the students' mean comfort level in dealing with people who have dementias was higher after the program than before the program. This study was not an experiment, however, so we do not know whether the program itself was causally responsible for the change. Perhaps any program that provided the students with the time to interact with such older adults would provide the same benefit.

11-63. If the samples were equal in size and all 250 people were randomly assigned to groups, then it would appear that the independent-samples t test would have its inoculation against unequal variances and could be used in this study.

11-65. $H_1: \mu_{treatment} \neq \mu_{control}$. Our samples come from populations in which the mean baseline health-related quality of life is different for those in the treatment group versus those in the control group.

11-67. The treatment group appears to have slightly lower mean health-related quality of life than the control group. The treatment group has slightly more spread in its health-related quality of life scores than the control group. Assuming there was no prediction of a directional outcome, we can say that $p < \alpha$, and we can conclude that the mean health-related quality of life was significantly lower for the treatment group than the control group.

11-69. The alternative hypothesis is nondirectional, so we can compare the p value directly with alpha. Because $p < \alpha$, we reject the null hypothesis and conclude that there was a statistically significant difference in the means of systolic blood pressure for the forearm versus the upper arm, with the readings being taken while the patients were supine.

11-71. This information appears after "upper-arm and forearm systolic," so the means of systolic blood pressure are being compared for those two places on the arm, with the measures being taken while the head of the bed was inclined at 30°.

11-73. The mean diastolic blood pressure on the forearm versus the mean diastolic blood pressure on the upper arm while the patient was supine.

11-75. The paired t test is negative, so the larger mean was subtracted from the smaller mean to obtain the numerator of the test statistic.

11-77. Question 11-75 said, "The paired t test was equal to -7.6." Because -7.6 is more extreme than the lower critical value of -2, we reject the null hypothesis. We can conclude that there was a significant difference in the mean diastolic blood pressure readings on the forearm versus the upper arm when the patient was inclined. Because the mean forearm reading was higher, we can say that the mean diastolic blood pressure was significantly higher when taken on people's forearms than on their upper arms.

Chapter 12

12-15. Independent variable.

12-17. Extraneous variable, which should be controlled via randomization.

12-19. We have one categorical independent variable with more than two levels, we have a quantitative outcome variable, and we are interested in comparing the group means.

12-21. $H_0: \mu_1 = \mu_2 = \mu_3$, where $1 =$ the group hearing speeded-up music, $2 =$ the group hearing slowed-down music, and $3 =$ the group hearing normal-speed music. Our samples come from populations in which the mean distance in the middle 20 minutes is the same for those hearing speeded-up music, slowed-down music and normal-speed music.

12-23. A multiple comparison procedure would be needed to determine how the means differ. A significant one-way ANOVA F test is not specific and can only say there is some difference in the means.

12-25. H_1: In terms of the simulation completion time, there is some difference in the means of the populations from which we drew our samples of first-year residents, last-year residents, surgeons who have led up to 40 procedures, and surgeons who have led 80+ procedures. H_0: Our samples came from populations (described above) that do not differ in terms of mean simulation completion time.

12-27. Table C does not have a listing for $df_W = 52$, so we would look at the next smaller $df_W = 50$. The F critical value is 2.79.

12-29. To look for possible outliers that could skew the means. We should be quite familiar with the data before computing inferential statistics.

12-31. We would need a multiple comparison procedure.

12-33. Six comparisons (Groups 1 vs. 2, 1 vs. 3, 1 vs. 4, 2 vs. 3, 2 vs. 4, and 3 vs. 4).

12-35. $.05 \times 6 = .30$.

12-37. The total sums of squares is the sum of the two numbers in that column, 13,284.085. The $df_B =$ number of groups minus $1 = 4 - 1 = 3$. The $df_W =$ total N minus the number of groups $= (14 \times 4) - 4 = 56 - 4 = 52$. The 3 goes in the first blank in the df column, and 52 goes in the second blank in that column. To get the total df, you add the numbers in that column: $3 + 52 = 55$. To get the mean square between, you will use the two numbers on the same line. Take SS_B and divide it by df_B to get 1,529.115. Do the same thing for the mean square within: $SS_W/df_W = 167.245$. Finally, to get the one-way ANOVA F test, take MS_B and divide it by $MS_W = 9.1429639 \approx 9.14$.

12-39. Because $p < a$, reject the null hypothesis.

Chapter 13

13-25. Sleep trouble score is the variable that we think is being affected by the texting, so sleep trouble is being treated here as a criterion variable.

13-27. $H_1: \rho \neq 0$. Nothing in the scenario suggests that we have a directional alternative hypothesis with a positive (or negative) linear relationship being predicted. The scenario says, "Suppose we want to know whether the number of text messages … will predict how much trouble they have with sleeping." The alternative hypothesis can be worded as follows: Our sample comes from a population in which there is some linear relationship between the number of text messages sent by college freshmen in a week and the amount of trouble they have with sleeping.

13-29. $H_1: \beta \neq 0$. Our sample comes from a population in which there is some nonzero regression slope when the number of texts is used to predict the amount of sleep trouble.

13-31. Predicted sleep trouble score $= 18 + 0.23X$, where $X =$ number of texts sent in a week.

13-33. The regression equation was calculated using a data set in which the mean number of texts in a week was 285. Corey's number of texts may be outside the range of the data for which the regression equation was created.

13-35. $H_1: \rho > 0$. Our sample comes from a population in which there is a positive linear relationship between left-nostril distances and cognitive abilities as measured by the MMSE. Those with shorter distances will tend to have lower MMSE scores, while those with longer distances will tend to have higher MMSE scores.

13-37. The t test for the slope $= .441/.1277 \approx 3.45$.

13-39. The Y-intercept is the point where the regression line crosses the Y axis. If X is zero, then the predicted score on the criterion variable equals the Y-intercept. Here, if $X = 0$, it means the gestational age is zero, which makes no sense.

13-41. There is a positive linear relationship between the variables because the slope is positive.

Chapter 14

14-29. We multiply the population proportion by (1 − population proportion). Then we divide by N, and then take the square root of our answer. The population proportion is hypothesized to be .14. We take $(.14 \times .86)/1{,}234 = .0000976$. After we take the square root of that number, we get the answer: .0098777.

14-31. The lower limit of the 95% confidence interval is the sample proportion minus the margin of error $= .2042139 - .0193603 = .1848536 \approx .185$. The upper limit is $.2042139 + .0193603 = .2235742 \approx .224$. So the confidence interval is [.185, .224]. This interval does not contain the hypothesized population proportion of .14. We may conclude that our sample of older Oklahomans comes from a population in which a significantly higher proportion of respondents consume 5+ fruit/vegetable servings daily.

14-33. In order: 0.04415808, 0.255432373, 0.02345898, 0.345621142, 0.014308943, 4.450199557, 0.0000443459, 0.023178123.

14-35. Because the observed test statistic is not more extreme than the critical value, we retain the null hypothesis and conclude that the blood drives produced a sample that fits the proportions that we hypothesized existed in the population, based on the information from a website.

14-37. Respondent's role may be considered a predictor variable.

14-39. The internal validity is extremely weak. We can draw no inferences about the professional's role having a causal influence on the responses.

14-41. H_1: Respondent's role is related to the response to the question about a nonpunitive reporting environment for nurses witnessing disruptive behavior from physicians.

14-43. $df =$ (number of rows $-$ 1) $\times$ (number of columns $-$ 1) $= (2-1) \times (3-1) = 1 \times 2 = 2$.

14-45. The first row's expected frequencies are 123.5609756, 453.0569106, and 19.38211382. The second row's expected frequencies are 29.43902439, 107.9430894, and 4.617886179. Notice that one of these expected frequencies is less than 5, which we said can be problematic to the robustness of the chi square for independence.

14-47. First row: 154.7293278, 226.7105559, and 6.853328045. Second row: same numbers.

14-49. Chi-square test for independence ≈ 10.95.

14-51. Our results support the alternative hypothesis, which said the respondent's role in the hospital was related to the judgment about whether a nonpunitive reporting environment existed for nurses.

14-53. $27/242 = .1115702 \approx .112$.

14-55. H_0: Consumption of 5+ fruit/vegetable servings daily (yes/no) is independent of age group. There are other ways of writing the null hypothesis. See chapter for examples.

14-57. Squared differences: each number is 351.3338786.

14-59. $11.32792131 \approx 11.328$.

14-61. Chi-square critical value $= 3.84$.

14-63. It means the two proportions differ significantly.

14-65. There are two categorical variables: phase of the moon and whether the patients have seizures. One might be tempted to use the chi square test for independence. But the next question should raise serious concerns about the appropriateness of this test statistic.

14-67. Expected frequencies, first row: 12.88571429, 15.71428571, 15.4. Second row: 28.11428571, 34.28571429, 33.6.

14-69. Observed minus expected, first row: 6.26, −6.26. Second row: −6.26, 6.26.

14-71. Squared differences divided by corresponding expected frequencies, first row: 0.820854629, 1.339289132. Second row: 0.865832965, 1.412674838.

14-73. The critical value is 3.84. Because the observed chi-square test for independence is more extreme than the critical value, we reject the null hypothesis.

14-75. Women in the highest quintile for fruit/vegetable consumption had a 32% lower risk of cardiovascular disease than women in the lowest quintile for fruit/vegetable consumption, after adjusting for age and randomized group membership.

14-77. The odds ratio = $(12 \times 147)/(44 \times 19) = 1764/836 = 2.1100478 \approx 2.11$.

14-79. After accounting for the severity of injuries and the location of the amputations for those in the study, the researchers' 95% confidence interval for the odds ratio did not contain 1. Therefore, cellulitis and TBI status remained significantly related, after adjusting for those extraneous variables.

Chapter 15

Scenario A.   a. Experiment. b. Treatment = independent variable. Shingles (yes/no) = dependent variable. c. Odds ratio and its associated CI.

Scenario C.   a. Observational research. b. Oral contraceptive use = predictor variable. Acute mountain sickness = criterion variable. c. Chi-square for independence.

Scenario E.   a. Experiment. b. Soothing method = independent variable. Number of seconds of crying = dependent variable. c. One-way ANOVA F test and a multiple comparison procedure with CIs for estimating population mean differences.

Scenario G.   a. Experiment. b. Treatment = independent variable. For this part of the scenario, the dependent variables are a history of osteoarthritis and duration of low back pain. c. If history of osteoarthritis was categorical (yes/no), then chi-square for independence. For number of months of low back pain, if sample sizes are equal, independent-samples t test and its associated CI for estimating the population mean difference.

Scenario I.   a. Observational research. b. Time of test = predictor variable. Test score = criterion variable. c. Paired t test and its associated CI for estimating the difference in the related population means.

Scenario K.   a. Observational research. b. Year = predictor variable. Satisfaction rating = criterion variable. c. AWS t test and its associated CI for estimating the population mean difference.

Scenario M.   a. Observational research. b. No predictor variable. Quality of life = criterion variable. c. One-sample t test and its associated CI for estimating the population mean.

Scenario O.   a. Observational research. b. Cortisol might be the predictor variable, and maternity blues might be the quantitative criterion variable, but it is possible that the blues led to the changes in cortisol. We cannot know because this study is not an experiment. c. Pearson's r, linear regression (if prior studies informed our decision about which variable should be the predictor and which variable should be the criterion), CI for the slope.

Scenario Q.   a. Observational research. b. No predictor variable. Birth weight = criterion variable. c. z test statistic and its associated CI for estimating the population mean.

Scenario S.   a. Experiment. b. Cost of the shot = independent variable. Rating of likelihood of catching the illness = dependent variable. c. Independent-samples t test and its associated CI for estimating the population mean difference.

Scenario U.   a. Descriptive research. b. Activity level of participants (moderately active versus sedentary) = predictor variable. Experience of cardiac-related event (yes/no) = criterion variable. c. Relative risk.

Scenario W.   a. Descriptive research. b. Month of delivery = predictor variable. Number of women with pregnancy-induced high blood pressure = criterion variable. c. Chi square test for goodness of fit.

Appendix

Table A.1 **Areas for the Standard Normal Distribution**

z	1	2	3	z	1	2	3	z	1	2	3
0.00	.0000	.5000	0.25	.0987	.4013	0.50	.1915	.3085
0.01	.0040	.4960	0.26	.1026	.3974	0.51	.1950	.3050
0.02	.0080	.4920	0.27	.1064	.3936	0.52	.1985	.3015
0.03	.0120	.4880	0.28	.1103	.3897	0.53	.2019	.2981
0.04	.0160	.4840	0.29	.1141	.3859	0.54	.2054	.2946
0.05	.0199	.4801	0.30	.1179	.3821	0.55	.2088	.2912
0.06	.0239	.4761	0.31	.1217	.3783	0.56	.2123	.2877
0.07	.0279	.4721	0.32	.1255	.3745	0.57	.2157	.2843
0.08	.0319	.4681	0.33	.1293	.3707	0.58	.2190	.2810
0.09	.0359	.4641	0.34	.1331	.3669	0.59	.2224	.2776
0.10	.0398	.4602	0.35	.1368	.3632	0.60	.2257	.2743
0.11	.0438	.4562	0.36	.1406	.3594	0.61	.2291	.2709
0.12	.0478	.4522	0.37	.1443	.3557	0.62	.2324	.2676
0.13	.0517	.4483	0.38	.1480	.3520	0.63	.2357	.2643
0.14	.0557	.4443	0.39	.1517	.3483	0.64	.2389	.2611
0.15	.0596	.4404	0.40	.1554	.3446	0.65	.2422	.2578
0.16	.0636	.4364	0.41	.1591	.3409	0.66	.2454	.2546
0.17	.0675	.4325	0.42	.1628	.3372	0.67	.2486	.2514
0.18	.0714	.4286	0.43	.1664	.3336	0.68	.2517	.2483
0.19	.0753	.4247	0.44	.1700	.3300	0.69	.2549	.2451
0.20	.0793	.4207	0.45	.1736	.3264	0.70	.2580	.2420
0.21	.0832	.4168	0.46	.1772	.3228	0.71	.2611	.2389
0.22	.0871	.4129	0.47	.1808	.3192	0.72	.2642	.2358
0.23	.0910	.4090	0.48	.1844	.3156	0.73	.2673	.2327
0.24	.0948	.4052	0.49	.1879	.3121	0.74	.2704	.2296

(*Continued*)

Table A.1 (*Continued*) Areas for the Standard Normal Distribution

z	2	3	z	2	3	z	2	3
0.75	.2734	.2266	1.15	.3749	.1251	1.55	.4394	.0606
0.76	.2764	.2236	1.16	.3770	.1230	1.56	.4406	.0594
0.77	.2794	.2206	1.17	.3790	.1210	1.57	.4418	.0582
0.78	.2823	.2177	1.18	.3810	.1190	1.58	.4429	.0571
0.79	.2852	.2148	1.19	.3830	.1170	1.59	.4441	.0559
0.80	.2881	.2119	1.20	.3849	.1151	1.60	.4452	.0548
0.81	.2910	.2090	1.21	.3869	.1131	1.61	.4463	.0537
0.82	.2939	.2061	1.22	.3888	.1112	1.62	.4474	.0526
0.83	.2967	.2033	1.23	.3907	.1093	1.63	.4484	.0516
0.84	.2995	.2005	1.24	.3925	.1075	1.64	.4495	.0505
0.85	.3023	.1977	1.25	.3944	.1056	**1.645**	**.4500**	**.0500**
0.86	.3051	.1949	1.26	.3962	.1038	1.65	.4505	.0495
0.87	.3078	.1922	1.27	.3980	.1020	1.66	.4515	.0485
0.88	.3106	.1894	1.28	.3997	.1003	1.67	.4525	.0475
0.89	.3133	.1867	1.29	.4015	.0985	1.68	.4535	.0465
						1.69	.4545	.0455
0.90	.3159	.1841	1.30	.4032	.0968	1.70	.4554	.0446
0.91	.3186	.1814	1.31	.4049	.0951	1.71	.4564	.0436
0.92	.3212	.1788	1.32	.4066	.0934	1.72	.4573	.0427
0.93	.3238	.1762	1.33	.4082	.0918	1.73	.4582	.0418
0.94	.3264	.1736	1.34	.4099	.0901	1.74	.4591	.0409
0.95	.3289	.1711	1.35	.4115	.0885	1.75	.4599	.0401
0.96	.3315	.1685	1.36	.4131	.0869	1.76	.4608	.0392
0.97	.3340	.1660	1.37	.4147	.0853	1.77	.4616	.0384
0.98	.3365	.1635	1.38	.4162	.0838	1.78	.4625	.0375
0.99	.3389	.1611	1.39	.4177	.0823	1.79	.4633	.0367
1.00	.3413	.1587	1.40	.4192	.0808	1.80	.4641	.0359
1.01	.3438	.1562	1.41	.4207	.0793	1.81	.4649	.0351
1.02	.3461	.1539	1.42	.4222	.0778	1.82	.4656	.0344
1.03	.3485	.1515	1.43	.4236	.0764	1.83	.4664	.0336
1.04	.3508	.1492	1.44	.4251	.0749	1.84	.4671	.0329
1.05	.3531	.1469	1.45	.4265	.0735	1.85	.4678	.0322
1.06	.3554	.1446	1.46	.4279	.0721	1.86	.4686	.0314
1.07	.3577	.1423	1.47	.4292	.0708	1.87	.4693	.0307
1.08	.3599	.1401	1.48	.4306	.0694	1.88	.4699	.0301
1.09	.3621	.1379	1.49	.4319	.0681	1.89	.4706	.0294
1.10	.3643	.1357	1.50	.4332	.0668	1.90	.4713	.0287
1.11	.3665	.1335	1.51	.4345	.0655	1.91	.4719	.0281
1.12	.3686	.1314	1.52	.4357	.0643	1.92	.4726	.0274
1.13	.3708	.1292	1.53	.4370	.0630	1.93	.4732	.0268
1.14	.3729	.1271	1.54	.4382	.0618	1.94	.4738	.0262

(*Continued*)

1	2	3	1	2	3	1	2	3
z			z			z		
or			or			or		
−z			−z			−z		
1.95	.4744	.0256	2.35	.4906	.0094	2.75	.4970	.0030
1.96	**.4750**	**.0250**	2.36	.4909	.0091	2.76	.4971	.0029
1.97	.4756	.0244	2.37	.4911	.0089	2.77	.4972	.0028
1.98	.4761	.0239	2.38	.4913	.0087	2.78	.4973	.0027
1.99	.4767	.0233	2.39	.4916	.0084	2.79	.4974	.0026
2.00	.4772	.0228	2.40	.4918	.0082	2.80	.4974	.0026
2.01	.4778	.0222	2.41	.4920	.0080	2.81	.4975	.0025
2.02	.4783	.0217	2.42	.4922	.0078	2.82	.4976	.0024
2.03	.4788	.0212	2.43	.4925	.0075	2.83	.4977	.0023
2.04	.4793	.0207	2.44	.4927	.0073	2.84	.4977	.0023
2.05	.4798	.0202	2.45	.4929	.0071	2.85	.4978	.0022
2.06	.4803	.0197	2.46	.4931	.0069	2.86	.4979	.0021
2.07	.4808	.0192	2.47	.4932	.0068	2.87	.4979	.0021
2.08	.4812	.0188	2.48	.4934	.0066	2.88	.4980	.0020
2.09	.4817	.0183	2.49	.4936	.0064	2.89	.4981	.0019
2.10	.4821	.0179	2.50	.4938	.0062	2.90	.4981	.0019
2.11	.4826	.0174	2.51	.4940	.0060	2.91	.4982	.0018
2.12	.4830	.0170	2.52	.4941	.0059	2.92	.4982	.0018
2.13	.4834	.0166	2.53	.4943	.0057	2.93	.4983	.0017
2.14	.4838	.0162	2.54	.4945	.0055	2.94	.4984	.0016
2.15	.4842	.0158	2.55	.4946	.0054	2.95	.4984	.0016
2.16	.4846	.0154	2.56	.4948	.0052	2.96	.4985	.0015
2.17	.4850	.0150	2.57	.4949	.0051	2.97	.4985	.0015
2.18	.4854	.0146	2.58	.4951	.0049	2.98	.4986	.0014
2.19	.4857	.0143	2.59	.4952	.0048	2.99	.4986	.0014
2.20	.4861	.0139	2.60	.4953	.0047	3.00	.4987	.0013
2.21	.4864	.0136	2.61	.4955	.0045	3.01	.4987	.0013
2.22	.4868	.0132	2.62	.4956	.0044	3.02	.4987	.0013
2.23	.4871	.0129	2.63	.4957	.0043	3.03	.4988	.0012
2.24	.4875	.0125	2.64	.4959	.0041	3.04	.4988	.0012
2.25	.4878	.0122	2.65	.4960	.0040	3.05	.4989	.0011
2.26	.4881	.0119	2.66	.4961	.0039	3.06	.4989	.0011
2.27	.4884	.0116	2.67	.4962	.0038	3.07	.4989	.0011
2.28	.4887	.0113	2.68	.4963	.0037	3.08	.4990	.0010
2.29	.4890	.0110	2.69	.4964	.0036	3.09	.4990	.0010
2.30	.4893	.0107	2.70	.4965	.0035	3.10	.4990	.0010
2.31	.4896	.0104	2.71	.4966	.0034	3.11	.4991	.0009
2.32	.4898	.0102	2.72	.4967	.0033	3.12	.4991	.0009
2.33	.4901	.0099	2.73	.4968	.0032	3.13	.4991	.0009
2.34	.4904	.0096	2.74	.4969	.0031	3.14	.4992	.0008

(*Continued*)

Table A.1 (*Continued*) **Areas for the Standard Normal Distribution**

z	2	3	z	2	3	z	2	3
3.15	.4992	.0008	3.22	.4994	.0006	3.45	.49972	.00028
3.16	.4992	.0008	3.23	.4994	.0006	3.50	.49977	.00023
3.17	.4992	.0008	3.24	.4994	.0006	3.60	.49984	.00016
3.18	.4993	.0007	3.25	.4994	.0006	3.70	.49989	.00011
3.19	.4993	.0007	3.30	.49952	.00048	3.80	.49993	.00007
3.20	.4993	.0007	3.35	.49960	.00040	3.90	.49995	.00005
3.21	.4993	.0007	3.40	.49966	.00034	4.00	.49997	.00003

Table A.1 was computed by William Howard Beasley.

Appendix

Table B.1 Critical Values for t Distributions

	Total α for a Two-Tailed Test					
	.20	.10	.05	.02	.01	.001
	Total α for a One-Tailed Test					
df	.10	.05	.025	.01	.005	.0005
1	3.078	6.314	12.706	31.821	63.657	636.619
2	1.886	2.920	4.303	6.965	9.925	31.599
3	1.638	2.353	3.182	4.541	5.841	12.924
4	1.533	2.132	2.776	3.747	4.604	8.610
5	1.476	2.015	2.571	3.365	4.032	6.869
6	1.440	1.943	2.447	3.143	3.707	5.959
7	1.415	1.895	2.365	2.998	3.499	5.408
8	1.397	1.860	2.306	2.896	3.355	5.041
9	1.383	1.833	2.262	2.821	3.250	4.781
10	1.372	1.812	2.228	2.764	3.169	4.587
11	1.363	1.796	2.201	2.718	3.106	4.437
12	1.356	1.782	2.179	2.681	3.055	4.318
13	1.350	1.771	2.160	2.650	3.012	4.221
14	1.345	1.761	2.145	2.624	2.977	4.140
15	1.341	1.753	2.131	2.602	2.947	4.073
16	1.337	1.746	2.120	2.583	2.921	4.015
17	1.333	1.740	2.110	2.567	2.898	3.965
18	1.330	1.734	2.101	2.552	2.878	3.922
19	1.328	1.729	2.093	2.539	2.861	3.883
20	1.325	1.725	2.086	2.528	2.845	3.850
21	1.323	1.721	2.080	2.518	2.831	3.819
22	1.321	1.717	2.074	2.508	2.819	3.792
23	1.319	1.714	2.069	2.500	2.807	3.768
24	1.318	1.711	2.064	2.492	2.797	3.745
25	1.316	1.708	2.060	2.485	2.787	3.725
26	1.315	1.706	2.056	2.479	2.779	3.707
27	1.314	1.703	2.052	2.473	2.771	3.690

(*Continued*)

Table B.1 (*Continued*) **Critical Values for *t* Distributions**

	Total α for a Two-Tailed Test					
	.20	.10	.05	.02	.01	.001
	Total α for a One-Tailed Test					
df	.10	.05	.025	.01	.005	.0005
28	1.313	1.701	2.048	2.467	2.763	3.674
29	1.311	1.699	2.045	2.462	2.756	3.659
30	1.310	1.697	2.042	2.457	2.750	3.646
35	1.306	1.690	2.030	2.438	2.724	3.591
40	1.303	1.684	2.021	2.423	2.704	3.551
45	1.301	1.679	2.014	2.412	2.690	3.520
50	1.299	1.676	2.009	2.403	2.678	3.496
55	1.297	1.673	2.004	2.396	2.668	3.476
60	1.296	1.671	2.000	2.390	2.660	3.460
70	1.294	1.667	1.994	2.381	2.648	3.435
80	1.292	1.664	1.990	2.374	2.639	3.416
90	1.291	1.662	1.987	2.368	2.632	3.402
120	1.289	1.658	1.980	2.358	2.617	3.373
100000	1.282	1.645	1.960	2.326	2.576	3.291

Table B.1 was computed by William Howard Beasley. Interpolation with respect to *df* should be done linearly in $1/df$.

Appendix

Table C.1 **Critical Values for *F* Distributions**

Denominator df	α	\multicolumn{8}{c}{Numerator df}							
		1	2	3	4	5	6	7	8
11	.05	4.84	3.98	3.59	3.36	3.20	3.09	3.01	2.95
	.01	9.65	7.21	6.22	5.67	5.32	5.07	4.89	4.74
12	.05	4.75	3.89	3.49	3.26	3.11	3.00	2.91	2.85
	.01	9.33	6.93	5.95	5.41	5.06	4.82	4.64	4.50
13	.05	4.67	3.81	3.41	3.18	3.03	2.92	2.83	2.77
	.01	9.07	6.70	5.74	5.21	4.86	4.62	4.44	4.30
14	.05	4.60	3.74	3.34	3.11	2.96	2.85	2.76	2.70
	.01	8.86	6.51	5.56	5.04	4.69	4.46	4.28	4.14
15	.05	4.54	3.68	3.29	3.06	2.90	2.79	2.71	2.64
	.01	8.68	6.36	5.42	4.89	4.56	4.32	4.14	4.00
16	.05	4.49	3.63	3.24	3.01	2.85	2.74	2.66	2.59
	.01	8.53	6.23	5.29	4.77	4.44	4.20	4.03	3.89
17	.05	4.45	3.59	3.20	2.96	2.81	2.70	2.61	2.55
	.01	8.40	6.11	5.18	4.67	4.34	4.10	3.93	3.79
18	.05	4.41	3.55	3.16	2.93	2.77	2.66	2.58	2.51
	.01	8.29	6.01	5.09	4.58	4.25	4.01	3.84	3.71
19	.05	4.38	3.52	3.13	2.90	2.74	2.63	2.54	2.48
	.01	8.18	5.93	5.01	4.50	4.17	3.94	3.77	3.63
20	.05	4.35	3.49	3.10	2.87	2.71	2.60	2.51	2.45
	.01	8.10	5.85	4.94	4.43	4.10	3.87	3.70	3.56
21	.05	4.32	3.47	3.07	2.84	2.68	2.57	2.49	2.42
	.01	8.02	5.78	4.87	4.37	4.04	3.81	3.64	3.51
22	.05	4.30	3.44	3.05	2.82	2.66	2.55	2.46	2.40
	.01	7.95	5.72	4.82	4.31	3.99	3.76	3.59	3.45
23	.05	4.28	3.42	3.03	2.80	2.64	2.53	2.44	2.37
	.01	7.88	5.66	4.76	4.26	3.94	3.71	3.54	3.41
24	.05	4.26	3.40	3.01	2.78	2.62	2.51	2.42	2.36
	.01	7.82	5.61	4.72	4.22	3.90	3.67	3.50	3.36
25	.05	4.24	3.39	2.99	2.76	2.60	2.49	2.40	2.34
	.01	7.77	5.57	4.68	4.18	3.85	3.63	3.46	3.32
26	.05	4.23	3.37	2.98	2.74	2.59	2.47	2.39	2.32
	.01	7.72	5.53	4.64	4.14	3.82	3.59	3.42	3.29
27	.05	4.21	3.35	2.96	2.73	2.57	2.46	2.37	2.31
	.01	7.68	5.49	4.60	4.11	3.78	3.56	3.39	3.26

(*Continued*)

Table C.1 (*Continued*) Critical Values for *F* Distributions

Denominator df	α	Numerator df 1	2	3	4	5	6	7	8
28	.05	4.20	3.34	2.95	2.71	2.56	2.45	2.36	2.29
	.01	7.64	5.45	4.57	4.07	3.75	3.53	3.36	3.23
29	.05	4.18	3.33	2.93	2.70	2.55	2.43	2.35	2.28
	.01	7.60	5.42	4.54	4.04	3.73	3.50	3.33	3.20
30	.05	4.17	3.32	2.92	2.69	2.53	2.42	2.33	2.27
	.01	7.56	5.39	4.51	4.02	3.70	3.47	3.30	3.17
32	.05	4.15	3.29	2.90	2.67	2.51	2.40	2.31	2.24
	.01	7.50	5.34	4.46	3.97	3.65	3.43	3.26	3.13
34	.05	4.13	3.28	2.88	2.65	2.49	2.38	2.29	2.23
	.01	7.44	5.29	4.42	3.93	3.61	3.39	3.22	3.09
36	.05	4.11	3.26	2.87	2.63	2.48	2.36	2.28	2.21
	.01	7.40	5.25	4.38	3.89	3.57	3.35	3.18	3.05
38	.05	4.10	3.24	2.85	2.62	2.46	2.35	2.26	2.19
	.01	7.35	5.21	4.34	3.86	3.54	3.32	3.15	3.02
40	.05	4.08	3.23	2.84	2.61	2.45	2.34	2.25	2.18
	.01	7.31	5.18	4.31	3.83	3.51	3.29	3.12	2.99
42	.05	4.07	3.22	2.83	2.59	2.44	2.32	2.24	2.17
	.01	7.28	5.15	4.29	3.80	3.49	3.27	3.10	2.97
44	.05	4.06	3.21	2.82	2.58	2.43	2.31	2.23	2.16
	.01	7.25	5.12	4.26	3.78	3.47	3.24	3.08	2.95
46	.05	4.05	3.20	2.81	2.57	2.42	2.30	2.22	2.15
	.01	7.22	5.10	4.24	3.76	3.44	3.22	3.06	2.93
48	.05	4.04	3.19	2.80	2.57	2.41	2.29	2.21	2.14
	.01	7.19	5.08	4.22	3.74	3.43	3.20	3.04	2.91
50	.05	4.03	3.18	2.79	2.56	2.40	2.29	2.20	2.13
	.01	7.17	5.06	4.20	3.72	3.41	3.19	3.02	2.89
55	.05	4.02	3.16	2.77	2.54	2.38	2.27	2.18	2.11
	.01	7.12	5.01	4.16	3.68	3.37	3.15	2.98	2.85
60	.05	4.00	3.15	2.76	2.53	2.37	2.25	2.17	2.10
	.01	7.08	4.98	4.13	3.65	3.34	3.12	2.95	2.82
70	.05	3.98	3.13	2.74	2.50	2.35	2.23	2.14	2.07
	.01	7.01	4.92	4.07	3.60	3.29	3.07	2.91	2.78
80	.05	3.96	3.11	2.72	2.49	2.33	2.21	2.13	2.06
	.01	6.96	4.88	4.04	3.56	3.26	3.04	2.87	2.74
100	.05	3.94	3.09	2.70	2.46	2.31	2.19	2.10	2.03
	.01	6.90	4.82	3.98	3.51	3.21	2.99	2.82	2.69
125	.05	3.92	3.07	2.68	2.44	2.29	2.17	2.08	2.01
	.01	6.84	4.78	3.94	3.47	3.17	2.95	2.79	2.66
150	.05	3.90	3.06	2.66	2.43	2.27	2.16	2.07	2.00
	.01	6.81	4.75	3.91	3.45	3.14	2.92	2.76	2.63
200	.05	3.89	3.04	2.65	2.42	2.26	2.14	2.06	1.98
	.01	6.76	4.71	3.88	3.41	3.11	2.89	2.73	2.60
400	.05	3.86	3.02	2.63	2.39	2.24	2.12	2.03	1.96
	.01	6.70	4.66	3.83	3.37	3.06	2.85	2.68	2.56
1000	.05	3.85	3.00	2.61	2.38	2.22	2.11	2.02	1.95
	.01	6.66	4.63	3.80	3.34	3.04	2.82	2.66	2.53
10000000	.05	3.84	3.00	2.60	2.37	2.21	2.10	2.01	1.94
	.01	6.63	4.61	3.78	3.32	3.02	2.80	2.64	2.51

Table C.1 was computed by William Howard Beasley. Interpolation with respect to *df* should be done linearly in 1/*df*.

Appendix

Table D.1 Critical Values for χ^2 Distributions

df	α for One-Tailed Test			
	.10	.05	.01	.001
1	2.71	3.84	6.63	10.83
2	4.61	5.99	9.21	13.82
3	6.25	7.81	11.34	16.27
4	7.78	9.49	13.28	18.47
5	9.24	11.07	15.09	20.52
6	10.64	12.59	16.81	22.46
7	12.02	14.07	18.48	24.32
8	13.36	15.51	20.09	26.12
9	14.68	16.92	21.67	27.88
10	15.99	18.31	23.21	29.59
11	17.28	19.68	24.72	31.26
12	18.55	21.03	26.22	32.91
13	19.81	22.36	27.69	34.53
14	21.06	23.68	29.14	36.12
15	22.31	25.00	30.58	37.70
16	23.54	26.30	32.00	39.25
17	24.77	27.59	33.41	40.79
18	25.99	28.87	34.81	42.31
19	27.20	30.14	36.19	43.82
20	28.41	31.41	37.57	45.31
21	29.62	32.67	38.93	46.80
22	30.81	33.92	40.29	48.27
23	32.01	35.17	41.64	49.73
24	33.20	36.42	42.98	51.18
25	34.38	37.65	44.31	52.62
26	35.56	38.89	45.64	54.05
27	36.74	40.11	46.96	55.48

(*Continued*)

Table D.1 (*Continued*) **Critical Values for χ^2 Distributions**

df	α for One-Tailed Test			
	.10	.05	.01	.001
28	37.92	41.34	48.28	56.89
29	39.09	42.56	49.59	58.30
30	40.26	43.77	50.89	59.70
40	51.81	55.76	63.69	73.40
50	63.17	67.50	76.15	86.66
60	74.40	79.08	88.38	99.61
70	85.53	90.53	100.43	112.32

Table D.1 was computed by William Howard Beasley. Interpolation with respect to *df* should be done linearly in 1/*df*.

Index

Bubble graph, 79
Buron (2010), 24–25, 30, 343–344, 352, 515, 526
Buron, B., *see* Buron (2010)
Buysse, Daniel J., 99; *see also* Buysse et al. (2008); Buysse, Reynolds, Monk, Berman, & Kupfer (1989)
Buysse et al. (2008), 90, 92, 107, 174, 188, 295, 296
Buysse, Reynolds, Monk, Berman & Kupfer (1989), 84, 108
Buzzy4shots.com, 298

C

Caffeine beliefs/alertness example, 15–19, 21
Cake, bleeding armadillo, *see* Bleeding armadillo cake
Cardiac arrest resuscitation/later quality of life study, 522
Cardiac event/exercise study, 525
Career satisfaction for physicians study, 29, 521
Careers in health sciences, 1–2
Carey et al. (2007), 94, 108
Carey, J.J., *see* Carey et al. (2007)
Carmody et al. (2011), 28, 30
Carmody, J.F., *see* Carmody et al. (2011)
Carney, C.E., *see* Carney, Edinger, Meyer, Lindman, & Istre (2006)
Carney, Edinger, Meyer, Lindman, & Istre (2006), 276, 286, 296
Case-control studies, 23, 302, 474, 475, 481
Cases, 476
Case studies, 6
Categorical data analysis, 440, 447, 448, 461–462, 485
Categorical variables, 10, 53–54
Causality, causal conclusions, causal relationships, 15, 26–27
CDC, *see* Centers for Disease Control and Prevention (CDC)
Cell phone apps/skin cancer study, 138–148, 151, 154–155
Cells in tables of frequencies, 138, 465–466
Cellulitis, 490–491, 511
Census Bureau, U.S., 58, 129

Centers for Disease Control and Prevention (CDC)
arthritis, 525
birth/death rates, 114
fruit/vegetable consumption, 440–441, 513–514
growth charts, 95, 105
hepatitis C, 190
obesity rates, 11, 58, 60, 61, 64, 110, 117, 129, 401, 404, 421–422, 426–427
smoking prevalence, 68, 127–128
Central Intelligence Agency, U.S., 121–122, 124, 126
Central Limit Theorem
assumptions, 224–225
confidence intervals for μ and, 184
example with small population, 168–170
explanation of, 165–167
need for, 168
online demonstrations, 171
rat shipment example, 193
shape of sampling distribution of M, 210, 279
statistical noteworthiness, 181
z test statistics and, 173–177, 208
Chaffin et al. (2012), 77, 81
Chaffin, M., *see* Chaffin et al. (2012)
Change across time, 344, 348
Characteristics of data sets, 34
Childhood immunizations, 320
Children's heart rates example, 249–250, 520
Children's medicines/flavorings example, 521–522
Children with ADHD/stability balls, 3, 8
Children with asthma in emergency room example, 6–9
Chi pronunciation, 448
Chi square, *see* Chi square tests for goodness of fit; Chi square tests for independence
Chi square tests for contingency tables, 462; *see also* Chi square tests for independence
Chi square tests for goodness of fit, 448–460
assumptions, 460
computation, 453, 456–457

T - #0302 - 071024 - C604 - 234/156/27 - PB - 9780367783532 - Gloss Lamination